Animal Behavior

Animal Behavior

Concepts, Methods, and Applications

Shawn E. Nordell
Saint Louis University

Thomas J. Valone
Saint Louis University

New York   Oxford
OXFORD UNIVERSITY PRESS

Oxford University Press is a department of the University of Oxford.
It furthers the University's objective of excellence in research, scholarship,
and education by publishing worldwide.

Oxford   New York
Auckland   Cape Town   Dar es Salaam   Hong Kong   Karachi
Kuala Lumpur   Madrid   Melbourne   Mexico City   Nairobi
New Delhi   Shanghai   Taipei   Toronto

With offices in
Argentina   Austria   Brazil   Chile   Czech Republic   France   Greece
Guatemala   Hungary   Italy   Japan   Poland   Portugal   Singapore
South Korea   Switzerland   Thailand   Turkey   Ukraine   Vietnam

For titles covered by Section 112 of the US Higher Education Opportunity Act,
please visit www.oup.com/us/he for the latest information about pricing and
alternate formats.

Published in the United States of America by
Oxford University Press
198 Madison Avenue, New York, NY 10016
http://www.oup.com

Library of Congress Cataloging-in-Publication Data
Nordell, Shawn E.
   Animal behavior : concepts, methods, and applications / Shawn E. Nordell, Saint
Louis University, Thomas J. Valone, Saint Louis University.
      pages cm
   Includes bibliographical references.
   ISBN 978-0-19-973759-8
   1. Animal behavior. I. Valone, Thomas J. II. Title.
   QL751.N74   2013
   591.5—dc23   2013026400

Printing number: 9 8 7 6 5 4 3 2

Printed in the United States of America
on acid-free paper

For

James and Geraldine Valone

and

Buck, Ernie, Kirby, Grace, and Max

Brief Contents

Contents

Chapter 4   Behavioral Genetics   56

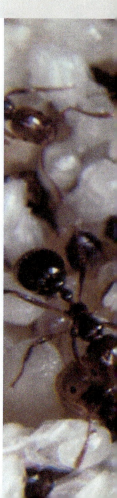

Chapter 5   Learning and Cognition   78

Chapter 6   Communication   112

Chapter 7   Foraging Behavior   142

Features

Chapter 8   Antipredator Behavior   170

Features

Chapter 11　Mating Behavior　254

Chapter 12   Mating Systems   286

Chapter 13   Parental Care   312

Chapter 14   Social Behavior   338

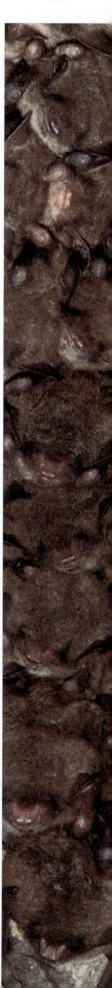

Preface

Our love of science is rooted in our undergraduate experiences, when we were both fortunate to first become immersed in the research process. We conducted projects that required us to develop research questions and hypotheses, consult and reference the primary literature, collect and analyze data, and present and discuss our conclusions. But not all undergraduates have such an opportunity. With this textbook, we hope to share the excitement of our own learning experiences. The narrative we present guides students through each step of the research process, from the development of the research question and hypothesis through tests of predictions, with ample (but not overwhelming) details on methodology and results. Our goal is for students to engage with the work of animal behavior research, just as we did as undergraduates.

This book is based on the growing mandate to shift science education from a pedagogy of rote memorization to one of critical thinking, emphasizing big-picture concepts and the nature of scientific inquiry. As instructors, we understand the need to provide not only extensive coverage of animal behavior but also the tools to create true learning opportunities. In addition, students increasingly seek relevance in their courses, asking, "How can I apply this?" To address all these needs, we go beyond merely presenting information: **we take a conceptual approach that highlights the process of science and the real-world applications of animal behavior research.**

The approach: concepts, methods, and applications

Our approach involves three major components. First, we organize each chapter around three to six major concepts. Second, we deconstruct research studies to emphasize the process of science. And third, we provide real-world applications in each chapter to tie the concepts to societal issues. Throughout the book, we use an accessible, question-driven style to engage students.

CONCEPTS

Each chapter is built around **broad organizing concepts**, such as "Evolution by natural selection favors behavioral adaptations that enhance fitness." The development of a conceptual understanding is crucial for students to be able to make connections, see broader implications, and apply their learning. We thus present major concepts as a framework for understanding and evaluating empirical research examples. These framing sentences synthesize and summarize foundational research on complex topics in animal behavior. By using concepts as an outline for each chapter, we offer students a clear learning progression, enabling them to scaffold their knowledge throughout the chapter and develop higher order thinking skills.

METHODS

We illustrate each concept using research from the primary literature, with an emphasis on the **methods of the featured studies** so that students can become immersed in the process of animal behavior research as a rigorous quantitative science. As scientists, we know that it is impossible to evaluate research without fully understanding how the research was conducted. Yet students often perceive science as a series of facts and do not have a full understanding of where our knowledge comes from. To counter their misperceptions, we clearly identify the research question, hypothesis, and prediction for featured studies and then illustrate how the methods allow the prediction to be tested. We present the resulting data in a way that shows students the individual variation present in all data. For example, we present means with standard errors in many of the results figures in the book. The book also offers a detailed chapter on standard methodology used in animal behavior research.

The featured studies have been carefully chosen to represent a broad range of taxa and include a combination of classic and contemporary research that is student accessible. We introduce the researchers from each study using their full names so that students can appreciate not only the diversity of taxa studied but also the diversity of researchers.

APPLICATIONS

Each chapter contains examples of how various people and groups are **applying the concepts** in animal behavior research to societal problems and issues. Students rarely have the opportunity to see how animal behavior research might be relevant to their own lives. Yet there are many applications of this field, such as how habitat selection behavior is being used to more effectively reintroduce species to restored areas or how crop damage can be mitigated by manipulating predation risk. Throughout the book, we highlight these examples in the "Applying the Concepts" feature.

Overview of chapters

The first three chapters lay the groundwork for understanding the science of animal behavior. In Chapter 1, we provide a brief review of the scientific method. Chapter 2 presents an overview of evolution, and Chapter 3 summarizes methods commonly used to study behavior and presents a historical review of the field. The next two chapters focus on the development of behavior through the examination of behavioral genetics (Chapter 4) and learning and cognition (Chapter 5). We then examine communication (Chapter 6), foraging (Chapter 7), and antipredator behavior (Chapter 8). Chapter 9 is devoted to animal movement, taking a look at both dispersal and migration. In Chapter 10, we cover habitat selection, territoriality, and aggression. The next three chapters focus on reproduction, examining mating behavior (Chapter 11), mating systems (Chapter 12), and parental care (Chapter 13). We end with a chapter devoted to social behavior (Chapter 14). Although the book contains chapters that differentially emphasize proximate and ultimate explanations of behavior, we infuse both approaches throughout the chapters by the incorporation of Tinbergen's four questions.

Teaching and learning features

- **"Scientific Process" boxes:** To further emphasize the process of science, each chapter contains one to three "Scientific Process" boxes. These present detailed research descriptions within a scientific framework, clearly and concisely laying out each step: research question, hypotheses, predictions, methods, results, and conclusions. Students can thus easily follow the research example at a glance, from its conception (the original research question) to the use of the scientific method (the creation of testable hypotheses, the experimental protocol used, the evaluation of data) and ultimately the findings of the work (conclusions of the study). The details contained in these examples illustrate and reinforce the rigorous nature of animal behavior research.
- **"Toolboxes":** These boxes explain essential skills or complex terms in the science of animal behavior. They do not appear in every chapter but are included as needed to build students' intellectual toolkit. They have two functions. First, these boxes show students how they can apply the scientific concepts to their own work. For example, one toolbox describes animal

sampling techniques because we know that many students will be asked to collect data as a part of this course. Second, toolboxes provide additional background information. For instance, many students may lack knowledge about phylogenies, which is crucial for employing the comparative method in behavior research. Therefore, one toolbox provides information describing phylogenies, how they are constructed, and how they are interpreted. Another toolbox explains how data are described and summarized to help students understand the data presented throughout the book.

- **"Applying the Concepts" boxes:** These boxes, which appear in each chapter, contain examples of how animal behavior research is being applied to real-life problems. This feature shows students the importance and relevance of "pure" scientific research to larger societal problems. By including these boxes in every chapter, we give students opportunities to see the broader implications and importance of research.

- **"Chapter Summary and Beyond":** At the end of each chapter, we provide a brief summary of the concepts covered. In doing so, we also point students to recent papers that further develop the ideas presented in the chapter. No textbook can be all-inclusive, so these papers are ideal for students or instructors seeking additional information about a concept.

- **Chapter questions:** At the end of each chapter, we provide a range of questions, including some that could be used as assignments, to promote critical thinking and foster class discussions. Answers and notes for even-numbered questions are included at the back of the text so that students can test themselves, and answers and notes for odd-numbered questions are included in the instructor's resources.

- **Diverse research examples:** The book contains research examples covering a diverse range of taxa from all over the world. We have consciously worked to include ample representation from major taxa (invertebrates, amphibians, reptiles, fish, birds, and mammals) in each chapter. Because most students have a limited knowledge of animal diversity, we have included an image of each featured research species in addition to brief natural history descriptions. (Furthermore, over 300 additional photos are available on the Instructor's Resource CD and the companion website, **www.oup.com/us/ nordell**, which we describe in the following section.) We also include data summary figures as they typically appear in the primary literature. Thus, students can see that variation in behavior is ubiquitous in research: individual data points are plotted on line graphs, and bar graphs contain means with standard error bars.

Support package

Oxford University Press offers a comprehensive ancillary package for instructors who adopt *Animal Behavior: Concepts, Methods, and Applications.* The following resources are available on the Instructor's Resource CD and the companion website, **www.oup.com/us/nordell**:

- **Video library and guide:** To help students better understand animal behavior research and visualize the science in each chapter, and to provide instructors with an easy-to-use resource for videos tied directly to the concepts covered in the text, we have assembled links to three types of short videos:
  1. Natural history videos that illustrate specific behaviors of the study organisms in the featured studies
  2. Process-of-science videos that illustrate the methodology described in featured research
  3. Interviews with the scientists whose research is featured in the book

We have carefully selected these videos because animal behavior is a visual endeavor, and yet students often have little experience with the diversity of taxa studied or their behavior. Each video link is accompanied by a brief description of its content and information on the video's length. We envision that instructors will incorporate many of these videos into their classes to help students better understand the behavior and research of animals.

- **Digital image library:** The image library includes electronic files in Power-Point format of every illustration, graph, photo, figure caption, and table from the text, in both labeled and unlabeled versions. Images have been enhanced for clear projection in large lecture halls. The library also includes more than 300 additional photos from researchers, depicting study organisms, research protocol set-ups, researchers collecting data, and habitats. For each chapter, we provide these photos and a guide to their content so that instructors can easily incorporate them into their courses.

- **Lecture notes for each chapter:** Editable lecture notes in PowerPoint format make preparing lectures faster and easier than ever. Each chapter's presentation includes a succinct outline of key concepts and featured research studies, and incorporates the graphics from the chapter.

- **Test bank:** We have created a test item file that includes over 250 multiple-choice questions suitable for exams (available only on the Instructor's Resource CD as a Word document).

- **Suggested further readings:** To assist instructors in identifying classic, contemporary, and relevant readings from the primary literature, we have assembled a list of over 25 suggested further readings taken directly from the journal *Behavioral Ecology*. These readings, like the video library, are tied directly to the concepts covered in the text. In addition, we provide a pedagogical guideline to help students develop their ability to critically evaluate the primary literature.

- **Answers and notes (for odd-numbered chapter questions)**

Contact your local OUP sales representative for a copy of the Instructor's Resource CD and access to the companion website.

Acknowledgments

This book is the product of our rather long collaboration in both the academic sphere and life outside of academia, but we have many people to thank for providing us with immense support. We thank our scientific mentors who helped us develop our love and understanding of science. For Shawn, this group includes her Master's advisor, Donald Thomson, who introduced her to the wonderful world of field research, and her Ph.D. advisor, Astrid Kodric-Brown, who taught her critical thinking without being critical. For Tom, this group includes his undergraduate advisor, Tom Caraco, who introduced him to the wonderful science of animal behavior, and his Ph.D. advisor, Michael Rosenzweig, who taught him how to be a scientist. Shawn would also like to thank her pedagogical mentor, Mary Stephen, director of Saint Louis University's Reinert Center for Teaching Excellence, who helped her develop her understanding of and research on the scholarship of teaching and learning. We would both like to thank our many, many wonderful students who graciously provided constructive feedback and input regarding our teaching and their learning of our favorite discipline. We also thank our Saint Louis University colleagues for their support throughout the process, particularly the Department of Biology, Don Brennan, and Joe Weixlmann. Finally, we thank Robin Carter, Scott Freeman, Sallie Marston, Manuel Molles, and Bob Ricklefs for sharing their wisdom and advice regarding the world of writing textbooks.

We also wish to express our deep appreciation to the wonderful team at Oxford University Press USA who guided us through this project: Jason Noe, senior editor, who somehow convinced us we should and could write this book and guided us graciously through the process; John Haber, developmental editor, who in a most affable manner taught us the fine art of clarity in writing and provided continuous good humor despite missed deadlines; Lauren Mine, developmental editor, who provided great insight and attention to details to ensure our vision was realized; Shelby Peak, production editor, for meticulously shepherding the manuscript through copy editing and typesetting; Melissa Rubes, Katie Naughton, Caitlin Kleinschmidt, and Andrew Heaton, editorial assistants, for conscientiously following up on so many details; Jason Kramer, marketing manager; Frank Mortimer, director of marketing; Patrick Lynch, editorial director; and John Challice, vice president and publisher. We also wish to thank those in production and design who worked so hard to make this such a beautiful book: Kim Howie, senior designer; Michele Laseau, art director; Lisa Grzan, production manager; and the team at Precision Graphics.

We could not have completed this project without the amazing and continued support and encouragement of our dear friends and family. We thank our friends in St. Louis (especially Vera and Joe Brandt, Bob and Caroline Cordia, Jean and Bill Curtis, Diane and Frank Lockhart, Sharon Matlock, Denise Mandle, and Chris Sebelski) and Tucson (especially Ellen Tuttle, Marcy Tigerman, Marcy Wood), for whom we express great love and gratitude. Their unfailing enthusiasm was appreciated more than we can ever say and helped us through many deadlines. We also thank our families both past and present, who supported our endeavors with love and appreciation. Shawn would like to thank her Academic Ladder Writing Club group members, who supported her writing on a daily basis.

Finally, we thank Buck, Ernie, Kirby, Grace, and Max, whose boundless energy and wagging tails kept us smiling and have taught us more about animal behavior than they will ever know.

We thank the following reviewers, commissioned by Oxford University Press, for providing thoughtful and constructive suggestions. The book benefited greatly from their skillful input:

Elizabeth Archie, University of Notre Dame

Suzanne Baker, James Madison University

Peter Bednekoff, Eastern Michigan University

Russell Benford, Northern Arizona University

Andrew R. Blaustein, Oregon State University

Joel Brown, University of Illinois at Chicago

Theodore E. Burk, Creighton University

Prassede Calabi, University of Massachusetts–Boston

Blaine Cole, University of Houston

Francine Dolins, University of Michigan–Dearborn

Richard Duhrkopf, Baylor University

Emily DuVal, Florida State University

Fred C. Dyer, Michigan State University

Janice Edgerly-Rooks, Santa Clara University

Miles Engell, North Carolina State University

Ann Fraser, Kalamazoo College

Sharon Gill, Western Michigan University

Harold Gouzoules, Emory University

Blaine D. Griffen, University of South Carolina

Sylvia L. Halkin, Central Connecticut State University

Jodee Hunt, Grand Valley State University

Valerie James-Aldridge, University of Texas–Pan American

Clint Kelly, Iowa State University

Astrid Kodric-Brown, University of New Mexico

William Kroll, Loyola University Chicago

David Lahti, Queens College

Tracy Langkilde, Pennsylvania State University

Susan Lewis, Carroll University

Catherine Lohmann, University of North Carolina at Chapel Hill

Jeff Lucas, Purdue University

Karen Mabry, New Mexico State University

John C. Maerz, University of Georgia

Tara Maginnis, University of Portland

Frank F. Mallory, Laurentian University

Mary Beth Manjerovic , University of Central Florida

Lauren Mathews, Worcester Polytechnic Institute

Marion McClary, Fairleigh Dickinson University

Kevin J. McGraw, Arizona State University

Daniela Monk, Washington State University

Cy L. Mott, Kentucky Wesleyan College

James Nieh, University of California, San Francisco

Brian Palestis, Wagner College

Daniel R. Papaj, University of Arizona

Stephen Pruett-Jones, University of Chicago

Rick Relyea, University of Pittsburgh

Lanna Ruddy, SUNY Geneseo

Mike Ryan, University of Texas

Debbie Schlenoff, University of Oregon

Toru Shimizu, University of South Florida

Joseph Sisneros, University of Washington

Donald Sparling, Southern Illinois University

Eric Strauss, Loyola Marymount University

Thomas Terleph, Sacred Heart University

Kevin Theis, Michigan State University

Jeffrey Thomas, Queens University of Charlotte

Robert M. Turnbull, University of Southern Mississippi–Gulf Coast

Al Uy, University of Miami

E. Natasha Vanderhoff, Jacksonville University

Sean Veney, Kent State University

Margaret Voss, Penn State University

David Westneat, University of Kentucky

Michele Jade Zee, Northeastern University

We also thank the following people for generously providing images and/or videos: Maria Abate, Mark Abrahams, Elizabeth Adkins-Regan, Ginger Allington, Esteban Alzate, Nick Barber, Anders Berglund, Thore Bergman, Jay Biernaske, Eric Bollinger, Thierry Boulinier, Alice Boyle, Jacob Bro-Jorgenson, Jason Brown, Valerie Bugh, Kevin Burns, Rhett Butler, Joanne Cable, Colin Chapman, Mark Chappell, Karen Cheney, Nikita Chernetsov, Aurelie Cohas, Aaron Corcoran, Isabelle Côté, Jillian Cowles, Susan Crowe, Herman Dierick, Niels Dingemanse, Hannah Dugdale, Jeffery Dunk, Doug Eifler, Josh Engel, Brad Fiero, Benjamin Fitzpatrick, Leonard Freed, Nicole Gerlach, Eric Gese, Matt Goff, James Grant, Kristine Grayson, David Green, Simon Griffith, Benoit Guénard, Beth Hahn, Jens Herberholz, Samantha Hilber, Chad Hoefler, Anne Houde, David Jamison, Julie Jaquiéry, Trevor Jinks, Jörgen Johnsson, Clement Kent, Alan Krakauer, Jens Krause, Ipek Kulahci, Kevin Laland, Jeffrey Lane, Bernd Leisler, Bill Leonard, John Lill, Adeline Loyau, Lauren Mathews, John McCormack, Mark McCormick, Randolf Menzel, Don Miles, Matthew Mitchell, Carson Murray, James Nichols, Justin O'Riain, Alvaro Palma, Luis Pardo, Lorna Patrick, Irene Pepperberg, Nigel Raine, Leeann Reaney, Diana Reiss, Raleigh Robertson, Helen Rodd, Kenneth Ross, Tiffany Roufs, Yvan Russell, Ralph Saporito, Gabriele Schino, Ingo Schlupp, Kenneth Schmidt, Peggy Sherman, Dominique Sigg, Hans Slabbekoorn, Marla Sokolowski, Verônica Thiemi Tsutae de Sousa, Geoffrey Steinhart, Bård Stokke, Paul Switzer, Ryan Taylor, Fabricio Barreto Teresa, Barbara Tiddi, George Uetz, Cock Van Oosterhuit, Michael Ward, Patrick Ward, Brandon Wheeler, Jan Wijmenga, Henry Wilbur, Gerald Wilkinson, Steve Yanoviak, and Mai Yasué.

We appreciate your constructive feedback. Please e-mail us your thoughts at shawn.nordell@gmail.com.

Shawn E. Nordell

Thomas J. Valone

Animal Behavior

Chapter 1

The Science of Animal Behavior

Figure 1.1. Giraffe behavior. An aggressive interaction between two male giraffes.

We can observe animal behavior everywhere. In backyards and parks, squirrels collect and bury nuts to eat later. Each spring, birds, frogs, and crickets sing to attract mates. Wasps make nests under the eaves of houses and other buildings, where the queen lays eggs and the colony cares for offspring. Farther from our home, male giraffes in Africa exchange blows by swinging their heads and necks against one another in competition for females (Figure 1.1) while fish in the Great Barrier Reef form large schools to minimize predation (Figure 1.2). And each fall and spring, millions of birds migrate from temperate to tropical regions.

This book will introduce you to the wonders of animal behavior. In this chapter, we begin by exploring the ways that animals and their behaviors are an important part of the world. We then introduce the science of animal behavior and the scientific method. We'll see how scientists test hypotheses to learn about the natural world. Most scientific hypotheses about behavior are developed in the context of evolution. In contrast, many nonscientists try to explain animal behavior based on the idea that animals possess human emotions. Such explanations are problematic, because they rarely make predictions that can be tested. Finally, we examine how scientists communicate their findings to others. One hallmark of scientific papers is that they are subject to review by other scientists, who evaluate the research before it is published.

Figure 1.2. Schooling fish. Goatfish and snapper form schools to minimize predation.

1.1 Animals and their behavior are an integral part of human society

Understanding the behavior of animals has always been important to people. Beginning in prehistoric times and continuing for tens of thousands of years, humans painted images of animals on cave walls all over the world. The drawings are detailed enough to allow us to identify different species (both extinct and extant), and many images depict animals exhibiting behaviors like eating, sleeping, and engaging in acts of aggression. Recent research indicates that these paintings likely aimed to present realistic depictions of animals and their behavior rather than symbolic connotations (Pruvost et al. 2011). Because humans in prehistoric times relied on animals for food, knowledge about animal behavior was important for survival (Shipman 2010).

Animals and animal behavior are still an integral part of modern society. Millions of people live and interact with animals daily. The American Veterinary Medical Association reports that in 2007, in the United States alone, over 80 million households contained a pet (American Veterinary Medical Association 2012). This number includes approximately 139 million freshwater fish, 90 million cats, 73 million dogs, 18 million other small mammals, 16 million birds, and 11 million reptiles. Most owners value their pets for companionship: we, for example, enjoy both watching our pets behave and interacting with them.

Many people work with animals. Cattle, chickens and turkeys, hogs, and sheep are important agricultural products; horses are used by ranchers, law enforcement officials, and the horse-racing industry; and dogs have long been used in both police work and the military (e.g., Lemish 1996) (Figure 1.3). In all of these cases, people manage animal behavior to accomplish a task. In a different way, the behavior of animals is also integral to medicine. It helps researchers assess and learn about sensory, motor, and cognitive functions. For instance, behavioral changes often reflect the effects of neurochemical agents, neurotoxins, or hormonal changes, which can be more easily studied in animals than humans. Research on memory, cognitive function, and learning often involves measuring and recording animal behavior.

In addition, animals provide entertainment for millions of people. According to the World Zoo and Aquarium Association, more than 600 million people visit zoos

Figure 1.3. Military working dogs. Dogs are used as trackers, scouts, sentries, and bomb detectors.

and aquariums worldwide each year, and visitors most value the ability to see animals behaving. Many movies, too, involve animals and their behavior, such as *March of the Penguins*, *Free Willy*, *Seabiscuit*, and *The Adventures of Milo and Otis*. Animal behavior is also featured on many popular television shows, such as *Meerkat Manor*, *The Crocodile Hunter*, and PBS's *Nature* series, and entire television networks, such as Animal Planet and National Geographic Wild, are dedicated to animals.

While these examples illustrate that animals and their behavior are an integral part of human society, the study of animal behavior involves much more than interacting with pets, working with animals, or watching animals in movies or on television. Rather, animal behavior researchers use the scientific method to understand the behaviors we observe. So what exactly do we mean by behavior? We explain this term in the following section.

Figure 1.4. Lizard thermoregulation. Ectotherms, like this chuckwalla (*Sauromalus varius*), move to cool locations as their body temperature rises.

Recognizing and defining behavior

We define **animal behavior** as any internally coordinated, externally visible pattern of activity that responds to changing external or internal conditions (Beck, Liem, & Simpson 1991; Levitis, Lidicker, & Freund 2009). Let us examine each part of this definition in turn.

Internally coordinated refers to internal information-processing such as endocrine signaling, sensory information processing, or the action of neurotransmitters. When two male giraffes meet during the breeding season, such processes coordinate their aggressive behavior. However, when a small caterpillar falls into a roaring river and is tossed about in the current, the animal is *not* exhibiting a behavior, because it has no control over its movement.

Externally visible activity refers to patterns that we can observe and measure. For example, we can observe a squirrel eating an acorn and can quantify this behavior. We cannot externally observe the variation in a lizard's heart rate.

We can, however, observe an animal's *behavioral response to changing conditions*. For example, male crickets, frogs, and birds vocalize in response to changes in day length, temperature, or moisture at specific times of the year. Similarly, during a summer day, a desert lizard moves from the top of a hot rock to the underside of a cool ledge to reduce its body temperature (Figure 1.4). Lizards are ectotherms and as such cannot regulate their body temperature internally but can do so behaviorally.

animal behavior Any internally coordinated, externally visible pattern of activity that responds to changing external or internal conditions.

Measuring behavior: elephant ethograms

Featured Research

Behaviors must be measurable—that is, we must be able to quantify our observations with numbers according to specific rules (Martin & Bateson 1993)—and different people must be able to recognize a behavior independently. Often, such characterization begins when a trained observer completes an **ethogram**, which is a formal description or inventory of an animal's behaviors. Ethograms typically list or catalogue defined, discrete behaviors that a particular species exhibits. Researchers can use an ethogram and measure how many times a behavior occurs (its frequency), the length of time of a behavior (its duration), the frequency of the behavior per unit time (its rate), or the vigor or forcefulness of the behavior (its intensity). Sometimes only certain behaviors, such as social behaviors (e.g., groom self, groom others, chase others), are quantified. Typically, a more complete list of behaviors that occur over a specific time period is recorded (e.g., sleep, groom self, groom others, walk, climb tree, eat food, drink). From these observations, a researcher can then determine both the total and the relative time that an animal engaged in each behavior—in other words, the

ethogram A formal description or inventory of an animal's behaviors.

TABLE 1.1 Elephant ethogram. This ethogram describes the behavior of captive elephants (*Source:* Rees 2009).

BEHAVIOR	DESCRIPTION
High-frequency behaviors	
Dusting	Collecting soil and throwing it over the body/rubbing it onto the skin (while standing still or walking), including digging into the soil for this purpose
Feeder ball	Feeding or attempting to feed at a metal feeder ball containing small quantities of food
Feeding	Collecting food with the trunk and placing it in the mouth while standing still or walking (does not include suckling or activity at the feeder ball)
Locomotion	Walking (except feeding or stereotyping)
Playing	Chasing another elephant/mock fighting with another elephant (but not as a result of an antagonistic encounter or as part of courtship)
Standing	Standing motionless (not while stereotyping or dusting)
Stereotyping	Repetitive behavior with no obvious purpose: weaving, head bobbing, pacing backward and forward or in an arc, walking in circles
Suckling	Calf suckling from mother or other female. Measured separately from feeding
Low-frequency behaviors	
Aggression	Hitting/pushing as a result of an antagonistic encounter (but not as a part of play)
Bathing	Standing/laying in pool/squirting water from pool over body with trunk
Digging	Digging in soil using the foot (but not as a part of a dusting behavior)
Drinking	Collecting water in the trunk and squirting it into the mouth
Lying down	Lying down on the ground (on its side or prone)
Rolling	Rolling in soil or mud (but not as a part of playing with another individual)
Sex	Courting or being courted/mounting another elephant or being mounted by another elephant of either sex

time budget A summary of the total time and relative frequency of different behaviors of an individual.

Figure 1.5. Captive Asian elephant. Ethograms are often constructed to describe and quantify the behavior of captive animals.

measure of the behavior divided by the overall time spent observing the animal. The resulting **time budget** indicates the total time and relative frequency of each behavior.

Ethograms are commonly constructed for animals in captivity and can be used as a baseline for healthy behavior. Paul Rees created an ethogram for captive Asian elephants (*Elephas maximus*) (Figure 1.5) at the Chester Zoo in the United Kingdom (Rees 2009). When elephants were in their outdoor enclosure, he recorded their behavior every five minutes for an entire day once a week for 11 months. His ethogram contains 15 behaviors, including feeding, standing, and digging (Table 1.1). In addition, he measured stereotypical behavior, or captivity-induced behavioral anomalies that are used as an index of the welfare of captive animals (Mason 2010). These behaviors often include repetitive behaviors that lack an apparent purpose, such as head bobbing or pacing. Rees found that captive elephants spent about a quarter of their time feeding, and that stereotypic behavior was negatively correlated with feeding behavior; in other words, the more time spent feeding, the less time spent pacing or head bobbing. Rees suggests that using widely spaced feeders to supply food slowly and at random times could reduce the frequency of stereotypic behavior.

We turn next to a discussion of the scientific method: the process scientists use to understand the behaviors we observe.

1.2 The scientific method is a formalized way of knowing about the natural world

When you think of science, you may think first of an impressive body of knowledge—the facts that you are asked to learn in class. Have you ever considered where that knowledge came from? It was obtained by human beings seeking to understand the natural world. Scientists engage in the **process of science**, which involves observing events, organizing knowledge, and providing testable explanations (Mayr 1982). The process of science is fundamental to our understanding of the natural world.

Scientific discoveries continue to occur today at a tremendous pace. Every week, new discoveries appear on television, in magazines, and on the Internet. As instructors, we look back at our courses from just five years ago and marvel at how much has changed in so short a time. You and your instructors may be involved in new discoveries yourselves.

Why does scientific progress occur so rapidly? What is different about scientific knowledge compared with knowledge in the humanities? As we see next, the process of science means constructing hypotheses that make testable predictions.

The importance of hypotheses

Some disciplines—like English, history, and philosophy—are part of the humanities. Others, like anthropology and sociology, are social sciences. Still others, like biology and chemistry, are natural sciences. In all these disciplines, scholars seek to gain a better understanding of the world. In the humanities, researchers are interested in understanding the human experience. The social sciences involve knowledge of human behavior and societies. And in the natural sciences, researchers strive to understand the natural world.

You may have noticed differences not only in the information presented in different courses but also in how you are expected to think about that information. For example, the humanities discuss art, literature, and how humans think and act. You might be expected to present your own perspective while critically evaluating the perspectives of others. Often, two people may arrive at very different perspectives using similar evidence.

Historians, for example, strive to understand why certain events happened by interpreting personal journals, letters, government publications, and newspaper articles. A historian forms a thesis, or narrative, to explain events and then interprets relevant documents to support that thesis (Hult 2002). For instance, a historian might be interested in what motivated Thomas Jefferson, John Adams, James Madison, and other Founding Fathers to declare independence from England and form their own system of government. Were economic considerations, such as preservation of individual wealth, the primary motive (Beard 1913), or convictions about individual liberty and rights (McDonald 1958)? By examining the Declaration of Independence, the U.S. Constitution, the *Federalist Papers*, and letters written by key figures, historians have developed these and other interpretations. In each case, research involves discovering, reevaluating, and interpreting the evidence.

Scientists, like other scholars, begin with questions, but they formulate these questions into hypotheses. Hypotheses are explanations that make predictions that can be tested. Because these tests can be repeated and confirmed by other scientists, the results are much less subject to debate. This is why science sometimes seems like just an accumulation of facts. Yet these facts are the results of scientific studies that have been confirmed over and over by the scientific community over many years.

The scientific method

Scientists use the scientific method to learn about the natural world. The scientific method is a formalized process that involves the testing of hypotheses (Figure 1.6). This process often begins with an observation of a single event or pattern that requires explanation. This observation forms the basis of a **research question**—a brief statement of something that we would like to understand. For example, suppose

process of science Observing events, organizing knowledge, and providing explanations through the formulation and testing of hypotheses.

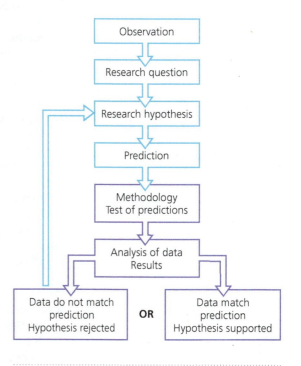

Figure 1.6. The scientific method. This flowchart summarizes the scientific method.

research question The first methodological step in the process of science. A formal statement of an unknown that one would like to understand.

research hypothesis an explanation based on assumptions that makes a testable prediction.

alternate hypothesis The statistical hypothesis that the proposed explanation for observations does have a significant effect.

null hypothesis The statistical hypothesis that observations result from chance. The hypothesis of no effect.

you were walking through your neighborhood and observed that some yards had many robins (*Turdus migratorius*) feeding in them, while other yards had very few. Robins, common songbirds found on lawns throughout North America, feed on invertebrates such as earthworms and beetle grubs. This observation might lead to the following research question (Scientific Process 1.1):

Research question: Why is there variation in the number of robins feeding in different yards?

The identification of patterns like this can be accomplished with careful observation and mere human curiosity. Throughout this book, you will see how researchers have identified different behavioral patterns that have led to a variety of research questions.

The next step in the scientific process is the formulation of a hypothesis, or, more formally, a research hypothesis. You may think of a hypothesis as an educated guess, but this definition is far too simplistic. Rather, a **research hypothesis** is an explanation based on assumptions that produces a testable prediction. Research hypotheses are evaluated using two statistical hypotheses that reflect the two possible outcomes. One outcome is that the proposed explanation for the observation *does* have a significant effect; this is the **alternate hypothesis**, or H_a. The other possible outcome is that the proposed explanation does *not* have a significant effect; this is the **null hypothesis**, or H_0, the hypothesis of no effect. These terms were coined by Sir Ronald Fisher (1966), a British geneticist, evolutionary biologist, and statistician. Together, the null and alternate hypotheses are known as statistical hypotheses.

For example, to explain why more robins feed in some yards than others, you might hypothesize that the quantity of food varies between yards and that this variation affects robin abundance. This leads to two statistical hypotheses:

Alternate hypothesis: The amount of food in a yard determines the number of robins feeding there.

Null hypothesis: The amount of food in a yard does not determine the number of robins feeding there.

The alternate hypothesis assumes that the amount of food in a yard is the only factor that determines the number of robins feeding there. Both the alternate and null hypotheses make predictions:

Alternate hypothesis prediction: Yards with more food will have more robins.

The null hypothesis makes a different prediction:

Null hypothesis prediction: There is no relationship between the amount of food in a yard and the number of robins in the yard.

The last step in the scientific process is to evaluate the research hypothesis by testing the prediction of the null hypothesis. One way to do this is to make new observations by collecting data from many different yards on the number of robins and the number of earthworms (a common food of robins) in each one. With these data, you can now test the prediction of the null hypothesis. There are only two possibilities. If there is no relationship between the amount of food in a yard and the number of robins, then you *fail to reject the null hypothesis*. When this occurs, you also *reject the alternate hypothesis*, because this hypothesis does not provide an adequate explanation for the pattern you observed. Alternatively, if you find that yards with more food have more robins, then you *reject the null hypothesis* and *fail to reject the alternate hypothesis*. You have thus found support for the alternate hypothesis.

When scientists find support for a research hypothesis, we do not conclude that the hypothesis has been "proven" to be correct. Scientists can never prove that a hypothesis is correct (Popper 1959) because random chance or an untested variable (i.e., a different alternate hypothesis) could also lead to the observed phenomenon. So all a scientist can do is to test the hypothesis repeatedly and always fail to reject it. As a result, all hypotheses are tentative explanations of observed phenomena. They may eventually be rejected and replaced by hypotheses that make more accurate predictions. Scientists can, and do, test multiple hypotheses simultaneously to

SCIENTIFIC PROCESS 1.1

Robin abundance and food availability
RESEARCH QUESTION: *Why is there variation in the number of robins feeding in yards?*

Hypothesis: The amount of food in a yard determines the number of robins feeding there.

Prediction: Yards with more food will have more robins.

Methods:

Researchers:
- counted the number of robins in each yard at the same time of day in 30 yards.
- quantified the amount of food (earthworms) available to robins in each yard by examining several 900 cm^2 sampling areas.
- mixed 40 g of yellow ground mustard seed into 4 L of water, then poured the mixture into each sampling area so that it was absorbed into the soil.
- counted the number of earthworms that emerged in each sampling area over a period of 10 minutes.

Results:

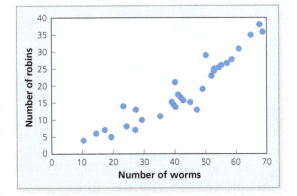

OR

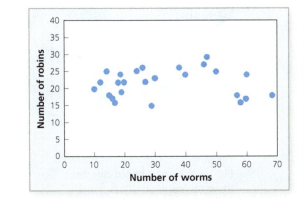

Figure 1. The highest number of robins were found in yards with the most worms.

Figure 2. There is no relationship between the number of robins and the number of worms in yards.

Conclusion 1: The hypothesis is supported. Yards with more earthworms have more robins.

Conclusion 2: The hypothesis is not supported. Yards with more earthworms do not have more robins.

explain observations of the natural world that can facilitate rapid scientific advancements (Platt 1964).

What do you do when you fail to reject the null hypothesis? When this occurs, you need to develop a new research hypothesis to explain the original observation—in this case, differences in the number of robins feeding in yards. This might involve a slight modification to the rejected hypothesis or something radically different. For instance, you might now postulate that robins avoid areas with predators like cats. Your new hypothesis would predict that yards with no predators will have high numbers of robins, while yards with predators will have few robins feeding there. This new hypothesis makes a completely different prediction than your original hypothesis—and this prediction, too, can be tested.

SCIENTIFIC PROCESS 1.2

Robin abundance and predators
RESEARCH QUESTION: *Why is there variation in the number of robins feeding in yards?*

New Hypothesis: The presence of predators affects the number of robins feeding in yards.

New Prediction: Yards with no predators will have a high number of robins, whereas yards with predators like cats will have few robins.

Methods:

Researchers:
- conducted an experiment. In half the yards, placed a cat (predator) in a cage in the center of the yard (treatment). In all other yards, placed an empty cage in the center of the yard (control).
- counted the number of robins in each yard at the same time of day for all yards.

Results:

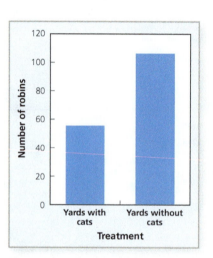

Figure 1. Yards without cats had more robins.

OR

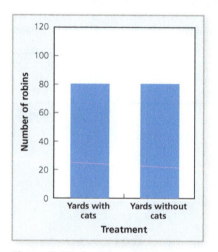

Figure 2. Yards with and without cats had similar numbers of robins.

Conclusion 1: The hypothesis is supported. Yards without cats have more robins.

Conclusion 2: The hypothesis is not supported. Yards without predators do not have more robins.

This time, you would gather data on the numbers of cats and robins in different yards (Scientific Process 1.2).

Correlation and causality

Let's say that the data do allow you to reject the null hypothesis. Suppose you found a positive relationship between the number of earthworms and the number of robins feeding in yards. Clearly, these data support your alternate hypothesis that the amount of food in a yard determines the number of robins feeding there. The data you collected represent a **correlation** between the two variables measured: they co-vary, or vary together, predictably. Correlations can be positive, as in this case, when both variables display either low or high values simultaneously, or negative, when one variable displays high values while the other displays low values. In addition, correlations may be zero when two variables vary independently of each other (Figure 1.7).

correlation Two variables that vary together predictably.

However, we still have work to do to test the hypothesis more fully. The reason is that correlation does not demonstrate causality. Even if two variables are correlated, one does not necessarily cause the other. There may be other reasons why the number of robins and the number of worms in yards co-vary positively. Perhaps some yards have shrubs that are excellent nesting sites for robins while others do not, and the yards with shrubs also have more earthworms. It would be unclear whether the robins preferred certain yards because of their earthworms, their shrubs, both, or some other factor entirely. Because correlation does not demonstrate causation, researchers often try to rule out plausible alternative explanations. They also try to find a mechanism that might underlie the correlation. Scientists place a premium on mechanistic explanations for patterns because mechanisms help to explain why the pattern occurs and so strengthen their conclusions. Because it is very difficult to rule out all other possibilities, it can be very difficult to establish causality through observation alone. Many scientists therefore test hypotheses using controlled experiments, as we will see in Chapter 3, but even experiments cannot prove a hypothesis is correct.

Hypotheses and theories

The more times a hypothesis has been tested and not rejected, the more confidence the scientific community has in the explanation. Hypotheses that make many predictions, have been tested hundreds or thousands of times by many different scientists, and have not been rejected come to be known as **scientific theories.** Scientific theories provide a conceptual framework that explains many phenomena and are well supported by observations and experimental tests. Examples include the cell theory of living organisms, the germ theory of disease, and Charles Darwin's theory of evolution by natural selection. In sum, scientific theories are well-substantiated explanations that form the basis of our understanding of the natural world (National Academy of Sciences 1998).

Note that the word *theory* has a very special meaning in science. For the general public, a theory often means a mere guess. You might have a "theory" about who the murderer is on a television show or who ate the chocolate chip cookies in your dorm. In contrast, scientific theories are hypotheses that have been tested repeatedly. They are more general than a specific hypothesis and tend to unify and simplify our knowledge by synthesizing information into a larger framework. For example, cell theory states that cells are the fundamental unit of all organisms.

Social sciences and the natural sciences

So far, we have discussed the humanities and the natural sciences. What about the social sciences, such as psychology? Social scientists create and test hypotheses using the scientific method to study human behavior and societies. Their methodology often includes collecting data from surveys, questionnaires, and public opinion polls, as well as conducting experiments to obtain information about human values, perceptions, and behavior. For example, social scientists might be interested in understanding why there is a higher smoking rate among adults in Missouri than among adults in Arizona. They might hypothesize that people in states with more antismoking education programs have a better understanding of the negative health effects of tobacco and so are less likely to smoke. To evaluate this hypothesis, the researchers would first need to collect data on the antismoking educational programs used in these states and the effect of these programs on a smoker's intention to quit, quitting success rate, and so forth. Alternatively, social scientists could conduct an experiment. Smokers might be randomly assigned to receive either antismoking educational material or no educational material (as a control). Again, the researchers would collect data on each smoker's intention to quit, quitting success rate, and so forth. Many other factors might also affect tobacco use, such as income or social factors, and these too would need to be considered.

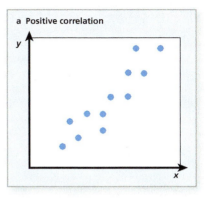

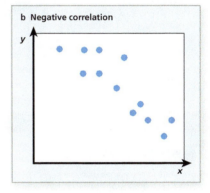

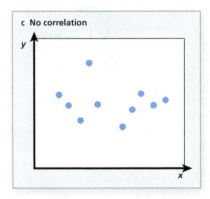

Figure 1.7. Correlations. Correlations can be (a) positive, when both variables (x, y) increase, or (b) negative, when one variable increases as the other decreases. In addition, there may be (c) no correlation between two variables when they vary independently.

■ **scientific theories** Hypotheses that make many predictions, have been tested many times by many different scientists and have not been rejected.

Where does the study of animal behavior fall in this framework? Animal behavior courses are typically offered by biology or psychology departments. Animals are a part of the natural world, and so it makes sense that animal behavior is a part of the natural sciences. Why, then, is animal behavior also studied in psychology departments? The answer is that psychologists have two ways to study human behavior: (1) directly or (2) indirectly, through the study of animals, in the same way that medicine uses animal models to understand human health. Psychological research on the evolution of human behavior often involves the study of our closest relatives. Animal behavior research, whether conducted by biologists or psychologists, uses the scientific method to formulate and test hypotheses.

1.3   Animal behavior scientists test hypotheses to answer research questions about behavior

To study the behaviors of animals, scientists must form and test hypotheses to answer research questions. These questions can be based on earlier observations, previous knowledge, prior research, a theory, or some combination of these. Let's examine how this process works using the case of wolf spiders.

Featured Research

Hypothesis testing in wolf spiders

Figure 1.8. Wolf spider. Male brush-legged wolf spiders wave and tap their forelegs to court females.

The brush-legged wolf spider (*Schizocosa ocreata*) is common in the leaf litter habitat of eastern North American forests. During the brief breeding season each spring, males search for females (Figure 1.8). Females typically mate only once, whereas males mate multiple times, and so unmated females become increasingly rare as the breeding season progresses. How do males find unmated females? Females are cryptic, meaning that their coloration lets them blend in well with the leaf litter. As a result, males search for females slowly, using chemical cues that females deposit on silk threads in the leaf litter. When males find receptive females, they court them with a multimodal display by walking rapidly in a jerky fashion while waving and tapping their forelegs (providing visual cues). They also transmit vibrations through the substrate by rubbing specialized organs together (providing seismic cues). These signals can be sensed by the female and may also be used by other nearby males to find a mate. George Uetz, his student Andy Roberts, and collaborator Dave Clark were interested in this mating behavior (Roberts et al. 2006). The research team investigated the following research hypothesis:

Research hypothesis: Male wolf spiders use the behavior of nearby males to find receptive females.

The null hypothesis is that males do not use the presence and behavior of other males as a cue to locate receptive females, while the alternate hypothesis is that they do. The alternate hypothesis predicts that the presence and behavior of a rival male will cause the searching male to head in the rival's direction, search for a female, and begin courtship tapping.

The research team tested these predictions by conducting an experiment. They collected juvenile spiders from a nearby forest and raised them in the lab until they were sexually mature. Focal males (the subjects of study) were tested for their reactions to (1) visual cues only (the sight of a demonstrator male tapping), (2) seismic cues only (the vibrations of another male), and (3) both cues simultaneously. The researchers recorded the focal male's behavior for three minutes prior to the stimulus, three minutes during the stimulus, and three minutes after the stimulus. This allowed researchers to characterize baseline behavior and examine changes in behavior after exposure to test stimuli. In each phase of a trial, the research team recorded two behaviors of the focal male: its chemoexplore behavior, or behavior to detect chemical signals (slow walking while rubbing the substrate with the pedipalp to detect chemical signals); and its courtship behavior, or behavior to find a mate (rapid walking in jerky movements while tapping the substrate). The researchers prompted the demonstrator males to tap by providing them with the silk of females that contained their chemical cues.

The researchers used 15 focal males in each treatment. In any study, it is important to collect data on many individuals to ensure that data are representative of the population or species. Individuals always exhibit variation, but hungry or sick individuals may exhibit particularly unusual behavior. The collection of data from many individuals helps prevent researchers from drawing conclusions that are based only on a few nonrepresentative individuals.

The research team found that the treatments had no effect on chemoexplore behavior (Figure 1.9). From these data, the team accepted the null hypothesis: male wolf spiders do not appear to use the visual or vibratory behavior of a courting rival as a cue to the location of a receptive female. Apparently, males just have to find females on their own.

Then a funny thing happened. A year or so later, Roberts, Uetz, and Clark were examining the mating behavior of these spiders in nature and observed males heading in the direction of other males who were tapping in courtship. This was exactly the behavior that they did *not* observe in their lab experiments. What was going on? They remembered that in their previous experiment, they had used a population of males raised in isolation in captivity that had had no experience with rival males. Could prior social interactions be an important prerequisite for this behavior?

To test this new hypothesis, the researchers conducted similar experiments using adult males caught in the wild. This time, they found that the focal males did display a greater frequency and duration of interactive behaviors (i.e., approach, courtship signaling) during and after exposure to the sight of another male tapping. The findings indicated that male wolf spiders can indeed use the tapping of rival males as a cue for finding females (Clark, Roberts, & Uetz 2012).

This story illustrates an important point. Research is an ongoing process of discovery. The behavior of males in the wild did not seem to match observations of laboratory-reared subjects, which led to a new hypothesis and experiments and a new understanding of spider behavior. This is an excellent example of how science progresses.

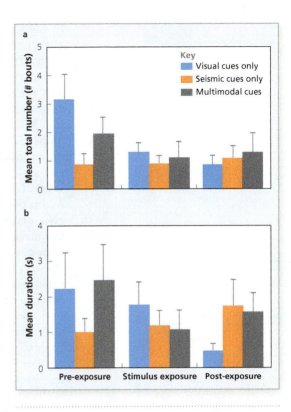

Figure 1.9. Spider response to male cues. There was no effect of experimental treatments on male behavior, as measured by (a) number of chemoexplore bouts or (b) duration of chemoexplore bouts before, during, or after stimulus exposure. The height of each solid bar represents the mean behavior, and the thin lines represent a measure of variation in the data, the standard error of each mean. (*Source:* Roberts et al. 2006).

Negative results and directional hypotheses

In the wolf spider example, the null hypothesis could not be rejected in the initial experiment. When scientists do not reject the null hypothesis (and thus reject the alternate hypothesis), we have **negative results**. Negative results indicate that the alternate hypothesis does not explain the behavior being examined. When this happens, scientists need to develop a new alternate hypothesis to explain the observation. This is an important part of the scientific process. It does not indicate a failure by the scientist, because we have learned that one possible explanation is incorrect.

Hypotheses can also be directional or nondirectional, depending on the specificity of the hypothesis. A directional alternate hypothesis predicts specifically how the variable under examination will affect a particular behavior—positively or negatively—whereas a nondirectional alternate hypothesis usually offers no specific prediction of how the variable will affect behavior. Researchers use a nondirectional hypothesis when they do not have a specific prediction about an animal's response. For example, when a moving individual determines that a predator is nearby, it could quickly flee to a safer location or it could become motionless to avoid detection.

negative results A situation in which one does not reject the null hypothesis (and thus one rejects the alternate hypothesis).

In one scenario, the speed of the individual's movement would increase as a result of detecting a predator, while in the other scenario, the speed of the individual's movement would decrease. In both cases, the speed of movement changes once the individual has observed a predator, but the direction of the change is different. Therefore, the appropriate hypothesis is nondirectional, since a specific prediction about movement (increase or decrease) may not be possible. The null hypothesis would state that the presence of a predator would not affect the speed of movement of the individual, while the alternate hypothesis would state that the presence of a predator would affect the speed of movement of the individual.

Generating hypotheses

At this point you might be wondering, "Where do hypotheses come from?" This is a great question. Recall that hypotheses are explanations for observed phenomena that make predictions. In other words, hypotheses are created by scientists to explain some observation, which for our purposes is a behavior exhibited by animals. Many scientific studies first begin investigating a system by describing it in order to identify patterns that exist. Once a system is described, hypotheses can be generated and tested to understand the observed patterns.

Researchers also construct hypotheses by considering how natural selection might have acted on the behavior to produce a behavioral adaptation. In this approach, they ask, "How does the observed behavior affect the fitness of the animal?" **Fitness** here is defined as the survivorship and reproductive success of an individual. (We will explore fitness further when we study evolution, starting in Chapter 2.) Such thinking can help generate alternate hypotheses that produce testable predictions.

In the wolf spider example, the hypothesis that males use rivals as a cue to locate females is based on evolutionary reasoning. It is beneficial for a male to be able to locate receptive females, because doing so allows the opportunity to court and mate. Any cue that reduces search time and increases encounters with receptive females should be advantageous.

Hypotheses from mathematical models

Another source of hypotheses and predictions are mathematical models. Scientists often use equations to clarify arguments and assumptions. The resulting mathematical models can generate predictions about which behaviors maximize an individual's fitness. All models about behavior are based on assumptions about the ecology and evolution of an organism. The advantage of mathematical models is that they allow scientists to easily manipulate their assumptions to produce new predictions about behavior.

For example, Hanna Kokko and Leslie Morrell constructed a model to understand variation in mate-guarding behavior—a behavior in which males guard a female before and after mating to ensure that she does not mate with other males (Kokko & Morrell 2005). In many species, females may mate with more than one male in a fairly short time period. This creates the potential for competition among the sperm of different males to fertilize the female's eggs. As a result, males cannot be sure of the paternity of offspring. What could males do to increase the probability that they will be the father of a female's offspring and thus increase their fitness? Mate guarding is one possible answer. However, the benefit of this behavior to the male comes with a cost: while mate guarding one female, he cannot be searching for or mating with other females. He thus faces a trade-off—and a question: How long should a male spend mate guarding a female? In other words, what amount of female mate guarding maximizes fitness for a male?

Kokko and Morrell created a mathematical model to predict how long a male should guard a female he has inseminated. The model was based on several assumptions. First, it assumed that male fitness increases with the number of eggs fertilized. This assumption seems reasonable, because the more eggs a male fertilizes, the

---

fitness The relative survivorship and reproductive success (ability to produce viable offspring) of an individual.

more offspring he is likely to sire. Second, the model assumed that some males are attractive to females, while others are unattractive, and females prefer to mate with attractive males because doing so increases their fitness. This is true for many animals, as we will see when we examine mating behavior in Chapter 11. Third, the model made a set of assumptions about possible female behavior. Once a female has mated with a male, that female could very rarely mate with other males; almost always try to mate with other males she encounters, even if she has already mated; or sometimes try to mate with other males, depending on whether the male with whom she just mated was attractive or unattractive. Kokko and Morrell included mathematical functions for each of these assumptions in their model.

A first set of predictions is drawn from the assumptions about female behavior. Kokko and Morrell's model predicts that if females either very rarely mate with other males or almost always try to mate with other males, then a male should not spend much time mate guarding (Figure 1.10). In the first case, the benefit of mate guarding is minimal, since the female will rarely mate with other males. In the latter case, the cost of mate guarding is so high that the benefit of the behavior will not be large enough to offset the cost.

A second prediction can be drawn from the third assumption about how male attractiveness affects female behavior. The model predicts that unattractive males should exhibit higher levels of mate-guarding behavior than attractive males when interacting with females that sometimes try to mate with other males, because females that mate with unattractive males may try to mate with a more attractive male in order to increase their fitness.

In short, mathematical models allow scientists to make different, testable predictions by varying their assumptions. These models are especially useful because a single model can be applied to many species and different assumptions about their behavior can be easily incorporated. Kokko and Morrell note that their model might best be tested on different species of birds.

Many hypotheses about animal behavior begin with the assumption that behavioral traits are adaptations that enhance fitness. These hypotheses must make testable predictions about behavior. In contrast, many people explain behavior by assuming that animals share the same emotions as humans. As we see next, such explanations can be problematic, because they rarely make testable predictions.

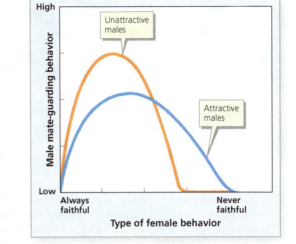

Figure 1.10. Mate-guarding model. This model predicts that mate guarding should be low for all males if females are always or rarely faithful. It also predicts that for females that tend to be faithful, unattractive males should guard more often than attractive males, whereas for females that tend to be less faithful, only attractive males should mate guard (*Source:* Kokko & Morrell 2005).

1.4 Anthropomorphic explanations of behavior assign human emotions to animals and can be difficult to test

Anthropomorphism is the tendency to attribute human motivations, characteristics, or emotions to animals. For example, a television documentary might describe a female coyote (*Canis latrans*) as "worried" when her mate is late returning to the den and "happy" when her mate returns with food. Another show might describe a female meerkat (*Suricata suricatta*) as "sad" when her grown offspring leave the den to establish their own territories. Of course, many cartoons depict animals as essentially humans in animal form—talking, singing, showing feelings, and scheming to achieve different goals.

Perhaps because anthropomorphism is so prevalent in the media, it is also very common among the public (Applying the Concepts 1.1). People often attempt to explain the behavior of animals by attributing their actions to anger, jealousy, or happiness. Maybe a bee stung your leg as you tried to swat it away. Using anthropomorphic thinking, you might believe that the bee stung you because you made it angry. This hypothesis may or may not be true, but how could you possibly test it?

■ **anthropomorphism** Attributing human motivations, characteristics, or emotions to animals.

APPLYING THE CONCEPTS 1.1

What is behind the "guilty look" in dogs?

Have you ever come home to find food missing off the counter or the garbage strewn all over the floor? Many dog owners have experienced this and have immediately scolded their dog for disobedience. They often report that their dogs displayed a "guilty look" because they "knew" they had been disobedient (Figure 1). These owners describe their dog's guilty look as avoiding eye contact, lying down and rolling on its side or back, wagging its tail low and quickly, holding its head down, licking or pawing, and moving away from the owner. But do dogs experience "guilt"? This is an anthropomorphic explanation of a behavior. Alexandra Horowitz is a cognitive scientist who studies dog behavior. She designed an experiment to examine the "guilty look" in dogs (Horowitz 2008).

Horowitz recruited 14 dog owners for her experiment. Each dog was exposed to four experimental treatments. At the beginning of each trial, the owner showed the dog a food treat and gave the dog a command not to eat it. The owner then left the room. The experiment had two conditions: either (1) the researchers gave the dog the treat (the dog ate the treat and disobeyed the owner's command), or (2) the researchers removed the treat (the dog could not eat the treat and disobey the owner's command). Once the treat was gone, the owners were called back into the room. They were told either that (1) the dog ate the treat or (2) the dog did not eat the treat (it had been removed), although this information may or may not have been true. Each dog experienced all pairwise combinations of the conditions. Owners were told to scold their dogs when told the treat had been eaten and to greet their dogs

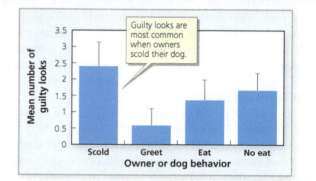

Figure 2. Dog responses. The mean (+ SE) number of guilty looks was highest when dogs were scolded by their owners (*Source:* Horowitz 2008).

in a friendly manner when told the treat had not been eaten. All trials were videotaped, and the researchers quantified the number of times each dog displayed behaviors associated with the "guilty look" (avoiding eye contact; lowering head, tail, or ears; moving away from the owner; and so forth).

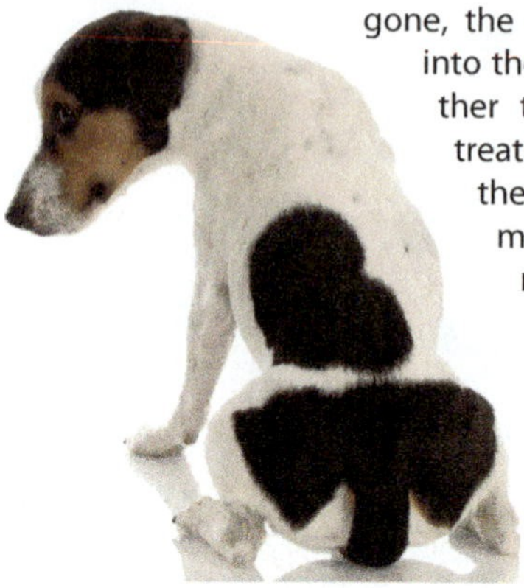

Horowitz found that the dog's obedience had no effect on its display of "guilty look" behaviors: the mean number of guilty look behaviors was similar whether or not the dog actually ate the treat. Owner behavior did have an effect on dog behavior, however, because "guilty look" behaviors were most common after owners scolded their dog (although in half of these trials, the dog had not disobeyed the command) and least common when they greeted their dog (Figure 2). Horowitz concluded that the dogs did not display more behaviors associated with the guilty look after disobeying a command. Horowitz speculates that the "guilty look" likely represents submissive behavior in response to owner scolding. However, she notes that her experiment does not rule out the possibility that dogs could experience guilt. Anthropomorphic explanations of animal behavior remain intriguing but difficult to test.

Figure 1. Guilty look.

What prediction does this hypothesis make? It is difficult, if not impossible, to characterize the emotional state of a bee.

One hypothesis to explain the behavior of the bee does make a prediction: the bee perceived your attempt to swat it as an attack, leading to antipredator behavior. In other words, bees sting potential predators to deter attacks. You could test this hypothesis by observing interactions between bees and their predators and then quantifying how often they use their stinger. As a control group, you could compare the interactions between bees and a nonpredator and quantify how often bees sting that species. The hypothesis would predict that bees will more often sting potential predators than they will nonpredators.

While anthropomorphic thinking rarely produces testable predictions, we do not want to dismiss the possibility that animals experience emotions. Indeed, one area of active research, known as cognitive ethology, attempts to understand animal cognitive, or thought, processes. Understanding animal thinking is challenging, but researchers are making progress in its study, as we will see in Chapter 5. In addition, for many animals, there is often a relationship between stress-related behaviors and physiology, as with the production of glucocorticosteroids in vertebrates (Blas et al. 2007). In these cases, we truly are making progress in understanding animal mental states and emotions. However, anthropomorphic hypotheses remain difficult to test and should be avoided until we are better able to understand animal emotions and cognition.

1.5 Scientific knowledge is generated and communicated to the scientific community via peer-reviewed research

Every day we hear about important scientific discoveries in areas ranging from human health to the environment. How do we evaluate all this information? How do we determine what is based on valid research and what is just personal opinion? The ability to evaluate scientific information critically and ascertain its validity is called **scientific literacy**. However, scientific literacy would not be possible if scientists could not evaluate and communicate their research in the first place. Even the most far-reaching discoveries remain unknown until scientists communicate their findings to the world.

scientific literacy The ability to evaluate scientific information critically and ascertain its validity.

The primary literature

Animal behavior research is conducted by investigators and their graduate and undergraduate students at universities, zoological parks, veterinary schools, government and private research institutions, and conservation organizations. After conducting research, scientists write papers and submit them to scientific journals to report their findings. These journals, known as the **primary literature**, are the primary or original source of scientific information (Figure 1.11). The editors of scientific journals use experts to help decide whether to accept or reject a paper for publication, a process known as **peer review**. These reviewers evaluate the importance of the research question and the validity of the hypotheses, methodology, analyses, and conclusions. They also comment on how clearly and concisely the paper is written; if a paper is poorly written, it will be difficult for the reader to evaluate the science. Why are a great many papers rejected? Reviewers scrutinize manuscripts for accuracy, and papers are rejected for such reasons as unsound methodology, incorrect analyses or statistics, poor experimental design, and faulty or inappropriate logic. As a result, the knowledge communicated in the primary literature has passed a rigorous, objective review.

primary literature The original source of scientific information, typically peer-reviewed scientific journals.

peer review A process in which editors of scientific journals use experts to help decide whether to accept or reject a paper for publication.

Each discipline has a variety of specialized journals (see Appendix 1). In addition, if the research presents results of broad scientific appeal, it may appear in one of the prestigious and less specialized scientific journals, such as *Science, Nature, Proceedings of the National Academy of Sciences, Proceedings of the Royal Society of London*, or the *Public Library of Science*.

Scientific papers follow a systematic format, which is summarized in Appendix 2 of this book. For instance, the "Methods" section of a paper contains detailed information about how the research was conducted. Such an explanation is important because other scientists must be able to replicate the work in order to test its validity. The paper may also contain figures and tables that summarize important findings. Most natural scientists follow the style manual of the Council of Science Editors (CSE), which describes how to cite references and handle other matters of style. Psychologists and other social scientists follow the American Psychological Association style format (APA), while medical journals tend to use the American Medical Association style format (AMA). Here we follow a modified version of the author-year system of the CSE style (e.g., Nordell & Valone 1998).

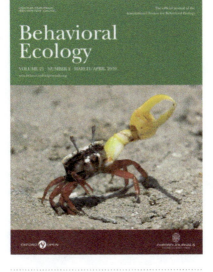

Figure 1.11. Scientific journal. *Behavioral Ecology* is a peer-reviewed journal devoted to animal behavior research.

TOOLBOX 1.1

Scientific literacy

When we hear or read about a scientific discovery, we tend to assume that the information is an accurate and unbiased representation of the original finding. In order to evaluate the accuracy of the report, it is important to be able to refer to the primary literature. How can you determine whether the information you are reading is peer reviewed (from the primary literature), an interpretation of the research (from the secondary literature), or simply someone's personal opinion? Answering this question is an important part of scientific literacy and is critical to being an informed citizen. This information can affect our health, the environment, and how we vote. Here are some of the key questions to ask.

1. **What is the name of the publication, and is it endorsed by a scientific society?**

   Most disciplines have scientific societies whose members are experts in that field. Many scientific societies publish a primary literature publication of research in their discipline.

2. **Who are the authors?**

   What are the researchers' institutional affiliations? Most primary research investigators will be affiliated with a college or university, medical/veterinary/research center, or zoological park or conservation organization. If the scientists are affiliated with an institute, it is important to identify who might be funding their research to determine if there are any conflicts of interest. Do the authors have scientific degrees such as a Ph.D., MD, or DVM? Most primary research investigators will have one (or more). Journals often do not list degrees, but you can often ascertain this information from additional websites.

3. **Is the article written in the standard scientific format, with an introduction, a methods section, a results section, a discussion or conclusion, and literature cited or references list?**

   Most primary research articles are written using this standard format. Not all journals require the term "introduction" as an actual heading, but they do usually require it as a portion of the paper. Secondary sources will not use this format.

4. **Does the article cite the primary literature in the references?**

   Many secondary literature publications do not cite references, whereas almost all primary literature articles do. Usually the publication will use the author-date system (author's last name, year of publication) within the text.

5. **Does the publication indicate that all articles are peer reviewed?**

   You can go to the website of the publication and see if there are instructions to the authors about how the submitted paper will be evaluated. If there is no information about peer review, the publication is probably secondary literature. Does the information on the journal website indicate that it publishes primary research in the discipline?

The secondary literature

secondary literature A report that summarizes and interprets the primary literature. Often reported in newspapers, magazines, and books.

If the results of scientific research are of interest to the public, they may also be reported in newspapers, magazines, and books, as well as on television and over the Internet. Reports like these are called the **secondary literature**, because the information has been presented before and often does not have the same validity as the primary literature.

Peer review is a hallmark of most research. It allows experts in a field to critically evaluate potential new knowledge. However, hypothesis testing remains the domain of the sciences and provides a means for the rapid advancement of our knowledge about the natural world.

Chapter Summary and Beyond

Animal behavior is important because it intersects with the lives of many people. We define such behavior as the internally coordinated activity patterns of animals that respond to changing external or internal conditions. Levitis, Lidicker, and Freund (2009) provide a nice review of the many definitions of behavior that scientists use. Modern animal behavior research involves more than simply observing animals; the science of animal behavior also involves testing hypotheses to explain behavior.

Science is a way of knowing about the natural world that involves using the scientific method. Scientists use the scientific method to construct hypotheses that make testable predictions. Researchers analyze data to determine whether they support or refute their hypothesis. Science never proves a hypothesis is correct. Rather, it is an ongoing process of hypothesis testing, and that process is crucial to scientific progress. A hypothesis that makes broad predictions and stands up to repeated testing may eventually become a theory.

Scientists must consider the limitations of studies that report correlations, because correlation does not demonstrate causation. In addition, anthropomorphic hypotheses assign human emotions to animals and are difficult to test. For a review of the history and problems associated with anthropomorphism in the scientific study of animal behavior, see Wynne (2004b; 2007).

Scientific literacy is the ability to evaluate scientific information critically. Scientific literacy includes understanding how hypothesis testing drives the scientific process. It also includes understanding peer review, in which scientists determine whether research is clear, valid, and important enough to be published in the primary literature. Laine and Mulrow (2003) describe the value of peer review and provide insights into how the process works. Papers in scientific journals follow a consistent, systematic format. The public often reads about them in the secondary literature.

CHAPTER QUESTIONS

1. Spend 20 minutes quietly observing an active animal. This animal could be a squirrel or bird on campus, a captive animal at the zoo, or your pet. Write one to two paragraphs describing the animal's behavior in detail. Note the species and location of the animal.

2. For each of the following, determine whether the example describes a behavior and use the term's definition to explain why. (a) A spider spins a web to catch food. (b) Male fish construct nests to attract females. (c) A fiddler crab waves its claw to fight a rival. (d) A lion's mane moves away from its body because of a gust of wind. (e) A dog salivates just before it is fed dinner. (f) A butterfly moves from a shady to a sunny location.

3. Many disciplines use the word "science" to describe their work. Your college or university may have departments of computer science, political science, food science, or even family and consumer science. Describe one of these disciplines and then determine whether or not you could consider it to be a "science." In this chapter, we discussed how science differs from other disciplines, and you can use that information to develop criteria for this assignment.

4. Why is a hypothesis not simply an educated guess?

5. Find an article on animal behavior from the primary literature and another from the secondary literature. Describe how you determined why one is an example of the primary literature and the other an example of the secondary literature.

6. Many people feed birds because they believe it makes the birds happy. How could you test this hypothesis?

7. Suppose you took your dog to the veterinarian because it was having intermittent diarrhea. One possible explanation for this symptom is that the dog has parasites. Many parasites go through life cycles in which they are only shed from the intestinal tract of the host every few days. The veterinarian has a fecal sample analyzed for parasites. The results come back negative—that is, the laboratory did not find any parasites in the fecal sample. Using the scientific method, can you now rule out parasites as the cause of the diarrhea? What is the rationale for your answer?

Evolution and the Study of Animal Behavior

Figure 2.1. Birds at a feeder. Birds feeding are at risk of predation.

We enjoy watching birds at the feeders in our backyard. Finches, sparrows, chickadees, cardinals, and jays are regular visitors (Figure 2.1). Every so often, so too are Cooper's hawks (*Accipiter cooperii*), predators of small birds. One day, we were watching a group of finches at the feeder when all of a sudden a Cooper's hawk appeared from over the house and attacked. The finches scattered, fleeing quickly to dense shrubs. Three escaped, but one ended up in the talons of the hawk. Such scenarios play out every day in nature—some individuals survive while others do not. The same variation occurs during mating season: some individuals produce many offspring while others produce none. Variation in survivorship and reproduction is the raw material for adaptive evolution.

Evolutionary biologist Theodosius Dobzhansky once said, "Nothing in biology makes sense except in the light of evolution" (1973), because evolution is the foundation of our understanding of the natural world, and animal behavior is one important aspect of that world. In this chapter, we examine the process of evolution by natural selection. We start by describing natural selection in some detail and examine the conditions required for evolution to occur. Although natural selection results in changes in populations, it acts on individuals that collectively make up those populations, rather than on the populations themselves. Finally we introduce the concept of sexual selection, a form of natural selection that focuses on reproduction. Sexual selection explains many of the morphological and behavioral differences that often exist between the sexes within a species.

2.1 Evolution by natural selection favors behavioral adaptations that enhance fitness

Figure 2.2. Dog breeds. There is tremendous variation in the body size and morphology of these breeds. (a) Boston terrier; (b) Labrador retriever; (c) greyhounds.

natural selection Differential reproduction and survivorship among individuals within a population. The mechanism that results in adaptive evolution.

heritable A trait that can be passed from parents to their offspring because of heredity.

evolution Changes in allele frequency in a population over time.

fitness The relative survivorship and reproductive success (ability to produce viable offspring) of an individual.

If you have ever visited a dog park, you've likely seen a variety of breeds, such as Boston terriers, Labrador retrievers, and greyhounds (Figure 2.2). Each of these breeds belongs to the same species, *Canis lupus familiaris*, but each breed possesses unique trait features with respect to body size and shape, tail length, head shape, coat texture, and color. The American Kennel Club recognizes over 200 breeds of dogs today. Why are there so many, and where did they come from?

All dog breeds have descended from gray wolf (*Canis lupus*) ancestors as a result of human manipulation of the breeding of individuals over the past 14,000 years (Akey et al. 2010). Over many generations, human breeders have selected individuals that possess certain traits for a particular breed, called the breed standard. Many breeds were created for specific tasks: terriers to hunt vermin, retrievers to find and retrieve hunters' birds, and greyhounds to hunt swift prey. Breeders allowed reproduction only among individuals that possessed certain traits. Differences in breed standards over time eventually led to ever-greater differences among the dog breeds we see today. This process is known as artificial selection because it is done "artificially"—that is, by humans.

Charles Darwin's great insight was that a similar process also occurs naturally, a process that he called natural selection. **Natural selection** is the differential reproduction and survivorship among individuals within a population. It is the mechanism that results in adaptive evolution. Darwin was not the first to suggest that species evolve, but he (along with Alfred Russell Wallace) was the first to describe the plausible mechanism, natural selection, by which evolution can occur.

Natural selection occurs because there is variation in traits among individuals in a population, and some traits provide individuals with greater reproductive success. When these traits are **heritable**, they are passed from parents to offspring. Natural selection can result in changes in allele frequencies in a population over time—a process that we recognize as **evolution**.

In his book *On the Origin of Species* (1859), Darwin articulated three conditions required for evolution by natural selection:

1. Variation exists among individuals in a population in the traits they possess.
2. Individuals' different traits are, at least in part, heritable. Traits can be passed from parents to their offspring so that offspring resemble their parents in the traits they possess.
3. Traits confer differences in survivorship and reproduction, a measure we call **fitness**: individuals with certain traits will have higher fitness, while those with other traits will have lower fitness relative to one another. Therefore, the fitness of individuals is not random; it is based on the traits they possess.

Today, we know that gene alleles are the basis of phenotypic traits. Natural selection acts on heritable variation among individuals and can result in changes in allele frequencies and associated trait values in a population. Traits that confer high fitness increase over time, while those that confer low fitness decline.

When traits are heritable, parents will tend to pass on the traits they possess to their offspring via their shared alleles. Many behavioral traits that have been studied are in fact heritable, including mating behavior, feeding behavior, overall activity level, and aggression (Stirling, Réale, & Roff 2002). In the next section, we examine how researchers study the heritability of behavioral traits.

TOOLBOX 2.1

Genetics primer

Deoxyribonucleic acid (DNA) condenses into chromosomes during mitosis and meiosis (Figure 1). Specific stretches of DNA contain genes, which code for polypeptides (e.g., proteins) that have a variety of functions.

Alleles are different versions of a gene (Figure 2). The locus is the physical location of an allele on a chromosome. Diploid organisms have two copies of each chromosome, and each may have a different allele. If the alleles are the same, then the

organisms are homozygous for these alleles; if the alleles are different, then they are heterozygous. During sexual reproduction, each parent contributes only one of its alleles. Therefore, offspring have different allele combinations than their parents.

During meiosis, homologous chromosomes line up in pairs (Figure 3). During this time, they can exchange segments of DNA, known as crossing over. This process results in novel combinations of DNA that increase genetic diversity.

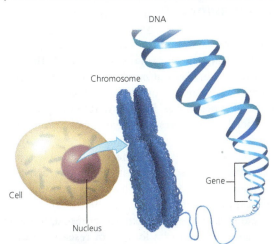

Figure 1. DNA. DNA is made up of two helical chains that condense into chromosomes during mitosis and meiosis. Genes are found in specific stretches of the DNA and code for polypeptides.

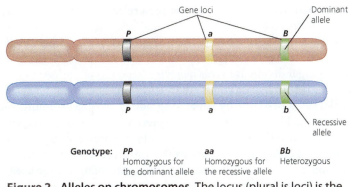

Figure 2. Alleles on chromosomes. The locus (plural is loci) is the physical location of a gene on a chromosome. Alleles at a gene locus can be different versions of the gene.

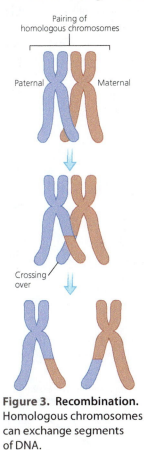

Figure 3. Recombination. Homologous chromosomes can exchange segments of DNA.

Measures of heritability

Traits must be heritable for natural selection to act on them. How do we determine the heritability of a trait? Two methods are commonly used:

1. **Parent-offspring regression** analysis examines the similarity between parents and their offspring in terms of the traits they possess. If a trait has a genetic basis, then the trait values of offspring should be similar to the trait

parent-offspring regression
A statistical technique used to examine the similarity between parents and their offspring in the traits they possess.

values of their parents: there should be a positive relationship between offspring and parent trait values. In this method, offspring trait values are plotted against parent trait values, and the slope of the resulting regression indicates the heritability of the trait.

2. In the **selection experiment** method, different groups of individuals are subjected to differential selection on the trait in question. If artificial selection acting on a trait results in changes in that trait value in subsequent generations, then the trait has a genetic basis.

Let's examine one case study in which both methods were used to determine the heritability of a behavioral trait—exploratory behavior.

selection experiment
An experiment in which different groups of individuals are subjected to differential selection on a trait.

Featured Research

Great tit exploratory behavior

Niels Dingemanse and his colleagues examined the heritability of exploratory behavior in free-living great tits (*Parus major*) (Dingemanse et al. 2002). The great tit is a small, colorful passerine bird found throughout much of Europe and Asia. It is a common inhabitant of woodlands, parks, and gardens, where it feeds on insects and seeds. It nests in tree cavities as well as nest boxes. Because of its widespread distribution, abundance, and use of artificial nesting sites, the great tit has become a favorite study subject for behaviorists.

Previous work indicated that individuals exhibit differences in their exploratory behavior when placed in novel environments: some actively explore their new environment quickly (bold individuals), while others are more reticent and slower to explore (shy individuals) (Verbeek, Drent, & Wiepkema 1994; Drent & Marchetti 1999). Are these personality traits heritable? The experimenters used birds from two populations in the Netherlands as part of a long-term research project. The birds bred in artificial nest boxes, which made it easier to capture them and determine relatedness (e.g., mother-offspring relationships). Birds were captured at the nest, uniquely color-banded, and, as adults, placed in an aviary that contained novel objects—five artificial wooden "trees." The research team recorded the number of flights and hops that birds took in the aviary in the first two minutes as an index of exploratory behavior. Individuals were then released back into the wild (Scientific Process 2.1).

There was a significant positive correlation between a mother's exploratory score and that of her offspring, demonstrating a genetic basis for differences among individuals in this behavior.

Next, the research team conducted a selection experiment for fast and slow exploratory behavior by separating adults into two groups. One group contained birds with the lowest exploratory behavior scores, while the other contained those birds with the highest scores. Subsequent juveniles from these groups were chosen as the initial breeding generation for two selection lines. In each selection line, nine pairs of birds were used for breeding, and this process was repeated for four generations. Researchers tested individuals from each generation to obtain their exploratory scores. The nine males and females with the highest (fast line) and lowest (slow line) scores of each generation were then selected for breeding.

The research team found strong changes in exploratory behavior of the two lines in the selection experiment over four generations. By the fourth generation, the average exploratory score for individuals in the high line was over four times higher than that for individuals in the slow line (Drent, van Oers, & van Noordwijk 2003) (Figure 2.3). This result indicates that exploratory behavior is heritable. Together, the results of these two experiments conclusively demonstrate that exploratory behavior in great tits has a genetic component. This conclusion was based

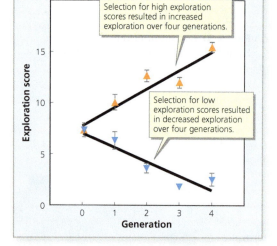

Figure 2.3. Selection experiment. The mean (± SE) exploratory score of individuals in the artificially selected lines for fast (orange) and slow (blue) exploration scores (*Source:* Drent, van Oers, & van Noordwijk 2003).

SCIENTIFIC PROCESS 2.1

Heritability of great tit exploratory behavior
RESEARCH QUESTION: *Is exploratory behavior in great tits a heritable trait?*

Hypothesis: Exploratory behavior has a genetic component.

Prediction: There will be a positive correlation between a parent's exploratory behavior and that of its offspring.

Methods: The researchers:

- placed individuals in an aviary (4.0 × 2.4 × 2.3 m) that contained five artificial wooden trees.
- recorded the number of flights and hops in the aviary in the first two minutes as an index of exploratory behavior.

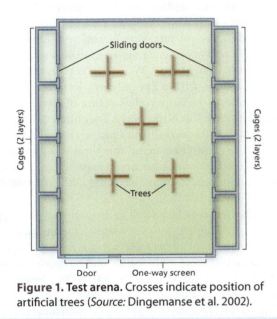

Figure 1. Test arena. Crosses indicate position of artificial trees (*Source:* Dingemanse et al. 2002).

Results:

- The parent-offspring regression on exploratory behavior was positive.

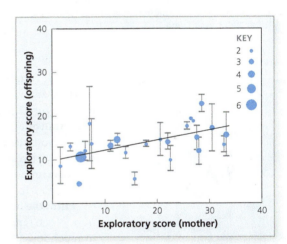

Figure 2. Parent-offspring regression. There was a positive correlation between parent and offspring (mean ± SE) behavior. Size of circles indicates number of individuals (*Source*: Dingemanse et al. 2002).

Conclusions: Exploratory behavior has a genetic component and is a heritable trait.

on two pieces of evidence: (1) offspring resemble their parents in this behavior, and (2) artificial selection on exploratory behavior produced significant differences in the two artificially selected lines.

Variation within a population

Through the process of natural selection, populations evolve. Such evolution requires variation among individuals in the traits they possess. Why do individuals in populations vary in behavior?

First, as we just saw, individuals differ in their genetic composition. Each generation introduces new genetic variation into populations through gene recombination, the immigration of new alleles into a population, and mutations (Hartl 2000). Because individuals in any population differ genetically, they will also tend to vary in their behavior. Furthermore, changes in environmental conditions can change the

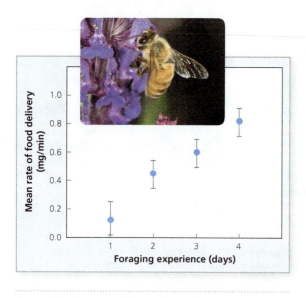

Figure 2.4. Honeybee food delivery. The mean (± SE) rate of food delivery for bees with different experience. Each day, honeybees' rate of food delivery increased, indicating that honeybees gain experience through trial-and-error learning (*Source:* Dukas 2008). *Inset:* honeybee.

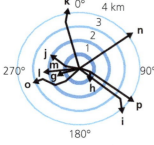

Figure 2.5. Owl dispersal. (a) Screech owl. (b) Each line represents the dispersal direction and movement distance of one individual. Individuals (letters) dispersed in all directions (*Source:* Belthoff & Ritchison 1989).

fitness of different traits and so maintain much variation in the frequencies of different alleles.

Second, many behaviors develop as a consequence of both genetic and environmental effects. Thus, even close relatives (with similar genes) often exhibit very different behavior as adults when they are exposed to different environmental conditions as juveniles. We will examine this issue in more detail in Chapter 4.

Third, many complex behaviors require learning and so are modified with experience. Because individuals will differ in experience over the course of a lifetime, we will observe differences in their behavior as well. For example, bees need to learn how best to extract nectar and pollen from flowers and then transport them efficiently back to their colony. Dukas (2008) showed that the feeding performance of honeybees (*Apis mellifera*)—that is, their rate of food delivery to the colony—increases dramatically over the first four days of foraging activity as a result of trial-and-error learning. An individual bee delivered food to the colony at a rate almost four times higher on its fourth day of activity than its rate on its first day (Figure 2.4).

Fourth, there might be little or no variation in fitness over a wide range of behaviors. One example is dispersal behavior. Dispersal is the process of moving away from the natal area, or place of birth, to find an adult breeding area or territory. For many species in many habitats, individuals will experience the same fitness whether they disperse north, south, east, or west. When this is true, we expect to see much variation in dispersal direction within a population. For example, James Belthoff and Gary Ritchison examined the dispersal direction of Eastern screech owls (Figure 2.5a) (*Otus asio*) (Belthoff & Ritchison 1989). They fitted nestlings with radio transmitters and located individuals several times per week as the birds dispersed from their natal territory. The researchers found that there was no particular pattern with respect to dispersal direction, which we would expect if characteristics that determine territory quality and fitness are not associated with compass direction relative to the natal location (Figure 2.5b).

Fifth, the fitness of a trait (behavior) may be related to its frequency in a population. When rare, the behavior may yield high fitness, but when common, it may result in much lower fitness. Such frequency-dependent selection can maintain different behaviors in a population. For example, some male fish defend territories to attract and mate with females, while other males hide at the edge of these territories and attempt to fertilize eggs of females attracted to the territory holder. Such "sneaker" male behavior is frequency dependent: it results in high fitness when most males in a population defend territories (a scenario that produces many opportunities to fertilize eggs) but much lower fitness when few males defend territories (producing fewer opportunities for reproduction).

Finally, individuals in all populations typically differ in size, nutritional status, health, and other traits. These differences can lead to significant variation in behaviors. For instance, many male birds sing complex songs. While song production has a strong genetic component, it also requires learning and can be physically demanding. Thus, in several species, vocal performance increases with age: older males that are in better condition produce different songs than younger males (e.g., Ballantine 2009).

For all these reasons, we expect much variation in any particular behavior within a population. Indeed, evolution by natural selection requires such variation.

Fitness and adaptation

As we have just seen, there is variation among individuals in a population in the traits they possess. Those traits might be the particular morphology, physiology, or behavior of an individual. In general, some traits confer higher fitness (survivorship and reproduction) than others. Traits that confer a selective advantage are called **adaptations**. Individuals with these traits will tend to survive better and leave more

TOOLBOX 2.2

Frequency-dependent selection

In frequency-dependent selection, the fitness of a trait (behavior) depends on its frequency in a population relative to other phenotypes. In positive frequency-dependent selection, the fitness of a phenotype increases as it becomes more common (Figure 1); in negative frequency-dependent selection, fitness declines as a phenotype increases in frequency (Figure 2). Consider a recent study by Benjamin Fitzpatrick and two high school students, Kim Shook and Reuben Izally, who examined

birds foraging on salamanders (Fitzpatrick, Shook, & Izally 2009) (Figure 3). The researchers noted that numerous species of small, slender, cryptic salamanders that live on the forest floor throughout North America all have a similar polymorphism. Within each species, some individuals have a dorsal stripe, while others do not. Why might all these different species have a similar polymorphism?

The research team investigated whether frequency-dependent selection as a result of bird predation maintains this polymorphism. They created a model salamander and then used it to make a mold to reproduce the model using modeling clay. The models did or did not possess an ocher dorsal stripe. They attached a food reward (half a peanut) to the underside of each model with edible library paste. They put the models on trays filled with leaf litter and used a random number generator to determine what type of model went on each tray. The trays and models were placed in an open area next to a woodlot.

For the first six days, the ratio of striped to unstriped models was 5:45, followed by one day of equal numbers of both morphs. For the next six days, the ratio was reversed (45:5), followed by one final day of equal abundances. Each day, the researchers quantified survival by counting the number of model salamanders that still had their food reward at the end of the day.

The team observed several blue jays foraging on the models, so they could confirm that the models were preyed upon. Fitzpatrick and colleagues found that when a morph occurred at low relative frequency it had higher relative survivorship (Figure 4), and vice versa. The rare form always had a survival advantage. Therefore, frequency did affect survivorship and was negatively frequency dependent. This may explain how this morphological polymorphism is maintained within a species.

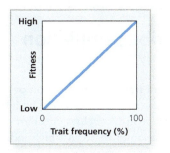

Figure 1. Positive frequency-dependent selection. The fitness of a trait increases as it becomes more common.

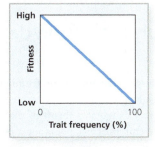

Figure 2. Negative frequency-dependent selection. The fitness of a trait declines as the trait increases in frequency.

a

b

Figure 3. Predator and prey. (a) Blue jay predator; (b) models of striped and unstriped prey.

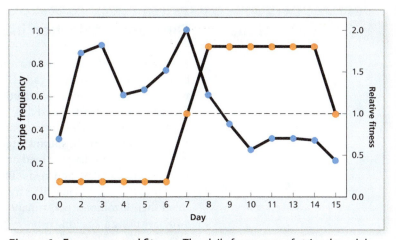

Figure 4. Frequency and fitness. The daily frequency of striped models (orange) and their relative fitness (blue) (*Source:* Fitzpatrick, Shook, & Izally 2009).

adaptation A trait that enhances fitness (survivorship and reproduction). Also an evolutionary process that results in a population of individuals with traits best suited to the current environment.

offspring than individuals whose traits do not confer a selective advantage (i.e., yield lower fitness), and so adaptations are identified relative to existing traits in a population (Reeve & Sherman 1993).

The most direct measure of the fitness of an individual is the number of progeny (that go on to reproduce) that it produces over its lifetime. However, few studies, particularly those that focus on behavior, measure fitness in this way. For one thing, this approach can be both logistically difficult and time consuming. Instead, most studies use indirect measures of fitness. Behavioral researchers often estimate fitness by quantifying parameters such as survivorship, number of matings, body size, growth rate, and feeding efficiency. Such indirect measures are typically positively correlated with more direct fitness measurements (e.g., Ritchie 1990; Taylor et al. 2008).

In this section, we've seen how evolution by natural selection favors behavioral adaptations that enhance fitness. Next, we examine in more detail how traits in populations change over time.

2.2 Modes of natural selection describe population changes

directional selection A situation in which individuals in a population with an extreme trait value at one end of the spectrum possess the highest fitness.

disruptive selection A situation in which individuals in a population with extreme trait values on both ends of the spectrum have the highest fitness.

stabilizing selection A situation in which individuals in a population with intermediate trait values have the highest fitness in a particular environment.

Natural selection can cause a change in the frequency of traits in a population over time. Evolutionary biologists have identified three modes of natural selection, which describe how populations change as a result of the relative fitness values of different trait values. To illustrate this process, consider a population of snails in which individuals exhibit continuous variation in color, ranging from white to pink to red. **Directional selection** occurs when individuals with one extreme trait value possess the highest fitness (Figure 2.6). In this case, let's say red individuals have the highest fitness in a particular environment; as a result, the frequency of red individuals in the population will increase. **Disruptive selection** occurs when individuals with either of two extreme trait values have the highest fitness (Figure 2.7). In this case, red and white individuals both have higher fitness than pink individuals in the same environment. Hence, the frequency of red and white individuals will increase, while the frequency of pink individuals will decline. Last, **stabilizing selection** occurs when individuals with intermediate trait values have the highest fitness in a particular environment. In this case, pink individuals have higher fitness than red or white individuals (Figure 2.8). When stabilizing selection is at work, the frequency of pink individuals will increase over time. Let's examine one example of each mode of selection.

Featured Research

Directional selection in juvenile ornate tree lizards

For many animals, the ability to move quickly affects survival: faster individuals can better evade predators and can also capture mobile prey more effectively. Therefore, we might expect directional selection to act on the speed of movement. Donald Miles examined this idea by studying selection on locomotor performance in juvenile ornate tree lizards (*Urosaurus ornatus*) (Miles 2004). Ornate tree lizards are found in a variety of habitats in southwestern North America. These small lizards (less than 6 cm in snout-vent length) feed on a diverse array of insects and are preyed upon by larger lizards, snakes, and birds (Figure 2.9).

Miles collected 45 juveniles from one population near Tucson, Arizona, and quantified their locomotor performance in the laboratory. All individuals were tested for two days using a 2 m long raceway with a sand floor. Using gentle prodding, individuals were induced to run down the length of the track. Eight infrared photocells, spaced at 25 cm intervals, recorded their movement and allowed the calculation of mean velocity (average speed over 2 m) and burst velocity (fastest speed over a 25 cm interval). Miles repeated this test eight times for each individual, giving each lizard one hour to recover between trials. He also recorded the body size and mass of each individual,

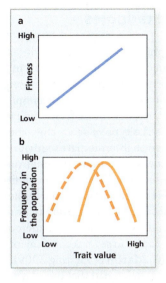

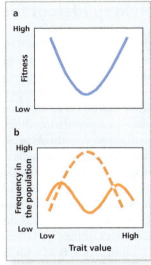

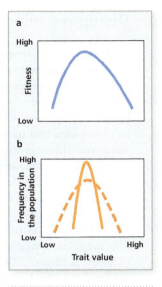

Figure 2.6. Directional selection. (a) High trait values have higher fitness than low trait values. (b) The dashed line is the population before selection. The solid line is the population after selection.

Figure 2.7. Disruptive selection. (a) Low and high trait values have the highest fitness. (b) The dashed line is the population before selection. The solid line is the population after selection.

Figure 2.8. Stabilizing selection. (a) Intermediate trait values have the highest fitness. (b) The dashed line is the population before selection. The solid line is the population after selection.

as well as its stride length, which was measured by the distance between successive footfalls in the sand. All individuals were uniquely marked and later released back into the wild at their capture site. Miles then recorded the survival of these individuals over the next six months by visiting the site repeatedly and recapturing surviving individuals. He used survivorship over this time period as a measure of fitness.

Miles found that larger, heavier individuals had high survivorship, as did those with higher mean velocity, higher burst velocity, and longer stride length (Table 2.1). Body size and speed are often correlated, with larger individuals tending to move faster, so which trait was more important in this case? Miles used a regression analysis to determine that locomotor performance was more important in affecting survival than body length. Locomotor performance is a function of stride length (Bonine & Garland 1999). He concluded that in this population, there was strong directional selection on limb length in juveniles: longer limbs produce larger strides that allow individuals to achieve greater velocities, which in turn enhances their survivorship.

Figure 2.9. Ornate tree lizard. These lizards are common in southwestern North America.

TABLE 2.1 Lizard traits. Mean ($\pm$ SE) trait differences between survivors and nonsurvivors (*Source:* Miles 2004).

TRAIT	SURVIVORS	NONSURVIVORS
Snout-to-vent length (mm)	36.6 $\pm$ 3.5	33.6 $\pm$ 4.2
Mass (g)	1.36 $\pm$ 0.43	1.13 $\pm$ 0.46
Initial velocity (m/s)	0.36 $\pm$ 0.16	0.35 $\pm$ 0.15
Burst velocity (m/s)	0.62 $\pm$ 0.18	0.48 $\pm$ 0.14
Mean velocity (m/s)	0.39 $\pm$ 0.14	0.29 $\pm$ 0.11
Stride length (mm)	89.4 $\pm$ 14.6	71.8 $\pm$ 13.1

Disruptive selection in spadefoot toad tadpoles

In disruptive selection, individuals with intermediate phenotypes have lower fitness than individuals with either extreme. This situation can occur when individuals in a population specialize on different resource types and there is competition among individuals for those resources, known as intraspecific competition. For example, imagine a population of individuals that feed on seeds that range in size from small to large. Assume that individuals in a population differ in the type of seeds they consume: one-third eat only small seeds, one-third specialize on intermediate seeds, and one-third consume only large seeds. If intermediate seeds are rare in the environment, then individuals that specialize on intermediate seeds will have the lowest fitness due to increased competition for food. Individuals that eat the extremes (either small or large seeds) will have higher fitness and should increase in the population. David Pfennig and colleagues, who have studied disruptive selection in spadefoot toad tadpoles (*Spea multiplicata*), examined a similar scenario in several experiments (Pfennig & Pfennig 2005; Pfennig, Rice, & Martin 2007; Martin & Pfennig 2009).

Spadefoot toads are small amphibians (less than 6 cm in length) that live in a variety of habitats in North America. In arid regions of southwestern North America, they breed in temporary rain-filled ponds created by intense summer rainstorms. Once a pond fills with water, eggs hatch and the emerging tadpoles feed in the ponds while they grow and develop over a span of three weeks. Tadpoles consume both detritus (decomposing organic material) and small invertebrates that are typically found in such ponds. Within a single pond population, individuals exhibit a wide range of variation in feeding morphology and behavior. Some individuals (detritus morphs) specialize on detritus: they have round bodies, smooth mouthparts, and small jaw muscles (Figure 2.10). Other individuals (carnivore morphs) specialize on invertebrates: they have narrower elongated bodies, notched mouthparts, and large jaw muscles. A third group (intermediate morphs) are generalists that consume both detritus and invertebrates: they possess morphology that is intermediate between that of the two specialists.

Pfennig and colleagues hypothesized that disruptive selection may be operating on tadpole morphology and feeding behavior. Previous work demonstrated that tadpoles compete for food and that each morph has the highest fitness when it is rare, because rare individuals experience reduced competition (Pfennig 1992). Specialized morphs should experience lower competition, because they compete for food with only a subset of the population—their own morphs and intermediate morphs, but not other specialist morph individuals. In contrast, intermediate morph individuals compete with all individuals in the population for food.

To evaluate their hypothesis, Pfennig's research team tested two predictions:

Prediction 1: Detritus morphs feed more efficiently than intermediate morphs on detritus; carnivore morphs feed more efficiently than intermediate morphs on invertebrates.

Prediction 2: Detritus and carnivore morphs have higher fitness than intermediate morphs.

To test the first prediction, the researchers examined the feeding efficiency of tadpoles forced to consume only one food type. In one experiment, they placed one individual in a small wading pool within a cage and allowed it to feed on detritus for eight days. The researchers measured the size of each individual (snout-vent length) at the beginning and end of the experiment to determine its growth rate, which was used as a measure of fitness. The morphology of each individual was quantified (smooth or notched mouth and the width of jaw muscles) to provide a continuous morphology index that ranged from "detritus morph" through "intermediate" to "carnivore morph." In a second experiment, the

Figure 2.10. Spadefoot toad tadpoles. Detritivore morph (top) and carnivore morph (bottom).

research team characterized the feeding performance of individuals on invertebrates by placing a single tadpole in a small container with ten fairy shrimp (*Thamnocephalus* sp), a common invertebrate food item for tadpoles. The researchers measured the amount of time that individuals required to capture and eat shrimp, an important aspect of feeding efficiency. They measured the morphology of each individual in this experiment to create a morphology index.

To test the second prediction, the research team marked over 500 individuals from one pond 11 days after it had filled with water and spadefoot toad eggs had hatched. They measured each individual to quantify a morphology index and then categorized individuals into one of three types: detritus morph, carnivore morph, or intermediate morph. Individuals of each morph type were marked with a different color elastomer tag, which was injected under the skin, and were then returned to the pond to continue their development. Eight days later, the researchers collected 1,500 individuals from the pond (including many marked individuals) and counted the number of each color of morph to obtain an estimate of each morph's survivorship (one measure of fitness). To obtain a second measure of fitness, the research team also examined 301 unmarked tadpoles from the 1500 collected. For each, they quantified its morphology index and recorded its size (snout-vent length). Body size in developing tadpoles is correlated with fitness: larger individuals have higher survivorship and fitness (Pfennig & Pfennig 2005).

The researchers found that the feeding performance of each specialized morph was higher than that of intermediate morphs, as they had predicted. For individuals feeding on detritus, detritus morph individuals exhibited higher growth rates than intermediate morphs, and carnivore morphs more efficiently captured invertebrate prey than did intermediate morphs (Figure 2.11). In addition, both measures of fitness indicated that intermediate morphs have lower fitness than either specialized morph. Intermediate morphs had a lower probability of survival (indicated by a significantly lower recapture rate) than both detritus and carnivore morphs. Among the unmarked individuals, intermediate morphs were smaller than both specialized morphs (Figure 2.12).

The research team concluded that their results supported their hypothesis. Intermediate morph individuals have lower fitness (lower survival and smaller body size) than either detritus morph or carnivore morph individuals, likely because of competition. Individuals with specialized morphology have fewer competitors and are more efficient feeders than individuals with intermediate morphology.

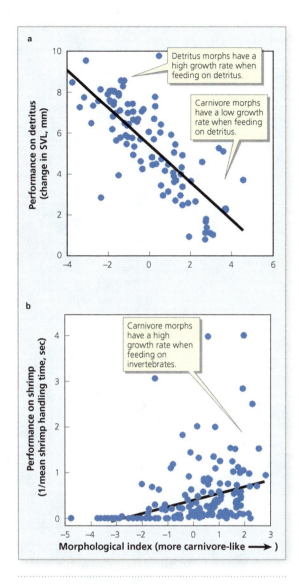

Figure 2.11. Tadpole feeding performance. (a) Growth rate of individuals feeding on detritus as a function of their morphology. Individuals with more carnivore-like morphology have slower growth. (b) Capture efficiency of individuals feeding on shrimp as a function of their morphology. Individuals with more carnivore-like morphology have higher capture efficiency (*Source:* Martin & Pfennig 2009).

Stabilizing selection in juvenile convict cichlids

Under stabilizing selection, individuals with intermediate trait values have the highest fitness. For example, consider territorial defense behavior. Many animals defend an area called a territory to protect the food it contains or to attract mates. However, defense of a territory requires time and energy to fight off potential intruders, and these costs increase as territory size increases. Thus, one might anticipate that there would be a territory size of intermediate value that would lead to the highest fitness. Jason Praw and James Grant examined this hypothesis in juvenile convict cichlids (*Amatitlania nigrofasciata*) (Praw & Grant 1999).

Convict cichlids are native to Central America and are a common aquarium fish. Individuals frequently defend small territories and feed on a variety of foods, including invertebrates, small fish, plants, and algae. In order to examine territory defense,

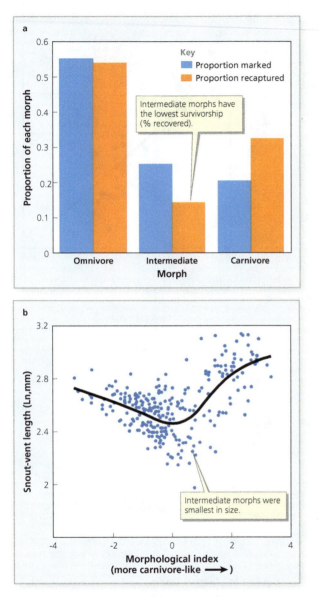

Figure 2.12. Tadpole survivorship and size. (a) The proportion of each morph type recovered in the mark-recapture study. Intermediate morphs had the lowest survival. (b) Unmarked individuals of intermediate morphology were the smallest size, a measure of fitness (*Source: Martin & Pfennig 2009*).

optimal trait value The trait value that confers the highest fitness in a population in a particular environment.

cost-benefit approach A method used to study behavioral adaptations in which the fitness benefits and costs of different traits are examined to determine which has the highest net benefit (benefit – cost).

Praw and Grant used aquaria that contained a single food patch that fish could defend. Food patches consisted of different numbers of ice cube tray cells. Six patch sizes were used, ranging from one cell up to 121 cells. Commercial fish food pellets were evenly distributed across a patch, which meant that larger patches contained more food than smaller patches (Scientific Process 2.2).

The researchers allowed a single large, dominant fish to defend a territory around the food patch. These individuals always attempted to defend the entire food patch, which varied in size. Next, the researchers introduced four small, subordinate intruder fish into each aquarium. Intruders attempted to feed from the food patch while the dominant fish defended its territory. For each dominant fish, Praw and Grant recorded the amount of food it ate, the number of intruder chases, and its growth rate over a ten-day period. Because all fish were juveniles (and did not exhibit mating behavior), the researchers focused on growth rate as a measure of fitness. In general, juvenile fish with high growth rates have higher fitness.

The researchers found that fish that defended very small or very large territories had lower growth rates than fish that defended intermediate-sized territories. This occurred because fish that defended very small territories had less food to eat, while those that defended very large territories spent much more time and energy chasing intruders. Individuals that defended intermediate-sized territories had more to eat than fish that defended small territories (and so grew faster) but spent less time (and thus energy) chasing intruders than fish with large territories. From this experiment, Praw and Grant concluded that selection would favor individuals who defend territories of intermediate size.

Studying adaptation: the cost-benefit approach

Under all modes of selection, the trait value that confers the highest fitness in a particular environment is called the **optimal trait value**. If the trait under selection is heritable, and if the selection regime remains constant, then populations will evolve toward the optimal trait value (or values) over successive generations. Thus, we see that natural selection is an optimizing process that allows us to understand adaptations by describing the fitness benefits and costs of different phenotypes. This **cost-benefit approach** illustrates a common method used to study behavioral adaptations: identify fitness costs and benefits of different traits to determine which trait has the highest net benefit.

We will see the cost-benefit approach used throughout the rest of this book. In fact, we have already seen it in Chapter 1, when Kokko and Morrell used this approach to predict how long a male should mate guard a female to maximize his fitness.

2.3 Individual and group selection have been used to explain cooperation

In the lizard, spadefoot toad, and fish examples described previously, natural selection was assumed to act on individuals. As noted, individuals that have the highest fitness are more likely to pass their genes and behavior on to the next generation. However, some behavior is not so readily explained by **individual selection**—natural selection at the level of individuals. Many social animals display cooperative

SCIENTIFIC PROCESS 2.2

Stabilizing selection on territory size in cichlids
RESEARCH QUESTION: *How does territory size in cichlids affect fitness?*

Hypothesis: An intermediate territory size will optimize fitness.

Prediction: Individuals that defend an intermediate-sized territory will have the highest growth rate.

Methods: The researchers:

- created different sized food patches.
- placed one patch in an aquarium with a single dominant (focal) fish.
- added four intruders to each aquarium and recorded the territorial defense behavior of the dominant fish and its growth rate.

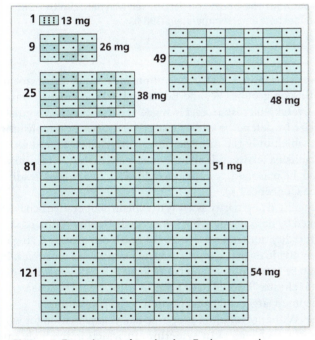

Figure 1. Experimental territories. Each rectangle represents a cell; food is indicated by dots. Patches contain 13–54 mg of food (*Source:* Praw & Grant 1999).

Results:

- The focal fish defended the entire food patch from intruders.
- Defense behavior increased as territory size increased.
- Fish defending intermediate-sized territories had the highest growth rate.

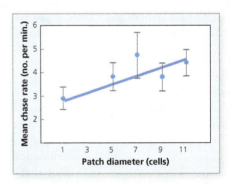

Figure 2. Patch defense. The chase rate increased with territory size (*Source:* Praw & Grant 1999).

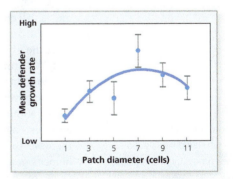

Figure 3. Growth rate. Fish on intermediate sized territories had the highest growth rate (*Source:* Praw & Grant 1999).

Conclusions: Selection will favor individuals that defend intermediate-sized territories, because they will have the highest fitness.

behavior, helping others to survive and reproduce. For example, many ant and bee colonies contain many sterile workers and a single queen who reproduces. The workers defend the colony and deliver food to the queen and her offspring. Why should these individuals forgo their own reproduction and help another reproduce? This question troubled even Darwin. He speculated that selection may act on the entire colony rather than on individuals alone.

■ **individual selection** Natural selection acting on individuals.

TOOLBOX 2.3

Game theory

The cost-benefit approach predicts that a behavior that maximizes fitness, irrespective of how other animals behave, will continue to be passed down from parent to offspring. However, in many situations, the fitness of a behavior will depend on how others behave. Determining the behavior that maximizes fitness in such cases requires the use of game theory, a cost-benefit approach first developed in economics. **Game theory** models involve finding the best solution to an interaction or game between two players (Parker & Hammerstein 1985). It takes into account the players, the strategies they adopt, and the payoffs for each one. In economics, the players are people, the strategies are different behaviors, and the payoffs are typically monetary. John Maynard Smith and George Price were the first scientists to apply these economic models to animal behavior (Maynard Smith & Price 1973). In these cases, the players are animals and the strategies are again different behaviors, but the payoffs are described in terms of

fitness. The solution to biological games is the **evolutionary stable strategy** (ESS); this strategy cannot be beaten by any alternative strategy, and no player will adopt a different strategy because doing so will result in lower payoffs. Oftentimes, the behavior that maximizes fitness will be frequency dependent. For example, in some species, males can use two behaviors to mate: they can (1) call to attract females or (2) wait silently near a calling male and intercept females attracted to the caller. The fitness of the two behavioral strategies can depend on the frequency of each; for example, if no males in the population call, the second strategy will have very low fitness. Game theory is used to determine the behavioral strategy that maximizes fitness in such circumstances—this strategy may be some mixture of each behavior in the population. We will see several examples of the use of game theory throughout the book. For details, see Maynard Smith (1982) and Riechert and Hammerstein (1983).

---

group selection Selection that favors particular groups of individuals over other such groups of the same species.

Vero C. Wynne-Edwards proposed that cooperation may result from **group selection**, or selection acting on groups (Wynne-Edwards 1962). Wynne-Edwards asked why animals do not overexploit resources. He suggested that they can regulate breeding and population size by defending widely spaced territories or establishing dominance hierarchies in which only the most dominant individuals reproduce—those that successfully establish a territory. If all individuals attempted to maximize their fitness by reproducing, he reasoned, resources would be overexploited, leading to extinction (Applying the Concepts 2.1).

Group selection quickly gained popularity as a way to explain cooperative behavior and is still commonly invoked by nonscientists, but is it needed? George Williams said no (Williams 1966). Territoriality does not need to be the result of a group's trying to limit the use of resources. Individuals may simply compete for available resources. In fact, Williams argued, group selection is based on faulty logic. An individual that did not limit its reproduction (a "cheater") would have a fitness advantage. The "cheater's" offspring would quickly become a larger and larger portion of the population.

kin selection A form of natural selection in which individuals can increase their fitness by helping close relatives, because close relatives share the helper's genes. Often used to explain altruism.

At about the same time, William Hamilton proposed a new explanation for why animals engage in cooperative behavior: **kin selection**, in which individuals can increase their fitness by helping close relatives, because close relatives share the helper's genes (Hamilton 1964; Maynard Smith 1964). Thus, individual selection can explain cooperation, through **inclusive fitness**—both individual fitness and the fitness obtained by helping close relatives. For example, in many social insects among which cooperation is common, females have very high inclusive fitness as a result of their sex-determining system, as we will see in Chapter 14.

inclusive fitness A combination of individual fitness and the fitness obtained by helping close relatives.

Not everyone was ready to dismiss group selection, however. For one thing, not all cooperating insects display a close genetic relationship among workers, often because two or more queens produce offspring (e.g., Kümmerli & Keller 2007). In response, David Sloan Wilson and others have developed a new theory of group selection called **multilevel selection**. Selection on groups, they argue, may be stronger than selection on individuals, but only in the right circumstances: groups must be small, and there must be minimal movement of individuals between groups (e.g., Wilson & Wilson 2007). Work in the laboratory has supported this model. For instance, when individual *Pseudomonas* bacteria chemically detect the presence of other *Pseudomonas*, they secrete a substance that enhances nutrition and the immune system for the entire

multilevel selection A form of natural selection that involves both selection on groups and selection on individuals.

APPLYING THE CONCEPTS 2.1

Do lemmings commit suicide?

Biologists may look skeptically at group selection, but you might not know that from documentaries and television. The idea has even become something of a legend. One common myth is that lemmings commit suicide to prevent overpopulation. Lemmings are small, herbivorous rodents common in Arctic regions and are one of the subjects of an early Disney nature movie, *White Wilderness*. As the movie describes, approximately every four years, the lemming population increases dramatically. In response, lemmings disperse into new habitats, a behavior that can result in mortality. They are seen swimming across rivers and lakes and even traveling on sea ice floats. In one memorable scene in the movie, individuals appear to jump off a cliff into the Arctic Ocean and swim off to their death. Could such behavior benefit the species as a whole by preventing overpopulation and the exhaustion of resources? If so, is it an example of group selection?

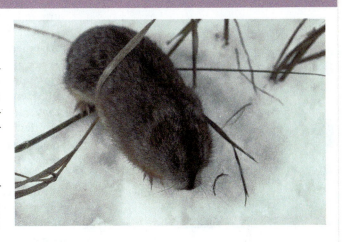

Research indicates that lemming populations do exhibit dramatic population fluctuations in the Arctic and that individuals disperse during periods of high abundance (Chitty 1996). However, there is no evidence that individuals deliberately commit suicide. Such an explanation, in fact, runs up against problems with the logic of group selection. Individuals that did not cooperate in committing suicide would have higher fitness than those that did, and "cheating" individuals would quickly dominate the population.

In reality, lemmings are simply dispersing to reduce intraspecific competition. Those that disperse onto sea ice floats are not intentionally committing suicide, but rather are moving to find a better habitat. The dramatic "suicide" in the movie was actually staged for the camera (Chitty 1996).

population (Sachs & Bull 2005). Empirical support from multicellular organisms is sparse, however, and the debate about the importance of multilevel selection for understanding the evolution of cooperation continues (Leigh 2010; Kokko & Heubel 2011).

2.4 Sexual selection is a form of natural selection that focuses on the reproductive fitness of individuals

Natural selection acts on heritable traits that affect survivorship *and* reproduction. **Sexual selection** is a form of natural selection that acts on heritable traits that affect reproduction. Individuals vary in their ability to compete for mates or attract individuals of the opposite sex, which leads to differential reproduction within a sex.

Darwin himself proposed that many traits observed in males and females of a species are due to sexual selection and often result in **sexual dimorphism**, or morphological differences between the sexes. Which traits are involved in sexual selection? We will look at this issue in more detail in Chapter 12; for now, consider the case of house finches.

sexual selection A form of natural selection that acts on heritable traits that affect reproduction.

sexual dimorphism Morphological differences between the sexes.

Sexual selection in house finches

Featured Research

House finches (*Carpodacus mexicanus*) are small, granivorous birds found in a variety of habitats in North America. Females have a dull brown overall plumage, while males are more colorful. Male head, body, and tail feathers can range in color from dull yellowish-red to bright red (Figure 2.13), a result of carotenoid pigments in their diet. The intensity of red of an individual's carotenoid colored feathers is determined entirely by diet. Coloration may therefore indicate a male's ability to obtain a nutritious diet, and females can use this trait to select the best males to mate with in a population.

Geoff Hill conducted a set of experiments to determine whether red coloration in male house finches is the result of sexual selection by females (Hill 1990). He examined whether females prefer to mate with males that contain the greatest amount of red coloration in their plumage using wild-caught house finches in laboratory experiments. The basic experiment consisted of placing a single female in a central compartment that was surrounded by four males in separate compartments. The female could see each male by moving to a perch next to his cage, but the males could not see each other.

For each male, Hill quantified the plumage by scoring the intensity of coloration on a standardized set of body locations (head, underbelly, and tail). He added these scores together to obtain a plumage index. Higher numbers indicated more intense red. A female was then allowed to associate with the males for two hours. Hill conducted all experiments during the breeding season, when females actively seek mates. He recorded the amount of time the female spent next to each male and used this as an indicator of female preference. In one experiment, Hill selected four males that naturally differed in red coloration. In another experiment, to control for other potential confounding factors, he artificially varied each male's body coloration with dyes. Some bright red birds were made dull, while some dull birds were made brighter red. In a second control experiment, he examined how females would react to a set of birds that all lacked red coloration. In this case, he presented a female with four females that had no red coloration at all. In each experiment, at least 14 females were tested for the particular set of four treatment individuals.

Hill found that the females in his experiments spent the most time next to those males with the most intense red coloration (Figure 2.14). When given the choice of four individuals that lacked red coloration, females exhibited no significant preference for any particular individual. These results support the hypothesis that red plumage color in males is a sexually selected trait. Additional work by Hill and colleagues has shown that plumage color correlates with the health and nutritional condition of a male (Hill & Montgomerie 1994; Hill, Inouye, & Montgomerie 2002). Thus, red coloration provides an accurate indicator of male condition. By choosing to mate with the reddest males, females are selecting the highest-quality mates. As we will see in Chapter 11, sexual selection is an area of great activity in behavior research.

Figure 2.13. Male house finches. Males range in color from (a) dull yellowish-red to (b) bright red, depending on their diet.

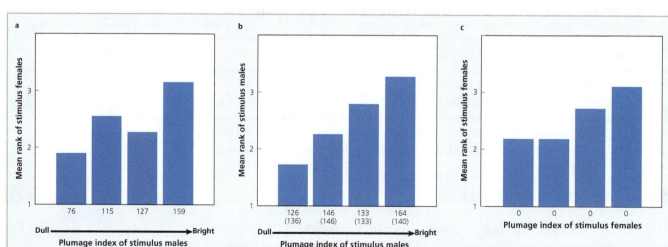

Figure 2.14. Female house finch preference. Mean preference ranks of the four stimulus birds. The number represents the plumage index score of each male. (a) Females preferred the male with the reddest color among males that varied naturally in color. (b) Females preferred the male with the reddest color among males whose color was manipulated. (Each male's pre-manipulated plumage index score is shown in parentheses.) (c) Females show no preference among four females lacking red color (Source: Hill 1990).

Chapter Summary and Beyond

The behaviors we observe in animals result from evolution by natural selection, which favors adaptations that confer high fitness. Evolution by natural selection requires that (1) individual differences in traits exist within populations, (2) traits can be passed from parents to offspring, and (3) traits confer differences in survivorship and reproduction on individuals. Populations evolve according to three modes of selection, which describe how trait values differ in terms of fitness. Directional selection results in an increase in one extreme trait value. Disruptive selection leads to an increase in two extreme trait values and a decrease in intermediate values. Stabilizing selection leads to an increase in intermediate trait values over time. However, for a trait to evolve, it must be heritable.

Researchers commonly determine the heritability of traits in two ways. They compare trait values of offspring with those of their parents, or they conduct artificial selection experiments to select for a particular trait value. If a trait is heritable, parent-offspring trait value correlations will be positive, and selection experiments will document changes in the trait value over successive generations. Stirling, Réale, and Roff (2002) review studies that indicate that many behaviors are heritable.

Natural selection acts on individuals and causes populations to evolve. Wynne-Edwards proposed that group selection might be an important evolutionary force to explain the dominance hierarchies and territorial behavior of many species. He envisioned that group selection would prevent the overexploitation of resources. However, his idea requires cooperation by all individuals to limit their reproduction and has been largely discredited. Recent theoretical and empirical work has identified some circumstances in which multilevel selection, which includes selection that acts on groups, may be stronger than individual selection. Wilson and Wilson (2007) and Wilson, Van Vugt, and O'Gorman (2008) provide details about multilevel selection and its relationship to earlier discussions of group selection.

Sexual selection is a form of natural selection that focuses on differential reproduction among individuals. It has been an important mechanism to explain the evolution of many forms of sexual dimorphism in animals as a result of mating preferences. Clutton-Brock (2007) and Johnstone (2008) provide detailed reviews of sexual selection and mate choice in animals.

CHAPTER QUESTIONS

1. Individuals in a species of moth vary in wing color from white to black, but all individuals prefer to land on the white bark of birch trees. White wing color and the behavioral preference for landing on white birch bark are adaptations, because the wing color of the moth matches the white color of the birch bark and predators have trouble finding white-winged moths on these trees. As a result, white-winged individuals experience lower predation than dark-winged individuals when landing on birch bark. Now consider a population of these moths that live near a power plant that burns coal and emits black soot that coats the bark of birch trees, turning them black. These moths still prefer to land on birch trees, but now white-winged individuals suffer very high predation because the color of their wings contrasts with that of the soot-covered black bark. Dark-winged moths now experience very low predation because they are difficult to see on the dark bark. Is the behavior of landing on birch trees still a behavioral adaptation in this population of moths?

2. A study found tremendous variation in the aggressive behavior of a lobster population off the coast of Maine, with the most aggressive lobsters tending to have the highest reproductive success. The researcher concluded that selection increases aggressive behavior of individuals in this population. Is this conclusion valid?

3. Imagine you are studying a population of individuals that differ in body size. Individuals that weigh less than 20 g or more than 30 g have lower fitness than those that weigh 25 g. What mode of selection is operating on the population and what evolutionary change would be predicted based on your data? What assumptions did you make to generate your prediction?

4. Consider a population of birds in which males compete with each other to establish territories. Only 50% of adult males successfully establish a territory, and females will only mate with males that possess a territory. In this population, half the adult males do not breed each year. Wynne-Edwards (1962) argued that such territorial behavior and the existence of many nonbreeders could best be explained by group selection. Explain these observations based on individual selection.

5. Describe the similarities and differences between natural selection and sexual selection.

Chapter 3

Methods for Studying Animal Behavior

Figure 3.1. *Cephalotes* ant. This species of ant can direct its gliding behavior, as ecologist Stephen Yanoviak discovered.

3.1

3.2

3.3

3.4

When we were graduate students, we quickly found out that research does not always go as planned. Once, during a study of the mating behavior of round stingrays (*Urolophus halleri*) in the Gulf of California, observations ended abruptly after a large storm created unfavorable conditions and all the stingrays left the study area. Luckily, when the weather stabilized, the stingrays returned, and we were able to continue (Nordell 1994). More recently, we set up an experiment to study squirrel foraging behavior. We were ready, but the squirrels apparently were not. Although we put out an abundance of a preferred food (hazelnuts) for several weeks, squirrels never came to our study site location. So we had to find a new location.

Many research projects begin with detailed planning. Yet sometimes scientific research begins almost as an "Aha!" moment, as it did for ecologist Stephen Yanoviak (2006). It all started during a project to study deforestation and the spread of mosquito-borne tropical diseases in Peru. Yanoviak was collecting mosquitoes in the rainforest canopy 30 m (over 100 ft) off the ground when he was attacked by ants. He noticed that after he brushed some of the ants off his body, rather than falling to the ground, they appeared to glide back to the tree trunk as if they could somehow direct their fall. Later, he made additional observations of this phenomenon by painting the ants' rear legs with white nail polish so that he could better observe them as they fell. He again saw that they seemed not only to glide and direct their movements toward the tree trunk, but even to make 180° turns in the air.

When he headed home, he asked an ant ecologist, Michael Kaspari, and an expert on flying and gliding animals, Robert Dudley, to collaborate with him to examine this behavior more thoroughly.

Several years passed before all three of them could get back to Peru and conduct more "drop" studies. When they did, they found that about 85% of the individuals in the ant species *Cephalotes atratus* seemed to have the ability to direct their gliding flight behavior (Yanoviak, Dudley, & Kaspari 2005) (Figure 3.1). This was an exciting finding, as no one had known that ants could control their descent or direct their gliding flight. So in this case, scientific discovery began quickly and accidentally.

In this chapter we begin by exploring a fundamental characteristic of animal behavior research: it looks at both the proximate mechanisms that generate behavior as well as the ultimate reasons for why it evolved. We then discuss how animal behavior researchers collect data using observational, experimental, and comparative research methods. Essentially, we examine the nuts and bolts of how animal behavior research is conducted. We then discuss how these methods have been used historically, showing how animal behavior has been examined from a variety of perspectives. We close the chapter with a discussion of important ethical issues in the study of animal behavior.

3.1 Scientists study both the proximate mechanisms that generate behavior and the ultimate reasons why the behavior evolved

What kinds of research questions can we ask about animal behavior? Consider the migration behavior of birds. Many birds travel long distances between summer breeding and winter nonreproductive grounds. The seasonal movement of individuals between these locations is known as migration behavior. We can ask several questions about this behavior. For example, what causes migration? The behavior often is initiated by seasonal changes in day length. These changes trigger hormonal responses that lead to increased feeding, fat deposition, and the onset of long-distance movement (Ramenofsky & Wingfield 2007). In some species, migration behavior is genetically determined and requires no learning, while in others, migration routes must be learned by following experienced individuals (e.g., Lishman et al. 1997) (Figure 3.2).

We can also ask, "Why do birds migrate?" Migration allows individuals to track resources and thus avoid places where resource availability is greatly limited in some seasons; in short, it promotes survival (Cox 1985). To understand the evolution of migration behavior for a particular population or species, researchers often examine how migration patterns differ among different populations of a species or between closely related species (Outlaw et al. 2003; Milá, Smith, & Wayne 2006).

Niko Tinbergen (1963) summarized the different types of research by outlining four basic questions that can be asked about animal behavior:

1. What is the mechanism that causes the behavior?
2. How does the behavior develop?
3. What is the function of the behavior?
4. How did the behavior evolve?

Tinbergen's four questions provide a framework for the study of behavior. Answers to Questions 1 and 2 are often referred to as **proximate explanations**, because they focus on understanding the immediate causes of a behavior (Mayr 1961). These

proximate explanation An explanation that focuses on understanding the immediate causes of a behavior.

Figure 3.2. Migration. Sandhill cranes (*Grus canadensis*) learn migration routes by following experienced individuals.

explanations often incorporate studies of genetics, sensory systems, neurons, hormones, and learning. Studies of the hormonal changes that trigger the migration behavior of birds and examinations of, the genetic basis of and learning required for migration are examples of proximate explanations. Answers to Questions 3 and 4 are known as **ultimate explanations**, because they require evolutionary reasoning and analysis (Mayr 1961). Understanding the effects of migration on survivorship and why migration evolved are examples of ultimate explanations. Researchers have argued that a complete study of behavior requires answers to all four questions. Throughout this book you will read examples of both proximate and ultimate research studies.

ultimate explanation An explanation of behavior that requires evolutionary reasoning and analyses.

observational method An approach in which scientists observe and record the behavior of an organism without manipulating the environment or the animal.

3.2 Researchers use observational, experimental, and comparative methods to study behavior

When conducting research, a scientist must determine the best method to address research questions and test their hypotheses. Three common methods used in behavioral research are the observational, experimental, and comparative methods (Altmann 1974; Lehner 1998; Bateson & Martin 2007). It is important to note that numerous protocols or data collection techniques can be used with each method and that researchers often employ more than one method in their studies (Applying the Concepts 3.1). In addition, data must be analyzed to see if the patterns observed support the prediction.

The observational method

In the **observational method**, scientists observe and record the behavior of an organism without manipulating the environment or the animal (Figure 3.3). This method is commonly used both to test hypotheses and to describe behavioral patterns. For example, the observational method is used to construct ethograms, as we saw in Chapter 1. Researchers studying animals in zoological parks frequently use the observational method, because behavior is often an important indicator of well-being or level of stress (Marriner & Drickamer 1994). For instance, changes in the behavior of an animal may indicate changes in its reproductive condition or health (Lindburg & Fitch-Snyder 1994).

Figure 3.3. Observational method. A primatologist observes behavior in the field.

APPLYING THE CONCEPTS 3.1

Project Seahorse

Amanda Vincent began studying seahorses because she wanted to do research outdoors and in the sea. Later, she became fascinated with the animal, particularly its unusual reproductive behavior. For example, only male seahorses get pregnant (Figure 1). Once she started her research, she became engrossed with them. Now Vincent works not only to research seahorses' behavior but also to conserve them. When she began studying seahorses, Vincent's work involved spending many hours conducting observations. These animals had rarely been observed in the wild, and very little was known about their behavior. This meant many months of work and many hours each day sitting very still while scuba diving in cold water. Her research began with basic questions about the mating behavior of seahorses. Are they monogamous? What are the behavioral differences between the two sexes of these species? Do males defend territories? To answer these questions, she had to find the seahorses during their reproductive periods. These animals can be very cryptic, or camouflaged, and so blend into their environment. She also had to develop techniques for tagging individuals so she could identify them later and design appropriate measures of behavior.

Because of the research conducted by Vincent, her students, and colleagues, we now know much more about the unique behaviors of seahorses. It turns out that some are truly monogamous, which is very rare in the animal world and even rarer among fish. Mated pairs greet each other in a daily ritual that appears to be important in maintaining that monogamous bond (Vincent 1995a). In addition, Vincent's research indicates that many populations of seahorses and pipefish are decreasing because they are heavily harvested for Chinese medicine, curios, and the hobby aquarium market (Vincent 1995b). In response,

Figure 1. Seahorse. This pregnant male big-belly seahorse, *Hippocampus abdominalis*, is one of many seahorse species studied by Dr. Vincent. It is found only along the coasts of Australia and New Zealand.

Drs. Amanda Vincent and Heather Koldewey founded Project Seahorse to protect seahorse populations while ensuring sustainable livelihoods for those dependent on seahorse fishing. Vincent continues to study the behavior (e.g., Vincent, Evans, & Marsden 2005) and work for the conservation of seahorses (e.g., Vincent 2008). For more information, see www.pbs.org/wgbh/nova/seahorse/vincent.html.

Featured Research

The observational method and reproductive energetics of chimpanzees

Research in the field often relies on the observational method. A long-term study of chimpanzees (*Pan troglodytes*) conducted by Carson Murray, Elizabeth Lonsdorf, Lynn Eberly, and Anne Pusey illustrates this method's uses (Murray et al. 2009). Murray's research team investigated the reproductive energetics of free-living female chimpanzees in Gombe National Park in Tanzania. Chimpanzees, medium-sized primates that live in equatorial Africa, prefer to feed on calorie-rich, ripe fruit but will eat a variety of foods. All female mammals face increased energetic demands both during gestation (pregnancy) and while feeding their young via lactation (Figure 3.4). In response, many pregnant and lactating females alter their diet to increase caloric intake. For instance, females of many small primate species (e.g., capuchins and lemurs) will alter their diet to include more nutrient-rich foods. However, prior to Murray et al.'s study, little was known about this kind of behavioral modification in great apes such as chimpanzees. Therefore, Murray's team asked the following question:

Research question: Do pregnant and lactating chimpanzees alter their feeding behavior to meet the increased energy requirements of reproduction?

In order to address this research question, they observed the behavior of three types of adult females: pregnant, lactating, and nonpregnant/nonlactating. They tested the following hypothesis:

Alternate hypothesis: Pregnant and lactating females will consume a diet that provides higher caloric intake than that consumed by nonpregnant, nonlactating females.

Null hypothesis: Pregnant and lactating females will *not* consume a diet that provides higher caloric intake than that consumed by nonpregnant, nonlactating females.

Prediction: Pregnant and lactating females will have a higher percentage of fruit in their diet (high-calorie food) than nonpregnant, nonlactating females.

The research team used data collected from one group of chimpanzees in the park to test their prediction. Over a period of 30 years, the group ranged in size from 38 to 64 individuals. Starting in 1973, researchers conducted full-day follows during which they tracked and recorded all the activities of one focal individual for an entire day. They recorded all of its feeding bouts (amount of time spent eating, type of food items eaten) and social interactions during that day. Each day, different individuals were followed, resulting in a large database of chimpanzee behavior. The researchers analyzed over 19,000 hours of data from over 2,300 "follows." From these data, they calculated the percentage of fruit in each individual's diet by dividing the time spent feeding on fruits by the total time spent feeding. Each adult female was classified as either pregnant (back-calculated from birthdates of offspring), lactating, or nonpregnant and nonlactating.

Murray's team found that the three types of adult females tended to differ in their feeding behavior. The diets of both pregnant and lactating females included more fruit (allowing more energy gain) than the diet of nonpregnant, nonlactating females (Figure 3.5). The research team concluded that pregnant and lactating females appear to alter their diet to meet the increased energy demands of reproduction. This conclusion was reached through research that relied solely on observations of animals, without any manipulations.

The experimental method

In the **experimental method**, scientists manipulate or change a variable to examine how it affects the behavior of the animal. The variable that is changed is called the independent variable, and it can be anything that we measure, control, or manipulate. It can be abiotic (e.g., temperature, humidity, wind) or biotic (e.g., habitat, food availability, social interactions). The researcher then measures changes in another variable, the dependent variable, that occur in response to changes in the independent variable. For example, suppose you were examining how food availability affects clutch size (the number of eggs laid by a female during one reproductive attempt). Your experimental treatment might consist of manipulating the amount of food (the independent variable) and then measuring the number of eggs produced (the dependent variable).

In order to determine whether or not your manipulation is actually influencing the dependent variable, it is essential to include a control group. The **control group** does not experience the manipulation but is treated similarly in all other aspects. In the simplest experimental design, only one factor differs between the experimental and control group. In essence, the control group represents the null hypothesis, and the experimental group represents the alternate hypothesis.

Figure 3.4. Female chimpanzee. Female chimpanzees face increased energy demands when pregnant and lactating.

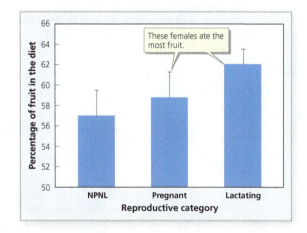

These females ate the most fruit.

Figure 3.5. Chimpanzee diets. Mean (+ SE) percentage of fruit in the diets of females with different reproductive conditions. Pregnant and lactating females ate more fruit than nonpregnant, nonlactating (NPNL) females (*Source:* Murray et al. 2009).

▪ **experimental method** An approach in which scientists manipulate or change a variable to examine how it affects the behavior of an animal.

▪ **control group** In an experiment, the group that does not experience a manipulation and thus provides a comparison to experimentally manipulated groups.

TOOLBOX 3.1

Animal sampling techniques

Murray et al. (2009) followed chimpanzees for many full days to collect data on their behavior. More typically, researchers collect a partial record of behavioral data using standardized sampling techniques. One common technique is **focal animal sampling**, which provides unbiased data from different individuals (Altmann 1974). In this method, a focal individual is randomly selected and observed for a specified time period, and all pertinent behaviors performed by the individual are recorded. After the time period has expired, a different individual is randomly selected and observed. Focal animal sampling has several advantages.

1. The behavior of different individuals is sampled. If individuals have unique markings or morphological aspects, the researcher can record the behavior of known individuals and identify the total number of unique individuals sampled. If it is difficult to identify individuals, the researcher can move to a different location before selecting a new individual. The researcher can also simply select an individual that is some distance away from the last focal animal observed. This ensures that a number of different individuals are sampled.

2. Because random individuals are selected, instead of individuals that exhibit either unusually high or low levels of a particular behavior, sampling is unbiased.

3. Because data are collected over a recorded amount of time, researchers can easily calculate the frequency of a behavior and the rate at which the behavior was displayed.

4. This technique requires little more than being able to keep track of time and observe and record behavior. Depending on the situation, these tasks might best be accomplished by a pair of researchers: a recorder, who keeps track of time and records the behavior, and an observer, who watches the animal and describes the behaviors displayed.

Another common sampling technique is known as **instantaneous** or **scan sampling**. Here, data are collected from individuals at regular time intervals. For instance, you might be interested in the frequency with which individuals in a group scan their environment (i.e., raise their head to observe their surroundings). You could use scan sampling at 30-second intervals to record the number of group members whose head is raised. Over a sufficiently long observation period, these data would provide information on the percentage of time individuals spend scanning the environment.

Featured Research ## The experimental method and jumping tadpoles

Verônica de Sousa and her colleagues used the experimental method to study an unusual behavior of an undescribed species of tropical tadpole (*Pseudopaludicola* sp.) (De Sousa, Teresa, & de Cerqueirra Rossa-Feres 2011). *Pseudopaludicola* are very small frogs (less than 20 mm snout-vent length) found throughout South America (Laufer & Barreneche 2008). During the rainy season, females lay eggs in the small puddles formed from the footprints of large mammals at the edge of ponds and swamps. The eggs develop into tadpoles that go through metamorphosis in these small puddles. Also in these puddles are aquatic juvenile dragonflies (naiads) that are voracious predators of tadpoles, fish, and invertebrates.

The researchers observed that tadpoles sometimes jump out of their puddles and wondered if this behavior could reduce the risk of predation. The puddles tend to be closely spaced and numerous, so a tadpole might be able to jump to a safer one. The researchers examined two questions: (1) Is tadpole jumping behavior a response to the presence of a predator? and (2) Does the jumping behavior increase survival? (Scientific Process 3.1)

The researchers collected tadpoles and dragonfly naiads and created small experimental arenas to simulate puddles. To examine the effect of a predator on jumping behavior, they conducted an experiment with three treatments that differed in the presence or absence of the naiad predator and then measured the frequency of tadpole jumping. They found that the frequency of tadpole jumping was significantly higher in the predator-present treatment than in either control treatment. Most jumps occurred in response to naiad movement.

SCIENTIFIC PROCESS 3.1

Jumping tadpoles
RESEARCH QUESTION: *Why do tadpoles jump out of puddles?*

Hypothesis: Jumping is an antipredator behavior.

Prediction: Tadpoles will jump most often from puddles that contain a predator.

Methods: The researchers:

- collected tadpoles and dragonfly naiad predators from the field.
- acclimated tadpoles in individual aquaria in the lab for four to five days.
- created experimental arenas with sandy bottoms to simulate puddles (9 cm diameter, 4.5 cm depth).

Figure 1. Predators and prey. Juvenile dragonflies (naiads) feed on tadpoles. Adult dragonflies are much larger than naiads.

Experiment 1: 41 tadpoles each experienced three treatments:

1. Predator treatment (PRE): one naiad in the arena
2. Predator control treatment (CPRE): an inanimate object (similar in size to naiad) in arena to control for the presence of a novel object
3. Control (CON): no object or predator in the arena

Each trial lasted 45 minutes and was videotaped. Researchers recorded jumping behavior (number of jumps/tadpole)

Results: Jumping behavior was most common in the predator-present treatment.

Experiment 2: 21 tadpoles were randomly assigned to one of two treatments:

1. Screen treatment: a screen was placed over the arena to prevent the tadpole from jumping out
2. Nonscreened treatment: no screen was placed over the arena, allowing the tadpole to jump out

Each trial lasted three hours or ended when the tadpole was consumed.

Results: Survival was highest in the nonscreened arena treatment.

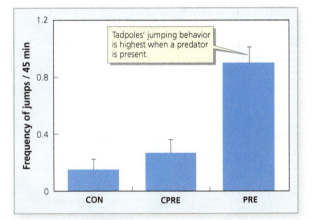

Tadpoles' jumping behavior is highest when a predator is present.

Figure 2. Jumping behavior. The mean (+ SE) frequency of jumping was higher for the predator treatment than both controls (*Source:* de Sousa, Teresa, & Rossa-Feres 2011).

Conclusion: Tadpole jumping behavior is an effective antipredator behavior that enhances survivorship.

To examine how jumping behavior affects tadpole survival, they conducted another experiment with tadpoles and naiads using two treatments that differed in whether tadpoles could get away from the predator. For this experiment, they measured the number of tadpoles that survived and found that the survival rate was much higher in the treatment in which tadpoles could escape the predator.

These experimental results indicate that naiad predators kill tadpoles and that tadpole jumping is an effective antipredator behavior that may have significant fitness benefits. Since tadpoles are aquatic, wouldn't jumping out of the puddle cause them to desiccate? Observations in the field confirm that tadpoles do spend quite a bit of time out of puddles, and yet they survive. There is often a thin film of water on the substrate that likely reduces the risk of desiccation.

comparative method An approach that examines differences and similarities between species to understand the evolution of behavioral traits.

ancestral trait A trait found in the common ancestor of two or more species.

derived trait A trait found in an organism that was not present in the last common ancestor of a group of two or more species.

phylogeny Hypothesized evolutionary ancestor-descendent relationships among a set of organisms.

Featured Research

sister species Two species that are more closely related to one another than to any other species.

The comparative method

In the **comparative method**, scientists examine differences and similarities between species to understand the evolution of behaviors. Often, sets of closely related species share similar behavioral adaptations because their shared ancestor possessed the behavioral trait. By comparing the behavior of closely related species, researchers can try to understand whether a behavior is ancestral or derived. **Ancestral traits** (also called plesiomorphic traits) are found in a common ancestor of two or more species, while a **derived trait** (or apomorphic trait) is found in a more recently evolved species and was not present in the common ancestor.

To understand the evolution of behavioral traits, researchers often require knowledge of the evolutionary relationships among taxa (taxonomic groups) such as species. Evolutionary relationships are provided by a **phylogeny**, a hypothesized evolutionary history of a group of species. For any set of species, a phylogeny illustrates our best understanding of the patterns of descent, showing ancestor-descendant relationships that reflect their evolutionary history. For example, two species that are more closely related to one another than to any other species share a recent common ancestor and are known as **sister species**. In a phylogenetic diagram these species are represented by two lines that share a node (the common ancestor).

The comparative method and the evolution of burrowing behavior in mice

How do researchers use the comparative method and phylogenies to examine the evolution of behavior? Consider work conducted by Jesse Weber and Hopi Hoekstra (Weber & Hoekstra 2009). These researchers were interested in understanding the evolution of burrowing behavior of mice species in the genus *Peromyscus*. Species of these small mice exhibit a wide range of burrowing behaviors. Weber and Hoekstra first constructed a phylogeny of seven species using genetic data (Figure 3.6). Next,

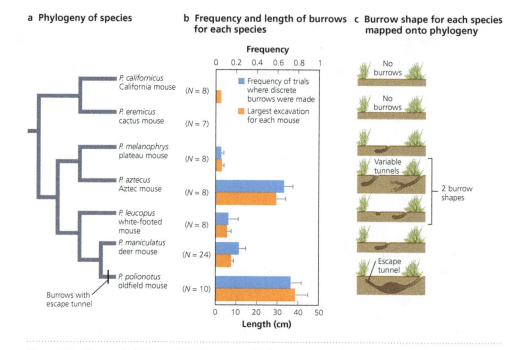

Figure 3.6. Mouse phylogeny and burrowing behavior. (a) The molecular phylogeny of seven species of mice identifies sister species. (b) The mean (+ SE) frequency and length of burrow construction for each species demonstrates variation among species. (c) Species range from no burrowing behavior to constructing complex burrows (*Source:* Weber & Hoekstra 2009).

TOOLBOX 3.2

Creating phylogenies

You probably learned about the **Linnaean classification system** in introductory biology (Figure 1). Carl Linnaeus, a Swedish scientist, designed the scientific classification system of naming organisms in 1735. The Linnaean system classifies organisms in a nested hierarchical system ranging from the largest groups of kingdoms into smaller and smaller groups that move from phylum down to class, order, family, genus, and finally species (Figure 1). Originally, organisms were classified based on morphological and ecological similarities, because these were the only characteristics available at the time. Today we have much more detailed information about organisms, because we have access to molecular and genetic data, which allow us to evaluate the evolutionary history of organisms. Many scientists use the **phylogenetic classification system**, also called **cladistics**. This system also produces a nested hierarchy that classifies organisms down to the species or even subspecies level, but it also includes domains and kingdoms. In addition, the phylogenetic classification system identifies **clades**, which are groups of organisms that are all descended from a common ancestor. Phylogenies are created using character traits, which can be anatomical morphology, physiological processes, behaviors, or genetic sequences. Scientists use cladistic computer programs to analyze such data and produce a **phylogenetic tree**, which is a hypothesis about the evolutionary relationships among a set of organisms (Figure 2). Just like all hypotheses, these can be supported or rejected as new and better data become available.

Cladistic analyses start with three basic assumptions:

- The characteristics of organisms change over time. The original state of the characteristic is known as ancestral and the changed state is known as derived (Figure 3).
- All organisms are related by descent back to a common ancestor.
- When a lineage splits, it tends to create a branching pattern. Splits result in two groups and not three or more lineages at once (although some exceptions do exist).

To create a phylogeny, researchers follow five steps:

1. Choose a set of organisms. This might be several populations of a species, several species in a genus, or groups of higher taxonomic levels.
2. Determine the character traits for the analysis. These might include morphological, molecular, or behavioral data (e.g., Blackledge et al. 2009). This critical step will be the basis of the analysis. The character traits must all be **homologies** (i.e., they must be similar as a result of coming from a common ancestor and not because of convergent evolution). It is also important to select character traits that are not readily affected by environmental conditions. This is one relative advantage of molecular and genetic traits.
3. Determine whether each character trait is ancestral or derived (i.e., determine the polarity) for each taxa. This is typically done by comparing the focal set of species to an **outgroup**, which is a species distantly related to the focal group. Ancestral traits are often shared with the outgroup, while derived traits are acquired by a recent common ancestor of the taxa under consideration.
4. Group the taxa by their shared derived traits (i.e., **synapomorphies**). Groups that share derived traits are hypothesized to form a clade.
5. Look for the simplest phylogenetic tree, a process known as **parsimony**. The most parsimonious tree requires the fewest evolutionary changes.

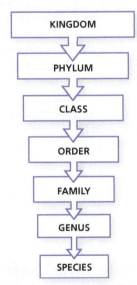

Figure 1. Linnean classification system. The classical Linnaean system is a hierarchy.

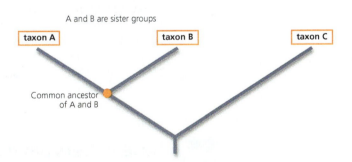

Figure 2. Sample phylogenetic tree. In the phylogenetic classification system, phylogenetic trees represent evolutionary relationships between organisms or taxa.

continued...

	Lancelet (outgroup)	Shark	Tuna	Salamander	Turtle	Chimpanzee
Character						
Vertebral column	no	YES	YES	YES	YES	YES
Bony skeleton	no	no	YES	YES	YES	YES
Four limbs	no	no	no	YES	YES	YES
Amniotic egg	no	no	no	no	YES	YES
Hair	no	no	no	no	no	YES

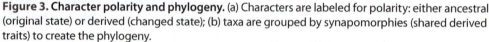

Figure 3. Character polarity and phylogeny. (a) Characters are labeled for polarity: either ancestral (original state) or derived (changed state); (b) taxa are grouped by synapomorphies (shared derived traits) to create the phylogeny.

they conducted behavioral studies of each species to measure different aspects of burrow construction (e.g., entrance length, burrow depth, and total length). They placed a mouse in a large chamber (1.22 m × 1.52 m × 1.07 m) that contained a sandy loam soil for 48 hours and allowed it to construct a burrow. Afterward, they removed the mouse and filled the burrow with polyurethane foam that expanded to create a cast of the burrow, which they then measured.

They found much variation in burrowing behavior of the seven species studied: two species did not construct burrows, three constructed simple burrows, and two made very complex burrows. Finally, the researchers mapped the burrowing behavior of the species onto their phylogeny.

The patterns in the behavioral studies and the phylogeny provided information about the evolution of burrowing behavior in these species. The analysis indicated that the ancestral trait is the absence of burrowing and the construction of large, complex burrows is a derived trait. They found that sister species often exhibited significant differences in burrowing behavior (e.g., the plateau mouse and Aztec mouse exhibited different behaviors, as did the deer mouse and the oldfield mouse). Because burrowing behavior was so varied among these closely related species, the researchers concluded that variation in this behavior most likely evolved independently in the different species. This conclusion implies that genetic differences among species produce the observed differences in burrowing behavior. Weber and Hoekstra are currently continuing their work to examine the genetic aspects of burrowing behavior in these mice.

As we've just seen, animal behavior can be examined using a variety of methods. Next, we examine how the study of behavior has been approached from a variety of different perspectives.

3.3 Researchers have examined animal behavior from a variety of perspectives over time

From Darwin forward, animal behavior research has been based on evolutionary thinking. Over time, a large number of rich and diverse research foci have addressed the study of animal behavior in unique ways by emphasizing different approaches and methodologies. In this section, we briefly review how research has changed over time.

Darwin and adaptation

Darwin pioneered the scientific study of behavioral adaptation in three books: *On the Origin of Species* (1859), *The Descent of Man and Selection in Relation to Sex* (1871), and *The Expression of the Emotions in Man and Animals* (1872). In the first book, Darwin discusses instinct, or behaviors neither learned nor requiring experience. One intriguing example he describes is cuckoo reproductive behavior. Many species of cuckoo lay their eggs in the nests of other species, who then care for the young (Figure 3.7). How could such behavior evolve by natural selection? Darwin noted that while most female birds lay an egg every day, female cuckoos do not: they lay one egg every several days. He reasoned that if a female cuckoo were to raise her own clutch of many eggs, she would be faced with a prolonged period of incubation and a clutch of chicks whose ages would differ greatly. Such a scenario would be costly for both the female and the chicks. Darwin hypothesized that selection would favor an individual that laid her eggs in the nests of others, circumventing this problem.

Figure 3.7. Cuckoo behavioral adaptation. A dunnock feeds a cuckoo chick. Many species of cuckoo lay their eggs in the nests of other species.

Darwin's work on animal behavior was wide ranging, both taxonomically (e.g., ants, birds, and bees) and topically (e.g., reproduction, aggression, and instinct). Although some of Darwin's descriptions were based on anecdotes and others were anthropomorphic, he always explained behavior using evolutionary reasoning and in terms of adaptation. His work represents the beginning of the evolutionary basis of animal behavior research and a tradition that has continued through to today.

Early comparative psychology

Georges Romanes, a contemporary of Darwin, approached the study of animal behavior from the perspective of a comparative psychologist. Psychologists study the mental processes and behavior of humans. Comparative psychologists, like Romanes, study animal behavior in order to understand human behavior. In *Animal Intelligence*, Romanes systematically examined the reasoning ability of a wide variety of animal taxa, ranging from protozoans to mammals (Romanes 1882). For him, an animal displayed a "mind," or conscious action, rather than mere reflex reaction, when it used previous experiences to learn and modify its behavior in an adaptive manner. For example, if a protozoan encountered a barrier and moved around it, Romanes interpreted this behavior as evidence of a conscious or intelligent choice, as opposed to a reflex movement.

Romanes's work stimulated other researchers concerned with examining animal consciousness. He eventually published a "mental tree" to describe a hierarchy of reasoning among taxa (Romanes 1888). While Romanes's construction is simplistic, the patterns it summarizes are surprisingly congruent with current research regarding which animals display higher stages of intellectual development. These include tool use in monkeys and communication in Hymenoptera (one of the largest orders of insects) (Burghardt 1985).

Many of the examples in Romanes's book were anecdotal or anthropomorphic, and they were often provided by nonscientists. So although it was one of the first attempts to systematically examine behavior throughout the animal kingdom, it was heavily criticized. C. Lloyd Morgan, for one, emphasized the need for "accurate observation" along with the methods and principles of modern psychology (Morgan 1894). Morgan, too, used animals as a way to understand human consciousness. However, he urged a more cautious approach, proposing that the simplest psychological process possible should be used to interpret an animal's behavior—an idea that came to be known as **Morgan's canon** (Morgan 1903; Burghardt 1985; Costall 1993). Although Morgan's canon became a cornerstone of comparative psychology, Morgan may not have meant it to be literally interpreted as such. Morgan was emphasizing the need for a more scientific

Morgan's canon The idea that the simplest psychological process possible should be used to interpret an animal's behavior.

approach to the study of the mental evolution of animals, but he acknowledged that one should not necessarily exclude more complex explanations of behavior (Costall 1993).

Comparative psychology in North America

In the 1930s, animal behavior research began to flourish in North America. Two comparative psychologists, E. L. Thorndike and Margaret Floy Washburne, were particularly noteworthy.

Thorndike, like Lloyd Morgan, stressed the importance of avoiding anecdotal evidence. He urged the use of repeated observations, under standardized conditions, to obtain appropriate sample sizes (Galef 1998; Stam & Kalmanovitch 1998). He thus pioneered the use of standardized methodology and the experimental method in the study of animal learning. He is best known for the use of "puzzle boxes," from which an animal could see food that lay outside its reach. These boxes had levers, pulleys, and treadles that had to be manipulated in order to escape the box. Thorndike observed and quantified how animals learned by trial and error to escape the apparatus to obtain food. Cats, for instance, quickly learned how to escape and escaped more quickly with repeated testing (Thorndike 1911). Thorndike called this increased ability a time curve, today known as a learning curve.

Margaret Floy Washburn was the first woman to earn a Ph.D. in psychology, and the second woman to be inducted as a fellow into the National Academy of Sciences. In addition to her extensive research on the motor theory of consciousness (Russo & Denmark 1987), she wrote a groundbreaking textbook that emphasized the experimental method for comparative psychology. *The Animal Mind* (Washburn 1908) systematically summarized research on a wide variety of taxa and helped to train a generation of comparative psychologists. For example, it summarized all the known research on sensory discrimination and spatial perception in organisms ranging from amoebas to vertebrates.

Behaviorism

In the early part of the twentieth century, John Watson spearheaded the development of behaviorism, a field of comparative psychology that studies behavior independent of animal mental states or consciousness. Behaviorists focused on what they could actually observe (Watson 1913; Watson 1924), including the specific stimuli that provoked a response. In this stimulus-response paradigm, both stimuli and responses could be external (behavior) or internal (physiology). Because Watson emphasized the benefits of using variables that could be controlled, he promoted a strong emphasis on laboratory work.

Two of the most influential behaviorists were Ivan Pavlov, a Russian digestive physiologist, and B. F. Skinner. Pavlov studied salivary gland secretions and the role they played in digesting food (Pavlov 1927). He noticed that dogs would often salivate profusely just prior to being fed. He discovered that this behavior could also be produced in the absence of food if he first paired the presentation of food with a novel stimulus, like the ringing of a bell. After several such pairings, dogs associated the novel stimulus with food and would salivate after hearing it, a behavior he called a conditional reflex. **Classical** or **Pavlovian conditioning** occurs whenever a novel stimulus (like the ringing of a bell) is paired with an existing stimulus (here, the food) and elicits a particular response (salivation). Eventually the novel stimulus alone elicits the same response as the existing stimulus.

Skinner also worked on laboratory animals in highly controlled conditions. He developed novel tools and procedures to study the development of behavior. Most famously, he is known for the creation of operant chambers, or **Skinner boxes**, which housed a single animal (Figure 3.8). Skinner examined how the behavior of animals could be modified by pairing a particular behavior with either a reward or a punishment (Skinner 1938). Behaviors could be enhanced if animals were given food rewards (a positive reinforcement) for performing the behavior. Alternatively, these behaviors could be reduced if they resulted in a painful shock (a positive punishment). In this process, called **operant conditioning**,

classical conditioning A type of learning in which a novel neutral stimulus is paired with an existing stimulus that elicits a particular response. Eventually the novel neutral stimulus alone elicits the same response as the existing stimulus.

Skinner box Operant chamber to study behavioral conditioning. Also known as an operant chamber.

operant conditioning A learning process in which an animal learns to associate a behavior with a particular consequence.

an animal learns to associate a behavior with a particular consequence (either positive or negative).

Classical ethology

In direct contrast to behaviorists, classical ethologists studied the behavior of wild animals in nature by observation and experimentation. In general, classical ethologists were not psychologists interested in understanding human minds and behavior, but rather were interested in studying animals and their behavior for their own sake. Karl von Frisch, Konrad Lorenz, and Niko Tinbergen shared the Nobel Prize in Physiology and Medicine in 1973 "for their discoveries concerning organization and elicitation of individual and social behaviour patterns" (Karolinska Institutet 1973). Von Frisch studied honeybee sensory perception and communication (Frisch 1956). He discovered that European honeybees (*Apis mellifera*) have a complex chemosensory system that they use for finding food. He found that they use vision, smell, and taste while foraging on flowers. For example, they orient from a patch of flowers to their hive using the position of the sun in the sky and can respond to polarized lightwaves from the sun. Even more astonishingly, he found that bees communicate the distance and direction of distant food sources to other individuals in the hive by performing a "waggle dance."

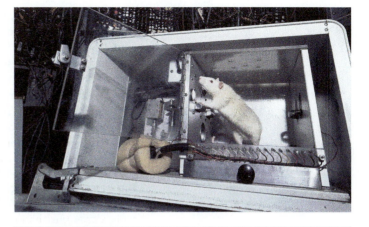

Figure 3.8. Skinner box. A rat is tested in a Skinner box.

Lorenz studied instinctive behavior in birds, particularly **imprinting**, in which a young animal learns the characteristics of its parent. Many birds, such as goslings, "imprint" on the sight and sound of their mother or other objects they see at hatching (Lorenz 1950). Lorenz found that there was a restricted period of time during which imprinting could occur.

Tinbergen studied instinct—behaviors that are under strong genetic control. Instinctive behaviors termed **fixed action patterns** are invariant and unlearned; once initiated, they are brought to completion. Much of his work attempted to understand the mechanisms, or **releaser stimuli**, that initiated a fixed action pattern. One classic example was egg-rolling behavior in graylag geese (*Anser anser*) (Tinbergen 1951). A bird incubating eggs will attempt to retrieve any egg that has been displaced from the nest. Tinbergen demonstrated that the sight of any egg-shaped object near a nest initiates egg retrieval.

Interdisciplinary approaches

Psychologists and ethologists have conducted animal behavior research increasingly as an interdisciplinary endeavor over the last few decades. For instance, we've seen the development of evolutionary psychology and cognitive ethology, disciplines that both emphasize the study of animal thought processes and cognition using an evolutionary framework. Evolutionary psychologists seek to understand human thinking and behavior, while cognitive ethologists focus on understanding the behavior of animals and often integrate information from neuroscience. Both fields assume that natural selection has shaped brain architecture and thought processes in an adaptive manner. Evolutionary psychologists, such as John Tooby and Leda Cosmides, suggest that we can best explain human behavior by understanding how natural selection shaped neural mechanisms in the past (Griffiths 2008).

Cognitive ethologists also seek to understand how natural selection has acted on mental processes and cognition in order to better understand the behavior of animals. In doing so, they often conduct comparative studies using broad taxonomic comparisons, frequently carrying out research in natural environments. Donald Griffin is credited with the development of cognitive ethology thanks to his book *The Question of Animal Awareness* (Griffin 1976). Today, cognitive ethologists study topics such as learning and memory; counting ability; symbol recognition; and vocal communication in organisms as diverse as bees, crows, rodents, and primates (Bekoff, Allen, & Burghardt 2002).

imprinting Rapid learning in young animals through observation. Individuals are typically attracted to objects they imprint on.

fixed action pattern Behaviors that are invariant and unlearned. Once initiated, they are brought to completion.

releaser stimuli Stimuli that initiate a fixed action pattern.

Behavioral ecology, which developed in the 1960s and 1970s and continues today, focuses on the evolution of behavior by studying its function. Behavioral ecologists sometimes use mathematical or computer models (frequently from economics) to examine the fitness benefits and costs of a behavior. What they learn from this process helps us understand how natural selection has molded that behavior, as we saw in Chapter 1 (Davies, Krebs, & West 2012).

You can already see that various perspectives have placed different emphases on Tinbergen's four questions. For example, classical ethologists tend to focus on proximate questions, while behavioral ecologists often emphasize ultimate explanations of behavior. You will see examples of many of these approaches and methods throughout this book. No matter which methods are used, all research also involves important ethical considerations. We examine those next.

3.4 Animal behavior research requires ethical animal use

In the previous chapters, we examined the process of science and described how scientific information is communicated. That said, scientists are not infallible. As we close this chapter, we examine two ethical issues related to the study of animals: scientific misconduct and animal care.

scientific misconduct Violation of ethical behavior in science.

Scientific misconduct includes the falsification or fabrication of data, purposefully inappropriate analysis of data, and plagiarism. These can occur in all branches of science and most often are identified in peer review, so papers with such problems are rarely published. If a paper does manage to pass through the peer-review system and is published, misconduct is typically then identified when other scientists cannot successfully replicate the findings.

Researchers studying animals must also consider how their work affects the individuals they study. The better we understand the behavior of animals, the better we can manage and care for them. However, research can also have negative impacts on animals. These can arise from simple observations of wild animals, inappropriate housing of animals in laboratories, and the various procedures performed. Scientists and governments have therefore developed ethical standards for animal research.

How research can affect animals

Observing animals in their natural environment often involves the lowest level of invasiveness and stress for individuals. However, even observational field research can negatively affect the animals studied. For example, repeatedly observing individuals can habituate them to humans. If individuals lose their fear of humans, they can become subject to increased predation from poachers or hunters. Manipulations of the animal's environment, such as providing supplemental food, can lead to still other unintended consequences. A large addition of food to the environment could result in a substantial increase in the population, which might result in overcrowding and increased stress (Robb et al. 2008).

Other kinds of behavioral research require more direct contact with animals (Figure 3.9). For example, field researchers may need to collect blood or tissue samples, quantify parasite loads, measure reproductive condition or morphology, or uniquely mark individuals. These procedures can increase animals' stress levels because they require researchers to trap, restrain, or anaesthetize individuals. For animals in captivity, there are additional considerations, including appropriate housing and husbandry (feeding and care), as well as how the animals will be procured. Researchers also need to determine the fate of the animals when the research is completed. Some animals may

Figure 3.9. Animal care. A zoo veterinarian takes measurements of a chameleon.

need to be sacrificed for examination of their body organs. For example, researchers studying how mating behavior affects neurobiology may need to examine the brains of their study organisms (e.g., Huang & Hessler 2008).

Sources of ethical standards

Most nations have set guidelines for the use of animals in research and teaching. In the United States, a variety of federal agencies and organizations have formulated specific guidelines that scientists must follow to maintain funding. These include the National Institutes of Health Office of Animal Care and Use (NIH-OACU), the Department of Agriculture Animal and Plant Health Inspection Service (USDA-APHIS), the Association for the Assessment and Accreditation of Laboratory Animal Care (AAALAC), and the American Association for Lab Animal Science (AALAS). In Canada, the Canadian Council on Animal Care sets these standards, while in Europe they come from the Council of Europe Conventions on the Protection of Animals and the European Biomedical Research Association.

Scientific societies also have animal care guidelines. These include the Animal Behavior Society (2006), the American Society of Mammalogists, the American Fisheries Society, the Ornithological Council, and the American Society of Ichthyologists and Herpetologists. In addition, the National Research Council's (NRC) Institute of Laboratory Animal Research Commission on Life Sciences publishes a free online book on animal care (Institute for Laboratory Animal Research 1996). In Appendix 3, you'll find further information about the network of agencies that provide guidelines for researchers.

The three Rs

Although there are differences among guidelines in different countries, most follow the **3 Rs**: replacement, reduction, and refinement. **Replacement** means using computer modeling, videotapes, or other approaches in place of actual animals. **Reduction** refers to limiting the number of animals subject to disturbance in research or teaching. Such an effort requires much thought and perhaps pilot studies in order to design a study that uses the fewest possible individuals. Finally, **refinement** involves improving procedures and techniques to minimize pain and stress for animals. Almost all animal care guidelines apply only to vertebrates—with one notable exception. In the United Kingdom, animal care guidelines cover the common octopus, *Octopus vulgaris*.

Governmental, research, and educational facilities that receive U.S. federal funding are required to have their research monitored by an Institutional Animal Care and Use Committee (IACUC). Before conducting research on animals, each scientist must submit an animal care protocol to the committee, which is composed of institutional scientists as well as members of the outside community. The protocol must describe why a species is appropriate for the research and whether nonanimal alternatives are available. For example, is it possible to use videotaped footage from previous work? Could computer modeling replace live animals? The protocol must also justify the number of individuals required, how they will be procured, and specific conditions for animal care (e.g., housing enclosure, food provided, temperature, and light/dark regime). In addition, the scientist must describe procedures that will be used to minimize stress to the animal. Finally, the scientist must address what will be done with the animals after the study is completed. Can they be used for future research? Will animals be released back into the wild at their capture sites, or will they be euthanized?

The IACUC reviews the research protocol to make sure that all animals will receive appropriate care and that the research is well designed, has not been conducted before, and complies with all animal welfare regulations. The committee also assesses the importance of the research question and the quality of the research before allowing it to be conducted. As you can see, ethical considerations are an important factor in animal behavior research.

replacement An ethical guideline that encourages the use of computer modeling, videotapes, or other approaches in place of actual research animals in the laboratory.

reduction An ethical guideline that promotes limiting the number of animals subject to disturbance in research or teaching.

refinement An ethical guideline that involves improving procedures and techniques to minimize pain and stress for animals.

TOOLBOX 3.3

Describing and summarizing your data

Researchers collect data and then use statistics to describe and summarize them. **Statistics** is a branch of applied mathematics that uses probability theory to produce the best conclusions from the available data (Sokal & Rohlf 2012). Researchers use statistics to plan the research protocol, summarize and interpret the data, and even predict future events. For centuries, statistics has been an important tool for scientists, as well as people in such fields as business, sports, and government.

Statistics are used to analyze data collected from a sample of the population being studied. A **sample** is a subset of a population or group that is chosen to be representative of the entire population. **Descriptive statistics** are used to summarize and describe measurements. For example, there are measures of central tendency, such as the mean, median, and mode, and measures of dispersion, such as the standard deviation, standard error, and variance. We use **inferential statistics** in testing hypotheses to determine if there is a statistical difference between two or more sets of data.

MEASURES OF CENTRAL TENDENCY

A **measure of central tendency** is a number that indicates the centrality of the data values. The most common measure is the mean, or average. The **mean** of a sample (often abbreviated as $\bar{x}$) is the sum of all the data divided by the number of data points. This measure indicates the average value of the data.

$$\text{Mean} = \bar{x} = \sum x_i / n$$

where $\sum x_i$ is the sum of all data values (1 to n).

The **median** is the middle value of a set of ordered data. In order to calculate the median, you would order your data from the smallest to the largest values and then find the value in the middle. If you have an even number of data points, then the median is the average of the two middle values. The median is a useful measure for data that are not normally distributed (i.e., they do not resemble a normal or bell-shaped curve). In these cases, there may be a few very small or very large data values that are far from the mean. These outlying data points can substantially inflate or decrease the mean, which means that they have a very large effect on the mean. For example, imagine analyzing the test scores of a group of students. If most scored in the 70%–80% range but two students scored a 100%, you might want to use the median to describe the central tendency of the scores, because the mean would be inflated by the two students who scored 100%. The median would better describe the centrality of the data.

Another measure of a set of data is the **mode**, or the most common value. If you wanted to examine and describe the number of classes that all freshmen take at your school, you might want to use the mode to describe such data.

MEASURES OF DISPERSION

In addition to summarizing the central tendencies of data, it is also important to describe variation using **measures of dispersion**. For example, imagine you gave two exams to a class of nine students. On Exam 1, all students scored between 75% and 85%. On this test, the scores (data) are tightly dispersed within a relatively small range. However, on Exam 2, students scored between 38% and 98%. This time, the test scores are widely dispersed, indicating a great deal of variation in the data. The means, however, of Exams 1 and 2 might actually be the same, say 80%. Therefore, reporting only the mean would fail to provide important information on differences in the distribution of the two exams' data. To accurately summarize data, you need to summarize the amount of dispersion or variability within the dataset.

There are several measures of dispersion. Often, the most useful are the range, standard deviation, standard error, and variance. The **range** is the difference between the highest and the lowest measurements. This is a very simple measure of data dispersion, but it can be a useful tool for understanding the full extent of variability in the data. For example, if you wanted to understand the age at which the symptoms of Huntington's disease (a hereditary disorder of the central nervous system) could appear, you would examine the range of ages at which the symptoms appeared in individuals with the disease. Then you would know the earliest and latest ages that these symptoms appear.

The **variance** (S^2) is a nonnegative number that provides information on spread in the data. The larger the variance, the more dispersion there is in the data. Calculating the variance requires squaring the deviations of each data value from the mean. Therefore, the variance is expressed in units squared (which are not the same as the units of the data) and tends to be a rather large number, which makes it difficult to graph with the mean or median. For these reasons, you will usually see the standard deviation reported instead of the variance. The **standard deviation** (S) is the square root of the variance; it is expressed in the same units as the data and is smaller than the variance, which makes it more easily represented on a graph.

$$S^2 = \sum (x_i - \text{mean})^2$$

Many scientific papers report dispersion in the data as the standard error (SE) of the sample mean. The **standard error** of a sample mean is simply the standard deviation of the data divided by the square root of the sample size (n):

$$SE = S / \sqrt{[n]}$$

The standard error provides information about the precision of the mean of a sample. This value is also used to construct confidence intervals around the sample mean. For example, a 95% confidence interval around the sample mean contains all data values that are ±1.96 times the standard error of the mean. As one collects more samples, we see that the standard error will decline, because we will obtain greater precision in our estimate of the sample mean. In contrast, the standard deviation of the data is unaffected by an increase in the number of samples collected.

Means and standard errors of the mean are commonly reported in figures that summarize data. For example, in a bar chart, the height of a bar indicates the mean of the sample and the vertical line at the end of the bar indicates the standard error of the sample mean. Throughout this textbook, you will see summaries of data from research studies that contain both the mean of a sample of individuals and the standard error of the mean.

Chapter Summary and Beyond

Explanations about animal behavior address either proximate or ultimate research questions, as summarized by Tinbergen. While emphasis on proximate and ultimate explanations has varied, together they provide complementary information and a more complete understanding of behavior. Laland et al. (2011) examine contemporary biological debates in light of the proximate-ultimate dichotomy. Animal behavior researchers commonly use observational, comparative, and experimental methods to test hypotheses.

Beginning with Darwin, researchers have studied behavior from many perspectives. Houck and Drickamer (1996) offer a detailed description of the history of the study of animal behavior, including seminal papers. Dewsbury (1985) provides autobiographies of many important figures in the history of animal behavior research.

The Association of Zoos and Aquariums, in collaboration with the Wildlife Conservation Society, has produced a DVD that illustrates research methods for studying animal behavior (*Methods for Animal Behavior Research*). In all research that involves animals, it is imperative to minimize stress and suffering. Research on animals is monitored and must be approved by institutional (and sometimes federal) animal care and use committees. The three Rs—replacement, reduction, and refinement—are often used as guidelines. Researchers must consider the possibility of replacing animals with computer models, reducing the number of animals required, and refining their techniques to minimize stress and disturbance to animals. For a more complete treatment of this subject, see Dawkins (2008) and Allen and Bekoff (2007).

CHAPTER QUESTIONS

1. A researcher is interested in studying food color preferences of cardinals, *Cardinalis cardinalis* (a common songbird). She offers a population of cardinals an equal number of small black seeds and large yellow seeds and finds that the birds eat significantly more of the large yellow seeds. From this experiment, the researcher concludes that cardinals prefer to eat yellow seeds rather than black seeds. Critique this experiment. Can you improve on its design?

2. Select three papers about animal behavior from the primary literature. Determine whether the researchers used observational, comparative, or experimental methods in the study. Does the research address proximate or ultimate questions about behavior? Briefly justify your answers.

3. The United Kingdom has added an invertebrate, the common octopus, to its list of animals included in the animal care guidelines. Why are invertebrates in general not included in animal care guidelines? Why do you think the British government included this particular invertebrate in its animal care guidelines? The following references might be useful in your analysis: Boal et al. (2000), Fiorito & Scotto (1992), Kuba et al. (2006), and Mather (2008).

4. In the phylogeny of felids, identify all sister species.

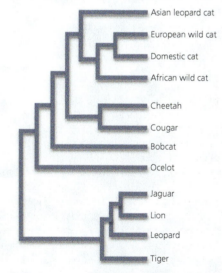

5. Imagine you are conducting an observational study examining the feeding time of individual mallard ducks in small flocks. You want to use focal animal sampling to determine if there are differences in male and female feeding times. How would you randomly select individuals from flocks?

Chapter 4

Behavioral Genetics

Figure 4.1. Herding behavior. Several breeds, such as this Border Collie, have been bred to herd livestock.

Our neighbor's Australian shepherd, Ryker, occasionally escapes from his yard to herd the sheep that live down the road. No one trained Ryker to do this, but he displays herding behavior (Figure 4.1). We have had Labrador retrievers for over 20 years now. The breed is very aptly named: these dogs tirelessly retrieve balls, sticks, and toys and are adept in the water. Our Labs, Max and Grace, race out each morning to combine retrieving balls with swimming in the pool (Figure 4.2). Why do dog breeds exhibit these differences in their behavior? These animals have a genetic predisposition to exhibit different behaviors—that is, their genes influence their behavior (Spady & Ostrander 2008). Over the past few centuries, Australian shepherds were bred for their herding behavior, while Labradors were bred for retrieving. These differences in behavior among dog breeds came about as a result of the natural variation in early dogs. Breeders could use this natural variation to select dogs with the appropriate traits to breed (artificial selection). Individuals that performed well on specific tasks were selected for future breeding, and because these behaviors have a genetic component, they increased in frequency over time in each breed.

However, genes do not tell the whole story. The environment, too, can strongly affect the behavior of animals. For example, the early social environment of puppies can be a critical factor in their behavioral development. In fact, the American Veterinary Society of Animal Behavior (AVSAB 2008) recommends that owners enroll puppies as early as eight weeks of age in socialization classes to minimize the risk of behavioral problems such as fear and aggression emerging in adulthood. In addition, although retrievers and sheepdogs naturally exhibit herding and retrieving behaviors, they need to be trained to fully develop these skills.

Figure 4.2. Retrieving. Retrievers were bred to retrieve game animals for hunters, but, as this one demonstrates, will retrieve almost anything.

phenotype (P) The observed traits of an individual.

genotype (G) All the alleles of an individual.

environment (E) The conditions an individual has experienced and their effect on its behavioral phenotype.

gene-environment interaction (GEI) Interactions between a genotype (G) and an environment (E) that influence the behavioral phenotype.

instinct Behaviors that are performed the same way each time, are fully expressed the first time they are exhibited, and are present even in individuals raised in isolation.

reflex One kind of innate behavior; an involuntary movement in response to a stimulus.

How, then, do genes and the environment affect animal behavior? Can we ascertain their relative importance in determining variation in behavior among individuals? In this chapter, we see how behavioral geneticists study the genetic and environmental influences on behavior that help to answer these questions.

4.1 Behavioral variation is associated with genetic variation

As we discussed in Chapter 2, individuals within and across populations display natural variation in their **phenotype**, or their observable traits, such as behaviors and morphology. This variation is due to differences in their **genotypes** (genetic makeup) and their environment. Behavioral genetics examines how genes and the environment contribute to these differences in behavior by partitioning these effects.

An individual's behavior, or its behavioral phenotypic value (P), is the result of three factors:

1. its genotype (G) at all loci that affect the behavior;
2. the **environment (E)** it has experienced; and
3. any interactions between them—more formally, **gene-environment interactions (GEI)**.

In the rest of this chapter, we will examine each of these factors, but we begin with the genetic basis of behavior.

The search for a genetic basis of behavior

Some of the earliest evidence for the genetic basis of behavior came from captive animals. Early researchers noted that many laboratory strains of mice and rats often exhibited consistent differences in behavior (e.g., Yerkes 1913). Because individuals in captivity are usually reared in similar conditions, differences in behavior most likely result from differences in their genotypes (e.g., Plusquellec & Bouissou 2001; Augustsson & Meyerson 2004). Other evidence came from studies of **instinct**, or innate behaviors—behaviors that are performed the same way each time, are fully expressed the first time they are exhibited, and are present even in individuals raised in isolation. Because all individuals in a species exhibit innate, nonlearned behaviors, these behaviors must have a genetic basis—which also means that they are heritable. Innate behavior includes **reflexes**—involuntary and often immediate behavioral responses to an external stimulus. One example is the blink reflex: the eyelid of many species, including humans, closes when an object moves quickly toward the eye. This reflex is common in vertebrates, as we learned firsthand when our dog Buck suffered facial nerve paralysis on his right side. The veterinarian confirmed the diagnosis by showing us that Buck's blink response was absent for his right eyelid but fully functioning for the left.

Konrad Lorenz and Niko Tinbergen described many innate behaviors. They noted that adult graylag geese (*Anser anser*), which lay their eggs in nests on the ground, will extend their neck and use their bill to gently roll displaced eggs back into the nest in a very fixed manner (Lorenz & Tinbergen 1957) (Figure 4.3). Tinbergen, a pioneer of modern animal behavior (see Chapter 3), also examined the innate escape response of newly hatched goslings (Tinbergen 1951). He noted that when

Figure 4.3. Graylag geese egg retrieval.
Egg retrieval behavior is a fixed action
pattern behavior initiated by the sight of
an egg outside the nest (*Source*: Lorenz &
Tinbergen 1957).

they observed a predator silhouette, they always responded by assuming a characteristic crouching position or running away. In both of these cases, the animals responded with a **fixed action pattern**—a behavior that displays almost no variation
and, once started, cannot be stopped until completed. These observations led
researchers to hypothesize that such behavior must be genetically based.

fixed action pattern Behaviors that
are invariant and unlearned. Once
initiated, they are brought to
completion.

Behavioral differences between wild-type and mutant-type fruit flies

Featured Research

John Paul Scott and Margaret Bastock, two researchers who each studied fruit flies
(*Drosophila melanogaster*) in the laboratory, produced more direct evidence of a genetic influence on behavior (Scott 1943; Bastock 1956). In fact, Bastock, a graduate
student of Tinbergen, helped open the door to the new field of behavioral genetics.

In her dissertation, Bastock examined behavioral differences between a **wild
type**, or the typical form in nature, and a mutant form of *D. melanogaster* (Bastock
1956). Genetic studies had previously identified a mutant fly, called "yellow" because

wild type The typical form of an
organism or gene that occurs in nature.

its body is yellow instead of the normal gray. This mutation
was rarely observed in nature but arose frequently and reproduced successfully in laboratory stocks. Why aren't
yellow forms more common in nature? Bastock wondered
whether the gene mutation in yellow flies might induce a
behavioral change that resulted in very low reproductive
success in the wild.

Fruit fly courtship begins with the male orienting
toward and then following the female. The male taps her
with his foreleg. The foreleg has pheromone receptors—
molecules on the surface of cells (or the nucleus) that receive chemical information—that identify sex and species.
Next, the male begins a wing vibration, a form of courtship
song, which is followed by the male licking the female. If the
female is receptive, they copulate (Figure 4.4).

Bastock interbred wild-type and yellow flies for seven
generations to create inbred flies whose genomes were very
similar except for one thing: the "yellow" gene. She conducted mating trials with both types of males and wild-
type females, recording their mating success and courtship
behaviors. She found that the wild-type males had overall
higher mating success and mated much sooner than the

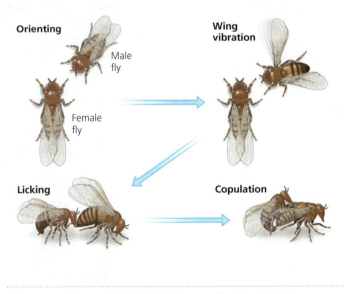

Figure 4.4. Fly courtship. Male fruit fly courtship behavior follows a sequence of events that includes orienting toward the female, vibrating the
wings (song production), and licking the female genitalia, which can lead
to copulation (*Source:* Sokolowski 2001).

yellow males. In addition, yellow males exhibited significantly less courtship behavior—in particular, wing vibration and licking—than wild-type individuals. Finally, Bastock conducted additional experiments to reject alternate hypotheses that females were responding to differences in the color or pheromones of the yellow males.

She concluded that important differences in courtship behavior resulted in the observed differences in the mating success of yellow individuals: females mate less with males that produce weak courtship songs. This experiment was one of the first examples of linking variation in genotype with variation in behavioral phenotype. While her work was underappreciated at the time (Cobb 2007), it set the stage for the behavioral genetic research of the future. Bastock used the phenotype in order to study genotypes. It took decades, and many technological advances, before researchers could more directly study behavioral genotypes, as we will see next.

Major and minor genes

Understanding the genetic basis of behavior requires unraveling the relationship between specific genes and behavior. Individual genes that are responsible for the majority of phenotypic variation are known as **major genes**, while those that contribute to small amounts of variation are known as **minor genes** (although the combination of many minor genes can have a large influence on a phenotype). The existence and identification of major genes provides insight into important sources of variation in behavior and is the focus of much work in behavioral genetics. Indeed, it has been said that the holy grail of behavioral genetic research is "to unearth a locus that harbors genetic differences that *causes* a change in behavior between individuals, populations, or species in nature" (Hoekstra 2010).

QTL mapping to identify genes associated with behavior

The genome of most animals is so large that searching for the specific genes that affect a behavior is often like looking for a needle in a haystack. For example, the mouse genome consists of over 23,000 genes on 20 chromosomes. How do researchers find specific genes? One approach is to examine **quantitative trait loci (QTL)**, which are stretches of DNA that either contain or are linked to genes influencing a trait such as behavior. **QTL mapping** is a statistical technique that combines genetic information with trait information to determine which regions of the genome contain the genes that influence the trait. This procedure can provide information on not only the number of genes involved but also their location on chromosomes. Such information can lead to the identification of **candidate genes**, major genes suspected of contributing to a large amount of the phenotypic variation in a specific trait. Let's look at one example.

major gene Individual gene that is responsible for a large fraction of phenotypic variation.

minor gene Individual gene that contributes to small amounts of variation in the phenotype.

quantitative trait loci (QTL) Stretches of DNA that either contain or are linked to genes influencing a trait such as behavior.

QTL mapping A statistical technique that combines genetic information with trait information to determine which regions of the genome contain the genes that influence the trait QTLs.

candidate genes Major genes suspected of contributing to a large amount of the phenotypic variation in a specific trait.

Featured Research

QTL mapping for aphid feeding behavior

Many insects feed on specific plant species and use a complex suite of feeding behaviors to assess these plants. Marina Caillaud and Sara Via used QTL analysis to examine the number of genes involved with plant choice in pea aphids (*Acyrthosiphon pisum*) (Caillaud & Via 2012). Pea aphids (Figure 4.5) have sucking mouthparts, called stylets, which they use to pierce the soft tissues of plants, such as leaves or stems, to feed on the sap in the phloem. In a population in New York, Caillaud and Via identified two genetically distinct races of pea aphids that feed on different plant species: one specializes on alfalfa, while the other specializes on clover. How do individuals select the correct plant species for feeding? Aphids explore the plant surface and subepidermic tissues, probe deeply to find phloem sap, and evaluate it prior to ingestion (Caillaud & Via 2000). The duration of these behaviors will differ greatly depending on whether an aphid is on a preferred or nonpreferred plant species.

For the QTL analysis, the researchers created F_1 and F_2 generations that resulted from crossing individuals of the two races. Next, they characterized the behavioral phenotype of plant choice behavior of individuals from the F_1 and F_2 generations, as well as that of each parental race. The researchers measured total time spent searching for a feeding site (exploration), the time spent on a plant before penetrating it, the latency to inject saliva into the plant to begin the digestive process, and the amount of time spent digesting sap from each plant. A strong preference for a plant was indicated by a rapid time to feed. They assessed the genotype of each individual using 116 amplified fragment length polymorphism (AFLP) markers (Vos et al. 1995). In this technique, genomic DNA is cut into pieces, or fragments. Known genetic markers (pieces of DNA) are then attached to each piece to create a **linkage map** showing the position of markers relative to each other, but not their specific location on the chromosome.

Caillaud and Via did not find any QTL for latency to inject saliva in the plant, which probably indicates that this behavior is influenced by numerous minor genes. However, for each of the other behaviors measured, they found from one to three QTLs, with the proportion of the behavioral variation in plant acceptance behavior explained by each QTL ranging from 7% to over 50%. Given the small number of QTLs and the large proportion of variance for which they accounted, the researchers proposed that perhaps only a few genes are involved with plant selection behavior.

In this example, the researchers tried to determine how many genes influence behavior. Once a major gene or genes are identified, researchers often proceed to examine specific genotypic influences on behavior and then characterize the gene product, as we see in the next study.

Figure 4.5. Pea aphid feeding. Pea aphids feed on plants by extracting sap.

linkage map A genetic map of the relative positions of genetic markers on chromosomes.

Fire ant genotype and social organization

Featured Research

Solenopsis invicta, the red imported fire ant, is native to South America but was accidentally introduced in the 1930s to North America in the cargo of ships sailing from South America. Not only do these ants bite, but they also inflict a painful sting. This exotic pest species has colonized the southern United States and continues to expand its range (Ascunce et al. 2011). The United States now spends about $6 billion annually on control, medical treatment, and damage to property related to this species.

Fire ants are social (i.e., group-living) insects that live in large colonies (Figure 4.6). Each colony has one or more reproductive queens and many sterile workers and soldiers, which find food and protect the colony. Fire ants exhibit two distinct kinds of social organization: some colonies possess a single reproductive queen (a monogyne colony), while others contain multiple queens (a polygyne colony). What is the proximate explanation for whether a colony is monygyne or polygyne? Do genes affect this variation in social behavior? Kenneth Ross's (1997) work suggests the answer is yes.

Ross collected ants from monogyne and polygyne colonies located in both their introduced range in the United States and their native range in Argentina. He used starch gel electrophoresis analysis to determine that the genotype and allelic frequencies at one locus, *Gp-9*, differed between the two types of colonies. He found that the *Gp-9* genotype had only two alleles, *Gp-9*^{B} and *Gp-9*^{b}. The monogyne colonies all had the *Gp-9*^{BB} genotype and thus only the *Gp-9*^{B}

Figure 4.6. Fire ant colony. Large queen and smaller workers tending larvae.

TOOLBOX 4.1

Molecular techniques

Molecular techniques for studying genes and behavior have become increasingly available. Three of the most prevalent are QTL mapping, creating knockout organisms, and microarray analysis.

QTL MAPPING

Quantitative trait loci (QTL) are portions of the genome that influence certain phenotypic traits. A QTL analysis correlates regions of the genome that have genetic markers with associated behavioral variation among individuals in a family. One way to create a family with behavioral variation is to cross (or mate) individuals that have different behavioral phenotypes and then make quantitative measures of the behavior of the offspring (Figure 1). We also need information on the individuals' genotypes. We can examine the genotype by cutting the DNA into fragments and attaching **genetic markers** (short sequences of known DNA) to identify each fragment (Figure 2). Several standard genetic markers are used such as restriction fragment length polymorphisms (RFLPs), where fragments differ in length; simple sequence repeats (SSRs), which are repeating sequences of only a few base pairs (also known as microsatellites); and single nucleotide polymorphisms (SNPs), sequences that differ in only a single base pair nucleotide.

These genetic markers must be scattered evenly throughout the genome.

Statistical analyses are used to determine if differences in behavior between individuals correlate with differences in marker genotypes. If there is a correlation, then the QTL that is associated with a behavior is assumed to be linked to the genetic marker.

CREATING KNOCKOUT ORGANISMS

One way to study the effect of a gene is to examine the behavior of animals with and without an active copy of the gene. This technique is often used in mice when the sequence but not the function of a gene is known. By examining behavioral and physiological differences in animals, researchers can gain understanding of the function of the gene.

Knockout mice are **transgenic**: they have had additional DNA added to their genome that stops a particular gene from functioning. This DNA is created by harvesting embryonic stem (ES) cells from early-stage mouse embryos. These cells are **totipotent** and so can grow into any type of cell. DNA that will inactivate the gene is inserted into the nuclei of the ES cells, and the cells are then grown in vitro in the lab for several days. Typically, another "marker" gene is also inserted that allows rapid identification of individuals with the new mutation. For example, this marker gene might produce a unique coat color in mice. The engineered cells are then injected into early-stage mouse embryos, which are implanted in the uterus of a female mouse.

The pups that are born are **chimeras**, meaning that they develop from two kinds of stem cells—one that contained the new mutation and those that did not. Chimeras are useful only if the new mutation was incorporated into the germ cells, which allows it to be passed to offspring. These mice are then

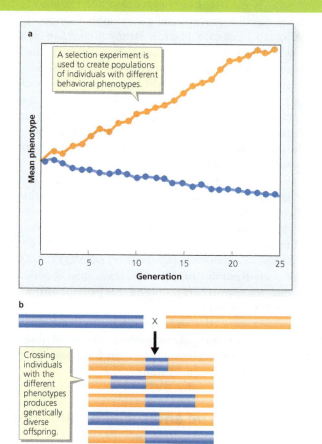

Figure 1. QTL analysis. Protocol for QTL analysis. (a) Individuals with different behavioral phenotypes are (b) crossed to create family groups with much behavioral variation (*Source:* Mackey 2001).

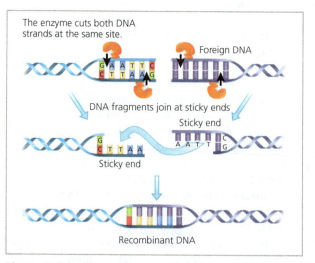

Figure 2. Genetic markers. Genetic markers are attached to DNA fragments for each individual using recombinant DNA techniques.

continued...

crossed with wild-type individuals to produce offspring with one copy of the mutation. Subsequent inbreeding produces offspring that are homozygous for the mutation. These are knockout individuals.

MICROARRAY ANALYSIS

Microarray analysis, or **gene expression profiling**, is another method used to identify genes that influence behavior. Not all genes are activated, or expressed, at all times, so they are not always undergoing transcription (DNA to mRNA) or translation (mRNA to proteins). Microarray analysis measures the amount of mRNA produced as an indicator of gene activation. This technique allows us to examine hundreds if not thousands of genes at once by quantifying which genes are activated in different individuals or tissues.

DNA from different tissues (such as different parts of the brain) or from different individuals (such as rover or sitter flies) is extracted, and then the mRNA from those cells is collected. Reverse transcriptase, an enzyme, is used to create complementary cDNA for each mRNA. The cDNA is labeled by color with a fluorescent tag to indicate where it came from. Equal amounts of the labeled cDNA are then placed onto a DNA chip, which is a modified microscope slide that contains hundreds or thousands of known genes of the species, placed in specific locations. The cDNA will hybridize only with its complementary DNA. The chip is then scanned with a laser to quantify all the genes that contain any colored fluorescent dyes. (If more than one type of cDNA is present, the location will appear as an additional color.) The intensity of the dye indicates the intensity of gene expression, with more dye indicating greater expression of that gene.

allele, indicating that this allele is fixed in monogyne colonies. The polygyne colonies, however, displayed more variation in their genotype and allelic frequencies. In particular, the reproductive queens were all heterozygotes, Gp-9^{Bb}. Thus, in this case, the alleles at one locus appear to regulate the social organization (number of reproductive queens) of a colony.

What does the Gp-9 gene code for? Genes code for proteins, biochemical compounds that have many functions. Proteins are called the "building blocks" of life because they can be structural components of cells, tissues, and organs. They also function as enzymes by catalyzing chemical reactions, and they are an important component in cell signaling as neurotransmitters, hormones, or receptors (see Chapter 5). In this way, proteins play a key role in initiating physiological changes in cells and organ systems that affect the brain and resulting behaviors.

Michael Krieger and Kenneth Ross sequenced the Gp-9 gene and found that it codes for a pheromone-binding protein (Krieger & Ross 2002). Pheromones are chemicals that trigger a response in another individual. Chemical recognition is an important cue that allows ants to determine whether individuals are members of their colony or a potential intruder. Further study determined that individual workers regulate the number of queens in their hive by accepting queens that produce the appropriate chemical signals and rejecting those that do not (Ross & Keller 2002). The Gp-9 protein likely plays a role in that recognition.

Experimental manipulation of gene function: knockout studies

So far we have examined how researchers find genes that influence behavior and then determine their products and effects on behavior. Another approach used to understand how major genes influence behavior is to disable a gene and examine the effect on behavior, a procedure known as a **knockout technique**. This technique requires that a researcher know the location of a gene and its DNA sequence in order to alter the gene so it cannot function. Individuals with an experimentally inactivated copy of a gene are known as knocked out (or KO) individuals (see Toolbox 4.1), and their behavior is compared with that of wild types. Mario Capecchi, Martin Evans, and Oliver Smithies received the 2007 Nobel Prize in physiology and medicine for their development of this technique. Let's examine a recent study that used the knockout technique in mice.

knockout technique A procedure in which a single gene is rendered nonfunctional.

Anxiety-related behavior and knockout of a hormone receptor in mice

Featured Research

Knockouts are often created by disabling a gene that codes for a receptor for a specific gene product. Arginine vasopressin (AVP) is a peptide hormone, a short polymer of amino acids linked by peptide bonds. This hormone is stored in and released from the posterior pituitary in the brain into the bloodstream, where it functions in fluid homeostasis and blood pressure control. AVP is also released into the brain, where it affects social behavior (e.g., social recognition, pair bonding, and paternal behavior) and behavior under stressful conditions. AVP affects only those cells that have a functional AVP receptor. Isadora Bielsky and colleagues used the knockout technique to study the function of one particular AVP receptor, AVPR1A (Bielsky et al. 2004). The V1aR gene in mice codes for AVPR1A, and so the researchers created a knockout for this gene.

TOOLBOX 4.2

Methods for studying behavior under stress

Stress affects most people and can have strong effects on behavior as well as mental and physical health. In order to study these effects, psychologists have developed a variety of test protocols for animal models. Since mice and rats are common study animals, these tests are designed to examine situations relevant to them. Wild mice and rats are nocturnal animals subject to predation by owls and carnivores. They seek safety from predators in underground burrows, logs, and dense vegetation—that is, dark, tight spaces. Open, well-lit areas are dangerous, because they do not allow individuals to hide from predators; as a result, such locations are stressful for rodents.

Three common laboratory tests provide individuals with areas that offer both safe and stressful locations. They include placing a subject in (a) an elevated chamber with both open and walled areas, (b) an open arena or field, and (c) an apparatus that has both well-lit and dark areas (Figure 1). Each test situation contains a region where individuals are highly exposed (open rather than walled elevated areas, the center rather than the edge of an open arena, and a bright versus a dark area) and areas where individuals are less exposed to potential predators. The former regions are more stressful and are normally avoided by control subjects. Researchers typically measure the amount of time spent in high- versus low-stress areas during each test to quantify how animals behave under different experimental treatments (e.g., when different genes have been knocked out or when individuals have been given different drugs).

a Elevated plus maze

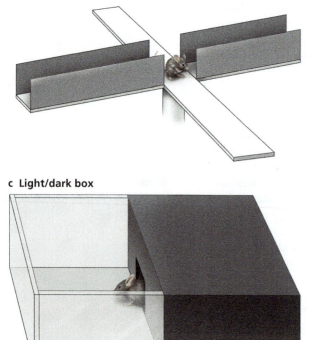

b Open field test

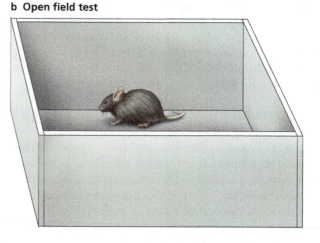

c Light/dark box

Figure 1. Stress tests. Three tests used to examine the behavioral response to stressful conditions: (a) time spent in open versus walled areas, (b) time spent in the center of an arena versus near the wall, and (c) time spent in a light versus a dark area.

The research team examined responses to stress- or anxiety-related behaviors. Mice are nocturnal and are therefore averse to bright light. Exposure to these conditions produces stress and stimulates movement to dark locations if they are available. In addition, mice display **thigmotaxis**, or a preference for physical contact, and therefore an avoidance of open areas, probably because such environments carry a high predation risk for wild mice. The team used several standard tests to examine the mice's behavior under stress, including their avoidance of both open spaces and brightly lit areas.

The researchers compared the behavior of wild-type and knockout mice on the same stress tests. In all three tests, the knockout mice spent more time than the wild-type mice in the light, in the center of the arena, and on the open arms of the elevated plus maze (Figure 4.7). From these data, the researchers concluded that the V1aR gene plays an important role in affecting the behavior of animals placed in stressful situations. Lack of proper gene function modifies behavior so that individuals spend more time in risky locations: these mice do not flee from dangerous conditions. Thus, this gene can strongly affect fitness by affecting habitat choice and movement behavior.

These examples demonstrate a variety of techniques used to examine the association between behaviors and genes. Next, we examine how environmental variation affects gene expression and behavior.

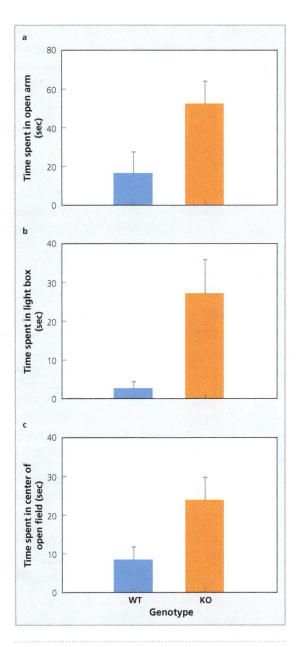

4.2 The environment influences gene expression and behavior

Up to this point, we have focused on genetic influences on behavior. Recall, however, that variation in behavior among individuals results from both variation in gene alleles and variation in environments. How can we characterize the effects of each?

Heritability

We can quantify the relative contributions of genetic and environmental variation to behavioral variation using statistical measures. **Heritability** (**h²**) is the proportion of phenotypic variation in a population that is due to genetic variation. We write

$$h^2 = \frac{V_G}{V_P}$$

Figure 4.7. Knockout mice behavior. Mean (+ SE) behaviors of wild-type (WT) and knockout (KO) mice. Knockout mice whose AVP receptor was nonfunctional spent more time in the (a) open arm, (b) well-lit areas, and (c) center of the open field compared to wild-type mice (*Source:* Bielski et al. 2004).

where V_G is the variance in the genotype and V_P is the variance in the phenotype. This is also known as broad-sense heritability, because it includes all genetic effects on the phenotype. Behavioral variation resulting from genotypes (V_G) is calculated by averaging the phenotypic values of one genotype across many different environments. An example of this calculation would be to measure the phenotype of clones or a highly inbred line across many environments. However, this approach is logistically very difficult and so is rarely taken. In order to know how the genotype affects the phenotype, we must partition all the genetic effects of each allele on a phenotype. These effects can be broken up into three factors:

1. **additive effects, A**, or the average effect of individual alleles on the phenotype;
2. **dominance effects, D**, or the interaction between alleles at one locus; and
3. **epistasis, I**, or the interaction between genes at different loci (Conner & Hartl 2004).

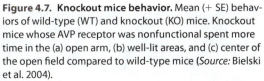

thigmotaxis A tendency to seek out physical contact with objects or enclosed spaces. The avoidance of open areas.

heritability (h²) (broad-sense) The proportion of phenotypic variation in a population that is due to genetic variation.

additive effects (A) The average effect of individual alleles on the phenotype.

dominance effects (D) The interaction between alleles at one locus.

epistasis (I) Interactions between genes at different loci.

narrow-sense heritability The proportion of phenotypic variance solely as a result of additive genetic values.

Dominance refers to the interaction of alleles at one locus in which one allele can mask the expression (phenotype) of the other. Epistasis occurs when the effect of one gene is modified by another gene: one gene can mask the expression of another, or they can act together to produce a new phenotype. We can now define the variance of the genotype as

$$V_G = V_A + V_D + V_I$$

and so heritability is

$$h^2 = \frac{V_A + V_D + V_I}{V_P}$$

Broad-sense heritability is a useful measure for clonal species, or species in which the offspring and parents have essentially identical genotypes. However, many species are not clonal. We thus need another estimate of heritability for the proportion of phenotypic variance that is due solely to additive genetic values, which we call **narrow-sense heritability**:

$$h^2 = \frac{V_A}{V_P}$$

This is the measure of heritability most commonly reported.

Narrow-sense heritability can be determined by examining the similarity of behavior between closely related individuals, such as parents and their offspring. As we saw in Chapter 2, heritability is commonly determined using parent-offspring regressions, in which the mean trait values of parents are regressed against the mean trait values of their offspring. The slope of the regression can range from zero to one: the higher the slope value, the more offspring resemble their parents. Higher slope values indicate that a greater proportion of the phenotypic variance is additive variance, or variation passed from parents to offspring. Lower slope values indicate that less of the phenotypic variance is additive genetic variance and so is not transmitted to offspring. This means that more nonadditive and environmental variance is expressed among offspring.

Not surprisingly, a variety of narrow-sense heritability values have been reported for behavior. These, for example, from low values for oviposition behavior in weevils (0.05) (Tanaka 2000) to moderate values for defense behavior in snakes (0.41) (Garland 1994) to much higher values for courtship behavior in crickets (0.72) (Hedrick 1994). This wide range illustrates how genetic and environmental factors differentially affect behavioral variation across traits and species (Ruefenacht et al. 2002) (Applying the Concepts 4.1). Next, we examine studies on fish, fruit flies, and birds to illustrate how environmental variation can generate behavioral variation among individuals.

APPLYING THE CONCEPTS 4.1

Dog behavior heritability

In captive breeding programs, individuals are selected as parents for the next generation in order to pass on desired traits. It is crucial, however, to know whether the trait has a heritable component, because if behaviors are affected only by the environment, artificial selection will not be effective. Historically, breeders simply assumed that traits were heritable.

A variety of behavioral tests have been developed to characterize the disposition or temperament of dogs (e.g., Wilsson & Sundgren 1997). For example, behavioral tests are used to assess breeds that will be used as service dogs or in hunting. Individuals must demonstrate the correct disposition for the situations they will encounter: service dogs must exhibit high levels of self-confidence in a wide variety of environments, while hunting dogs must not react negatively to gunshots. Silvia Ruefenacht and colleagues used 25 years of standardized behavioral field tests on over 3000 dogs with known pedigrees to determine the heritability of seven different behaviors in German shepherds (Ruefenacht et al. 2002), including self-confidence, reactions to different stimuli, reaction to gunfire, and play behavior. Heritability scores ranged from 0.09 to 0.24, with the highest score being reaction to gunshots. This finding confirms that although selection for specific behavioral traits can be fruitful, the environment also exerts a great deal of influence on behavior.

Environmental effects on zebrafish aggression

How do we study environmental influences on behavior? If closely related individuals exposed to different environments exhibit different behavior, it is likely that their behavioral differences are the result of environmental factors. Christopher Marks and colleagues adopted this approach to understand how water conditions affected aggressive behavior in zebrafish (*Danio rerio*) (Marks et al. 2005) (Scientific Process 4.1).

Zebrafish live in a variety of habitats ranging from oxygen-rich fast-flowing streams to oxygen-poor stagnant pools throughout the Himalayas. Oxygen depletion, or hypoxia, is a common stress for aquatic organisms that can affect their development and behavior. Marks and colleagues raised closely related zebrafish in water that had either low or high oxygen levels (hypoxic and normoxic environments, respectively) and then examined their aggressive behavior as adults when placed in each type of environment.

Individual males and females were bred to produce several clutches of full siblings (siblings that share 50% of their gene alleles). Within two hours of fertilization, the eggs of each clutch were split into the two developmental environments. Normoxic tanks had high levels of dissolved oxygen, while hypoxic tanks had low dissolved oxygen levels. Fish were raised in these environments with unlimited food for 75 days until they were adults.

After the fish had matured, they were tested for aggressive behavior in one of the two developmental environments. Aggression assays were conducted by placing a small mirror on one wall of the test chamber. Fish respond aggressively to their mirror image because they treat the image as they would an intruder. The researchers videotaped the behavior of each fish for two minutes after the mirror was put in place and recorded the amount of time the fish spent butting or nipping the mirror. Prior to the test, fish were allowed to acclimate for 16 hours in the test environment.

Hypoxic-reared fish displayed higher levels of aggression in the hypoxic test chamber, and normoxic-reared fish exhibited higher levels of aggression in the normoxic test chamber. Thus, fish were most aggressive in their rearing conditions. Because the experimental design used full siblings reared in different environments, the results demonstrate how the environment during development can affect behavioral phenotype. Marks suggested that aggressive behavior is energetically costly—and this cost may become particularly high when fish experience a novel oxygen environment. Fish reared in normoxic environments probably experience an oxygen deficiency when in hypoxic ___ ions and so are incapable of high levels of activity. It is less clear why fish reared in __ environments and acclimated for 16 hours in normoxic conditions exhibited ___ 's of aggression, and so further research on this phenomenon is needed.

___ example illustrates how environmental influences can result in very different ___ closely related adults. In the next section, we examine one proximate mecha-___ an explain such observations: environmental influences on gene expression.

___vironment and gene expression in fruit flies

___ ___es do not produce specific behaviors but rather code for a diverse array of molecules that affect brain function, through which behaviors are expressed. Animals acquire sensory inputs about the environment: both abiotic aspects and biotic aspects, including the presence of others, such as conspecifics or predators (see Chapter 6). These inputs are integrated in the brain and can result in differential **gene expression** . In other words, genes and their products are expressed at different times depending on sensory inputs.

gene expression The process by which gene products are produced.

Research examining environmental influences on gene expression often manipulates the social environment of individuals. Social environment can refer to the presence, type, and number of nearby individuals, as well as interactions with these individuals. How might differences in the social environment affect behavior? Consider that behavioral interactions between two conspecifics often depend on the sex of the participants: in male-male interactions, aggressive behaviors may be displayed, while male-female interactions often involve courtship behaviors. Lisa Ellis and Ginger

SCIENTIFIC PROCESS 4.1

Environmental effects on zebrafish aggression
Research Question: *How does environmental variation affect aggression in zebrafish?*

Hypothesis: Both developmental environment (DE) and behavioral (test) environment (BE) can affect aggression.

Prediction: (a) If only DE affects behavior, aggression will be unaffected by BE, and aggression will be highest in normoxic (high-oxygen) water; (b) if DE and BE affect behavior but do not interact, aggression will be higher for fish in a normoxic BE and will be highest for fish whose DE was normoxic; (c) if DE and BE interact, aggression will be highest for fish whose BE and DE match.

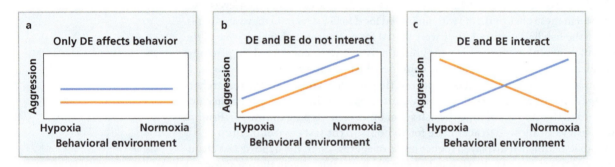

Figure 1. Predictions. The blue line represents fish raised in a normoxic (high-oxygen) development environment, and the orange line represents those raised in a hypoxic (low-oxygen) development environment (*Source:* Marks et al. 2005).

Methods: The researchers:

- collected eggs from 12 full-sibling clutches (all individuals shared 50% of gene alleles, because members of a clutch have the same parents).
- divided the eggs into two developmental environments (normoxic and hypoxic). The embryos were raised until adulthood (about 75 days of age) with unlimited food. Normoxia development treatment tanks had continuously oxygenated water (high dissolved oxygen content = 6.8 mg/L); hypoxia development treatment tanks had nitrogen gas bubbled into the water (low dissolved oxygen content = 0.8 mg/L).
- tested fish in either a normoxic or a hypoxic test chamber after a 16-hour acclimation period.
- measured aggression as time spent biting or nipping a mirror image.

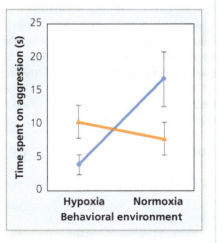

Results:

Hypoxia-raised fish had higher levels of aggression in the hypoxic test chamber, and normoxia-raised fish had higher levels of aggression in the normoxic test chamber.

Figure 2. Results. The blue line is normoxic and orange line is hypoxic development environment (*Source:* Marks et al. 2005).

Conclusion: Zebrafish display more aggressive behavior in the environment in which they were raised, indicating an interaction between environmental and behavioral environments.

Carney used **microarray analysis** to examine gene expression in male fruit flies exposed to different social environments (Ellis & Carney 2011). In particular, they wanted to determine whether flies possess specific genes for male-female interactions and other genes for male-male interactions—or whether they have instead a set of "socially responsive" genes that are expressed in all social interactions.

Ellis and Carney raised male flies and exposed them to one of two treatment groups or controls. One group was allowed to court a single female, while the other interacted with a rival male for 20 minutes. Control males did not interact with another fly. Microarray analysis was performed using DNA from the heads of flies in these two groups to examine differences in gene expression in the brain and important sensory organs, such as antennae.

The researchers identified hundreds of genes that were socially responsive. Of these genes, 505 responded to male-male interactions and 281 responded to male-female interactions. However, many of these genes were expressed in both treatments and so are presumed to be socially responsive genes involved in all conspecific interactions. The analysis also identified 240 genes that were responsive only to the male-male interaction treatment. Only 16 genes were uniquely responsive to male-female interaction. Thus, the researchers identified 16 candidate genes associated only with male courtship behavior, while hundreds of other genes were involved in male-male interactions. Why there was such a difference in the number of genes involved in these different social interactions is unclear, but clearly, environmental conditions strongly affect gene expression, and resulting behavior, in male flies.

Social environment and birdsong development

A second focus of environmental effects on gene expression examines birdsong. All birds produce simple vocalizations, known as calls, that are innate. Many birds, particularly songbirds in the suborder Passeri (of the order Passeriformes), also produce complex vocalizations known as songs, which are used to defend a territory and attract a mate (Figure 4.8). In many songbird species, only males sing (although in tropical species, often both sexes sing). Avian vocalizations are produced by the syrinx, which vibrates like human vocal cords when air passes through it.

Environmental exposure to conspecific song is a critical factor for proper song development in birds (e.g., Beecher & Brenowitz 2005). Song learning occurs in several stages, and species are classified as closed- or open-ended learners. **Closed-ended learners** must hear a tutor sing its conspecific song shortly after hatching. During this critical period, which may last for several weeks, the bird apparently learns the song but does not attempt to reproduce it. After a few weeks, singing begins, and birds produce vocalizations known as a subsong, which initially differs greatly from, but then converges to, the song it learned. Finally, a bird's song becomes crystallized and does not change. In contrast, a fixed critical period does not exist for **open-ended learners**, which can acquire new song elements throughout life.

Early work on song learning relied on socially isolated birds hearing a tape tutor. This approach maximized control of the learning environment and showed that birds deprived of a tutor failed to develop proper song (e.g. Marler & Peters 1977). Additional work revealed that song learning is affected by the *type* of tutor. For white-crowned sparrows (*Zonotrichia leucophyrs*), the critical period lasts longer when a subject is paired with a live tutor rather than a tape tutor (Baptista & Petrinovich 1984). In addition, while the songs of live-tutored starlings (*Sturnus vulgaris*) largely matched those of wild birds, tape-tutored birds produced much simpler songs, with a repertoire only about half the size of that of wild birds (Chaiken, Böhner, & Marler 1993). These results have spurred better understanding of the role of environmental influences on song learning (e.g., Beecher et al. 2007). We also now know much about the neural physiology and genetics of song learning, as we see next.

microarray analysis Measurement of the activity of many genes by quantifying gene products.

closed-ended learners Individuals that must hear a tutor sing its conspecific song shortly after hatching in order to learn the song correctly.

open-ended learners Individuals that can acquire new song elements throughout life.

Figure 4.8. Zebra finch. Zebra finches learn their songs from tutors.

Social environment and gene expression in birds

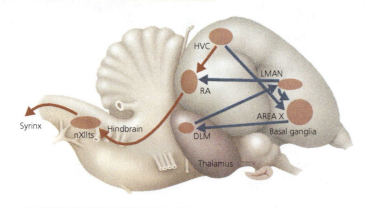

Figure 4.9. Avian song circuit. There are two neural circuits in the songbird brain. The red arrows indicate neural connections within the posterior circuit, which is involved in song production. The blue arrows indicate connections in the anterior circuit, which is involved in song learning. Abbreviations: HVC = neostriatal nucleus; RA = robust nucleus of the arcopallium; DLM = medial nucleus of the dorsolateral thalamus; LMAN = anterior neostriatum (*Source:* Clayton, Balakrishnan, & London 2009).

knockdown technique A procedure that reduces the expression of a gene.

spectrogram (sonogram) Two-dimensional representations of sound that allow researchers to characterize the acoustic structure of vocalizations.

The brains of songbirds differ greatly from those of non-songbirds, with songbird brains having enlarged and interconnected areas involved in song memory and production. These areas of the brain, called the song system, consist of a posterior nucleus, which controls sound production (air flow and the syrinx), and an anterior nucleus involved in song learning. An important localized region in the anterior nucleus is Area X (Figure 4.9). Exposure to birdsong, particularly conspecific song, causes numerous genes to be expressed in different regions of the song system. The most important appear to be *FoxP2* and ZENK (an acronym for the genes *Zif-268*, *Erg-1*, *NGFI-A*, and *Krox-24*) (Mello, Vicario, & Clayton 1992; Bolhuis et al. 2000; Clayton, Balakrishnan, & London 2009).

One way these genes were identified was by impairing their function and observing the effects on song development. Sebastian Haesler and colleagues used this approach to determine the role of *FoxP2*, which encodes the Forkhead Box protein P2, in zebra finches (*Taeniopygia guttata*) (Haesler et al. 2004). They found that *FoxP2* undergoes increased gene expression in Area X of a bird's brain, both when young birds learn to sing and when open-ended learning adults change their song. This suggests that *FoxP2* plays a role in song learning.

Haesler's research team reduced *FoxP2* levels in Area X before young zebra finches began to learn their song (Haesler et al. 2007). They used a **knockdown technique**, which involves using a virus to insert short sections of RNA into the *FoxP2* gene at two different locations to reduce its expression. (They used a knockdown procedure that reduces gene expression, because a knockout technique had not yet been developed for songbirds.) Control birds had a short section of RNA inserted into a noncoding region of DNA that has no effect on the gene's expression. Each subject was kept in a sound-isolation chamber with an adult male tutor during the critical period when young finches learn the songs of a tutor. At this time, the songs of both the tutor and the subject were recorded.

Spectrograms (or **sonograms**) are two-dimensional representations of sound that allow researchers to characterize the acoustic structure of vocalizations (Figure 4.10).

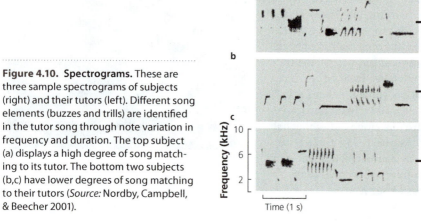

Figure 4.10. Spectrograms. These are three sample spectrograms of subjects (right) and their tutors (left). Different song elements (buzzes and trills) are identified in the tutor song through note variation in frequency and duration. The top subject (a) displays a high degree of song matching to its tutor. The bottom two subjects (b,c) have lower degrees of song matching to their tutors (*Source:* Nordby, Campbell, & Beecher 2001).

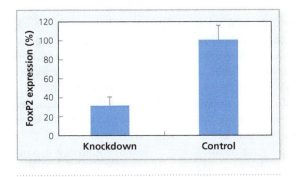

Figure 4.11. FoxP2 expression. Mean (+ SE) FoxP2 expression in knockout and control birds. The expression of the FoxP2 gene in Area X of the brain was significantly lower in knockdown birds compared to controls (*Source:* Haesler et al. 2007).

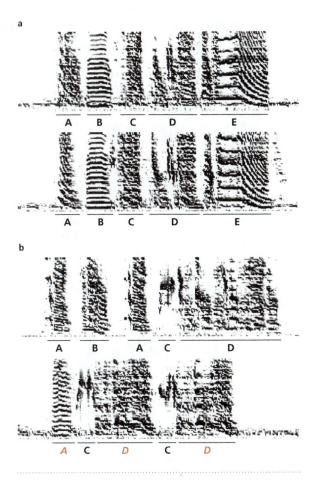

Figure 4.12. FoxP2 knockdown spectrograms. (a) The spectrograms of a control tutor (top) and subject song display a high degree of song matching. (b) The spectrograms of Fox P2 knockdown subjects do not match those of their tutor (top) well (*Source:* Haesler et al. 2007).

Spectrograms can reveal the number of song elements and their characteristic features (e.g., maximum frequency, maximum duration, and intervals between elements). They are commonly used to quantify the similarity of acoustic elements between the vocalizations of two individuals, such as subject and tutor (e.g., Nordby, Campbell, & Beecher 2001).

In knockdown birds, *FoxP2* expression was reduced by approximately 70% (Figure 4.11), which affected their song development: acoustic similarity between the tutor and the knockdown subject was much lower than it was in the controls. Knockdown birds tended to omit specific syllables and failed to copy accurately the duration of some song syllables (Figure 4.12). Because birds with experimental down-regulation of the *FoxP2* gene in Area X inaccurately learned their tutor song, the researchers concluded that *FoxP2* is required for normal song development.

These examples illustrate how the environment influences gene expression and behavior. Next, we examine gene-environment interactions.

Gene-environment interactions

As we've seen, both genes and the environment can have significant effects on the variation we see in behavioral phenotypes. Often, when a single genotype is reared in different environments, we may observe different behavioral phenotypes, a **reaction norm**. If the environment has a greater effect on one genotype than others, we say that there is a gene-environment interaction (GEI). Recall that an individual's behavior is the result of its genotype, the environment, and any interactions between them. As such, phenotypic variation in behavior within a population (V_P) results from variation in genotypes (V_G), variation in environments (V_E), and variation that results from their interaction (V_{GEI}):

$$V_P = V_G + V_E + V_{GEI}$$

Understanding GEI allows us to better understand phenotypic variation in populations.

reaction norm The range of behaviors expressed by a single genotype in different environments.

Rover and sitter foraging behavior in fruit flies

Featured Research

One now-classic example of a behavior known to be influenced by GEI comes from Marla Sokolowski's work on the foraging behavior of larval fruit flies (*Drosophila melanogaster*). Fruit fly eggs develop during a series of larval stages (Figure 4.13). For several days, the larvae crawl to food sources, eat, and then molt as they increase in size.

rover Larval rovers have longer foraging trails than sitters in the presence of food and are more likely to leave a food patch.

sitter Sitters have shorter foraging trails than rovers in the presence of food and are less likely to leave a food patch.

The larvae of *D. melanogaster* exhibit two behavioral polymorphisms, or variants, known as **rover** and **sitter**. Rovers have longer foraging trails than sitters in the presence of food and are more likely to leave a food patch (Figure 4.14). Sokolowski found that this difference in feeding behavior is due to different alleles at a single gene, *foraging* (the *for* gene) (Sokolowski, Pereira, & Hughes 1997). About 70% of individuals in natural populations have the dominant rover allele (*for*^{R}) and exhibit the rover phenotype, while 30% are homozygous for the sitter allele (*for*^{s}) and exhibit the sitter phenotype. Both of these genetic variants are considered wild type because they occur in natural populations. The polymorphism is maintained by frequency-dependent selection: the rover allele has higher fitness in crowded conditions, where it can benefit by traveling farther to find food, while the sitter allele has higher fitness in less-crowded environments. However, in the absence of food, rovers behave like sitters, probably because there is no benefit to increased movement (Graf & Sokolowski 1989). This pattern suggests a significant GEI, because rover larval behavior changes strongly with differences in food abundance, while sitter larval behavior does not.

Subsequent work examined GEI in adult flies. Sokolowski's research team examined the behavior of groups of 25 to 30 adult rovers and sitters exposed to different levels of food availability (Kent et al. 2009). Half were placed in holding vials

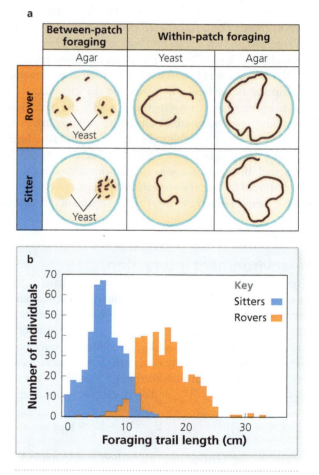

Figure 4.14. Rover-sitter foraging trail lengths. (a) Between-patch foraging shows that rovers tend to leave yeast (food) patches while sitters do not. Within-patch foraging shows that without food (agar only) there is no difference in behavior while in the presence of food (yeast) rovers make longer foraging trails than sitters. (b) Histogram of trail lengths in the presence of food for hundreds of larvae. Sitters (blue bars) tend to have much shorter foraging trails than rovers (orange bars) (*Source:* Sokolowski 2001).

Figure 4.13. Drosophila life cycle. Adults lay eggs that hatch within 24 hours. Larvae grow for several days and molt twice. Larvae then encapsulate into a puparium and undergo a four-day metamorphosis prior to emerging as adults.

that contained food (fed environments) for 16 to 18 hours prior to behavioral testing. The other half were placed in a food-deprived environment for the same time prior to behavioral testing. The researchers then examined the movement behavior of the flies using a horizontal Plexiglas maze (Figure 4.15). They placed the flies in a vial with food and allowed them to feed for 15 minutes before allowing them to leave the patch and move through the maze. Empty glass vials were placed at all maze exits. The team recorded the proportion of flies in the collecting vials after three minutes, which they called the food-leaving score.

The behavior of fed versus food-deprived rovers and sitters was quite different, indicating a significant GEI (Figure 4.16). Fed rovers had a much higher food-leaving score than food-deprived rovers. However, there was little difference between the behavior of fed sitters and that of food-deprived sitters. Why might this be? The researchers removed the heads from the flies and used mass spectroscopy to determine the compounds stored there. They found that rovers and sitters store energy differently (Figure 4.17). Fed rovers and fed sitters stored the same levels of lipids, but in the food-deprived environment, sitters stored more lipids. The researchers also found a large

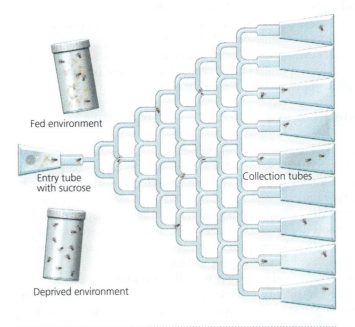

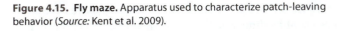

Figure 4.15. Fly maze. Apparatus used to characterize patch-leaving behavior (*Source:* Kent et al. 2009).

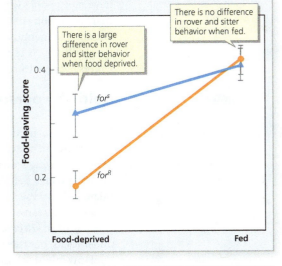

Figure 4.16. Adult rover and sitter food-leaving behavior. Mean (± SE) food-leaving score for fed and food-deprived flies. Note the difference in the change in sitter (blue) and rover (orange) food-leaving scores between fed and food-deprived flies (*Source:* Kent et al. 2009).

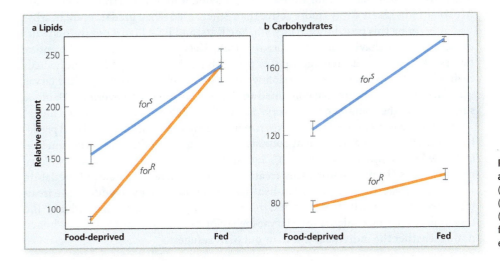

Figure 4.17. Energy storage in rovers and sitters. Change in the heads of rovers (orange) and sitters (blue) in the mean (± SE) relative amount of (a) lipids and (b) carbohydrates between fed and food-deprived individuals (*Source:* Kent et al. 2009).

difference between the carbohydrate storage of rovers and sitters in both environments. Sitters always store more carbohydrates than rovers. Carbohydrate levels for both were lower in the food-deprived environments, but they were lowest for rovers. This finding indicates that GEIs affect both behavioral and metabolic traits. Sokolowski's work demonstrates a GEI in both larval and adult fly behavior, showing that rover genotypes exhibit greater changes in behavior across different environments.

Up to this point, we have seen how variation in genotypes and the environment are used to explain differences in behavior. GEI is a third source of variation and is increasingly documented in behavioral studies (e.g., Bonte, Bossuyt, & Lens 2007; Sambandan et al. 2008). Studying these interactions allows us to understand the enormous behavioral variation that exists in nature. But the behavioral phenotype of an individual has limits. We examine these limits next in a discussion of animal personalities.

4.3 Genes can limit behavioral flexibility

personalities Consistent relative differences in behavior among individuals over time or across different environmental contexts.

So far we have seen how genetic factors, environmental factors, and their interaction produce variation in behavior. Genes also appear to limit behavioral variation. We all have met people who are consistently shy, while others are outgoing. Animals, too, display **personalities**—consistent differences in behavior over time or across different environmental contexts. Animal personality traits include variation in boldness (the willingness to take risks and explore novel objects), overall level of activity, and aggressiveness (Sih, Bell, & Johnson 2004; Stamps & Groothuis 2010).

 Featured Research

Bold-shy personalities in streamside salamanders

Streamside salamanders (*Ambystoma barbouri*) live primarily in the deep pools of shallow streams in eastern North America. The pools are separated by riffles, shallow areas of faster flowing water. Salamanders deposit eggs in these streams. Developing larvae face two issues before they undergo metamorphosis and leave the stream. First, the salamander larvae must maximize feeding time—to shorten their development time—before the streams and pools dry up. Second, they must avoid predators such as the green sunfish (*Lepomis cyanellus*). When predators are present, larvae can seek shelter under rocks, but this behavior reduces time spent feeding.

How might salamanders respond to these two challenges? One might expect that the best solution would be to modify behavior given the situation. An individual might, say, maximize feeding time when predators are absent, while minimizing feeding time when predators are present. That flexibility would produce a negative correlation between behaviors, because the two are mutually exclusive: the more time spent hiding, the less time can be spent feeding. But if an individual's personality is heritable, as we saw in Chapter 2 for exploratory behavior, it might constrain its behavior. Are some individuals always more active than others even when predators are nearby?

To examine this possibility, Andy Sih, Lee Kats, and Erik Maurer examined salamander larvae behavior in the laboratory (Sih, Kats, & Maurer 2003) (Scientific Process 4.2). They collected egg masses from which they obtained four individuals each, which they assumed were full siblings and thus closely related. Each of the larvae was placed in an aquarium that contained an elevated, opaque disk covering 50% of the surface area of the container, which served as a refuge under which larvae could hide.

Each larva was exposed to two experimental treatments that manipulated predation risk. The "fish cue" treatment consisted of adding water from an aquarium that contained four sunfish to the subject's aquarium. Because salamanders can perceive chemical cues from predators, this treatment should increase perceived predation risk for the larvae. In the control "no-fish" treatment, tap water was added. The treatments were separated by two days and the order randomized across individuals. The researchers recorded salamander larvae behavior every 30 minutes for ten hours, noting whether the individual was in or out of its refuge.

SCIENTIFIC PROCESS 4.2

Salamander personalities

Research Question: *How do salamander larvae respond to predation risk?*

Hypothesis: Individuals will vary their activity level in response to the level of threat.

Prediction: All individuals will exhibit high activity level when predation risk is low and low activity level when predation risk is high.

Methods: The researchers:

- collected ten egg masses (sibships) to obtain four individual siblings from each.
- placed each larva in an aquarium that contained a refuge: an elevated, opaque disk that covered 50% of the surface area of the container.
- exposed each larva to two experimental treatments:
  - high predation risk, "fish cue" treatment: water from an aquarium containing four sunfish was added to the subject's aquarium.
  - low predation risk, "no fish" (control) treatment: tap water was added to the subject's aquarium.
- recorded salamander larvae behavior every 30 minutes for ten hours, noting whether the individual was in or out of its refuge.

Results:

- Larvae in the control treatment spent more time out of the refuge eating than did larvae in the high-risk treatment.
- For both treatments, there was a significant positive correlation with the amount of time each larva spent out of the refuge.
- Siblings that spent more time in the open in the control also spent more time in the open with the predator treatment.

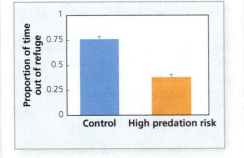

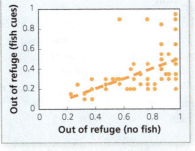

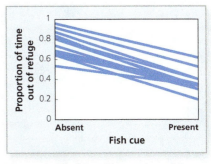

Figure 1. Refuge use. The mean (+ SE) proportion of time outside the refuge for control (blue) and high-risk (orange) treatments (*Source:* Sih, Kats, & Maurer 2003).

Figure 2. Predator effect. There was a positive relationship for time spent out of refuge with and without fish predator cues. Each circle is one individual (*Source:* Sih, Kats, & Maurer 2003).

Figure 3. Sibships. Proportion of time out of refuge for each sibship without (left) and with (right) fish predator cues. Points connected by a line represent one sibship (*Source:* Sih, Kats, & Maurer 2003).

Conclusions: Genetic differences across sibships explain the consistent relative differences in activity level across treatments, indicating that salamanders exhibit behavioral personalities.

On average, larvae spent about 75% of the time out of the refuge eating in the no-fish control treatment but only 40% of the time when fish cues were present. Individuals clearly responded to increased predation risk by spending more time hiding, as one might expect. However, examination of their individual behavior revealed another story as well. When plotting the amount of time each larva spent out of the refuge for both treatments, a significant positive correlation appears, indicating that an individual that spent much time out of the refuge when no fish cues were present also spent much time out of the refuge when fish cues were present. These larvae were acting "bold" in both situations, while "shy" individuals spent little time away from the refuge in both treatments. None of the salamanders exhibited what might be considered "optimal" behavior: none spent much time away from the refuge when no fish cues were present and little time away from the refuge when fish cues were present.

The differences between bold and shy behavior can clearly be seen when data from each sibling group (sibships) are combined into one value for each treatment. All sibships reduced time out of the refuge when fish cues were present. Still, those that spent the most time out of the refuge when no fish cues were present also spent the most time away when fish cues were present. Other sibships spent much less time away in both treatments. The consistency within but strong differences across sibships suggests that genetic differences underlie the observed behavioral differences. More important, we see how genotype can limit behavioral plasticity.

The existence of animal personalities suggests that behavior is not completely flexible and that genes play a role. Why do personalities exist? And how are they maintained in a population? Recent modeling has made some headway on this question, as we will see.

Featured Research

Animal personalities: a model with fitness trade-offs

Max Wolf and colleagues argue that most personality traits reflect aspects of risk-taking behavior. Boldness, high levels of exploratory behavior, and high overall activity time should increase risk of mortality from predation but also lead to increased resource acquisition. In contrast, shyness, low levels of exploratory behavior, and low overall activity should, in general, be less risky and less rewarding. In this way, animal personalities are tied to a fitness trade-off between current and future reproduction. Effort allocated to current reproduction comes at the cost of future reproduction (we will discuss this trade-off in depth in Chapter 13).

Wolf and colleagues developed a model of risky and shy personalities. They considered a model organism that lives for two years and can reproduce each year. High- and low-quality resources exist in this organism's environment, but high-quality resources are difficult to find. Individuals that do find high-quality resources have higher reproductive success than those that do not. The researchers assumed a trade-off in reproduction between Years 1 and 2 that is mediated by exploratory behavior. If individuals engage in high exploration in Year 1, they will have a low probability of reproducing in Year 1 but a high probability of finding high-quality resources (and thus having high reproductive success) in Year 2. Individuals that spend little time exploring will have a high probability of reproducing in Year 1 but will likely only find low-quality resources.

Wolf and colleagues evaluated the lifetime fitness for individuals that exhibit different levels of exploratory behavior through a series of simulations. The model shows that an individual with high exploration can have the same fitness as one with low exploration behavior, because they represent different allocations to current and future reproduction (Wolf et al. 2007). Individuals that explore thoroughly invest more heavily in future reproduction, while those that explore only superficially invest primarily in current reproduction.

Next, the researchers ran a series of simulations to examine the effects of competition for resources and predation risk on the lifetime fitness of individuals. These simulations revealed a general principle with two parts: (1) individuals that invest most heavily in reproduction in Year 2 should be more cautious (shy) to enhance their probability of surviving to reproduce, and (2) individuals that invest more in reproduction in Year

1 should be more bold (aggressive) to enhance their ability to compete for resources. The model suggests that bold, aggressive behavior allows greater short-term resource acquisition, which can be allocated to current reproduction, while shy, reserved behavior allows greater allocation to future reproduction. Because these behaviors produce equal fitness, they will both be maintained in a population. Future tests of this model will help us better understand the evolution of animal personalities.

Behavioral genetics holds great promise for understanding variation in behavior among individuals, populations, and species. Researchers have made a great deal of progress in identifying major genes and GEIs that affect behavior. Recent work has also revealed how genes can limit behavioral flexibility. In the coming years, we can expect progress in behavioral genetics to accelerate as we obtain entire genome sequences of additional species.

Chapter Summary and Beyond

Behavioral genetics is the study of the effects of genetic and environmental influences and their interactions on behavior. Anholt and Mackay (2010) provide a comprehensive overview of this field. An important aspect of behavioral genetics is the identification of genes that contribute to large amounts of variation in behavior (Fitzpatrick et al. 2005; Pennisi 2005). Both QTL mapping and microarray analysis have been used to identify genes that influence behavior (Flint 2003). The knockout technique (in which a single gene is rendered nonfunctional) has greatly expanded our understanding of how single genes affect behavior. This technique is often used in animal models to gain insight into human pathologies (Van Der Staay 2006).

While genes affect behavior, so too does the environment. The social environment is often manipulated to examine gene expression profiling (Robinson, Fernald, & Clayton 2008), most notably in birdsong. Recent work has begun to reveal how genotypic differences affect selective learning of the song of different tutors (e.g., Mundinger 2010). The recent sequencing of the zebra finch genome should make it possible to use avian knockouts and microarray analysis to understand the effect of specific genes on the singing behavior of birds (Warren et al. 2010). Multidisciplinary research has found behavioral, neural, and genetic parallels between human language development and birdsong development (Bolhuis, Okanoya, & Scharff 2010). Studying GEIs should help us better understand variation in behavior (Bendesky & Bargmann 2011).

Work on animal personalities has revealed how genotypes can limit behavioral variation among individuals (Stamps & Groothuis 2010). Sih et al. (2004) expand this finding to a discussion of behavioral syndromes, or suites of correlated behavioral traits across a population. Recent work focuses on the fitness consequences of different personality traits (Smith & Blumstein 2008) and the search for genetic aspects of personality in humans (Arbelle et al. 2003).

CHAPTER QUESTIONS

1. A researcher examined the behavior of adult minnows from two isolated populations. One population lived in a warm stream with an average temperature of 20°C, while the other lived in a cold stream with an average temperature of 14°C. Individuals from the warm stream population were significantly more aggressive than individuals from the cold stream. The researcher concluded that the difference in aggression was mainly because of environmental conditions. Evaluate this conclusion.

2. Propose a study to determine the heritability of personality in the salamanders studied by Sih, Kats, and Maurer (2003).

3. Briefly describe the information QTL mapping, microarray analysis, and knockout techniques provide to behavioral geneticists.

4. Identify one research study from the chapter for which you might reasonably conclude that phenotypic variation is more strongly associated with genetic variation than environmental variation.

5. In the salamander system studied by Sih, Kats, and Maurer (2003), explain how the existence of animal personalities indicates that genes limit behavioral plasticity.

6. Heritability values for behavioral traits are typically lower than those for morphological traits. Propose an explanation for this difference.

Chapter 5

Learning and Cognition

Figure 5.1. Mexican jay. These birds learned to associate a whistle with the presence of food.

5.4

5.5

5.6

5.3

5.2

5.1

Many years ago, during an ornithology class field trip to the Chiricahua Mountains of southeastern Arizona, we heard a loud police whistle nearby. Almost instantly, we began to hear vocalizations of a flock of Mexican jays (*Aphelocoma wollweberi*) that appeared to be moving toward the location of the whistle (Figure 5.1). We followed the jays to find a researcher who was busy collecting data on aggressive behavior.

This flock was being studied by Jerram Brown and his students. They had uniquely color-banded individuals and were investigating the behavior and ecology of these social birds. In Brown's early work, researchers placed small piles of food in the center of a flock's territory to attract many individuals, who then competed for access to the food. One year the researchers blew a police whistle a few times just before putting out the food, and soon the jays learned to associate this novel sound with the presence of food at a specific location. Over time, the researchers noticed that yearlings learned the association from older birds, because unmarked young birds would quickly respond to the sound even though no whistle had been blown for almost a year (Brown 1997b).

The ability to learn about their environment helps animals survive and reproduce. At our house, ruby-throated hummingbirds (*Archilochus colubrus*) appear not just to learn the location of feeders we set out but also to remember them from one year to the next. We know it is time to put the feeders up each spring when we see hummingbirds hover at the location of last summer's feeders. We watch gray squirrels (*Sciurus carolinensis*) bury nuts in the ground each fall and then revisit those locations during winter when food is less available. The squirrels are adept spatial learners—they remember hundreds or even thousands of locations where they buried food.

In this chapter we examine the broad array of ways that animals learn, ranging from simple habituation to complex problem solving. We also examine the physical changes in the brain that are associated with learning. We then discuss individual and social learning and see how the latter can lead to animal traditions and even culture and end by discussing animal cognition.

5.1 Learning allows animals to adapt to their environment

learning A relatively permanent change in behavior as a result of experience.

Learning is a relatively permanent change in behavior as a result of experience. It is a process by which animals modify their behavior, or adapt to their environment, in ways that allow them to experience increased fitness.

Featured Research ▶

Learning as an adaptation in juncos

Is there evidence that learning is adaptive? Yes, as we will see throughout this chapter. One of the simplest approaches to answer this question is to compare the behavior of individuals that differ in age and experience. If learning improves fitness, then older, more experienced individuals should outperform those that are less experienced. Kimberly Sullivan tested this hypothesis in a simple experiment with yellow-eyed juncos (*Junco phaeonotus*) (Figure 5.2), small sparrows that live in the mountains of southwestern North America (Sullivan 1988). Because juncos breed throughout the spring, their chicks hatch at different dates, and so juveniles vary in age at the end of summer. Previous work allowed Sullivan to age birds into four categories: recently fledged juveniles (four to seven weeks after leaving the nest, or fledging), young juveniles (eight to ten weeks after fledging), older juveniles (11 to 14 weeks after fledging), and adults (greater than 1 year old).

Then she asked a simple question: Does age affect feeding efficiency? Juncos mandibulate, or maneuver, food in their bill for several seconds to prepare it for consumption. Sullivan predicted that as birds age, they gain more experience handling food items and so become more efficient foragers. She tested this prediction by providing food that required different levels of manipulation: mealworms (*Tenebrio molitor* larvae) were cut into small pieces (0.009 g dry weight) that were easy to handle or large pieces (0.026 g dry weight) that were more difficult to handle. Birds had access to one food type for several days to be sure they had some exposure to the food type. Each day over the following two weeks, Sullivan recorded the handling time of each food item (the time from first contact until the item was consumed) for each bird. She found that for both prey types, adults had the lowest handling time, while recently fledged birds had the highest. She also calculated energy gain for the foraging birds. The difference in handling times led to differences in energy intake rate: it was highest for adults and lowest for recently fledged birds (Figure 5.3). Sullivan attributed these patterns to differences in experience. Older birds with more experience had learned how to best handle insect prey and thus became more proficient at the task.

Evolution of learning

To humans, the benefits of learning seem obvious. We learn how to talk, read, and drive a car. We learn to play sports and musical instruments. Animals do none of these things. What do they learn and why? A long-standing answer was that animals modify their behavior to deal with dynamic, changing environments. Recent work, however, suggests a more complex explanation.

Theory indicates that two factors affect the evolution of learning: environmental stability and the usefulness of past experience (e.g., Mery & Kawecki 2002; Dunlap & Stephens 2009). Consider two possible types of worlds: one is fixed and nothing ever changes (e.g., no new predators or parasites ever appear), while the other is dynamic

Figure 5.2. Yellow-eyed junco. This small bird is found throughout the mountains of the southwest United States and Mexico.

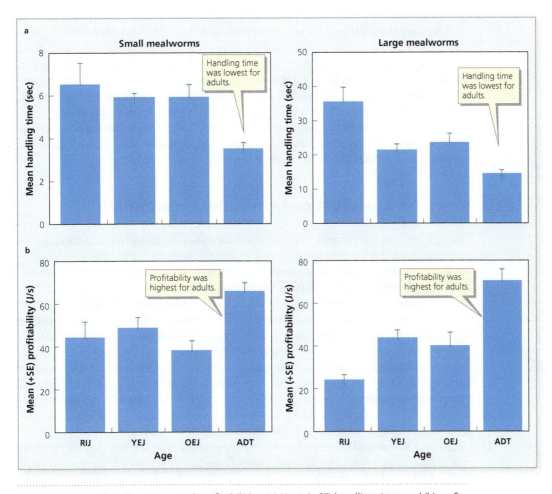

Figure 5.3. Junco handling times and profitabilities. (a) Mean (+ SE) handling times and (b) profitabilities for small (left) and large (right) mealworms for each age class. RIJ = recently independent juveniles; YEJ = young experienced juveniles; OEJ = old experienced juveniles; ADT = adults (*Source:* Sullivan 1988).

and changes unpredictably (e.g., new predators and parasites may appear in the middle of the breeding season). Now imagine two habitats (A and B) that differ in quality for reproduction: breeding in one leads to high fitness, while breeding in the other leads to reproductive failure. In the fixed world, the high-fitness habitat (A) will always be the best place to reproduce. In the dynamic world, the high-fitness habitat can change. Sometimes A will be best and at other times B will be best, thanks to the unpredictable nature of predators and parasites. Let's assume that in this changing world, each habitat has a 50% chance each breeding season of being high quality or causing reproductive failure (Figure 5.4).

In which world should learning evolve? Neither. This answer may not be intuitive, so let's examine why. In the fixed world, individuals that breed in the high-fitness habitat would quickly outcompete those that reproduce in the other habitat. If habitat choice is influenced by genes, the world would soon be full of individuals that selected only the high-fitness habitat, and there would be no need to learn about the other habitat or the differences between them. On the other hand, in the dynamic world, there is nothing to learn, because learning is useful only if individuals can benefit from their experience. In other words, there must be a predictable relationship between experience and the best option now available. In the dynamic world, the habitats change unpredictably, and so both habitats always have a 50% chance of being the best option, no matter what happened in the past. Again, there is no benefit to learning.

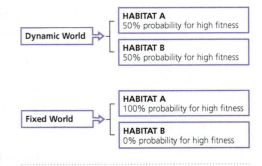

Figure 5.4. Dynamic and fixed worlds. In a dynamic world, reproducing in each habitat can lead to high fitness; in a fixed world, only one habitat leads to high fitness.

Learning theory thus demonstrates that two factors affect the evolution of learning: the regularity of the environment and the reliability of past experience. These factors are independent of each other but are not mutually exclusive. As environmental regularity increases, learning will become less favored, because in a completely regular world, evolution will fix behavior. As the reliability of experience increases, learning will be strongly favored, because individuals that learn will have higher fitness than those that do not (Dunlap & Stephens 2009). The real world usually falls between these two extremes, and so we expect learning to evolve. Indeed, most—if not all—animals that have been studied show the capacity to learn.

Green frog habituation to intruder vocalizations

habituation The reduction and then lack of response to a stimulus over time.

dear enemy hypothesis Territory owners will show reduced aggressive interactions toward territorial neighbors, compared to strangers.

Perhaps the simplest form of learning is **habituation**, the reduction and then lack of response to a stimulus over time. An environmental stimulus is anything in the environment (biotic or abiotic) that an individual can perceive, and any reaction to that stimulus is a response. Habituation to stimuli that are irrelevant to survival and reproduction, such as leaves blowing in the wind, is useful because it allows time and effort to be spent in more productive ways. For example, the **dear enemy hypothesis** (Temeles 1994) predicts that many songbirds will show reduced aggressive interactions with territorial neighbors, compared to strangers, as a territory owner becomes more familiar with its neighbor. The mechanism for this reduction in aggression is habituation. Strangers represent potential threats as new competitors for territorial space, while a neighbor represents a much lower threat, because it has already established a nearby territory. A territory owner can save time and effort by essentially ignoring a neighbor while aggressively interacting with a stranger.

Patrick Owen and Stephen Perrill investigated whether the dear enemy hypothesis explains aggression in territorial green frogs (*Rana clamitans*) (Owen & Perrill 1998). Males defend territories around ponds, where they vocalize to attract females. Territorial males respond to strangers with both aggressive vocalizations and physical attacks. Aggressive vocalizations can be distinguished from advertisement calls used to attract females by their lower dominant frequency. Owen and Perrill predicted that if males habituate to familiar stimuli, there should be a decrease in their response to a new rival's vocalization after an initial aggressive response.

The researchers studied males at four ponds near Indianapolis, Indiana. They simulated intruders by synthesizing the calls of two males and played them from a speaker placed 1 to 2 m away from a focal calling male. Synthesizing the calls standardized their length and intensity. One intruder's dominant frequency was set at 350 Hz, while the second intruder's dominant frequency was set at 450 Hz. For each focal calling male, the researchers first recorded his advertisement call. Next, they played one of the synthetic intruder calls at five- to 15-second intervals for up to an hour. During playback, the researchers recorded the focal frog's movements toward the speaker and its aggressive vocalizations. When the focal frog stopped moving and began to produce advertisement calls again, the researchers stopped the playback and started a 15-minute rest period, after which they initiated the same intruder call once more. The researchers again recorded the focal male's movement toward the speaker and aggressive calls. When movement again ceased and advertisement calls were being produced, the researchers played the second synthetic call to simulate a new intruder and recorded the vocalizations of the focal frog.

Habituation to the intruder calls was observed in both the focal frog's movement and vocalizations. At first, focal frogs made several movements toward the speaker, but this behavior declined over time (Figure 5.5). Movement toward the speaker increased following the rest period and re-initiation of the synthetic call, but at a lower level. Calling frogs produced advertisement calls prior to the playback of the synthetic intruder but switched to aggressive calls when the synthetic call began. Eventually, focal frogs switched back to advertisement calls until a new synthetic call was played (Figure 5.6).

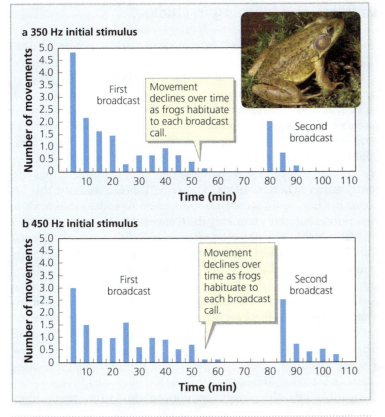

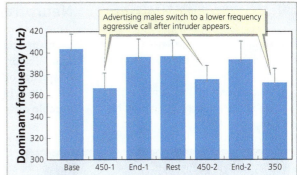

Figure 5.6. Advertisement and aggressive calls. Mean (+ SE) dominant frequency of calls during different periods. Focal frogs start by producing advertisement calls with high frequency to attract females ("base"). When challenged by an intruder vocalization ("450-1" and "450-2"), they switch to lower frequency aggressive calls for a short time until they habituate and then return to using advertisement calls ("end-1" and "end-2"). When a new intruder appears ("350"), they again return to producing lower frequency aggressive calls (*Source:* Owen & Perrill 1998).

Figure 5.5. Movement in response to calls. Number of movements over time as conspecific calls are broadcast. (a) The rest period between broadcasts begins at 65 minutes and the second broadcast begins at 80 minutes. (b) The rest period begins at 70 minutes and the second broadcast begins at 85 minutes. Individuals habituate to each broadcast call and show less movement over time (*Source:* Owen & Perrill 1998). *Inset:* green frog.

These results support the dear enemy hypothesis and demonstrate habituation to a stranger: both physical and vocal responses to the simulated intruder were initially high, but over time, both responses declined as the intruder became more familiar and was perhaps perceived to be less of a threat.

These examples illustrate how learning allows individuals to adapt to their environment. How do animals learn? We examine that question in the next two sections. First, we look inside the brain to see proximate changes in neurons associated with learning. Then we examine how animals learn by making associations.

5.2 Learning is associated with neurological changes

How are experiences translated into changes in behavior? Information from experiences and environmental stimuli are perceived via sensory receptors and relayed to the central nervous system through nerves. Nerves are composed of neurons, cells that receive and transfer electrical and chemical signals. The junction between two neurons, the synapse, is believed to play an important role in learning and memory. Two aspects of the nervous system have garnered much attention in research on learning: both changes in neurotransmitters and the number of synapses between neurons are associated with learning. A typical approach to understanding such proximate mechanisms of learning is to characterize synapse characteristics before and after a learned experience to determine what changes occur. This is an approach that we will see in the next two studies.

Neurotransmitters and learning in chicks

imprinting Rapid learning of a phenotype in young animals.

filial imprinting The learning of the phenotype and identity of parents by offspring.

Figure 5.7. Lorenz and geese. A clutch of geese imprinted on Lorenz's boots.

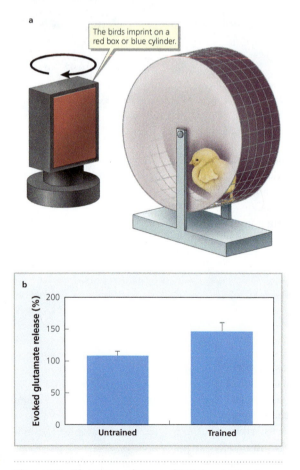

The birds imprint on a red box or blue cylinder.

Figure 5.8. Glutamate release and training. (a) Training protocol (*Source:* Horn 1998); (b) mean (+ SE) glutamate release in different chicks. More glutamate was detected in trained than in untrained birds (*Source:* Meredith et al. 2004).

One important form of learning is **imprinting**, rapid learning that occurs in young animals during a short, sensitive period and has long-lasting effects. For example, offspring often learn the phenotype and identity of their parents through **filial imprinting**, because their parents are the first objects they encounter. Vulnerable offspring benefit from this learning by being able to quickly identify their parents and remain close to them. Konrad Lorenz made imprinting famous when he showed how graylag geese (*Anser anser*) hatchlings would imprint on his boots when these were the first objects they saw. In the absence of their parents, they simply followed him (his boots) around (Lorenz 1935) (Figure 5.7). More recently, this behavior has been used for the reintroduction of endangered birds to their former habitats (Applying the Concepts 5.1).

Many birds, like domestic chickens (*Gallus gallus domesticus*), visually imprint on a stimulus when they hatch. In their brain, the intermediate and medial parts of the hyperstriatum ventrale (IMHV) appear to play an important role in memory for imprinting; earlier work demonstrated that lesions in this area of the brain prevent imprinting, as does the blocking of postsynaptic neurotransmitter receptors (Horn 2004). Rhiannon Meredith and colleagues studied another imprinting mechanism in chicks by investigating whether the release of neurotransmitters from the presynaptic neuron is also associated with such learning (Meredith et al. 2004).

In this study, the researchers divided chicks into two groups. Half were trained by exposure to a visual imprinting stimulus (a red box or a blue cylinder). The other birds were used as a control group: they were not trained and so had no visual stimulus for imprinting. During training, birds were placed on a running wheel; the birds typically attempted to move toward the stimulus, and the wheel recorded these movements. Less than ten minutes after training, the researchers measured the strength of imprinting, or preference score, by sequentially placing two objects in front of the chick while it was on the running wheel. One was the imprinted object, while the other was a novel object. The preference score was calculated by dividing the amount of running toward the imprinted object in a fixed time

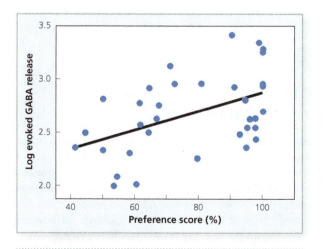

Figure 5.9. GABA release. Increased GABA was associated with higher preference scores (strength of imprinting) (*Source:* Meredith et al. 2004).

APPLYING THE CONCEPTS 5.1

Operation migration and imprinting

Many birds follow long migration routes from their summer breeding areas to their winter feeding grounds. Birds typically learn these routes from their parents. However, the hand-reared birds of endangered species that are being reintroduced into the wild do not have this option. How can they learn their natural migration routes? William Sladen and colleagues hypothesized that they could use imprinting to train young birds to follow a human flying an ultralight aircraft (Sladen et al. 2002).

Work began with a common species, the Canada goose (*Branta canadensis*). The researchers took eggs from parents, hatched them in an incubator, and hand-reared them. Birds did indeed imprint on the human handlers and followed them as they flew the aircraft, first over short distances, but eventually over longer flights. The same procedure was then used to train endangered species raised in captivity, such as trumpeter swans (*Cygnus buccinator*) and whooping cranes (*Grus americana*) (Figure 1) (Ellis et al. 2003). This success is particularly impressive given the cranes' 1,800 km migration. This research inspired the movie *Fly Away Home*, in which captive Canada geese imprint on a young girl. She trains the birds to follow their historic migration route all the way from Ontario, Canada, to North Carolina.

Figure 1. Imprinting on aircraft. Whooping cranes following an ultralight aircraft.

period by the total amount of running in response to both objects. A score of 50 showed no preference for the imprinted object, while higher scores showed greater preference and a higher strength of imprinting. At the end of the experiment, chicks were sacrificed, the IMHV tissue dissected, and assays conducted to measure the release of several amino acid neurotransmitters, including glutamate and gamma-aminobutyric acid (GABA).

The research team found higher glutamate in the IMHV tissue of trained chicks compared to that of the controls (Figure 5.8), but no difference in GABA. However, among the trained birds, chicks with the highest preference score—those that most strongly imprinted on the test object—had higher levels of GABA in their brain (Figure 5.9).

These results suggest that neurotransmitters play a role in imprinting. One neurotransmitter, glutamate, was more strongly released in the brain of birds that visually imprinted on an object, while another, GABA, was correlated with the strength of imprinting. This work illustrates that neurotransmitter release in presynaptic neurons played a role in imprinting over the short timescale of the experiment. Next, we examine recent research that investigates long-lasting memory.

Dendritic spines and learning in mice

Featured Research

Memory is the retention of a learned experience and is a critical factor in animals' ability to utilize their prior experiences. How is memory stored in the brain? We are

memory Retention of learned experiences.

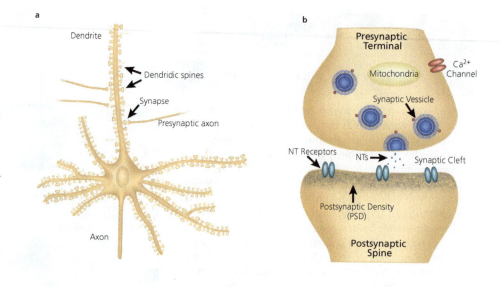

Figure 5.10. Neuron and dendritic spines. (a) Dendritic spines are small protuberances on the dendrites of neurons in the brain. (b) Each spine forms a synapse with another neuron (NT:neurotransmitter) (*Source:* Smrt & Zhou 2010).

> **neural plasticity** Structural changes in the brain, especially in the number of synapses and the strength of chemical synapses between neurons.

> **dendritic spines** Small protuberances on a dendrite that typically receive synaptic inputs.

just beginning to understand the answer to this question, and changes in the brain appear to be involved.

Structural changes in the brain, especially in the number of synapses and the strength of chemical synapses between neurons, is a phenomenon known as **neural plasticity** (Hering & Sheng 2001). Synaptic connections in the brain are dynamic, owing in part to the formation and elimination of **dendritic spines**, small protuberances on a dendrite that typically receive synaptic inputs (Figure 5.10). The dynamic nature of dendritic spines means that the number of synaptic connections between neurons can vary. Recent work suggests that the dynamic nature of spines plays a role in learning.

Guang Yang, Feng Pan, and Wen-Biao Gan examined spine formation associated with learning in mice (Yang, Pan, & Gan 2009). One group of mice was trained to learn a new motor skill: running on a rotating rod (a rotorod) suspended above the cage floor (Figure 5.11), an activity that is analogous to humans "logrolling" in water. The mice had to learn novel motor coordination skills and balance in order to stay on the rod as its speed increased. Another group of mice received no training and served as controls. The researchers used transcranial two-photon microscopy to examine the fluorescent-labeled dendritic spines of living subjects. This technique involves surgically thinning the cranium to allow for high-resolution imaging of dendritic spines.

After one to two days of training, both young and old mice that learned the new motor skill showed significantly higher levels of dendritic spine formation than controls (Figure 5.12). In addition, the performance of mice on the rotorod (the revolutions per minute they could achieve) was associated with the number of new spines formed: mice that had developed more new dendritic spines performed the task better. Interestingly, over the course of the following two weeks, most of the new spines disappeared, so that the net result was essentially no difference in the total number of spines between trained and control

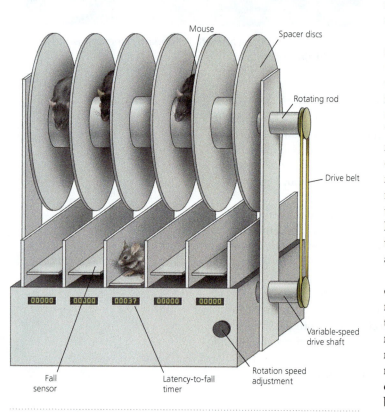

Figure 5.11. Rotorod. Mice are placed on a rotating rod that increases in speed as a novel motor skill. Researchers measure the ability of mice to stay on the rod (*Source:* Carter, Morton, & Dunnett 2001).

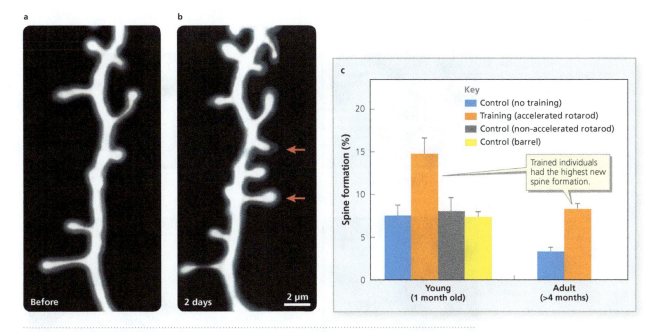

Figure 5.12. Dendritic spine formation. Image of dendrite (a) before and (b) after training, showing new spine formation (red arrows). (c) Mean (+SE) spine formation. Both young and adult animals that were trained and learned a new behavior had more spine formation (*Source:* Yang, Pan, & Gan 2009).

mice. In fact, the researchers estimated that less than 0.1% of the new spines would persist for life.

How do these results relate to lifelong memory? One possible explanation is that lifelong memory may be associated with both the formation of new spines during learning and the pruning of synapses that, in essence, "remodel" the brain following learning. Further studies are required to examine this hypothesis.

Such mechanistic studies of brain structure allow us to understand how the brain changes with learning. We can also ask evolutionary questions about learning and memory. Are there brain differences between species associated with differences in learning and memory? The next study provides an answer.

Avian memory of stored food

Many animals, like squirrels, **cache** (or store) food such as hard nuts for later use. Caching provides food for the future but is only useful if that food can be relocated. What is the mechanism that allows animals to find their caches? Some rely on local landmarks, like rocks and logs (Jones & Kamil 2001), but most use their memory. Memory of a specific object, place, and time ("what-where-when" memory) is called **episodic memory**.

Caching is particularly common in two families of birds, Corvidae (crows and jays) and Paridae (chickadees and titmice), but caching propensity varies across species (Figure 5.13). Some, such as Clark's nutcrackers (*Nucifraga columbiana*) and willow tits (*Poecile montanus*), cache heavily and rely on stored food for survival. Others, such as scrub jays (*Aphelocoma coerulescens*) and marsh tits (*Parus palustris*), cache occasionally, while species such as jackdaws (*Corvus monedula*) and blue tits (*Parus caeruleus*) rarely, if ever, cache food.

If cache recovery is accomplished through memory, then species that cache heavily should rely more heavily on a well-developed memory. Examining the brain structure of such species could help confirm this hypothesis. In many vertebrates,

Featured Research

cache Food stored in a hidden location for later retrieval.

episodic memory Memory of a specific object, place, and time.

Corvids   Parids

Cache heavily

Clark's nutcracker   Willow tit

Cache moderately

Scrub Jay   Marsh tit

Rarely cache

Jackdaw   Blue tit

Figure 5.13. Corvids and parids. There is variation in the level of caching within each group. (a) Clark's nutcracker, (b) willow tit, (c) scrub jay, (d) marsh tit, (e) jackdaw, and (f) blue tit.

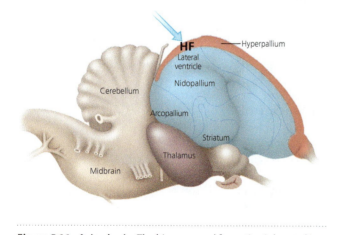

Figure 5.14. Avian brain. The hippocampal formation is located in the temporal medial lobe and labeled HF.

spatial memory is associated with the hippocampus (also known as the hippocampal formation in birds), which is located in the medial temporal lobe (Figure 5.14). Does increased caching behavior correlate with increased hippocampus size? Early work found that across families, the hippocampus is larger in species of birds that often cache food (Krebs et al. 1989). However, species in different families may differ in brain structure for a variety of reasons. Jeff Lucas and colleagues (Lucas et al. 2004) examined how species within the same family vary in food-hoarding propensity and brain structure (Scientific Process 5.1). They gathered data from studies on 13 corvid and ten parid species. The degree of food caching ranged from noncaching species to those that rely on cached food for survival. They found a strong association across species within both families: those that relied heavily on food caches indeed had a larger hippocampus than those that rarely cached food, suggesting a link between hippocampus size and function (memory of cache locations) in these species.

Each of these examples illustrates how learning is associated with changes in the brain, ranging from rapid changes in dendritic spine formation associated with learning a new skill to evolutionary changes in brain structure associated with long-term memory. Next, we examine how animals learn by making associations.

SCIENTIFIC PROCESS 5.1

Brain structure and food hoarding

RESEARCH QUESTION: *Does brain structure differ between birds that do and do not need to remember the location of cached food?*

Hypothesis: Hippocampal formation (HF) size in the brain correlates with spatial memory capability.

Prediction: Species that rely more on caching will have larger HF size.

Methods: The researchers:

- assembled data on HF size for 13 species of Corvids and ten species of Parids, corrected for differences in body and brain size.
- characterized each species as: (1) a nonhoarder, (2) a nonspecialized hoarder that caches food occasionally, or (3) a specialized hoarder that relies heavily on cached food for survival.

Results: In both families, specialized food hoarders had larger HF size than species that rely less heavily on cached food. Species that do not cache food exhibited the smallest HF.

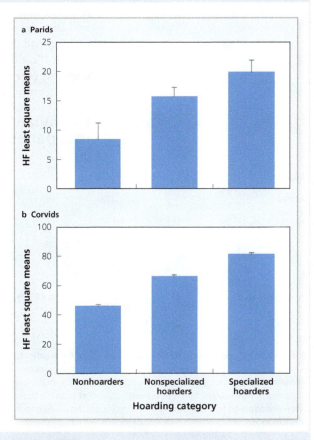

Figure 1. Hippocampal formation size. Mean (+SE) HF size for species of (a) Parids and (b) Corvids that differ in food hoarding (*Source:* Lucas et al. 2004).

Conclusion: HF appears to play an important role in facilitating memory of cache locations for future food retrieval.

5.3　Animals learn stimulus-response associations

Learning often requires making associations. For example, when we pick up a leash, our dogs associate that gesture with going for a walk. One common type of learning involves making an association between an environmental stimulus and a behavioral response—a stimulus-response association. Many animals make such associations, allowing them to learn about their environment, prepare for any similar future events, and respond accordingly.

Classical conditioning

The researcher most often associated with stimulus-response learning is Ivan Pavlov. Pavlov was studying the physiology of digestion when he observed that dogs often began salivating when they could smell food. More interestingly, they would also begin salivating when they saw the technician who normally fed them. They had formed an association between the technician and the upcoming delivery of food (Pavlov 1927).

innate (instinct) Behaviors that are performed the same way each time, are fully expressed the first time they are exhibited, and are present even in individuals raised in isolation.

classical conditioning (Pavlovian conditioning) A type of learning in which a novel neutral stimulus is paired with an existing stimulus that elicits a particular response.

This ability to learn new associations between a stimulus and an **innate**, or unlearned, response is called **classical conditioning** (or **Pavlovian conditioning**). Classical conditioning begins with an innate response to a stimulus, such as salivating in response to the sight of food. Pavlov called the food the unconditional stimulus (US) and salivating the unconditional response (UR). Next, Pavlov paired a neutral stimulus, such as the ringing of a bell, with the presentation of food. Such "conditioning" was done repeatedly until the animal learned to associate the bell with the arrival of food. In Pavlov's terms, the bell was a conditional stimulus (CS), and salivation in response to the bell was a conditional response (CR). Subsequently, a dog would salivate after simply hearing the bell, even in the absence of food: a new predictable relationship had developed between a stimulus (bell) and a response (salivating). Often the word "conditioned" is used instead of "conditional" when referring to a stimulus or response. Next, we examine how such conditioning can affect fitness.

Featured Research

Pavlovian conditioning for mating opportunities in Japanese quail

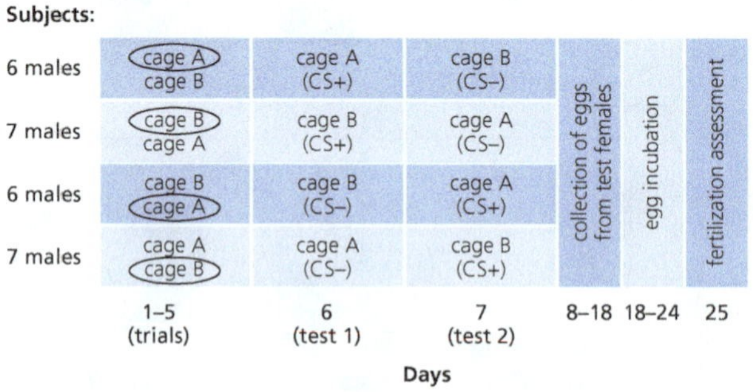

Classical conditioning allows individuals to become better prepared for future events by learning new associations. How does such learning affect fitness? Elizabeth Adkin-Regan and Emiko MacKillop studied how classical conditioning affects mating behavior in Japanese quail (*Coturnix japonica*), medium-sized Asian birds commonly raised in captivity for egg production.

The researchers conditioned adult males and females in the laboratory to two different mating situations using cages that differed in size, location, and appearance (wire or Plexiglas construction) (Adkin-Regan & MacKillop 2003). Each bird learned that mating always occurred in one cage condition (conditioned stimulus, or CS+) but not the other (CS−) (Figure 5.15). For half the birds, a mate was added after two minutes only when they were placed in the Plexiglas cage, while for the other half, a mate was added only when they were in the wire cage. For the focal males, once the female was added, mating ensued. For females, no mating was allowed during conditioning, because female Japanese quails

Figure 5.15. Conditioning design for males. Cage A is made of wire and Cage B of Plexiglas. For training trials 1–5, mating only occurred in the CS+ cage (circled). During tests 1 and 2, mating occurred in both cages. Females were given similar training but were not allowed to mate during training (*Source*: Adkin-Regan & MacKillop 2003).

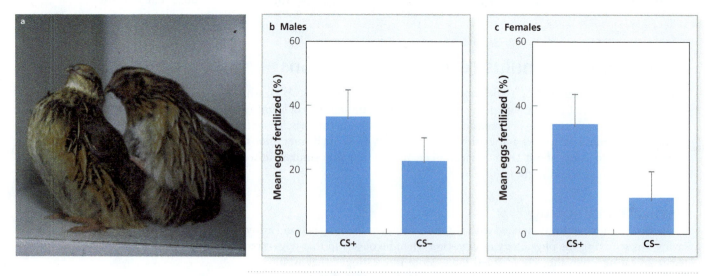

Figure 5.16. Percentage of eggs fertilized. (a) A male copulating with a female. (b) Mean (+ SE) percentage of eggs fertilized by males and (c) females in the CS+ and CS− cage (*Source*: Adkin-Regan & MacKillop 2003).

can store sperm, which could have biased the results. The male was thus kept behind a wire mesh screen. Training occurred twice a day for five days.

After training, the focal individuals were tested in both cages: researchers presented these individuals with a mate and then counted the number of inseminations that occurred. Eggs were collected, incubated, and allowed to develop for one week before being examined for embryos. Males fertilized more eggs in the cage where they had been conditioned to expect a mate (the CS+ condition). Females also had more eggs fertilized in the CS+ condition (Figure 5.16). Both sexes thus achieved greater reproductive success in the condition where they had learned that mating opportunities occur. These results show that Pavlovian conditioning can affect fitness. What caused these differences in fertilization success? The researchers hypothesized that males may transfer larger or more effective ejaculates and that females may behave differently to allow more sperm transfer under the conditioned stimulus. Although the exact mechanism by which this occurred is not yet understood, it is clear that fertilization success was affected by the learning of conditions.

Fish learn novel predators

Featured Research

Another important component of fitness is predator avoidance. Many animals face the threat of predation, but how do they know to avoid predators? One possibility is that a prey can innately identify its predators. However, many animals move large distances when migrating or searching for appropriate habitats and so may encounter many different predators. Can they identify all possible predators innately, or do they learn about them?

Matthew Mitchell and colleagues examined the ability of juvenile lemon damselfish (*Pomacentrus moluccensis*) to learn about novel predators (Mitchell et al. 2011) (Scientific Process 5.2). Lemon damselfish larvae develop in open water for about a month, after which they settle on the Great Barrier Reef off the coast of Australia. Settlers are small (1 cm in length) and suffer high predation.

Many fish, including damselfish, respond to conspecific **chemical alarm substances** released from damaged epidermal cells when, for example, a fish is wounded or killed by a predator. Fish that perceive chemical alarm substances respond with innate antipredator behaviors: they reduce feeding, increase vigilance behavior (scanning the environment for predators), and may spend more time in shelter. The researchers investigated whether fish learn their predators by association with the presence of the chemical alarm.

chemical alarm substances
Chemicals released from damaged epidermal cells in fish that function as an alarm cue for others.

The experimenters created a "cocktail" of chemical odors using adults of four fish species caught from the reef: two fish predators and two nonpredators. Individuals of each species were placed in separate tanks and left for six hours—enough time for the tank water to accumulate species-specific odors. The researchers mixed water from each of their separate tanks into a "cocktail" of chemical odors.

Mitchell's team collected damselfish recruits that had not yet settled using light traps set 50 to 100 m from the reef. Since predators of these fish are only found on reefs and not in open waters, the damselfish recruits were assumed to be naïve in regard to these predators. On Day 1, half the damselfish were conditioned to the cocktail odors plus a conspecific alarm substance. The other half were conditioned to the cocktail odors plus seawater as a control. The next day, damselfish were tested by exposing them either to odors from one of the cocktail species or to the combined odors of two novel species, only one of which was a fish predator. The researchers recorded the amount of time the test fish spent feeding, their distance from the shelter, and the time spent in shelter.

When the odor of each species was added, damselfish conditioned with the cocktail of species and alarm substance reduced their feeding time significantly more than did the controls. The test fish had apparently learned a new association between the odor of each cocktail fish and predation risk, as indicated by the chemical alarm substance. Individuals tested with the odor of the cocktail fish paired with saltwater did not respond with antipredator behaviors. Individuals also did

SCIENTIFIC PROCESS 5.2

Fish learn predators

RESEARCH QUESTION: *How do fish learn about predators?*

Hypothesis: Damselfish make associations between conspecific chemical alarm cues and the odor of a heterospecific fish to learn their predators.

Prediction: Individuals will reduce feeding and increase antipredator behavior when they detect odors that have been associated with chemical alarm cues.

Methods: The researchers:

- created two cocktails of odors: one from four species of known fish, A, B, C, and D (A = *S. dermatogenys*, a predator; B = *P. fuscus*, a predator; C = *C. plebeius*; D = *R. aculeatus*); and one from two species of novel fish, E and F (E = *C. batuensis*, a predator; F = *A. steinitzi*).
- paired the cocktail with a conspecific chemical alarm substance (treatment).
- paired the cocktail with seawater (control).
- tested individuals by placing them in a tank with a shelter (a pot).
- added food and recorded behavior for five minutes.
- added a stimulus odor (A-F) and recorded the amount of time test fish spent feeding, their distance from the shelter, and the time spent in shelter for five minutes.

Results: Fish exposed to the cocktail plus the chemical alarm substance exhibited less feeding behavior than did control fish and fish exposed to a novel odor.

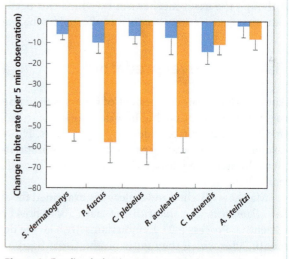

Figure 1. Feeding behavior. Mean (+ SE) change in feeding behavior for treatment fish (orange) and controls (blue) when exposed to the odor of six fish species. From left to right, these represent fish A, B, C, D, E, and F (*Source:* Mitchell et al. 2011).

Conclusion: Fish learn associations between the odor of a fish and chemical alarm substances.

not respond to the odors of the two novel fish species. Interestingly, there was no difference in shelter use between the two groups, perhaps because there were no predators in sight.

These results demonstrate that lemon damselfish can rapidly learn associations between fish odors and the risk from unfamiliar predators. In addition, because damselfish did not respond to the odor of a novel predator, there was no evidence that they have innate knowledge of predatory species; instead, they must learn to identify predators. The ability to make these important (and rapid) associations allows damselfish to minimize predation risk in new environments.

Operant conditioning

operant conditioning (instrumental conditioning) A learning process in which an animal learns to associate a behavior with a particular consequence.

As we just saw, classical conditioning involves learning novel associations that affect innate behavior. In contrast, **operant conditioning** involves learning associations between learned behaviors and outcomes. Edward Thorndike and B. F. Skinner pioneered the study of operant conditioning, also known as instrumental conditioning (see, e.g., Thorndike 1911; Skinner 1938). From Skinner's perspective, much of behavior is learned from its consequences: reinforcements (positive outcomes) increase a behavior, while punishments (negative outcomes) decrease it. Any of four types of operant conditioning can occur, according to whether a stimulus is added or removed:

1. In positive reinforcement, a behavior increases because it is associated with the addition of a desired stimulus, such as food.
2. In negative reinforcement, a behavior increases because it is associated with the removal of an aversive stimulus, such as pain.
3. In positive punishment, a behavior declines because it is associated with an aversive stimulus, such as pain.
4. In negative punishment, a behavior declines because a desired stimulus is removed, such as a food reward.

Skinner revolutionized the study of behavior in the mid-twentieth century, and his work is still used extensively to train animals (Applying the Concepts 5.2). Much of Skinner's work involved understanding how and when each operant condition should be applied to produce the desired behavioral outcome. He also invented novel research protocols and apparatuses to study operant conditioning. A case in point is the **operant chamber**, or "Skinner box," which typically contains a lever that a rat or pigeon can press. Operant chambers themselves can be very useful tools to study animal problem solving.

Operant conditioning is also known as **trial-and-error learning**, because behavior often changes incrementally as the animal "makes progress" toward solving a problem. Skinner developed a graphical representation of such progress, such as a decline in errors over time, called the individual's **learning curve**. Learning curves allow researchers to quantify learning and characterize differences in learning ability.

operant chamber An enclosure used to study behavioral conditioning.

trial-and-error learning Learning through repetition that results in rewards or progress toward a goal.

learning curve A graphical representation of a change in learning over time.

APPLYING THE CONCEPTS 5.2

Dog training

Have you ever rung a friend's doorbell only to hear uncontrolled barking? Have you had a dog run full force at you through the front door? Dog owners face behavioral problems like these all the time. Usually the owner simply yells at the dog: "Get away from the door," "Sit down," "Go away," or some other command that may be ineffective. Sophia Yin is a veterinarian who specializes in studying domestic animal behavior and helping owners with their problem animals. She uses operant conditioning with positive reinforcement to train dogs with problem behaviors.

To address dogs' problems with greeting people, Yin and colleagues turned first to professional dog trainers and guided them to train dogs to remain on a small mat for one minute, despite loud distractions (Yin et al. 2008). The trainers began with two exercises. First, the six dogs learned to eat from a remote-controlled food-reward dispenser. Once the dogs ate readily from the food dispenser, the trainers played a tone, which they followed by giving each dog a food reward. After the dogs learned to look for food after hearing the tone, the researchers added a condition: the dog had to look at the trainer before the food would be dispensed. In the second

exercise, the dogs learned to walk toward a "target" and touch it with their nose, which again resulted in the tone and a food reward. Once the dogs had mastered these two exercises, the real training protocol began.

The protocol had three stages: (1) a "down-stay" on a rug-covered platform; (2) running to the platform and doing a down-stay; and (3) the down-stay with distractions (such as a doorbell, loud knocking, or people walking). Training sessions lasted up to 30 minutes, depending on the dog's motivation. The trainers taught the dogs to complete the behaviors for each stage separately and then put them together in sequence. After eight days in the laboratory, dogs remained in a down-stay for a full minute despite distractions, whereas before training, they did so for only five seconds.

Dog owners used these same techniques with similar success: barking decreased from 19 to two barks per minute, while jumping decreased from eight to almost zero jumps per minute after training. The authors concluded that positive reinforcement is a powerful technique for modifying dog behavior. Dog trainers and owners are using these same techniques all over the world (Miller 2001).

Learning curves in macaques

One way to understand how quickly animals learn is to train them to overcome an innate preference. For example, when given a choice between a small and a large quantity of food items, many primates, including humans, strongly prefer the larger quantity (Boysen, Berntson, & Mukobi 2001). This innate preference makes sense because food intake affects fitness, and so more is usually better. Elisabeth Murray, Jerald Kralik, and Steven Wise studied the strength of this preference in rhesus macaques (*Macaca mulatta*) (Murray, Kralik, & Wise 2005). They offered six subjects a choice between one and four peanut halves that were placed in the experimenter's open hands. If a test subject reached for the hand with one food item, it received four; when a macaque reached for the hand with four food items, it received one. Here, one action provided four times more rewards than the other and therefore represented a more positive reinforcing outcome. Twenty trials were conducted each day, with a 20-second delay between them. How quickly did individuals learn to select the hand that contained only one food item?

As expected, each subject showed a strong initial preference for the hand with four food items (the "incorrect" choice). Macaques did learn to select the hand with one food item in order to receive four, demonstrating trial-and-error learning (Figure 5.17), but there was tremendous variation in their learning curves. One macaque learned rather quickly, after about 340 trials, but another macaque took over 2,700 trials to attain a low error rate. In fact, a wide variation in learning curves is common in many species (Boutin 2007) (Applying the Concepts 5.3).

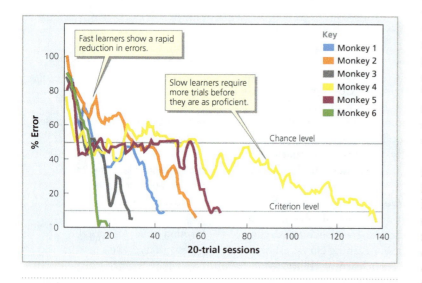

Figure 5.17. Macaque learning curves. Individual macaques (different lines) learned to select the hand with one food item instead of the hand with four at different rates. Those macaques that learned the fastest had the fewest errors in the least number of trials (*Source:* Murray, Kralik, & Wise 2005).

Trial-and-error learning in bees

Following Skinner's work and the development of operant chambers, many learning studies, like the previously described work on macaques, were conducted in laboratory settings designed to control as many factors as possible. Can operant learning be studied in more natural settings? Given that individuals differ in their learning rate, how does learning ability affect fitness?

Nigel Raine and Lars Chittka set out to answer those questions (Raine & Chittka 2008). Bumblebees (*Bombus terrestris*) use a variety of cues to learn which flowers provide nectar and pollen food rewards. Bees feed on many types of flowers but have an innate preference for the color blue, because blue flowers tend to have more nectar (Raine & Chittka 2007). Using operant conditioning, Raine and Chittka examined the learning curves of bees trained to associate the color yellow with a food reward in the laboratory.

The researchers worked with 12 colonies that contained uniquely marked workers. All bees were first allowed to feed from artificial flowers that contained sugar water. Each flower was multicolored—both blue *and* yellow. Once bees were used to feeding from artificial flowers, the researchers began training trials. Bees were placed in a flight arena where the artificial flowers were either all blue or all yellow, in equal proportion. Only the yellow flowers contained sugar water, whereas the blue flowers were empty. Thus, bees had to learn that they would be rewarded for probing yellow flowers and not rewarded for probing blue flowers.

APPLYING THE CONCEPTS 5.3

Are you smarter than a pigeon?

Operant chambers have been used extensively to study problem solving not just in animals but also in people. One example is the Monty Hall dilemma, named after the original host of the television game show *Let's Make a Deal*. Contestants in this game are offered the choice of three doors and told that a valuable prize, such as a new car, is hidden behind one door, while joke prizes, like a goat, are hidden behind the other two doors. The players get to keep the prize that is hidden behind the door of their choice.

The contestant first selects one door, and then the host reveals the joke prize behind one of the other doors. Now the contestant gets a second chance to choose a door. And there lies the dilemma: Stay with the original door choice or switch doors? (Remember: the door that the contestant first chose is still closed.) What would you do? Which decision would give you the best chance of winning a valuable prize?

Switching doors, it turns out, is the best option because it gives you a two-thirds chance of winning the valuable prize, while staying with your first choice gets you the prize only one-third of the time (Gill 2011). This answer is not terribly intuitive, however, so how does it work?

The probability of picking the correct door at the start of the game is one-third, since there are three doors. If you stay with that choice even after seeing one (wrong) door opened, your probability of winning the prize will still be one-third (Figure 1), so your probability of having made the wrong choice on your first pick will thus be two-thirds. Let's suppose that you indeed selected the wrong door with your first choice. When Monty Hall then opens one door and reveals a goat, the door you did not select must have the prize (since you selected the wrong door on your first choice).

Why is it best to change doors? Because two-thirds of the time you will have selected the *wrong* door on your first pick. You can get the prize 100% of these times by switching doors, and so you can obtain the prize two-thirds of the time by switching. Want to try it yourself? To play the Monty Hall game online, see http://www.nytimes.com/2008/04/08/science/08monty.html.

Humans tend to do poorly in this game; most decide to stay with their original choice. Walter Herbranson and Julie Schroeder used an operant chamber to see how pigeons (*Columba livia*) would perform (Herbranson & Schroeder 2010). Could birds learn to make the correct decision more quickly than you? Pigeons

were first trained to peck keys to obtain food rewards. The experiment consisted of three keys, analogous to the three doors of the game show. A computer randomly selected one key to lead to the valuable prize (not a car, but access to food for a few seconds). The other two keys provided no reward.

Just as in the show, the birds first had to select one of three keys, which were illuminated with white light. Once they did so, one of the two unselected keys (a "no reward" key) went dark, while the selected key and the remaining unselected key were both illuminated with green lights. This cued the subject to make a second choice between the keys: either select the same key again or switch. After this second choice, the pigeon was rewarded if it selected the correct key or else received nothing. Each bird was presented with the problem at least ten times each day over a period of 30 days, and each time the researchers recorded whether the subject switched its choice or stayed with the original key when given a second chance.

One Day 1, pigeons switched only about 36% of the time, but by Day 30, they were switching over 96% of the time (Figure 2), resulting in many more rewards, as predicted. The researchers gave a similar test to undergraduates, who played for points rather than food (alas, again not a car). Each individual played the game 200 times. During the first 50 trials, humans did rather better than pigeons; they switched about 56% of the time. During the last 50 trials, however, humans still switched only 65% of the time. Thus, humans got better at making the correct choice, but they did not perform nearly as well as pigeons. The pigeons learned that switching results in a greater probability of being rewarded

Car hidden behind Door 3	Car hidden behind Door 1		Car hidden behind Door 2
Player initially picks Door 1			
Host must open Door 2	Host randomly opens either goat door		Host must open Door 3
Probability 1/3	Probability 1/6	Probability 1/6	Probability 1/3
Switching wins	Switching loses	Switching loses	Switching wins

Figure 1. Monty Hall problem. This dilemma represents the problem faced by humans and pigeons when a player initially chooses Door 1. However, players win more often if they switch doors in their next choice.

continued…

and so increased their switching behavior based on positive reinforcement. Undergraduates, on the other hand, did not learn the best option as readily based on their reward probabilities. Their poorer performance does not demonstrate poor reasoning abilities. More likely, the pigeons' performance was better because they experienced more trials to learn about the dilemma—and perhaps because pigeons valued food more than students valued points.

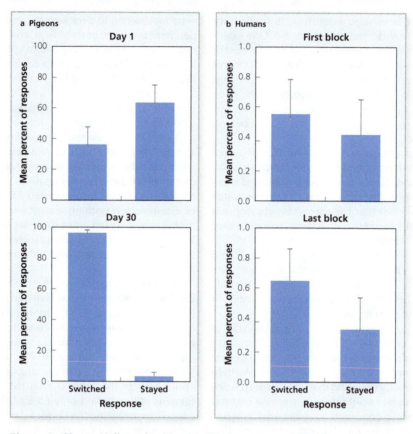

Figure 2. Monty Hall results. Mean (+ SE) percentage of responses. (a) Results from pigeon studies. (b) Results from human studies (*Source:* Herbranson & Schroeder 2010).

The researchers recorded over 100 flower visits per bee as bees searched for food. At first, bees in all colonies made mostly incorrect choices because of their innate preference for blue flowers, but all eventually learned to select mostly yellow flowers. However, the colonies differed in their learning rate: some learned quickly, while others did so much more slowly (Figure 5.18).

Next, the researchers investigated whether this variation in learning was associated with differences in fitness in a natural environment. They allowed bees from each colony to forage on wildflowers around Queen Mary's College in London. They measured foraging success based on a bee's weight when departing and returning to the hive, with the difference in weight being the mass of nectar that the bee was bringing to the hive. Bees were allowed to forage naturally for six days, providing data from thousands of foraging bouts.

Colonies differed in food delivery: those quick to learn in the laboratory delivered more food per hour to the colony than did slow learners, which researchers hypothesized would lead to enhanced survivorship (Figure 5.19). From these experiments, the researchers concluded that bees can learn where to feed by trial and error and that learning ability can affect fitness.

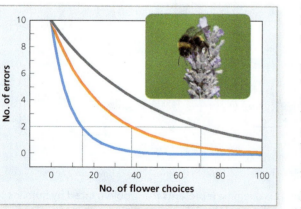

Figure 5.18. Bumblebee colony learning curves. Average learning curves for bees from three colonies. Note the variation in learning rate among colonies (*Source:* Raine & Chittka 2008).

These examples demonstrate that animals often learn by making associations. Many individuals may learn in isolation, as we saw in the case of the Japanese quail. But in many species, individuals live with conspecifics, which provide a rich social environment and enhanced learning opportunities, as we see next.

5.4 Social interactions facilitate learning

In social species, other individuals are a source of information for learning, a phenomenon known as **social learning**. In individual learning, such as trial-and-error learning, information is acquired through an individual's own activities; in social learning, individuals learn by observing others. Animals learn about food, mates, predators, and their environment through individual learning. They can also learn through social learning, and this approach reduces the time and energy costs of learning (Rieucau & Giraldeau 2011).

Young animals have much to learn about their environment, but two of the most important are what to eat and how to avoid being eaten. In the next two examples, we see how young birds and mammals learn their diet and how to detect and avoid predators.

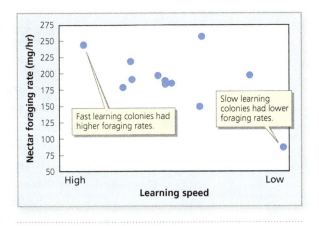

Figure 5.19. Learning speed and feeding rate. Mean nectar foraging rate for the 12 colonies. Colonies with a high learning speed in the experiment had higher mean feeding rates on wildflowers. (*Source:* Raine & Chittka 2008).

■ **social learning**   Learning by observing or interacting with other individuals.

Ptarmigan chicks learn their diet

Featured Research

Precocial birds, like chickens, are highly developed and mobile when they hatch, and so they quickly need to learn what to eat. Traci Allen and Jennifer Clarke examined how young precocial white-tailed ptarmigans (*Lagopus leucura*) learn their diet (Allen & Clarke 2005). Ptarmigans live in the Rocky Mountains (Figure 5.20), where they feed on the leaves, flower buds, flowers, and berries of shrubs and wildflowers. Their natural habitat contains dozens of plants, so how do birds learn the best plants to include in their diet? One possibility is trial-and-error learning. Alternatively, they might learn from their mother. Chicks follow their mother for several weeks as they mature, and hens with chicks often produce unique vocalizations when feeding, known as food calls. Do chicks learn to identify important food plants from their mother's food calls?

Figure 5.20. White-tailed ptarmigan. A common bird in alpine meadows in western North America.

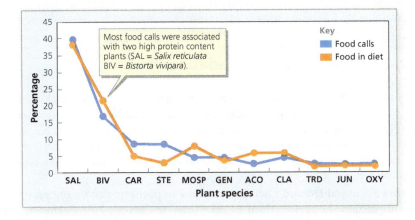

Figure 5.21. Food call frequency and diet. The percentage of food calls given (blue) by hens at a particular plant corresponds to the frequency of that plant in the diet (orange) of chicks (*Source:* Allen & Clarke 2005).

TABLE 5.1 Plants associated with food calls. Food calls were given near 11 plants that comprised a high percentage of the diet of chicks (*Source:* Allen & Clarke 2005).

FOOD CALL FOODS		NON–FOOD CALL FOODS	
PLANT SPECIES	% IN DIET	PLANT SPECIES	% IN DIET
Salix reticulata	38.3	*Trifolium nanum*	1.8
Bistorta vivipara	21.5	*Bistorta bistortoides*	0.9
Moss spores	7.5	*Poa spp.*	0.9
Trifolium dasyphyllum	5.6	*Potentilla nivea*	0.9
Acomastylis rossii	5.6	*Silene acaulis*	0.9
Carex jonseii	4.7		
Gentianodes algida	3.7		
Stellaria umbellata	2.8		
Claytonia lanceolata	1.8		
Juncus spp.	1.8		
Oxyria digyna	0.9		
Total percentage	**94.2**		**5.4**

Allen and Clarke observed 12 hens and their broods. Each hen and brood was observed for an average of six foraging bouts. During a bout, the research team recorded the food eaten by both the hen and her chicks each minute and all food calls given by hens. For each food call, they noted the food plant associated with the call and the subsequent behavior of the chicks. They also analyzed the nutrient content of each plant eaten by the birds.

Ptarmigans used 16 plant species as food. Hens produced a total of 47 food calls in association with only 11 plant species, but most of these calls were produced when hens fed on just two species: dwarf alpine willow (*Salix reticulata*) and alpine bistort (*Bistorta vivipara*)—two plants very high in protein content (Table 5.1). On hearing a food call, chicks invariably ran toward the female and began feeding on the plant near her. Over 94% of chick diets consisted of the 11 plants associated with the food calls. There was a strong relationship between the number of times a plant was associated with a food call and the proportion of that plant in a chick's diet (Figure 5.21). Allen and Clarke concluded that chicks learn from their mother's food calling and that hens facilitate learning by producing calls when feeding from the most nutritious plants.

Featured Research

Prairie dogs learn about predators

Many young animals are highly vulnerable to predators. For example, black-tailed prairie dogs (*Cynomys ludovicianus*) are large social rodents that live in arid grassland habitats in North America (Figure 5.22) that also contain predators such as black-footed ferrets (*Mustela nigripes*), coyotes (*Canis latrans*), bobcats (*Lynx rufus*), and a variety of hawks and rattlesnakes. How do young prairie dogs learn about their predators and how to avoid them?

Debra Shier and Donald Owings tested the hypothesis that young prairie dogs learn antipredator behaviors from adults (Shier & Owings 2007). They captured 36 juveniles with their mothers within two days of emergence from their burrows. Each mother and her litter were housed in a separate enclosure, with an artificial

Figure 5.22. Prairie dogs. Highly social rodents that live in large colonies.

underground burrow at one end. Each juvenile first went through a pretraining assessment of antipredator behavior with predator test stimuli: a black-footed ferret, a prairie rattlesnake (*Crotalus viridis*), a moving hawk model, or a cottontail rabbit (*Sylvilagus audubonii*), which served as a nonpredator control. The ferret and rabbit were presented within a mesh box, while the snake was presented behind a mesh barrier. The model hawk was flown over the enclosure. The researchers videotaped each juvenile for ten minutes to record its antipredator response to the test stimulus.

Next came a five-week training period where the young pups were randomly divided into two groups and exposed to the predators. One group spent five weeks with an experienced adult that had lived in the wild and had previous exposure to the predators. The other group either trained alone or with an inexperienced sibling. Following training, juveniles were given a post-training test in which their response to each test stimulus was recorded when they were alone. The researchers measured the juveniles' activity level, frequency of fleeing, antipredator vocalizations, and vigilance behavior.

The post-training tests indicated that the treatments strongly affected antipredator behavior. Prairie dog juvenile pups trained with an adult were less active, produced more alarm calls, exhibited greater levels of vigilance, and more often fled to the burrow when exposed to a predator (Figure 5.23).

Following these tests, all juveniles and adults were marked and released back into the wild in a new colony to assess their survivorship. One year after testing, the juveniles reared with experienced adults had higher survivorship (Figure 5.24) than those reared alone or with inexperienced siblings. Shier and Owings concluded that juveniles learn antipredator behaviors from experienced adults and that learning helps them survive.

Learning the location of food patches

Social learning is not restricted to juveniles. Adults, too, use information from the behavior of other individuals, called **social information**, to learn about the environment. This is especially true for information concerning food patches, particularly for animals that feed on mobile prey. Consider a tern, a bird that feeds on fish near the water's surface by diving into the water. A school of fish moves constantly, so a food patch is always moving as well. Because patches with food can be difficult to locate, animals will often use other foraging individuals as a cue to their location. This phenomenon is known as **local enhancement**, and it occurs when an individual's focus is directed to a particular part of the environment due to the presence of another.

Location is only one aspect of a food patch. Perhaps even more important is the quality of that patch, such as how much food it contains. How can an individual learn about food patch quality? One possibility is again local enhancement. The more food a patch contains, the more foragers will typically be present at that patch. An even better strategy for determining patch quality is to observe individual feeding rate in a patch. Feeding rates will be high in rich patches and low in poor patches. Knowledge obtained from others about the quality of a resource is called **public information** (Valone 1989).

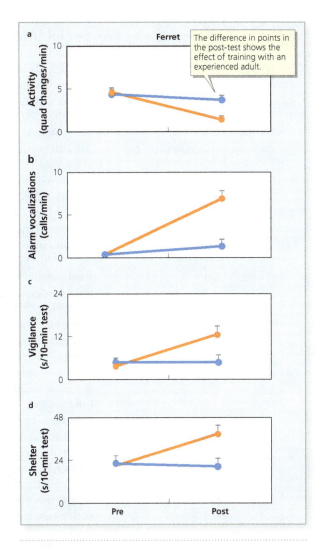

Figure 5.23. **Juvenile antipredator behavior.** Mean (+ SE) antipredator behavior of juveniles trained with an experienced adult (orange) and those not trained with an experienced adult (blue) for exposure to one predator type, a ferret. (a) Activity level, (b) alarm vocalizations, (c) vigilance rate, and (d) time in shelter. Note large differences in behavior in the post-training data (*Source:* Shier & Owings 2007).

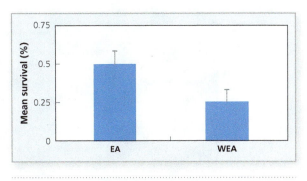

Figure 5.24. **Survivorship differences.** Mean (+ SE) survival probabilities. Juveniles trained with an experienced adult (EA) had higher survivorship in the wild after one year than juveniles trained without an experienced adult (WEA) (*Source:* Shier & Owings 2007).

Featured Research

Social information use in sticklebacks

social information Any information obtained from others about the environment.

local enhancement The direction of an individual's focus to a particular part of the environment by the presence of another.

public information Information obtained from the activity or performance of others about the quality of an environmental parameter or resource.

Do animals use public information to assess patch quality? Isabelle Coolen and colleagues examined this question in nine-spined sticklebacks (*Pungitius pungitius*) (Coolen et al. 2005). They predicted that if only local enhancement information is available, individuals will use it, because high numbers of fish should indicate the best patch. If fish could observe the feeding success of others, however, they should use that public information to select a food patch—regardless of the number of fish at the patch.

To test this prediction, the researchers divided a tank into three areas of equal size (Figure 5.25). Each endzone constituted a food patch and contained a feeder that dispensed food into the tank. The center area contained a compartment with a single observer, the test fish, who could see six conspecifics through a one-way mirror. The conspecifics served as "companions" so that the test fish was never isolated.

In a first experiment, a test fish could see conspecific "demonstrators" at each food patch. Six demonstrators were at one patch, and two were at the other. No food was provided from either feeder, so only local enhancement information was available—the number of demonstrators. After ten minutes, all demonstrators were removed and the observer was allowed to swim freely in the entire tank, although no food was actually present. The researchers recorded the mean percentage of time spent in each compartment.

The test fish spent considerable time in the central compartment adjacent to the companion fish, but with respect to the two food patches, test fish spent significantly more time at the patch that had previously contained six demonstrators (Figure 5.26). Sticklebacks were apparently using the number of fish at each food patch as a cue to its quality.

To determine whether sticklebacks also use public information to estimate patch quality, the researchers conducted a second experiment. Here, they manipulated the feeding rates of demonstrators at each patch. The experimental tank was the same as before, but now the test fish could watch demonstrators feed on bloodworms from each patch. The patch with two demonstrators was designated the "rich" patch, and food was released from the feeder there six times during the ten-minute trial. The patch with six fish was designated the "poor" patch, and food was released there only twice during the trial. Feeding rates therefore differed at the two patches.

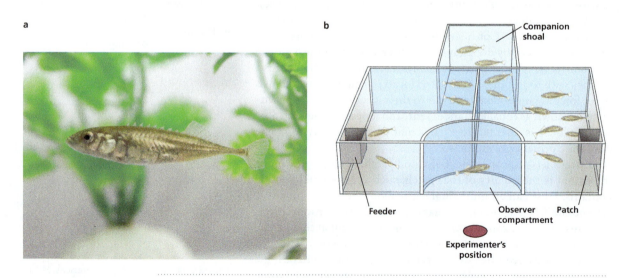

a

b

Figure 5.25. Stickleback experimental tank. (a) Stickleback. (b) Test aquarium. Individual test fish were placed in the observer compartment. Two or six demonstrators were placed in each side compartment, which contained a food patch, the feeder. The companion shoal of six individuals did not have access to food during demonstrations (*Source:* Coolen et al. 2005).

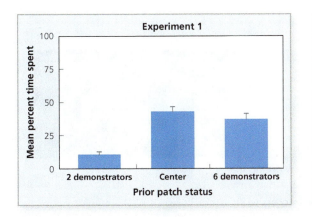

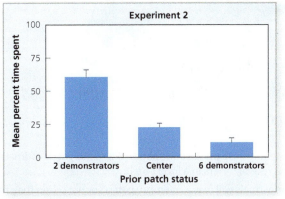

Figure 5.26. Percentage of time in each patch when seeing demonstrators that do not feed. Mean (+ SE) percentage of time spent in different locations. When demonstrators did not feed during the observation period, test fish spent more time near the patch that had six demonstrators compared to the patch that had only two demonstrators (*Source:* Coolen et al. 2005).

Figure 5.27. Percentage of time in each patch when seeing demonstrators that feed. Mean (+ SE) percentage of time spent in different locations. When feeding rates of demonstrators differed during the observation period, test fish spent more time at the rich patch that had two demonstrators compared to the poor patch that had six demonstrators (*Source:* Coolen et al. 2005).

After all demonstrators and food were removed, the test fish exhibited a strong preference for the rich patch, even though only two fish had been there (Figure 5.27). Coolen's team concluded that fish do use public information to select a patch. When local enhancement and public information provide contradictory information, as they did in the second experiment, fish rely on public information to select a food patch, likely because it provides more accurate information about patch quality.

Public information use in starlings

Featured Research

Public information, as we might expect, allows faster estimates of patch quality. The ability to assess quality rapidly is especially important when many patches contain little or no food. Obviously, individuals benefit by spending as little time as possible in such patches. Jennifer Templeton and Luc-Alain Giraldeau tested this prediction in starlings (*Sturnus vulgaris*) (Templeton & Giraldeau 1996) (Scientific Process 5.3). Starlings often feed in large flocks and probe the ground for invertebrate prey in the soil. The researchers captured wild birds and examined their feeding behavior in artificial food patches in the laboratory.

They trained birds to search a food patch that contained 30 holes. Each hole was covered by a piece of opaque latex, which prevented the birds from seeing the contents of the hole. A bird could search the hole by probing a slit in the latex with its bill and opening its mouth—a natural behavior for starlings searching for food. When a bird removed its bill, the cover returned to its original position, leaving no sign that the hole had been searched. Birds learned that the patch contained food in some feeding trials and was empty in others. The food items were large mynah pellets, which are easy to see when removed from a hole. Their size made it easy for birds to recognize successful probes by others. Templeton and Giraldeau measured how long a bird would search an empty patch, based on the number of holes searched, before leaving to feed on another.

Each focal bird was tested with a foraging partner. In one treatment, the partner bird did not sample the patch and so provided no public information. In a second treatment, the partner was trained to probe just three holes, and so provided very little public information. Finally, in the third treatment, the partner bird was trained to sample many holes, providing more public information. Templeton and Giraldeau predicted that the more public information available, the less time a bird would spend in the patch, because it could more quickly estimate that the patch had no food.

SCIENTIFIC PROCESS 5.3

Starlings use public information

RESEARCH QUESTION: *How do starlings estimate food patch quality?*

Hypothesis: Starlings use public information (PI) to aid patch estimation by noting the feeding success rate of others in a food patch.

Prediction: Individuals will leave an empty patch sooner when they have access to PI. The more PI they obtain, the sooner they will leave an empty patch.

Methods: The researchers:

- trained starlings to probe an artificial patch for food pellets. Focal starlings were paired with a partner bird in three treatments: (1) partner bird does not sample from patch (no PI), (2) partner bird samples at a slow rate (low PI), and (3) partner bird samples at a high rate (high PI).
- exposed five focal birds to all treatments.
- recorded the number of holes that the focal bird probed before leaving an empty patch.

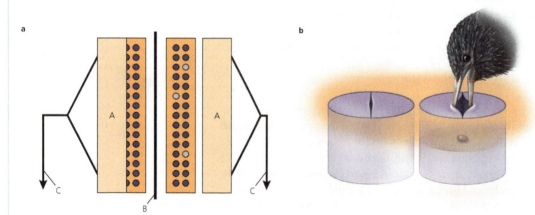

Figure 1. Food patch. (a) Experimental patch with holes. (b) A bird probing a hole for food (*Source:* Templeton & Giraldeau 1996).

Results: Focal birds that exploited the patch alone (no PI partner) sampled significantly more holes before leaving the patch than did birds with a PI partner. Focal birds paired with the high-PI partner sampled the fewest holes before leaving.

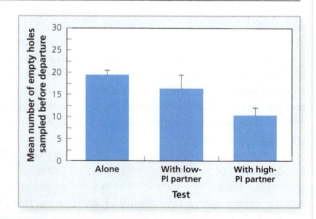

Figure 2. Results. Mean (+ SE) number of holes sampled by birds in each treatment (*Source:* Templeton & Giraldeau 1996).

Conclusions: The use of such PI allows birds to estimate patch quality faster than they could without access to this information. The use of PI reduced the amount of time focal birds spent in an empty patch.

Birds with a partner that provided no public information sampled about 20 holes before abandoning the patch. In contrast, birds abandoned the patch after sampling only 16 holes when paired with a partner that provided some public information and abandoned it after only ten holes when the partner provided more public information. Remember that none of the birds obtained food in this experiment, so only the partner's unsuccessful probes provided information to the focal bird. This public information reduced the time a focal bird spent on a poor patch, as predicted, and demonstrates how animals can benefit by using it.

These examples illustrate how sociality can promote social learning. Next, we examine one outcome of social learning: the development of animal traditions and culture.

5.5 Social learning can lead to the development of animal traditions and culture

We have seen that conspecifics can facilitate learning because they act as a source of information and that such learning is beneficial, so social learning is widespread. Social learning can also lead to variation among populations, as we see next.

Several subspecies of chimpanzees (*Pan troglodytes*) occur throughout tropical Africa as distinct populations. Some populations use stones to crack open nuts, while others do not. Why do these differences among populations exist? One explanation might involve genetic differences. Alternatively, the variation across populations might correspond to ecological differences, because behavior that is adaptive in one environment might not be adaptive in another. For instance, the use of stones to crack nuts requires the presence of both nuts and appropriate stones. If either is absent, the behavior will also be absent. A third possibility involves social learning, which could allow a trait to spread rapidly through one particular population. Differences in behavior among populations that are transmitted between generations through social learning are called **behavioral traditions**.

A behavioral tradition commonly observed in populations of birds and cetaceans (whales, porpoises, and dolphins) are local **song dialects**, characteristic differences in songs that vary geographically (Figure 5.28). Song is learned from an adult tutor (as we saw in Chapter 4), and so changes can spread rapidly (e.g., Noad et al. 2000; Wright, Dahlin, & Salinas-Melgoza 2008). Another common behavioral tradition involves tool use, or the manipulation of an object to achieve a beneficial outcome. Many birds and primates use stones and twigs as tools to obtain food (Figure 5.29).

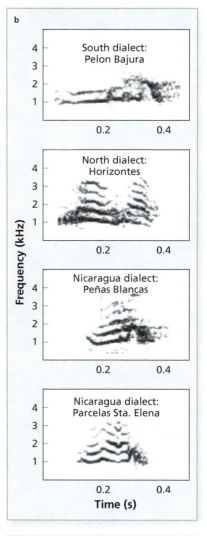

Figure 5.28. Birdsong dialects. (a) Yellow-naped Amazon parrot (*Amazona auropalliata*). (b) Sonograms vary greatly across research sites (*Source:* Wright, Dahlin, & Salinas-Melgoza 2008).

Figure 5.29. Tool use. (a) Bonobos (*Pan paniscus*) use stone tools. (b) New Caledonian crows (*Corvus moneduloides*) use twigs as tools.

behavioral traditions Differences in behavior among populations, transmitted across generations through social learning.

song dialects Characteristic differences in songs that vary geographically among populations.

Individuals place basket sponges on their rostrum to forage in rubble.

Figure 5.30. **Dolphin tool use.** Dolphins use marine sponges to dig up hidden prey on the seafloor, which is strewn with sharp rubble.

Bottlenose dolphins (*Tursiops* sp.) off the coast of Australia display a number of unique foraging strategies that include tool use. In particular, individuals remove marine sponges from the substrate and wear them on their rostrum or beak as they probe the seafloor to locate submerged prey in rough rubble (Figure 5.30). Janet Mann and her students noted that sponge tool-use behavior is found only in one population of dolphins near Shark Bay and tends to be most commonly observed in females. Do calves learn this behavior through social learning?

Mann and colleagues used 14 years of data on the behavior of mothers and their calves to examine this question. They found that this behavior appears to be transmitted primarily from mothers to their offspring, suggesting that social learning plays a role (Sargent & Mann 2009). Further work indicates that sponge tool-use behavior may function to minimize damage to the skin while probing the rocky substrate for prey (Patterson & Mann 2011).

Some species, like chimpanzees, exhibit behavioral traditions. For example, populations vary not only in their use of stones to crack nuts, but also in their use of twigs to fish for termites, and in specific grooming and courtship behaviors (Whiten et al. 1999). Differences in multiple traditions may be evidence of broader differences among populations, or **animal culture** (Whiten & van Schaik 2007).

animal culture Differences in multiple behavioral traditions among populations.

One way to test genetic, ecological, and social learning explanations for animal traditions is through experimental manipulation of populations. Manipulating populations can be difficult or impossible with primates and cetaceans, but it is feasible with smaller vertebrates, like fish. Robert Warner used this approach to study a behavioral tradition in coral reef fish.

Featured Research

Behavioral tradition in wrasse

Warner studied the mating behavior of bluehead wrasse (*Thalassoma bifasciatum*) off the coast of Panama (Warner 1988; Warner 1990). These small fish live year-round in coral reefs and spawn each day (Figure 5.31). Males defend territories along the vertical face of the reef (Figure 5.32), where they display to attract females. Females move to these territories and deposit their eggs for fertilization within them. Individuals live for a maximum of three years, but over the course of 12 years, Warner noticed remarkable consistency in the location of spawning sites: despite great variation in population size and many available sites, he never observed the

Figure 5.31. Bluehead wrasse. A species that exhibits a behavioral tradition in spawning site use.

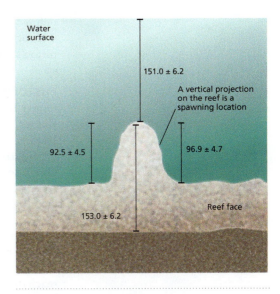

Figure 5.32. Bluehead wrasse spawning sites. The physical characteristics of a typical spawning site. Measurements in centimeters (*Source:* Warner 1988).

use of a new spawning site. Why did this population use the same 87 sites repeatedly? One explanation is that females simply used the best sites for reproduction. Alternatively, spawning sites might be learned from adults as part of a behavioral tradition.

To test these alternatives, Warner conducted a clever experiment. He first recorded spawning behavior in six reef patches. He then removed all adults from these patches and replaced them with adults from the same population that had used a different reef for spawning. Warner predicted that if fish assess a spawning site based on its quality, such as resource availability or predation risk, then the new fish should use the same sites used by the original fish. In contrast, the new fish might establish their own traditions by choosing new sites, and subsequent generations would stick to these choices. As a control, Warner briefly removed all adults from three other reef patches and returned them after a few hours.

Warner found that new individuals used some of the same sites as fish that were removed but also some new sites (Figure 5.33). Furthermore, their use of the sites was remarkably consistent over three more years—a new behavioral tradition had begun. In the control sites, the exact same spawning sites were still used.

How do traditions develop in these fish? Warner observed that females frequently arrived at the reef to spawn in small groups and that they often varied in size and thus age (Warner 1990). He suggested that young females learn the traditional spawning sites from older females. Warner proposed that all spawning sites are essentially equivalent in quality for reproduction, because sites do not directly affect the fitness of offspring: eggs develop into pelagic young far from the reef itself. As such, spawning sites could then be maintained by tradition.

We now understand that social learning can help explain variation in behavior among populations. Additional studies, like Warner's, are needed to understand how often animal culture underlies this variation. In the last section of the chapter, we examine how the mental capacity of animals is involved in learning.

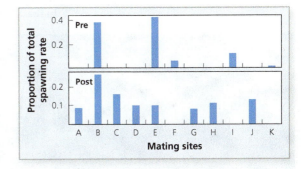

Figure 5.33. Bluehead wrasse spawning site use. The proportion of total spawning rates on one reef during the experiment, showing that new wrasse used different breeding sites. The letters represent different spawning sites used by old (pre) and new (post) fish (*Source:* Warner 1990).

5.6 Some animals can use mental representations to solve complex problems

We have seen that animals learn by making associations from their own experience or by observing others. But animals also solve problems. **Cognition** is broadly

cognition The ability to generate and store mental representations of the physical and social environment to motivate behavior or solve problems.

defined as the ability to generate and store mental representations of the physical and social environment to motivate behavior or solve problems. How is animal cognition studied?

History of animal cognition research

One of the first researchers to study the cognitive abilities of animals was Wolfgang Köhler. As we have seen, trial-and-error learning is one way to solve problems. Köhler was interested in whether animals could learn in other ways. We have all experienced that "aha" moment when "the light bulb goes on" and we suddenly find a solution to a problem. This mental process is called **insight learning**— spontaneous problem solving without the benefit of trial-and-error learning. Köhler wondered whether animals could learn to solve problems in the same way.

He worked with a small captive colony of chimpanzees at the Canary Island Anthropoid Station, where he developed a series of problems for the chimps to solve using food as motivation. In one experiment, he placed bananas outside of the chimp's enclosure and placed several sticks in the enclosure that they could use to reach the food. One chimp, Tschego, spent half an hour trying to reach the food with her arms before she quit. As a group of younger individuals approached the food, Kohler wrote, "suddenly, Tschego leap[ed] to her feet, seize[d] a stick, and quite adroitly pull[ed] the bananas till they [were] within reach" (Köhler 1925). In another experiment, a chimp named Sultan quickly figured out how to put two sticks together so that they were long enough to reach the food.

In these experiments, the animals had no prior experience with the problem and yet appeared to solve it quickly, as one would expect from individuals that possessed the capacity for insight learning and cognition. Determining if the animal used mental representations to solve the problem was, however, a challenge (Wynne 2004a). Did the chimps have a mental representation of sticks as an extension of their arms? Perhaps.

Other researchers have focused on teaching animals communication skills so that they can directly demonstrate cognitive abilities. Primatologists have long recognized the existence of complex social relationships between individuals in the wild. For example, males form alliances during aggressive interactions, and females form alliances to help one another in rearing offspring. Complex social relationships seem to require higher order mental skills associated with cognition, such as the ability to identify specific individuals and/or remember the outcome of past interactions (Cheney, Seyfarth, & Smuts 1986). These complex features of primate society led researchers to speculate about their mental capacity—in particular, whether chimpanzees can use language to communicate. Obviously, many animals communicate to share information. Birds sing to announce their presence in a territory, while wolves howl to maintain social bonds. Language, however, requires syntax—a set of rules for correctly stringing together signs or symbols to transfer complex information. The use of such rules is one aspect of cognition.

Several research teams have tried to teach chimpanzees to use a rudimentary vocabulary in order to determine whether the animals could string words or objects together to form a sentence (Gardner & Gardner 1969; Premack & Premack 1983; Rumbaugh 1977). However, most researchers agree that none of these studies succeeded in demonstrating that chimps could use the rules of a language

insight learning Spontaneous problem solving without the benefit of trial-and-error learning.

Figure 5.34. Elephant insight. This elephant, Kandula, moved the cube and stood on it to reach food.

(Hauser, Comsky, & Fitch 2002). For example, chimpanzees cannot imitate novel sounds created by someone else—a prerequisite for vocal language (Bolhuis & Wynne 2009).

Elephants and insight learning

While early work focused on primates, research on cognition today examines problem-solving skills in a variety of animals. Inspired by Köhler's work with chimpanzees, Preston Foerder and colleagues examined the capacity for insight learning in elephants. Elephants are highly social and thought to be intelligent, and so are good candidates for insight learning. The research team worked with Asian elephants (*Elephas maximus*) at the Smithsonian National Zoological Park (Foerder et al. 2011).

In the first experiment, the researchers placed food out of the elephants' reach but left sticks in the enclosure as possible tools. The researchers tested whether any of the elephants would use the sticks to reach the food. Unlike Köhler's chimps, none did.

In a second experiment, the researchers hung fruit over the animals' heads, out of reach, and left large moveable cubes in the enclosure. The youngest elephant, Kandula, quickly learned to maneuver a cube to the proper location so that he could stand on it and get the food (Figure 5.34). Because Kandula had no experience with the cube and did not use trial-and-error learning, the researchers concluded that the solution he devised was indeed insight learning and was therefore likely evidence of cognition.

Numerical competency in New Zealand robins

Along with studies of language and insight learning, another aspect of animal cognition involves numbers. Because "number" is an abstract concept rather than an inherent property of an object, the ability to count and assess "more" rather than "less" should indicate cognitive ability. In fact, many primates exhibit **numerical competency**: they can recognize numerical quantities (if only up to the number four). Nor is this skill restricted to primates. Irene Pepperberg demonstrated quite sophisticated numerical competency in an African gray parrot (*Pasittacus erithacus*), Alex, that was trained to vocally label objects and their color (Figure 5.35). Alex could correctly answer questions such as "How many blue keys?" and discriminate numbers up to nine (Pepperberg 2006).

numerical competency The ability to recognize numerical quantities.

Much of the research on numerical competency has been conducted in the laboratory. Simon Hunt, Jason Low, and Kevin Burns asked whether species in the wild could exhibit numerical competency and whether the development of this capability required extensive training. They studied a population of color-banded wild New Zealand robins (*Petroica australis*) in the Karori Wildlife Sanctuary in New Zealand (Figure 5.36). Because of their remote location, these birds are not wary of humans and so are easy to observe up close. The birds live in forests and regularly capture large invertebrates that they dismember into smaller pieces before eating. These food items are frequently cached for later consumption in depressions in tree branches. Birds rely on cached food to survive the winter, but caches may be pilfered by others. It may therefore be important for an individual to assign the caches relative importance: a cache with many food items should be more valuable than one with a single item. Such an assessment requires numerical competency.

The research team tested whether birds used numerical judgments when retrieving cached food (Hunt, Low, & Burns 2008). They presented mealworms (*Tenebrio molitor*

Figure 5.35. Parrot insight. Alex the parrot discriminates the color and number of objects in an experiment.

Figure 5.36. New Zealand robin. These birds display numerical competency.

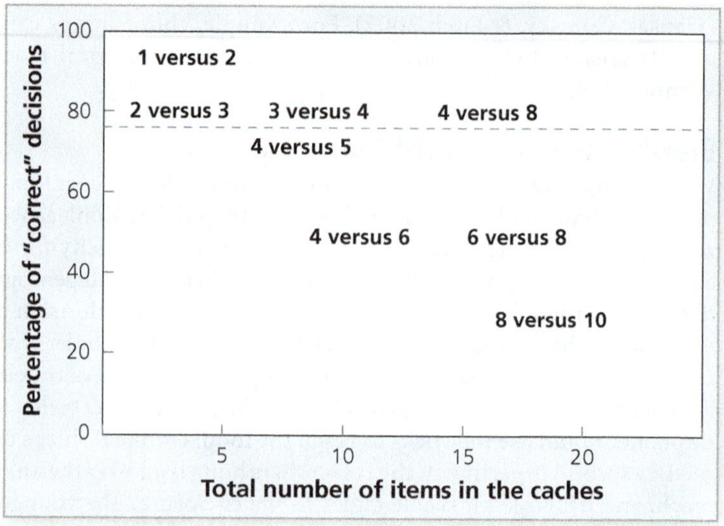

Figure 5.37. Percent correct choices. The percentage of time birds chose the cache with more food items in each of the treatments. Treatments are labeled with the number of food items hidden in each location. Treatments above the dashed line differ significantly from random choice (*Source:* Hunt, Low, & Burns 2008).

larvae) to 14 subjects on a tree branch that contained two cache sites. While a subject was observing, the researcher placed a number of mealworms in one cache location and then covered it with a piece of leather. They then placed a different number of mealworms in the other cache location and covered it with another piece of leather. Items were placed in caches one at a time, at five-second intervals, so that the birds could observe each addition of an item to the cache. The researchers then recorded which cache the robin selected. Robins frequently turn over leaves in search of food and readily learned to remove the leather lids to obtain cached items. Birds were allowed to consume the contents of only one cache, so there was incentive to select the larger cache with more food.

The research team varied both the total number of items in the two caches and the difference in food items between the caches. In general, the birds preferred the larger cache when the total number of food items was fewer than ten (Figure 5.37). When the caches contained more food items, there was no significant difference in cache selection. Additional experiments ruled out the use of scent, the volume of food, the time needed to add to a cache, or cache location bias as explanations of the results. Because the caches were covered, the robins could not visually compare the number of items in each cache. They must have counted items as they were placed in each of the two locations, compared the numbers, and then selected the larger cache. These results indicate that wild New Zealand robins have a sophisticated numerical sense when dealing with a small number of items that is not developed through extensive prior training.

Featured Research ▶

Cognition and brain architecture in birds

Earlier, we saw that the size of the hippocampus among bird species varies with caching behavior. Can we understand variation in cognitive abilities across species based on variation in brain size?

Brain size is related to body size, so larger species tend to have bigger brains, which can confound comparative studies. However, cognitive function is thought to reside in the forebrain, and so one way to account for variation in body size is to focus on relative forebrain size—the ratio of forebrain size to brain stem size. We can then

TOOLBOX 5.1

Neurobiology primer

Nerves are composed of neurons, cells that receive and transfer electrical and chemical signals. Neurons have many dendrites that receive a signal; a cell body (or soma), where the information is integrated; and an axon that conducts the electrical signal to the axon terminal, where the signal can be transmitted to other neurons, muscles, or organs (Figure 1). Neurons usually have a polarity, so information travels in only one direction.

Electrical signals transmit information through the neuron and involve the opening of receptor ion channels on the dendrite, which allow an influx of ions. The result is a change in ion concentration inside and outside the cell that leads to a difference in electrical voltage across the membrane, called a membrane potential. When the difference in electrical charge reaches a specific threshold, an action potential is generated that transmits the signal along the axon. Chemical signals transfer information at the synapse—the junction between two neurons. The signal is transmitted from the presynaptic neuron to the postsynaptic neuron through the release of chemical neurotransmitters in the synaptic cleft, where they bind to receptors on the postsynaptic membrane (Figure 2). Subsequently, the signal is taken up by the presynaptic neuron.

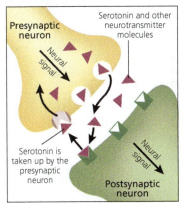

Figure 2. Synapse. The movement of neurotransmitters across the synapse. Electrical signals transform into chemical signals that move across the synapse.

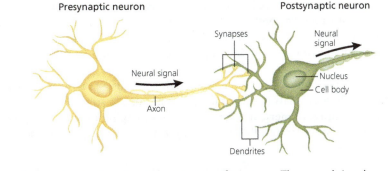

Figure 1. Neuron. The different parts of a neuron. The neural signal travels from the presynaptic to the postsynaptic neuron.

ask whether species with larger relative forebrains possess greater cognitive abilities. The challenge is to quantify cognitive ability across species. Louis Lefebvre has pioneered one way to do this by focusing on foraging innovations.

A foraging **innovation** is the ingestion of a new food type or the use of a new foraging technique. Examples include reports of a granivorous cardinal eating nectar from a flower, a mockingbird obtaining food from the mouth of a sea lion, and a diurnal warbler feeding at night on insects attracted to an artificial light (Lefebvre et al. 1997). For Lefebvre, foraging innovations provide insight into general cognitive abilities, because an animal is solving an ecologically relevant problem: how to obtain food. Since the behavior is recorded in the wild, this measure of cognition avoids the problems inherent in tests on captive animals. Innovations are also quantifiable and available for hundreds of species: one can simply tally the number of foraging innovations found in past studies and analyze them by species, family, or order.

Lefebvre and colleagues found reports of 322 feeding innovations for birds of the British Isles and North America (Lefebvre et al. 1997). They calculated the relative frequencies of foraging innovations for each of 16 orders to obtain the number of foraging innovations per species. They then correlated this number with the mean relative forebrain size. For British and North American birds, relative forebrain size varied across taxa. Passerines and parrots, for example, have larger relative forebrains

innovation A novel, learned behavior.

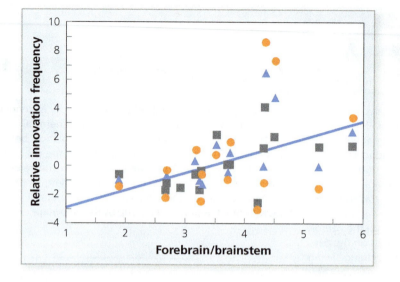

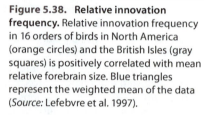

Figure 5.38.   Relative innovation frequency. Relative innovation frequency in 16 orders of birds in North America (orange circles) and the British Isles (gray squares) is positively correlated with mean relative forebrain size. Blue triangles represent the weighted mean of the data (*Source:* Lefebvre et al. 1997).

than doves and grebes. More important, the foraging innovation rate was positively correlated with relative forebrain size (Figure 5.38). These analyses support the hypothesized link between cognitive ability and the relative size of the forebrain in birds. This approach can provide unique insight into how cognitive abilities differ among species.

The study of animal cognition is challenging because it involves understanding whether and how animals use mental representations to solve problems. Examples of insight learning, numerical competency, and foraging innovations provide different avenues to understanding the mental capacities and problem-solving abilities of animals.

Chapter Summary and Beyond

Learning allows animals to modify their behavior to adapt to their environment. The simplest forms of learning involve imprinting, which allows animals to learn about their parents, and habituation, which allows them to lessen their response to nonthreatening stimuli. Recent work has examined imprinting involved in species recognition in birds in the wild (e.g., Hansen, Erik, & Slagsvold 2008). Learning is associated with changes in the brain, including increased neurotransmitter release and the formation and elimination of dendritic spines. Further research has examined how spine size and membrane neurotransmitter receptors are modified with learning (Kasai et al. 2010). Across species of birds, we also see a correlation between the size of the hippocampus and the propensity to rely on memory to find cached food. Roth et al. (2010) have examined variation in this relationship and note that a similar pattern occurs in rodents.

More sophisticated learning involves the formation of associations between a stimulus and a response. In classical (Pavlovian) conditioning, a new association is formed between a previously neutral stimulus and an innate behavioral response. In operant conditioning (trial-and-error learning), learned behaviors change based on positive and negative reinforcement or punishment. The formation of these associations involves trial-and-error learning, and learning curves can be used to characterize the proficiency of learning.

Social learning occurs when animals observe and learn from others. This type of learning is widespread across taxa (e.g., Valone 2007) and can enhance the speed of learning. Through social learning, populations can acquire traditions and culture—learned behaviors transmitted across generations. Researchers are actively debating definitions and examples of animal traditions and culture (Laland & Janik 2006).

Some animals have sophisticated cognitive abilities for problem solving, and this finding has spurred a great deal of research on insight learning, communication, numerical competency, and spatial memory. Recent work on primates has focused on their natural use of gestures to understand the roots of language (Balter 2010). Behavioral innovations, particularly novel feeding behaviors, vary across species and are correlated with relative forebrain size, which can provide insight into a species' cognitive abilities. More recent analyses of primates have found a similar positive correlation between behavioral innovations and relative forebrain size (Reader & Laland 2002; Lefebvre, Reader, & Sol 2004).

Work on animal cognition is particularly intense in cetaceans, primates, and dogs (e.g., Pak 2010; Seed & Tomasello 2010; Howell & Bennett 2011) and has led to ethical debates about the treatment of captive animals (e.g., Allen & Bekoff 2007; Grimm 2011). This area of research has also led to a discussion of animal consciousness, or whether or not animals "think" about objects or events (Griffin 1984; Griffin & Speck 2004). Ongoing work examines animals' neural capacity for consciousness and focuses on behavior suggesting animal self-awareness (Bekoff & Sherman 2004).

CHAPTER QUESTIONS

1. Kim Sullivan (1988) concluded that yellow-eyed juncos appear to learn how to handle insect prey as they mature. What pattern in the data would lead you to conclude that there was no evidence of learning?

2. You discover a fruit fly mutant that has no memory. How will this mutation affect its ability to learn?

3. In one experiment, juvenile chipmunks were housed in two large outdoor aviaries that contained large trees at one end and underground burrows at the other. In Aviary A, the researchers shook the tree branches violently for three seconds, which caused a loud sound twice a day. In Aviary B, the researchers also shook the tree branches violently for three seconds twice a day, making a loud sound—but here, several acorns, a favorite food of chipmunks, would fall to the ground five seconds later. In both aviaries, chipmunks were initially startled by the loud sound and ran to underground burrows. Predict how the behavior of chipmunks in the two aviaries might change after repeated trials.

4. Local enhancement and public information are two forms of social learning. How do they differ?

5. Explain how Warner's experiment (1988, 1990) with wrasse demonstrated a behavioral tradition of spawning site use. How could he rule out genetic or ecological factors to explain the observed behavior?

6. You have taught your puppy Max to lift a front paw when you say the word "shake." As an adult, Max still lifts a paw when you give this command. Describe the changes that likely occurred in the puppy's brain associated with this learned behavior.

7. Imagine two groups of primates. One group is known to learn socially by observation, while the other exhibits only individual trial-and-error learning. Imagine that one individual in each group learns a new foraging skill, such as using a rock to crack open hard nuts. What prediction can you make about the number of individuals in each group that will display this skill over time?

Chapter 6

Communication

CONCEPTS

FEATURES

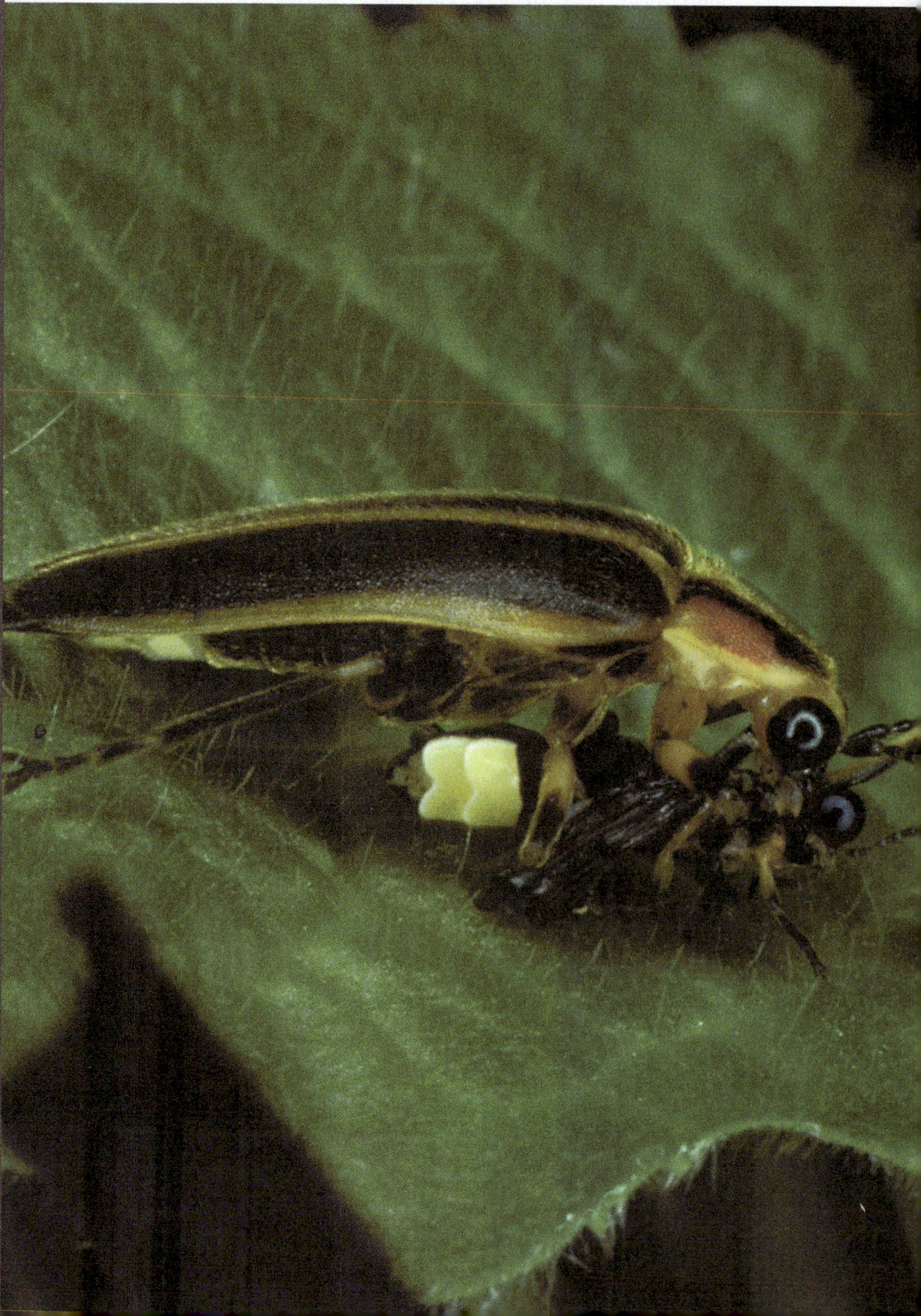

Figure 6.1. Firefly deception. Predatory female *Photuris* firefly eating a male *Photinus* firefly.

Our university operates a biological field station in the Ozark Mountains in Missouri. If you are lucky enough to visit in late spring or early summer, you will be treated to a remarkable sight. Just after dark each evening, the restored prairie comes alive, as thousands of fireflies (or lightning bugs) flash on and off as they fly low over the vegetation. Fireflies are beetles in the order Coleoptera that live in temperate or tropical habitats and are bioluminescent—that is, they produce light. If you watch carefully, you can observe differences in the duration and rate of flashes among the different fireflies. You are watching different species in the field, each with its own distinctive flash pattern for courting conspecifics. For example, the males of each *Photinus* species display a species-specific flash pattern to females, who respond with their own distinctive flash pattern. Males land nearby, while continuing to flash, and then mate. However, the field also contains predatory fireflies in the genus *Photuris*. These species mimic the flashes of *Photinus* females to lure males. Once a *Photinus* male lands nearby, the predatory *Photuris* attacks and eats it (Figure 6.1) (Lloyd 1975; Champion de Crespigny & Hosken 2007).

This amazing display is an example of animal communication. As we see in this chapter, animal communication involves the use of signals by one individual to influence the behavior of another. Signals are generally reliable indicators about conditions, like the courtship flash displays of male and female *Photinus* (Figure 6.2): they are accurate indicators about the species and sex of the signaler. However, signals are not always accurate indicators, as is demonstrated by the predatory *Photuris*. The evolution and accuracy of signals will depend on the benefits and costs of signal production and the fitness interests of those involved. While the simplest communication involves two individuals, third-party eavesdroppers can also intercept signals, which can result in behavioral modifications by both the eavesdropper and the signaler.

6.1 Communication occurs when a specialized signal from one individual influences the behavior of another

signaler An individual that produces a signal.

signal receiver An individual that detects a signal.

communication The process in which a specialized signal produced by one individual affects the behavior of another.

signal A packet of energy or matter generated from a display or action of a signaler that travels to a receiver.

waggle dance A behavior performed by a honeybee scout that recruits workers to a food source.

Figure 6.2. Firefly. Fireflies communicate using a bioluminescent flash display.

When a male firefly observes the flash pattern of a conspecific female, he stops flying, lands near her, and begins courting behavior. This behavioral sequence, which consists of a specialized signal produced by one individual, the **signaler**, that affects the behavior of another, the **signal receiver**, illustrates **communication**. Technically, a **signal** is a packet of energy or matter (e.g., light or sound waves, chemicals) generated from a display or action of a signaler that travels to a receiver (Hebets & Papaj 2004). Signals can be not only physiological traits such as the firefly flash but also morphological traits and behaviors. Let's start with an example of a behavioral signal.

Honeybees and the waggle dance

Karl von Frisch was awarded the Nobel Prize in 1973, in large part for his work on communication in honeybees (*Apis mellifera*). Honeybees live in colonies, and individual workers leave the hive in search of food. Like many other insects, these animals utilize visual and chemical signals. When an individual scout finds a rich food source, it flies back to the hive and recruits others to help exploit the food. Such communication allows the colony to rapidly exploit the food resource before competitors do. So how does this process work? In a series of studies, von Frisch manipulated the location of a distant food source and observed the behavior of both scouts and new recruits. He determined that the scout performs a specific behavior that he called the **waggle dance** (von Frisch 1967).

During a waggle dance, the scout moves in a figure-eight pattern on a vertical wall of the honeycomb. During the linear movement of the dance, the scout vigorously wags its body, and the duration of the wagging, von Frisch argued, indicates the distance to the food. Subsequent work indicates that every 75 msec of waggling translates into a distance of approximately 100 m from the hive (Seeley 1985) (Figure 6.3). The waggle dance also describes the direction of the food source, relative to an imaginary line that runs from the hive to the sun. For instance, if the sun is on the horizon (as at dawn) and the scout's linear movement is 30° left of vertical, then the food is 30° left of the sun (Figure 6.4).

Figure 6.3. Waggle dance. The waggle dance conveys the direction and distance to a food source to attending bees (*Source:* Grütter & Farina 2009).

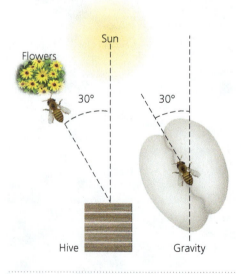

Figure 6.4. Linear movement of the waggle dance. The bees' linear movement indicates the direction of the food.

There has been some debate about the relative importance of odor cues versus the visual waggle dance signal used by recruits to find a food source (e.g., Vadas 1994). Von Frisch recognized that odor is used to locate nearby food resources, but he believed that the waggle dance provides the crucial signal, especially for food located large distances away—distances of more than, say, 500 m (von Frisch 1971). However, most recruits take much longer to fly to the food source than would be expected based on a straight-line route (e.g., von Frisch, Wenner, & Johnson 1967; Gould & Gould 1988). If the waggle dance provides an accurate signal, travel times should be much shorter. Adrian Wenner and Dennis Johnson proposed that the bees doing the waggle dance present the odor of the food source, not a precise direction and distance; this behavior then stimulates recruits to search for food on their own (von Frisch, Wenner, & Johnson 1967; Wenner, Well, & Johnson 1969). They suggested that new recruits fly downwind a few hundred meters before starting their search and use olfaction as a method of locating food by flying upwind, in a zigzag pattern, back toward the hive (von Frisch, Wenner, & Johnson 1967; Wenner 2002). If so, no wonder they take longer than expected. In fact, many insects that rely only on odor cues to find food do fly back and forth into the prevailing wind (Kennedy 1983).

Odor or the dance in bees

Featured Research

Which is more important, odor or the waggle dance? One way to find out is to track the actual movements of new recruits. Joe Riley and colleagues examined the flight paths of recruits that viewed the waggle dance from a hive placed in a large (1 by 1.5 km), flat, mowed field (Riley et al. 2005). The field contained few natural sources of nectar or pollen, so the only food available was located at a feeding station. The station was placed 200 m due east of the hive and contained a 0.2 to 1 M sucrose solution with no artificial odors. The researchers recorded wind speed and direction at ten-second intervals from four locations surrounding the field and at the hive to determine their effect on the flight of bees. Bees were required to move through a clear plastic tube to enter and leave the hive, which facilitated bee capture. This set-up allowed the researchers to mark most of the bees in the colony with small numbered tags on their backs. They could then record the identity and movement of bees into and out of the colony. At the start of the experiment, individual scouts were allowed to visit the feeder and their identity was recorded. Their waggle dance indicated that food was located due east of the hive. The researchers captured recruits who had never previously visited the feeding station and placed small transponders on their backs that transmitted continuous information about their location (Figure 6.5). These individuals were released either at the hive or at one of three locations 200 to 250 m southwest of the hive, and their movements were tracked.

Most of the recruits released at the hive almost immediately began flying east toward the food station (Figure 6.6). However, only two of the 19 recruits actually found the feeding station. After 200 m, most individuals began to follow a more circuitous flight path, which the researchers interpreted as searching behavior for the food. In addition, the 17 bees transported 200 to 250 m southwest of the hive and then released tended to fly due east for 200 m before also adopting a circuitous flight path. The wind data indicated that no odor from the feeding station was available to the bees at their release sites. There was no evidence that individuals flew downwind and then returned in a zigzag pattern to find the food. Instead, the bees had to fly *across* the prevailing wind to maintain a due east heading (Figure 6.7).

The flight paths of new recruits confirmed von Frisch's hypothesis that the waggle dance appears to signal the distance and

Figure 6.5. Bee with transponder. A honeybee with a transponder attached to its back can be tracked using radar.

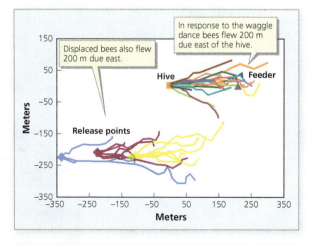

Figure 6.6. Bee flight paths. Each line represents the flight path of an individual (*Source:* Riley et al. 2005).

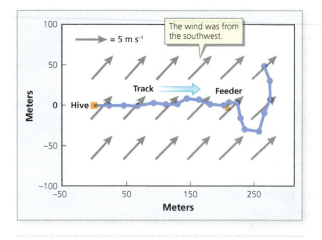

Figure 6.7. **Flight track in wind.** Flight track (blue) of an individual bee during a southeast wind (gray arrows). The bee did not simply fly downwind and in fact needed to adjust its flight direction to fly due east across the wind (*Source:* Riley et al. 2005).

Figure 6.8. **Vervet monkeys.** These African primates live in social groups.

direction of the food. In addition, the data demonstrate that recruits can travel to the area of a novel food source without odor cues. However, they also show that the dance signal is not sufficient to allow a recruit to locate a food source precisely. Additional work has demonstrated that new recruits use odors to pinpoint the exact location of the food (Grütter, Balbuena, & Farina 2008; Grütter & Farina 2009). In sum, the waggle dance is a behavioral signal that allows individuals to travel to the area of a distant food source, but bees then use local odor cues to find the food itself.

Auditory signals: alarm calls

Many birds and mammals vocalize when they spot a predator. These **alarm calls** are unique vocalizations produced when a predator is nearby. In a set of groundbreaking studies, Dorothy Cheney and Robert Seyfarth found that the alarm calls of vervet monkeys (*Chlorocebus pygerythrus*) (Figure 6.8) differ in how they affect receivers (Cheney & Seyfarth 1980; Cheney & Seyfarth 1985; Seyfarth, Cheney, & Marler 1980). Vervet monkeys are attacked by a variety of predators, including leopards (*Panthera pardus*), eagles, and snakes; when a vervet spots one of these animals, it gives an alarm call that alerts others in its group. However, individuals produce different alarm calls for different types of predator (Seyfarth, Cheney, & Marler 1980). Why? Each predator represents a threat that requires a different behavioral response (Figure 6.9).

Leopards, for example, attack from the ground. When a leopard is spotted, a "bark" call is given, and vervets escape attack by moving up into trees. "Cough" calls are given when an eagle is sighted. Eagles attack from above. Avoiding their attack requires moving down from treetops and into dense bushes. Snakes, in contrast, often hide in dense grass. When a python or cobra is observed, vervets give a "chutter" call that causes individuals to stand erect and look into grass clumps.

alarm call Unique vocalizations produced by social animals when a predator is nearby.

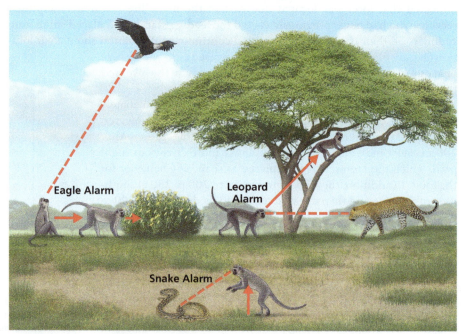

Figure 6.9. **Vervet alarm calls.** Vervets produce different alarm calls and then subsequent behavioral responses to different types of predators.

Titmouse alarm calls

Featured Research

Cheney and Seyfarth stimulated research on the alarm call systems of other species. Jason Courter and Gary Ritchison, for instance, examined alarm calls in the tufted titmouse (*Baeolophus bicolor*), a small songbird (Courter & Ritchison 2010). Titmice produce an alarm call in response to a perched avian predator. Vocalizations can be visualized in a sonogram, which characterizes frequencies of sound as a function of time (Figure 6.10). The titmouse call is composed of three basic notes, called Z, A, and D, which are typically produced in sequence. However, individuals often vary the number of D notes. This call recruits other birds (both con- and heterospecifics) to approach and mob the predator. Mobbing behavior, in which birds produce loud vocalizations, often harasses a predator and can drive it away from an area (see Chapter 8).

Titmice co-occur with several potential avian predators. Small raptors like Eastern screech owls (*Megascops asio*), sharp-shinned hawks (*Accipiter striatus*), and Cooper's hawks (*A. cooperii*) are known predators of titmice, while large raptors like great horned owls (*Bubo virginianus*) and red-tailed hawks (*Buteo jamaicensis*) rarely attack titmice. The size of a predator thus correlates negatively with risk to a titmouse. Courter and Ritchison conducted a study to determine whether the alarm calls differ in responses that correlate with the size and degree of threat posed by a perched avian predator (Courter & Ritchison 2010).

The experiment consisted of six treatments, each presenting a different predator, and a control. At several sites, the researchers established a feeding station that contained sunflower seeds, a food readily consumed by titmice. Lifelike models of six different birds were placed on platforms 1 m from the feeding station; one control presented nothing on the platform. Three models were high-risk predators (Eastern screech owl, sharp-shinned hawk, and Cooper's hawk), two were low-risk predators (great horned owl and red-tailed hawk), and one was a nonpredator control bird, a ruffed grouse (*Bonasa umbellus*). The researchers examined whether titmice would produce different alarm calls in response to these different models. As soon as the titmice were within 25 m of the feeding station, the models were uncovered. The researchers recorded titmouse behavior for six minutes and recorded all alarm calls.

Alarm calling varied across treatments. The number of D notes produced per titmouse was higher in response to small, high-risk predators compared to the number produced in response to large, low-risk predators and controls (Figure 6.11). Furthermore, in general, the smaller the predator, the longer the mobbing response (Figure 6.12): mobbing behavior lasted for approximately 250 seconds for the small-bodied, high-risk predators, compared to only 150 seconds for the larger, low-risk predators and less than 50 seconds for the controls. These results indicate that titmice produce different alarm calls that lead to differences in the behavior of receivers, which correlate with varying levels of threat. This finding illustrates the rapid speed and flexibility of auditory signals in response to changes in the environment.

Information or influence?

There is an important caveat here. In these and other examples, we do not know whether the signal actually encodes information about conditions, although this metaphor has previously dominated work in animal communication (e.g., Otte 1974). For example, vervet alarm calls may not mean "leopard present" or "snake present," but rather "climb

Figure 6.10. Titmouse alarm call. Sonogram of the tufted titmouse alarm call showing the structure of Z, A, and D notes (*Source*: Courter & Ritchison 2010).

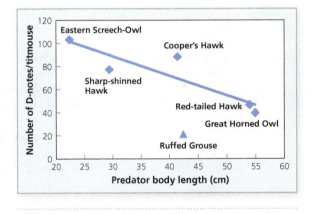

Figure 6.11. Alarm calls vary with predator size. Titmice produced more D notes per minute when presented with models of a small hawk compared to larger models or controls (*Source*: Courter & Ritchison 2010).

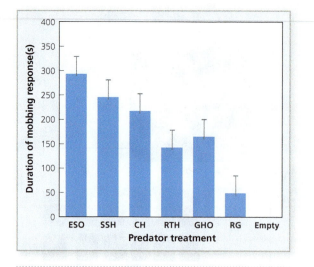

Figure 6.12. Duration of titmouse mobbing behavior.
Mean (+ SE) duration of mobbing behavior for each treatment. Titmice spent more time mobbing models of small predators than they did mobbing larger ones or controls (*Source:* Courter & Ritchison 2010). ESO = Eastern screech owl; SSH = sharp-shinned hawk; CH = Cooper's hawk; RTH = red-tailed hawk; GHO = great horned owl; RG = ruffed grouse.

a tree" or "look down" (Scott-Phillips 2008; Owen, Rendell, & Ryan 2010). Drew Rendall, Michael Owen, and Mike Ryan contend that the assumption that signals encode specific information is problematic for two reasons (Rendell, Owen, & Ryan 2009; Owen, Rendell, & Ryan 2010). First, it can imply a language-like meaning of communication that, as we saw in Chapter 5, has been challenging to document. Second, it can encourage attempts to characterize the information encoded in a signal, rather than focusing on factors that shape signal properties. These researchers suggest that it is more productive to study signals as traits that have evolved to influence the behavior of others.

Not everyone agrees. Others argue that removing the concept of information from the study of communication is unwarranted (Carazo & Font 2010; Seyfarth et al. 2010), and this discussion continues to stimulate further work. What we can say is that signals influence behavior *as if* they contained information about the location of a food source, the threat of a predator, or the phenotype of a signaler.

6.2 Signals are perceived by sensory systems and influenced by the environment

sensory receptor Nerve endings that respond to an internal or external environmental stimulus.

Animals receive signals through **sensory receptors**—nerve endings that respond to an internal or external environmental stimulus. These receptor cells then transmit information along axons to the central nervous system, where it is processed and an appropriate response is generated. Sensory systems vary greatly across taxa and have co-evolved with many kinds of signals, including chemical, visual, auditory, vibrational, and even electrical signals. These signals differ in many ways, particularly in terms of their movement through the environment. We illustrate some important differences by examining factors that shape the evolution of chemical, visual, and auditory signals.

Chemoreception

chemoreceptor A sensory receptor that detects chemical stimuli.

olfaction The detection of airborne chemical stimuli.

gustation The detection of dissolved chemicals, often within the mouth.

odorant A gaseous compound that is perceived as odorous.

volatile A chemical that can evaporate; that is, become gaseous.

pheromones Volatile organic compounds that are species-specific and affect the behavior of another individual of the same species.

The most primitive and universal sensory receptors are the **chemoreceptors** that detect chemical stimuli. Chemical signals are compounds secreted into the environment and detected by odor-binding proteins in sensory structures such as the antennae of invertebrates and the olfactory system of vertebrates. Chemoreception, the detection of chemical stimuli, includes **olfaction**, or the sense of smell, as well as **gustation**, or the detection of dissolved chemicals. Olfactory signals can be transmitted in water or air, are long-lasting, and can travel long distances. They can also be deposited on a substrate, as illustrated by the territorial scent marking of mammals and the foraging trails of ants. Because these signals can be produced in variable amounts, animals can vary their strength. In addition, chemical signals can travel around environmental barriers such as dense vegetation. Once transmitted, however, chemical signals cannot be modified.

There are two general classes of chemical stimuli: general odorants and pheromones. General **odorants** are **volatile**, or gaseous, compounds that are perceived as odorous. They allow animals to locate food and avoid predators and environmental problems such as fire. **Pheromones** are also volatile compounds, but they are species-specific—that is, these organic compounds affect the behavior of another individual of the same species. Pheromones can indicate the sex of a species, food trails, or predation (Applying the Concepts 6.1).

APPLYING THE CONCEPTS 6.1

Pheromones and pest control

Up to one-third of food production worldwide each year is destroyed by insect pests (Witzgall, Kirsch, & Cork 2010). One response to this problem has been the widespread use of chemical insecticides. However, these chemicals are not species-specific, and many kill nontarget species as well. They are also environmentally toxic and facilitate the evolution of resistance by the same pests they are designed to kill (Witzgall, Kirsch, & Cork 2010). Increasing attention to this issue has produced an alternative solution: the development and use of pheromones as a tool in pest management. Many insects use volatile sex pheromones to communicate to find potential mates. Pheromones can be effective in minute amounts, are species-specific, and are environmentally benign. One technique is to release these pheromones throughout a field or orchard to disrupt mating, because males cannot then locate females.

Jianhuamo Mo and colleagues studied the effectiveness of pheromones in controlling damage by light-brown apple moths (*Epiphyas postvittana*) in citrus orchards over the course of three years (Mo et al. 2006). They compared two orange groves: one control and one into which they released small quantities of an apple moth sex pheromone. Each orchard contained 30 to 60 single, constrained test females. After one week, females were examined to determine whether they had been inseminated by a male. The researchers found that females in the treatment orchard were almost never inseminated, compared to approximately 50% of those in the control orchard. In addition, examination of orchard fruits revealed that 20% had evidence of insect damage in the control orchard, whereas less than 10% of the treatment orchard fruits were damaged. Results like these are encouraging the development of pheromone-based management for a diversity of insect pests.

Ant pheromones

Featured Research

For many insects, pheromones are important for recognizing conspecific colony members, potential mates, or potential intruders who could steal nest resources. These pheromones are often composed of species-specific hydrocarbons found in the cuticle (outer covering). In ants, sentries at the nest entrance often use tactile cues such as antennal contact to determine whether individuals should be admitted to the nest or attacked. However, this type of signal means that an intruder could get to the nest entrance and perhaps even in the nest before being detected.

Andreas Brandstaetter and colleagues wondered whether cuticular hydrocarbons might be volatile; perhaps ants could detect them at some distance, instead of using tactile cues (Brandstaetter, Endler, & Kleineidam 2008) (Scientific Process 6.1). The researchers collected the hydrocarbon signal from the postpharyngeal gland of workers from several colonies of the Florida carpenter ant (*Camponotus floridanus*) and dissolved it in a solvent. They then placed the signal close to an ant, but at a distance from which it could not touch the signal. They observed that when the signal came from a non-nestmate, the ants responded with more aggressive behaviors, just as they do when they receive tactile cues. The researchers concluded that there does seem to be a volatile component to this signal.

photoreceptor A specialized neuron that is sensitive to light.

rod A type of photoreceptor, sensitive to low light levels.

cones. A type of photoreceptor for color vision under bright light conditions.

Photoreceptors

Animals often need to determine the directionality of light and detect spatial patterns. Vision receptors, or **photoreceptors**, have special pigment molecules that convert light energy into nerve impulses. In vertebrates, there are two types of photoreceptors: **rods**, highly sensitive in dim light, and **cones**, which are responsible for color vision. The number and ratio of rods and cones vary across species and depend on the light levels of the environment. This variation leads to a wide diversity of wavelengths of light that are detected across species (Figure 6.13).

Visual signals can be detected quickly, allowing a rapid response. In clear skies or clear water, they can also be visible at fairly large distances. However, their perception requires sufficient light levels. Thus, in deep ocean or murky water and at night, conditions favor the evolution of signals detected by other sensory systems. In addition,

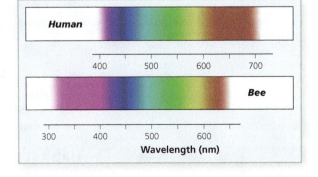

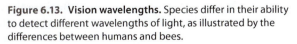

Figure 6.13. Vision wavelengths. Species differ in their ability to detect different wavelengths of light, as illustrated by the differences between humans and bees.

SCIENTIFIC PROCESS 6.1

Chemical signals in ants

RESEARCH QUESTION: *Can ants distinguish nestmates from non-nestmates using an odor signal?*

Hypothesis: Ants use volatile hydrocarbons to distinguish nestmates from non-nestmates.

Prediction: Individuals will respond differently to odors from nestmates and odors from non-nestmates.

Methods: The researchers:

- collected carpenter ant (*Camponotus floridanus*) colonies from two Florida islands.
- collected the hydrocarbon signal from the postpharyngeal gland of workers of each colony and dissolved it in a solvent, hexane.
- presented hydrocarbon signals from nestmates (NM), non-nestmates (n-NM), and controls (hexane only) to individuals on a magnetic stir bar (the "dummy"), placed 1 cm away.
- recorded the most aggressive behavior of the individuals during a period of 150 seconds. Behaviors were categorized as nonaggressive (categories 0–2) or aggressive (categories 3–5).

TABLE 1 Six-category behavioral index.

CATEGORY	DESCRIPTION
0	No antennal scanning
1	Antennal scanning not directed toward dummy
2	Antennal scanning directed toward dummy
3	Weak mandibular threat (mandibles slightly open)
4	Strong mandibular threat (mandibles widely open)
5	Body jerking (repeated, rapid, forward-and-back jerking with open mandibles)

Results: More workers displayed aggressive behaviors toward non-nestmate odors than they did toward nestmate or control odors.

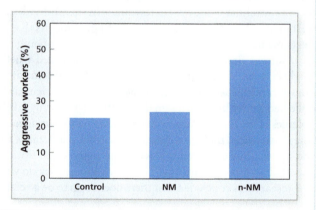

Figure 1. Aggression. Aggressive response to treatments (*Source*: Brandstaetter, Endler, & Kleineidam 2008).

Conclusion: Carpenter ants can use airborne chemical cues from a short distance of 1 cm to discriminate between nestmate and non-nestmate hydrocarbons.

visual signals can be blocked by obstacles, so again, they do not transmit well in dense media like murky water and thick vegetation. Therefore, visual signals are most often used by diurnal organisms that interact over relatively short distances in open habitats.

Featured Research

Habitat structure and visual signals in birds

Because the ability to perceive visual signals increases with brightness, variation in light levels should also affect their evolution. Karen Marchetti examined the evolution of bright plumage in Old World warblers that breed in the forests of Kashmir, India (Marchetti 1993). These birds live in a variety of forest habitats that vary in light intensity. She examined several species in the genus *Phylloscopus*, as well as the closely related goldcrest (*Regulus regulus*). These small birds are mostly dull green but differ in the number of bright color patches (from zero to four) that occur on their wings, crown, rump, and tail. Males compete for territories using display behaviors such as wing flashes, head tipping, and rump flashing. Marchetti found that these behaviors correspond well to the presence of different color patches; for instance, only birds with a crown stripe display head-dipping behavior. This movement allows the birds to make themselves temporarily more conspicuous. Marchetti hypothesized that color patterns were morphological signals used in competition for territories and that habitat brightness has influenced the evolution of these signals. She reasoned that because the amount of light influences visual signal perception, birds living in darker environments, like coniferous forests, should have more color patches than birds living in open, treeless areas. In other words, selection would favor the evolution of bright visual signals in darker habitats.

Marchetti recorded the species of birds present and their color patterns at five locations along an elevational transect. These locations varied in vegetation, from an open, bright habitat above the tree line to a relatively dark coniferous forest. At each location, she recorded the brightness of the habitat based on a still camera's automatic shutter speed, measured while holding the camera's aperture and film constant. Shutter speed is affected by light level and can be converted into foot-candles as a measure of illumination. Marchetti found a strong negative correlation between habitat brightness and the number of color patches on male birds, as predicted (Figure 6.14).

To further test her hypothesis, Marchetti manipulated the color patterns of one species, *P. inornatus*, which has two wing bars. She captured males while they were establishing territories before females arrived to breed and assigned them to one of three groups. "Control" birds had their wing bars colored with clear paint, "reduced" birds had their wing bars made smaller with green paint, and "enlarged" birds had their wing bars made larger with yellow paint. Marchetti then released these males and measured the effects of the manipulation on their territory size.

Males who had their wing bars experimentally enlarged obtained the largest territories, while males with reduced color patches ended up with the smallest territories (Figure 6.15). To discount the possibility that wing bar size rather than brightness of the color patch affects territory size, Marchetti conducted one final experiment: she added a novel color patch to the crown of males but did not enlarge their wing bars. These males too obtained the largest territories.

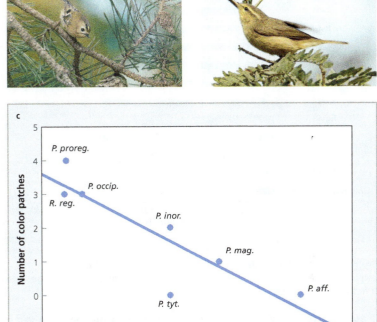

Figure 6.14. Plumage and habitat brightness. Compare (a) *R. regulus*, a species with three bright patches, and (b) *P. affinis*, a species with no bright patches. Each point on the graph (c) represents a species. Birds that live in darker habitats have more bright color patches (*Source:* Marchetti 1993).

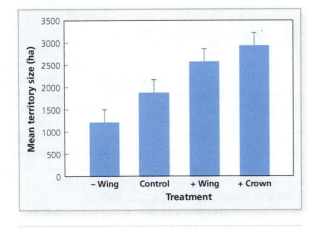

Figure 6.15. Experimental manipulation of plumage. Mean (− SE) territory size for each treatment. Individuals with experimentally reduced wing coloration (−Wing) established smaller territories than controls. Birds with experimentally added wing (+Wing) or crown (+Crown) coloration established the largest territories (*Source:* Marchetti 1993).

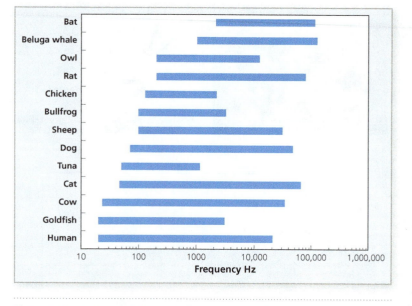

Figure 6.16. Auditory abilities of species. Species differ in the range of frequencies their auditory system can detect. Each bar represents the range of frequencies detected.

■ **auditory receptor** Specialized nerve cells that detect auditory signals.

Marchetti concluded that males rely on color patches as a visual signal in defending territories. Selection appears to have favored the evolution of multiple colorful patches in darker environments as a way to enhance signal transmission. If bright plumage increases signal transmission and aids a male's ability to defend a territory, then why aren't all birds selected to be brighter? One possibility is that increased brightness makes birds more conspicuous and so increases predation risk. We can thus think of the evolution of bright color patches as a trade-off between mating behaviors and predation risk.

Auditory receptors

Auditory signals are mechanical waves that cause the vibration of specialized nerve cells called **auditory receptors**, and, as with visual signals, taxa vary greatly in their ability to detect different frequencies (Figure 6.16). Two properties facilitate the evolution of auditory signals. First, these signals can bypass obstacles: sound waves can travel around objects. That means that auditory signals can sometimes be perceived by receivers when visual signals cannot. Second, auditory signals can be modified rapidly. They can be "turned on and off" depending on conditions (say, when a predator is nearby), and they can be produced at different amplitudes (decibel level) to overcome background noise. This feature makes them more versatile than chemical signals. For example, Jeffrey Cynx and colleagues recorded the songs of individual zebra finches (*Taeniopygia guttata*) while varying the amplitude

APPLYING THE CONCEPTS 6.2

Urban sounds affect signal production

Birds use auditory signals to defend territories and attract mates. Historically, these vocalizations occurred in their native rural habitat, but today, many birds live in noisy urban environments. Hans Slabbekoorn and Ardie den Boer-Visser investigated whether urban environments affect the songs of great tits (*Parus major*) (Slabbekoorn & den Boer-Visser 2006). Great tit songs are composed of two- to four-note phrases of different frequencies. The researchers recorded the songs of birds in ten large European cities and compared them to the songs of birds in matching forests outside each city.

Slabbekoorn and den Boer-Visser found that urban birds produced songs that had a higher minimum frequency and were shorter and faster than the songs of birds living in forests (Figure 1). The change in songs should allow for better transmission of the acoustic signal. The researchers suggest that the most likely explanation for the difference in minimum song frequency is learning: birds learn to modify their songs to compensate for low-frequency urban sounds, such as the noise of cars and trucks. This behavioral plasticity may be a critical element that allows species to adapt to urban life. These results add to a growing body of evidence that human alteration of the environment affects animal communication (Laiolo 2010).

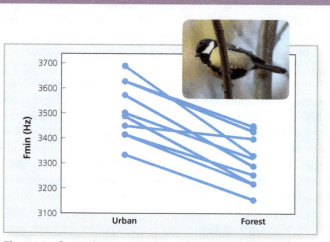

Figure 1. Great tit song. Minimum frequency in songs of ten urban and matched rural populations. Each line connects urban-forest population pairs (*Source:* Slabbekoorn & Den Boer-Visser 2006). *Inset:* great tit.

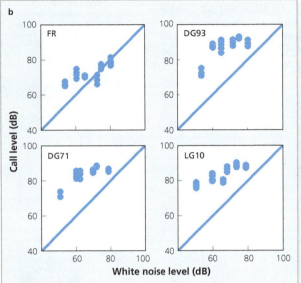

Figure 6.17. Call intensity varies with background noise level. (a) Zebra finches. (b) Individual finches increase the decibel level of their call as white noise decibel levels rise. On the line, the intensity of the call equals that of white noise. Note how most of the points are above the line. Note that each call produced (each point) tends to be louder than the white noise (*Source:* Cynx et al. 1998).

(loudness) of white noise played from a speaker (Cynx et al. 1998). They found that birds produced louder songs when the white noise was louder (Figure 6.17). Auditory signals can be detected at great distances in both air and water, as well as in habitats with different vegetation. Because they can be turned off quickly, they provide excellent and relatively safe long-distance communication. They can also be rapidly modified in response to current conditions (Applying the Concepts 6.2).

Habitat structure and bowerbird vocal signals

Featured Research

To be effective, signals must be able to be detected over background noise: they must stand out from their environment, as was seen in the previously mentioned research on zebra finch song production. We can expect that the environment will affect the evolution of signals, and particularly those that influence the behavior of individuals at some distance. For instance, the flash display of fireflies works as an effective signal over long distances because fireflies are nocturnal, and the luminous signal stands out in the darkness.

Auditory signals can also travel over long distances, but they degrade (lose quality) and attenuate (lose intensity) as they travel; furthermore, the rates of degradation and attenuation are affected by habitat structure. For example, higher frequencies attenuate more rapidly in dense vegetation (Blumenrath & Dabelsteen 2004), which should favor the use of lower frequencies, because selection should favor individuals whose vocalizations are best transmitted through their particular habitat (Morton 1975). James Nicholls and Anne Goldizen tested this hypothesis by comparing advertisement calls of satin bowerbirds (Nicholls & Goldizen 2006).

Satin bowerbirds (*Ptilonorhynchus violaceus*) live in a variety of habitats in eastern Australia, ranging from dense rainforests to open woodlands. Males construct a stick bower on the ground, from which they display to attract mates. They also produce a loud call to attract females, and previous work noted geographical differences in these calls (Tack et al. 2005). Nicholls and Goldizen wondered whether such variation was due to differences in the vegetation structure across populations.

To test the prediction that call frequency should be lower in more densely vegetated habitats, the researchers recorded male advertisement calls from 18 locations that varied in habitat and vegetation structure. They used sonograms to characterize the average call structure at each site, including the minimum frequency, maximum frequency, dominant (or loudest) frequency, and call duration

Figure 6.18. Satin bowerbird call sonograms. (a) Satin bowerbird. (b) Sonograms differ by location and habitat. Note differences in frequencies within the calls from different sites (*Source:* Nicholls & Goldizen 2006).

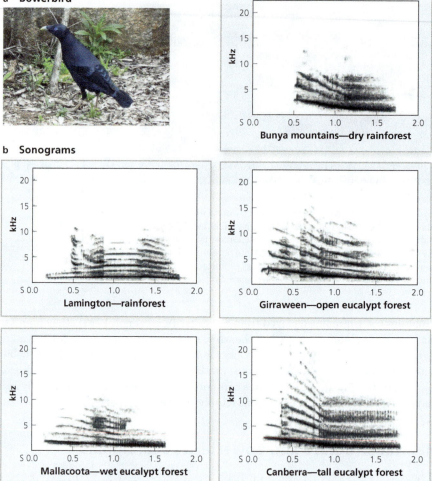

a   **Bowerbird**

b   **Sonograms**

Bunya mountains—dry rainforest

Lamington—rainforest

Girraween—open eucalypt forest

Mallacoota—wet eucalypt forest

Canberra—tall eucalypt forest

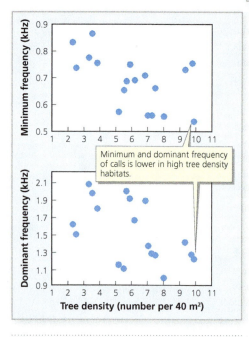

Minimum and dominant frequency of calls is lower in high tree density habitats.

Figure 6.19. Bowerbird call variation and habitat structure. The minimum and dominant frequencies of calls are lower in habitats with higher tree density. Each point represents the mean values for each location (*Source:* Nicholls & Goldizen 2006).

(Figure 6.18). To characterize vegetation, they counted the number of tree stems wider than 5 cm at 1 m above the ground that occurred along four 40 m transects. Bowerbirds vocalize at a height of a few meters off the ground, and stems of this diameter will scatter sounds to attenuate a call (Bradbury & Vehrencamp 1998).

Call structure varied across sites and was related to habitat type. Minimum frequency and dominant frequency were negatively correlated with tree density—that is, they were lower in sites with more trees (Figure 6.19). This finding appears to support the prediction that selection favors vocal signals that minimize attenuation. What produces these population differences? One possibility was that morphological differences across populations, such as body size, could account for this correlation, but after examining the data, the researchers ruled this explanation out. Another possibility is that the differences in calls are associated with genetic differences among populations. However, data indicate that populations in different habitats that produce different calls were not necessarily genetically divergent, and genetically divergent populations often produced similar calls. One final possibility concerns learning: juveniles may simply learn those habitat-specific calls that they hear best. Further work is required to test this hypothesis.

These examples show how the environment can affect the evolution of signals. Recall that the function of communication is to influence the behavior of another individual. As such, the evolution of signals is strongly influenced by the fitness interests of both signaler and receiver. This relationship is the focus of the rest of the chapter.

6.3   Signals can accurately indicate signaler phenotype and environmental conditions

For signals to evolve, they must increase the fitness of the signaler. Because signals affect the behavior of another individual, selection on signals will also be affected by receivers. Receivers should ignore signals that do not enhance their own fitness, as we saw in the discussion of habituation in Chapter 5. As a result, the fitness interests of signalers and receivers may differ, which will affect the evolution of a signal. In particular, divergent interests can lead to two outcomes. In the first, the signal will evolve to become an accurate indicator of signaler phenotype or the environment. In the second, the signal will evolve to become an inaccurate indicator of conditions. We start by examining the former outcome and then address the latter in the next section.

Signals as accurate indicators: theory

Three conditions favor the evolution of signals as accurate, or "honest" indicators (Maynard Smith & Harper 2003). First, the fitness interests of the signaler and the receiver may be similar. Both parties will then benefit from an accurate relationship between the signal and the signaler phenotype or environmental conditions.

Second, we can expect signals to be accurate indicators when they cannot be faked. Suppose, for example, that a signal is a function of body size. In contests over resources or territories, males often produce vocal signals. The frequency of a vocalization, or its pitch, is often negatively related to body size. Only the largest males can produce the lowest-frequency vocalizations, because they have the longest sound-producing structure—such as the larynx in frogs and mammals or the harp in crickets (Clutton-Brock & Albon 1979; Scheuber, Jacot, & Brinkof 2003). In these cases, the ability to produce low-frequency vocalizations is an accurate signal of male size.

Finally, and perhaps most commonly, signals may be accurate indicators because they are costly to produce or maintain (Zahavi 1975; Grafen 1990). For example, the long horns of beetles used in aggressive interactions (Emlen 1994), the wide eye-span of stalk-eyed flies (David et al. 2000), and the wattle size and color of pheasants used to attract mates (Ohlsson et al. 2002) are physiologically costly to produce: only males with access to abundant resources possess the largest, most extreme form of these traits (Figure 6.20). Other signals used in courtship include the bright plumage or body coloration of many birds and fish, and these are costly too. They make the male more visible (Magnhagen 1991) and are therefore found in those individuals most able to avoid predators. Because they are accurate indicators of male phenotype, females frequently prefer mates with extreme signals (see Chapter 11).

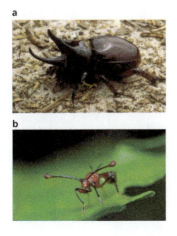

Figure 6.20. Costly traits. (a) Horn length in beetles, (b) eye span in stalk-eyed flies, and (c) wattle size and color (red) in pheasants are examples of traits that are costly to produce: each is larger and/or more intense when individuals have access to better diets.

We illustrate these ideas with several studies of predation, courtship, and aggression. In the first example, the fitness interests of predator and prey are similar. The latter two examples illustrate how the cost of a behavioral signal can lead to reliable indicators of signaler phenotype.

Aposematic coloration in frogs

Some species possess **aposematic coloration**—bright coloring on their bodies that makes them stand out from the environment and indicates that they contain noxious chemicals or poisons that make them unpalatable or dangerous prey. Such coloration is common in insects but is also found in fishes, amphibians, and snakes (Ruxton, Sherratt, & Speed 2005) (Figure 6.21). These species benefit if predators learn to avoid attacking them. Predators, too, benefit by learning such an association (see Chapter 5), because they will then not waste time hunting unpalatable prey.

Ralph Saporito and colleagues conducted an experiment to investigate whether the bright body coloration in the dendrobatid frog (*Oophaga pumilio*) functions as an aposematic signal. This frog commonly lives in leaf litter in the tropics and has a bright reddish-orange dorsal color with blue appendages. Dendrobatid frogs contain skin alkaloids that predators find distasteful (Daly & Myers 1967; Daly, Spande, & Garraffo 2005).

Saporito's team examined predation attempts on clay models molded to resemble the toxic *O. pumilio* and a nontoxic, brown leaf-litter frog similar to frogs of the genus *Craugastor* (Saporito et al. 2007) (Figure 6.22). Each model measured approximately 20 mm in length—the average size of *O. pumilio* in the study region in Costa Rica. To quantify predation, the researchers placed 800 frog models on either the forest floor or a white piece of paper at 5 m intervals along 40 transects. The white paper was used to take into account the effect of cryptic coloration, since half the brown models would then be obvious to predators. After 48 hours, the models were collected. The soft modeling clay retained impressions from predation attempts. Birds left U- or V-shaped attack marks, while mammals left teeth marks.

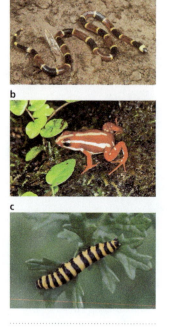

Figure 6.21. Aposematic coloration. (a) Coral snake, (b) poison dart frog, and (c) cinnebar moth caterpillar. All of these species are unpalatable to predators.

Figure 6.22. Clay models and frogs. Toxic *Oophaga pumilio* (a) and its clay model (b). Nontoxic *Craugastor fitzingeri* (c) and its clay model (d).

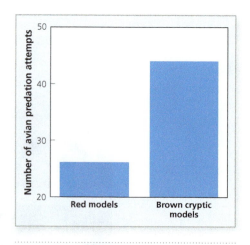

Figure 6.23. Predator attacks on frog models. Brown cryptic models were attacked more often than red (aposematic) models (*Source:* Saporito et al. 2007).

Over 12% of the models were attacked, mostly by birds. Although more models placed on the forest floor were attacked than those placed on white paper, the difference was not significant. However, the brown models were attacked by birds at almost twice the rate of the brightly colored models (Figure 6.23). These data support the hypothesis that bright coloration is a signal to predators that prey are unpalatable. Bright body coloration benefits the signaler when predators associate the colors with distasteful prey, allowing them to learn to avoid such items.

aposematic coloration Brightly colored morphology in a species that stands out from the environment and is associated with noxious chemicals or poisons that make them unpalatable or dangerous prey.

Courtship signaling in spiders

Chad Hoefler and colleagues studied the mating behavior of wolf spiders (*Pardosa milvina*) (Hoefler, Persons, & Rypstra 2008; Hoefler et al. 2009) to determine if males display accurate signals of their quality to females. In the presence of a female (or silk containing female pheromones), males engage in active displays that involve leg raises. Females prefer to mate with males that display the most intense courtship behaviors (Rypstra et al. 2003). Hoefler and his research team hypothesized that males in good condition should perform leg raises at a higher rate than those in poor condition (Scientific Process 6.2).

They tested this prediction by conducting two experiments. In the first, they collected juvenile spiders from the wild and reared them to maturity in the laboratory. Groups were fed according to one of two different regimes for three weeks to produce spiders in both good and poor condition. The research team then randomly paired males and females to produce four test groups in which male-female pairs differed in body condition: both high condition, both low condition, male in high condition and female in low condition, and male in low condition and female in high condition. They placed a female in a cylindrical arena (8 cm high by 19 cm diameter) to deposit silk for one hour and then placed her under a glass vial in the center of the arena. Each male was allowed to court the female and they measured all courtship behavior (number of leg raises) that occurred over a period of five minutes. They did this for three consecutive days.

As predicted, a male's condition affected his behavior. Males in good condition displayed leg raises at higher rates than males in poor condition. In addition, male behavior was consistent over all three days in three of the four test groups. In the group that matched males in good condition with females in poor condition, display rate declined over the three days.

A second experiment examined whether female preference for high-display males resulted in a fitness benefit. Here, the researchers collected a separate group of adult spiders. They divided males into two groups according to display rate and mated them randomly with females. The researchers then recorded the number of spiderlings that emerged from each mating and their subsequent survival in the absence of resources.

Females that mated with males with high display rates had higher fitness: they had more spiderlings than females that mated with males with low display rates, and their spiderlings lived significantly longer. Taken together, these results indicate that male display is an accurate signal of male quality: only males in good condition were able to maintain a high display rate, and display rate was consistent over time. Females obtained fitness benefits by mating with males that displayed at high rates. How can we interpret the behavior of good-condition males that decreased their display rate over time when courting females in poor condition? All males that displayed to a female whose condition was similar to or higher than their own displayed highly repeatable courtship levels over each day, indicating that they are capable of continuing with the high rate of leg raises. Perhaps it was not beneficial for males that perceived a female to be in poor condition to expend the energy to continue a high-intensity courtship. This hypothesis remains to be tested.

SCIENTIFIC PROCESS 6.2

Signaling in male wolf spiders
RESEARCH QUESTION: *Do male wolf spider leg raises provide an accurate signal to females?*

Hypothesis (1): Male condition affects leg-raising rate.

Prediction (1): High-condition males will display at a higher rate than low-condition males.

Hypothesis (2): High display rate in males indicates high quality.

Prediction (2): Females mated to males with high leg display rates will have higher reproductive success.

Methods (1): The researchers:

- collected 60 juvenile male and 60 juvenile female spiders and raised them in the lab for three weeks. Low-condition treatment (LC) spiders were fed one cricket twice per week. High-condition treatment (HC) spiders were fed four crickets twice per week.
- created four mate-pairing test groups:
  (1) HC male with HC female,
  (2) HC male with LC female,
  (3) LC male with HC female, and
  (4) LC male with LC female.
- measured male courtship rate.
- repeated the trial for the next three days using the same individuals to test for consistency.

Results (1): HC males displayed at higher rates than LC males. Display rate was consistent across the sample, except for HC males mated to LC females.

Methods (2): The researchers:

- mated females with males in two groups—males ($n = 21$) with high leg display rates (5.9 leg raises/min) and males ($n = 36$) with low leg display rates (3.0 leg raises/min)—and measured the number of spiderlings and their survival.

Results (2): Females mated to high-display-rate males had more spiderlings and higher offspring survival than females mated to low-display-rate males.

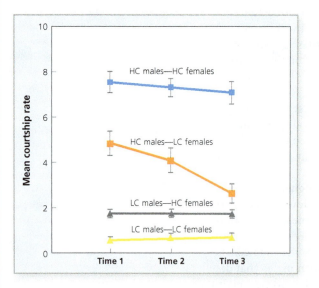

Figure 1. Courtship. Mean (± SE) courtship rate over trials for each treatment (*Source:* Hoefler et al. 2009).

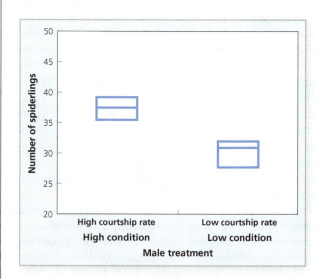

Figure 2. Reproductive success. Reproductive success of females mated to males that differ in courtship rate. Each box represents the middle 50% of the data. The median is each horizontal line (*Source*: Hoefler et al. 2009).

Conclusion: Mate condition affects male courtship display rate and is an accurate signal of male quality and fitness.

Aggressive display and male condition in fighting fish

Accurate signaling is common in a different context as well: aggressive interactions. As we see next, the cost of signaling also plays a role in selecting for accuracy.

Many animals, including fish, engage in aggressive contests over food and mates (see Chapter 10). When two fish fight, they often begin by engaging in a head-on opercular flare—an encounter in which the opercula, the bony flaps that cover the gills, flare out at the opponent. The frequency and duration of the display increase during a contest, and both are strongly correlated with winning: individuals that have high flare rates and long flare durations typically win.

Mark Abrahams, Tonia Robb, and James Hare hypothesized that the opercular flare display is an accurate signal of fighting condition because it is costly (Abrahams, Robb, & Hare 2005). The opercula play an important role in respiration. For the gills to function properly in gas exchange, they must be adequately ventilated by moving water. In calm water, ventilation is accomplished through opercular movements. Fish engaged in long flare displays in calm water must therefore sacrifice normal flapping movement and consequently water ventilation. Abrahams and colleagues hypothesized that only individuals in the best physiological condition would be able to sustain long opercular displays because of this cost. To test this hypothesis, they measured the ability of fish to maintain long opercular displays in normoxic (normal oxygen level) and hypoxic (low oxygen level) water. They predicted that flare displays would be significantly shorter under hypoxic conditions (Scientific Process 6.3).

The team examined the behavior of a popular freshwater species, Siamese fighting fish (*Betta splendens*), known for its aggressive behavior. Fish were exposed to four experimental treatments that varied in both dissolved oxygen level (normal or hypoxic) and the presence or absence of an intruder. To simulate an intruder, the researchers placed a mirror behind a removable opaque partition at one end of the tank. Fish appear to perceive their image in a mirror as an intruder and will display aggressively toward it.

Fish did not display when there was no "intruder" but readily displayed to the "intruder" that appeared in the mirror. In addition, opercular display behavior was significantly lower in hypoxic than it was in normoxic conditions, exactly as predicted. Abrahams and colleagues concluded that opercular displays are accurate signals of physiological condition because they are costly. They suggested that opercular displays cannot be faked: a fish flaring its operculum is analogous to a terrestrial vertebrate holding its breath. Only individuals in good physiological condition can maintain long opercular flare displays.

These examples show how, and under what conditions, signals can be accurate indicators of signaler phenotype or environmental conditions. However, not all signals are accurate indicators of phenotypes or conditions, as we see next.

6.4 Signals can be inaccurate indicators when the fitness interests of signaler and receiver differ

Not all communication involves accurate signals. What happens when the fitness interests of signaler and receiver conflict, as in our firefly example at the beginning of this chapter? The signaler can benefit by producing an inaccurate, or "dishonest," signal. The receiver that is able to discriminate accurate from inaccurate signals will also benefit. This produces a co-evolutionary arms race (Dawkins & Krebs 1979) between signalers and receivers: signalers are selected to produce signals that affect the behavior of receivers (e.g., by producing inaccurate signals that are difficult for receivers to discern), and receivers are selected for better discrimination abilities (Rice & Holland 1997; Holland & Rice 1998; Lozano 2009). Who wins in such a "race"? Neither. The evolution of the signal trait and its effect on receivers is an ongoing process: it is influenced by the fitness benefits and costs of signal accuracy in the signaler

SCIENTIFIC PROCESS 6.3

Fighting fish opercular display

RESEARCH QUESTION: *Does opercular display behavior vary with the condition of a fish?*

Hypothesis: Opercular displays are difficult to perform at high rates because they reduce water ventilation over the gills.

Prediction: Males in the best physiological condition will perform opercular displays at a higher rate than individuals in poorer condition.

Methods: The researchers:

- manipulated males by placing individuals in water that differed in oxygen level: normoxic water (at least 4 ppm dissolved oxygen) allowed fish to remain non-stressed; hypoxic water (3 ppm dissolved oxygen), created by bubbling nitrogen into the water, produced stressed individuals.
- recorded fish display behavior in ten-minute trials for two treatments: with or without an intruder (represented by a mirror image in the test tank).
- recorded the percentage of time a fish spent displaying (operculum flared out) over the course of 19 trials.

Results: Fish in normoxic water displayed significantly more than fish in hypoxic water.

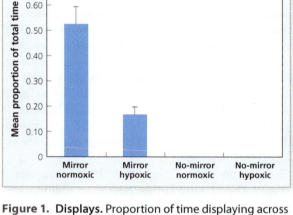

Figure 1. Displays. Proportion of time displaying across treatments (*Source*: Abrahams, Robb, & Hare 2005).

Conclusion: Only males in good physiological condition can display at a high rate. Opercular displays are an accurate signal of male condition.

---

and discrimination ability in the receiver (Krebs & Dawkins 1984; Dawkins & Guilford 1991; Johnstone & Grafen 1993; Rowell, Ellner, & Reeve 2006).

Inaccurate signals are common in interspecific signaling, such as between prey and predators. Here, the fitness interests of the species differ radically: prey fitness is enhanced by avoiding predation, while the predator benefits from consuming the prey. One well-known case of inaccurate signaling is mimicry—the adaptive resemblance of one species (the mimic) to another (the model), so that a third species (the receiver) is duped. In **Batesian mimicry** (Bates 1862), a palatable mimic resembles an unpalatable model that predators have learned to avoid. The predator perceives an inaccurate signal from the mimic and so does not attack it. In this case, the relative fitness benefit of the signaler favors the evolution of a signal that is difficult to discern. A poor signal that is easy to discern will lead to the death of the signaler, but an effective signal will only result in the loss of a meal for the predator. A second form of mimicry is **aggressive mimicry**, in which a predator mimics a nonthreatening model species in order to gain access to food. Again, the receiver is duped, but in this case, it is attacked by the aggressive mimic (as we saw in the *Photuris/Photinus* fireflies). Let's look at an example of each.

Batesian mimicry A palatable mimic resembles an unpalatable model that predators have learned to avoid.

aggressive mimicry A situation in which a predator mimics a non-threatening model.

Batesian mimicry and Ensatina salamanders

As we saw earlier, aposematism, or bright warning coloration, is commonly found in species with toxins. This signal evolved to convey accurate information to predators about the unpalatability of a prey species, and both signaler and receiver benefit from it. However, this signaling system is also ripe for "deceit," or inaccurate signals. Individuals of a third species that does not produce a toxin, the Batesian mimic, benefit if they resemble the toxic model species.

Batesian mimicry is found in many animal taxa. Shawn Kuchta, Alan Krakauer, and Barry Sinervo examined mimicry in Ensatina salamanders (*Ensatina eschscholtzii*), which live throughout central California (Kuchta, Krakauer, & Sinervo 2008). Most subspecies are cryptically colored, but one, the yellow-eyed salamander (*Ensatina eschscholtzii xanthoptica*), is brightly colored. As its name suggests, it has yellow eyes, and its legs and ventral region are bright orange. Is it a Batesian mimic? One hypothesis is that it mimics aposematic toxic newts in the genus *Taricha* (Stebbins 1949) (Figure 6.24). *Taricha* newts and the yellow-eyed salamander co-occur geographically, and both possess an orange ventral region and bold yellow patches on the iris.

To test this hypothesis, the research team conducted a series of feeding trials with western scrub jays (*Aphelocoma californica*). Scrub jays feed on a variety of insects and small terrestrial vertebrates, including salamanders. In the experiment, captive jays were housed individually in a large aviary. On the first day, all jays were presented with an edible salamander (*Batrachoseps attenuatus* or *B. luciae*) to ensure that they would eat salamanders. The next day, all jays were presented with a toxic newt (*Taricha torosa*). On the following day, each jay was presented with either a yellow-eyed salamander (mimic) or a palatable Ensatina subspecies (*E. e. oregonensis*). On the final day of the experiment, each jay was given the other subspecies of Ensatina. All salamander prey offered to the jays were first euthanized to eliminate behavioral differences and thus force jays to make feeding decisions based on morphology alone. Prey were presented in a lifelike posture in the middle of the aviary.

The researchers found that seven of the ten jays did not touch the toxic model. Three jays briefly poked at it but did not attempt to consume it. These ten jays had apparently learned to avoid the toxic prey in the wild. However, the behavior of the jays differed with respect to the two salamander subspecies. All ten birds contacted the palatable Ensatina subspecies (*E. e. oregonensis*) more quickly than they contacted the yellow-eyed salamander. The yellow-eyed salamanders were not even touched by three of the jays (Figure 6.25). They were consumed in only half the trials, whereas the palatable Ensatina subspecies (*E. e. oregonensis*) was consumed in nine out of ten trials. The researchers concluded that the yellow-eyed salamander effectively mimics the coloration of the toxic newt *T. torosa*, thereby obtaining a fitness

Figure 6.24. Salamander mimicry. Ensatina salamander mimic (right) and its toxic model (left). Note their yellow eye coloration and orange ventral surface.

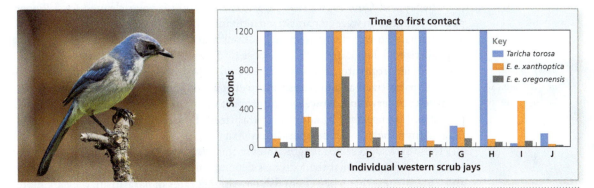

Figure 6.25. Jay attacks on salamander models and mimics. Time to first contact with models for ten birds. Scrub jays rarely contacted the toxic model (blue). Jays took longer to contact the mimic Ensatina salamander (orange) than they did to contact the cryptic salamander (gray) (*Source:* Kuchta, Krakauer, & Sinervo 2008).

benefit. The mimicry was not perfect: jays consumed more salamander mimics than they did toxic models. Nonetheless, attack intensity on mimics was less than on non-mimics, so we can see how selection would favor the evolution of Batesian mimicry as an inaccurate morphological signal.

Featured Research

Aggressive mimicry in fangblenny fish

Recall that in aggressive mimicry, a predator mimics a nonthreatening model so that it can closely approach its prey. Karen Cheney and Isabelle Côté examined aggressive mimicry in a coral reef system in Australia (Cheney & Côté 2007). Ectoparasites, such as gnathiid isopod larvae, infest many reef fish (Grutter 1994). These ectoparasites are also a food source for **cleaner fish**—small fish that feed on parasites and the dead skin of larger fish. In essence, the cleaner fish performs a service to the parasitized "client" fish, creating a mutualistic relationship. However, when client fish have few parasites, cleaner fish may feed on healthy tissue and can harm clients (Cheney & Côté 2005), so the mutualism depends on how heavily the client fish is parasitized. The juvenile bluestreak cleaner wrasse (*Labroides dimidiatus*) is a common cleaner fish found throughout the Indo-Pacific. Another species, the bluestriped fangblenny (*Plagiotremus rhinorhyncos*), closely resembles the bluestreak wrasse but consumes only the healthy scales, mucus, and dermal tissue of client fish. Both species have black bodies with neon blue stripes (Figure 6.26). Given the close morphological resemblance between the two, Cheney and Côté tested the hypothesis that fangblennies are aggressive mimics of juvenile cleaner fish. Clients with high parasite loads, they predicted, should seek interactions with cleaners and so may not be as vigilant about discriminating a cleaner from a fangblenny. On the other hand, clients with low parasite loads have less to gain by interacting with cleaners and so should be more vigilant to deception.

Staghorn damselfish (*Amblyglyphidodon curacao*) (Figure 6.27), a common reef client fish, were collected from a site in the Great Barrier Reef along with juvenile bluestreak cleaner fish and fangblennies. The experiment consisted of a treatment (damselfish parasitized) and a control (damselfish not parasitized). To parasitize damselfish, a fish was confined to a small section of the tank where approximately ten unfed gnathiid isopods were released. On average, approximately six isopods attached themselves to the fish, a typical parasite load for *A. curacao*. Unparasitized fish were free of isopods. During each trial, all three fish were allowed to interact for 15 minutes, and the researchers recorded the amount of time damselfish spent being cleaned by the cleaner fish, which fish (cleaner or client) terminated the interaction, the number of times the damselfish avoided an interaction with the cleaner fish, the number of successful attacks by the fangblenny on the damselfish, and whether the damselfish aggressively chased the fangblenny or cleaner fish.

Cheney and Côté found that parasitized damselfish were cleaned for a longer period of time than were unparasitized individuals. In addition, parasitized individuals terminated fewer cleaner fish interactions and avoided cleaner fish less than did unparasitized fish (Figure 6.28). All these findings indicate that parasitized damselfish benefited from interactions with cleaner fish, just as expected. Fangblennies also attacked parasitized and unparasitized fish at a similar average rate of 4.5 times per 15-minute observation. However, the proportion of successful attacks was higher on parasitized damselfish. In addition, parasitized damselfish retaliated less often to fangblenny attack than did unparasitized fish (Figure 6.28).

These data suggest that fangblennies in this system are aggressive mimics of juvenile cleaner fish, and that their success depends on the state of a client fish. Cheney and Côté suggest that when client fish have higher rates of parasitism, they are more likely to seek out interactions with cleaner fish. Because of the benefits of removing parasites, they are willing to incur the costs of mimics and so are less vigilant. This allows aggressive mimics like fangblennies to be more successful.

■ **cleaner fish** Fish that feed on ectoparasites and the dead skin of other fish.

a

b

Figure 6.26. Cleaner and mimic. This figure illustrates the close resemblance between the fangblenny mimic (a) and bluestreak cleaner (b) model.

Figure 6.27. Staghorn damselfish. A common reef fish in the western Pacific.

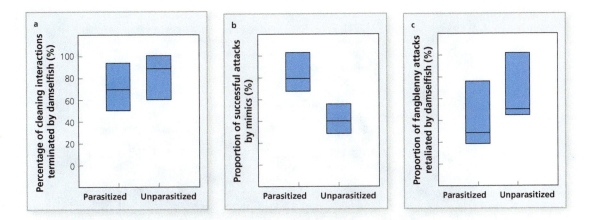

Figure 6.28. Interaction between damselfish and cleaner fish. Parasitized damselfish (a) terminated fewer interactions with cleaner fish, (b) were attacked more successfully by fangblennies, and (c) initiated fewer retaliatory chases than did unparasitized damselfish. In each plot, the median is represented by the horizontal line and the box represents the middle 50% of the data (*Source:* Cheney & Côté 2007).

Intraspecific deception: false alarm calls

So far, our examples have involved mimicry and interspecific "deception," or inaccurate signaling between species. Within-species, or intraspecific, deception also occurs but is less common for two reasons. First, intraspecific interactions are more common, and so individuals can learn to ignore inaccurate signals faster (Dawkins & Guilford 1991; Johnstone & Grafen 1993). We can say that inaccurate signals are subject to negative frequency-dependent selection, which keeps them rare, as we discussed in Chapter 2. Second, natural selection will favor receivers that can discriminate accurate from inaccurate signals. For instance, we have seen that male quality is often associated with a morphological or behavioral signal. Since low-quality males can benefit by providing an inaccurate signal, how do females avoid deception? They do so by using signals in males that are costly to produce or difficult to fake (Zahavi 1975; Grafen 1990). Inaccurate signals will then yield low fitness, because they will be ignored.

One type of intraspecific signal that involves deception is the production of false alarm calls—alarm calls that are produced when no threat is nearby. Let's look at two examples. In each case, the caller produces an inexpensive inaccurate signal and then takes advantage of the receiver's behavior.

Topi antelope false alarm calls

Featured Research

Topi (*Damaliscus lunatus*) are large antelope (weighing 75 to 150 kg) that live in the open savannahs of sub-Saharan Africa. When topi spot a predator—such as a lion or even humans—they emit snort vocalizations and then stare in the direction of the predator with their ears pricked up (Figure 6.29). Alarm snorts likely function to inform a predator that it has been spotted, because many predators can successfully capture prey only when they have the element of surprise (see Chapter 8).

During the breeding season, males defend individual mating territories from other males. Herds of nonterritorial females roam through male territories selecting mates during their one-day estrus (the sexually receptive, fertile period). A female will visit many territories and mate multiple times with several males—a behavior that can be used to determine when a female is fertile.

Jakob Bro-Jorgensen and Wiline Pangle noticed that males frequently gave alarm calls during visits by estrus females, even when no predators appeared to be nearby (Bro-Jorgensen & Pangle 2010). They proposed that these false alarms "deceive" a female into staying on a male's territory so that he can obtain additional matings. This **sexual deception hypothesis** predicts that males will give false alarm calls only when an estrus female is in their territory and only when she is trying to leave.

sexual deception hypothesis
Males will produce deceptive signals to females in order to enhance their own reproduction.

Figure 6.29. Topi. A male with a female in his territory gives an alarm snort and looks off to the distance.

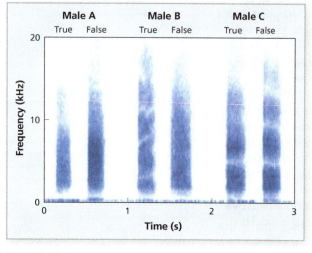

Figure 6.30. Topi alarm call spectrogram. True and false alarm calls for three males. Note their similarity (*Source:* Bro-Jorgensen & Pangle 2010).

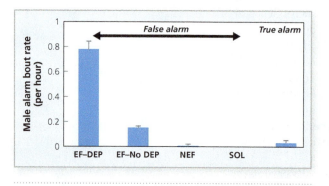

Figure 6.31. Male alarm call rate. Mean (+ SE) alarm call rates. Males give more false alarms when visited by estrus females (*Source:* Bro-Jorgensen & Pangle 2010). EF-DEP is estrous female attempting departure; EF-No DEP is estrous female not attempting departure; NEF is non-estrous female; SOL is solitary female.

To test this hypothesis, the researchers observed the behavior of 53 estrus and 20 non-estrus females between 2005 and 2009 in Masai Mara National Reserve in Kenya. Individuals were identified by their coat color and marking patterns. The researchers recorded the territorial location of the female, all matings, alarm snorts, and departure attempts (i.e., when a female attempted to leave a male's territory). Following each snort, the researchers scanned the area for predators. Predators, if present, were easy to locate because the habitat was open and topi tend to stare at a predator following a snort alarm. The researchers also recorded all alarm snorts to determine whether true alarm snorts—produced when a predator was actually present—and false alarm snorts differed in structure.

They found that alarm snorts given in the presence of predators were acoustically identical to false alarm snorts given in the absence of predators (Figure 6.30). The researchers observed that false alarms were most likely to be given when an estrus rather than a non-estrus female was in a male's territory, and especially when she was trying to depart the territory (Figure 6.31). Males often gave false snorts from the edge of their territory. After hearing a snort, females moved away from the male, which in turn tended to move her back toward the center of his territory. Following false alarm snorts, a male secured an average of 2.8 additional matings before the female left his territory.

From these results, the researchers concluded that male topi give inaccurate alarm signals to keep estrus females from leaving their territory. Females apparently cannot distinguish true from false alarms and so move away from the apparent danger—which leads them back to the center of the deceiver's territory. The male benefits by acquiring additional matings and siring more offspring (e.g., Dziuk 1996). Why do females respond to this inaccurate signal? Although the cost of producing the inaccurate signal is quite low for males, the cost to the female of ignoring an alarm signal can be very high.

Capuchin monkeys and inaccurate signals

In many social groups, dominant individuals—who may be older or larger than others—have preferential access to resources (see Chapter 10). Dominant individuals will often feed first, and if food is limited and easily defendable, they may prevent subordinates from feeding. Can inaccurate alarm calls allow subordinates to gain access to food? If so, then we can make several predictions:

1. These calls should be produced by subordinates and not by dominant individuals, who already have free access to resources.
2. The calls should be given when food is clumped rather than dispersed, because it is more difficult for subordinates to gain access to clumped resources.
3. The calls should be given when a caller is close enough to food to take advantage of the distraction.
4. The calls should benefit the caller by eliciting antipredator behavior in dominants, such as moving to a safer location (and most likely away from food).

Brandon Wheeler tested these predictions on a population of tufted capuchin monkeys (*Cebus apella nigritus*) in Iguazú National Park in Argentina (Wheeler 2009). Capuchins are medium-sized frugivorous (fruit-eating) primates that live in social groups of up to several dozen individuals (Figure 6.32). Within a group, dominant and subordinate individuals can be identified by their interactions (see Chapter 14). Wheeler studied a single group of about two dozen members over a period of 18 months. Individuals were identified by unique facial marks. Wheeler focused on the production of antipredator "hiccup" calls. Two or more of these indicate a high level of danger and often elicit antipredator behavior, especially "look" and "escape" reactions, in which receivers scan the environment and flee to a safer location.

Wheeler established food patches consisting of one to six platforms suspended in tree branches several meters off the ground—a typical feeding location for capuchins (Figure 6.33). Bananas, a preferred food, were placed on the platforms at the start of each day. The amount and distribution of food was manipulated. When only a few platforms

Figure 6.32. Tufted capuchin. Capuchins live in social groups.

Figure 6.33. Feeding platform. The feeding platform used to examine capuchin behavior.

were present, the food was clumped and thus easier to defend; when more platforms were included, the food was more dispersed and thus harder to defend.

At the start of each day, food was placed on the platforms. Upon arrival of the group, the researchers selected a focal individual and collected data using continuous sampling. The observers recorded all hiccup calls produced by the focal animal as well as its location with respect to the food platforms when the call was produced. In addition, the observer recorded the position of the focal individual every 30 seconds (less than 2 m from the platform or farther than 2 m from the platform). Hiccup calls were classified as accurate if a predator was observed nearby and inaccurate if no predator was observed.

Wheeler observed the behavior of capuchins at 321 feeding platform trials and collected data from 499 individual focal animal observations over a total of 45 hours. While only 14 individuals produced 25 inaccurate alarm calls, 24 of those were produced by a subordinate. Most were produced when the food patch consisted of no more than three platforms (i.e., when food was clumped); few were given when the patch consisted of five or six platforms (Figure 6.34). Individuals were almost always (85% of the time) within 2 m of the food patch when an inaccurate alarm call was emitted, and the false alarms produced antipredator behavior in 40% of receivers (Figure 6.35). The caller often benefited from the "deception," because in most cases the receiver abandoned the food platform, allowing the caller to feed.

These results are consistent with each of the researcher's predictions and imply that inaccurate alarm calls are produced to influence others to benefit the caller. Subordinate capuchins produce inaccurate alarm calls to distract others so that they can gain access to clumped food. Note, however, that such calls were given sparingly, probably because of frequency-dependent selection: subordinates who produce them too often will eventually be ignored.

Our discussion of signals to this point has largely assumed that communication occurs between one signaler and one receiver. In many situations, however, other individuals are nearby and so can intercept signals that are produced. We examine the implications of this in the last section of the chapter.

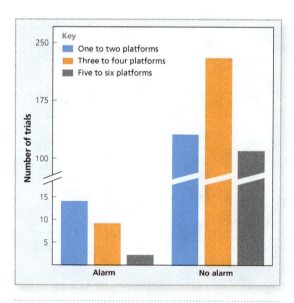

Figure 6.34. Alarm calls vary with food distribution. False alarms were given more often when food was clumped in one or two platforms rather than when it was more evenly distributed across five or six platforms (*Source:* Wheeler 2009).

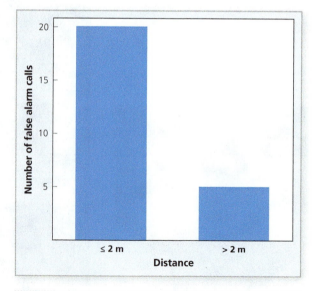

Figure 6.35. Alarm calls vary with distance to food. False alarm calls are most often given when a capuchin is less than 2 m from food (*Source:* Wheeler 2009).

6.5 Communication networks affect signaler and receiver behavior

While the simplest communication system involves two individuals, signals can also be intercepted by others. These **bystanders**, or **eavesdroppers**, are present but do not take part in the signaling (Matos & Schlupp 2005). Eavesdroppers can benefit by learning about the presence of competitors or predators. In turn, the presence of bystanders can influence the behavior of signalers, a phenomenon known as an **audience effect** (Zuberbühler 2008). We examine such communication networks in the context of foraging, predation, and aggressive interactions.

bystander (eavesdropper) A third-party individual that detects a signal transmitted between a signaler and a receiver. Also called an eavesdropper.

audience effect Occurs when the presence of bystanders influences the behavior of a signaler.

Squirrel eavesdropping

Featured Research

In the autumn and early winter, Eastern gray squirrels (*Sciurus carolinensis*) frequently cache (or store) food items like nuts for later retrieval. Food caches, however, are subject to pilfering: conspecifics or heterospecifics can steal food. In the deciduous forests of eastern North America, squirrels live with many potential cache pilferers, including other squirrels and birds. Blue jays (*Cyanocitta cristata*) also make food caches and retrieve them by remembering visual landmarks around the cache location (Bednekoff & Kotrschal 2002). Given that these species both cache and live in the same area, Kenneth Schmidt and Richard Ostfeld wondered whether squirrels might modify their behavior to avoid losing their food caches. A jay that observed a squirrel caching food could easily remember its location and return later to pilfer the contents. The researchers hypothesized that squirrels might alter their feeding preferences for different food items depending on whether jays were nearby or too far away to watch. They predicted that when jays are nearby, squirrels should reduce their preference for feeding on cacheable food items, which could be stolen after being cached.

To test this prediction, Schmidt and Ostfeld established pairs of food patches in eastern New York in early winter (Schmidt & Ostfeld 2008). Each food patch was filled with pea gravel and 15 hazelnuts. One patch contained unshelled hazelnuts, while the adjacent tray contained shelled hazelnuts. Squirrels frequently cache intact hazelnuts but rarely cache nuts without shells, presumably because unshelled nuts cached in the ground rot quickly. Food items were mixed into the pea gravel, forcing squirrels to dig through the gravel to find food. The first few nuts were relatively easy to find, but food discovery then became more and more difficult. This design ensured that a squirrel would leave a patch before all the food was found, because the last few items would be very difficult to acquire. The number of food items left, called the giving-up density, indicated how much effort the squirrel allocated to feeding in that patch (see Chapter 7).

Jays use vocalizations to communicate with conspecifics to defend territories and attract mates. Can squirrels use jay vocalizations to assess the threat of pilfering? Schmidt and Ostfeld recorded the amount of nuts left in each patch for three experimental treatments: (1) blue jay vocalizations from nearby speakers (less than 25 m away), (2) blue jay vocalizations from far-away speakers (greater than 125 m away), and (3) vocalizations of other common nonpilfering birds, such as northern cardinals (*Cardinalis cardinalis*) (control group). Food patches were available to the squirrels all day. At the end of each day, the researchers collected all the nuts that remained in each tray.

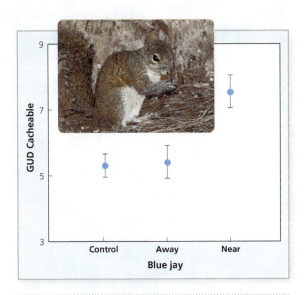

Figure 6.36. Giving-up densities of cacheable food. Mean (± SE) giving-up density (GUD) of cacheable food items for each treatment. Squirrels left more food items (higher GUDs) of cacheable nuts when jay vocalizations were played nearby than they did when vocalizations were played far away or for controls (*Source:* Schmidt & Ostfeld 2008).

In each case, squirrels consumed the same number of noncacheable nuts per food patch. However, significantly more cacheable nuts were left in the patch when jay vocalizations were played nearby (Figure 6.36). These results suggest that squirrels eavesdrop on heterospecific blue jay vocalizations to learn their presence and location. When jays appeared to be nearby, squirrels reduced the amount of effort they devoted to acquiring cacheable food, presumably because it was more likely to be pilfered if cached.

Eavesdropping in túngara frogs

During the breeding season, several species of frogs produce vocalizations to attract mates. However, predators can more easily locate a calling male, too. For example, in the tropics, predatory bats are known to identify the location of túngara frogs (*Physalaemus pustulosus*) (Figure 6.37) using their vocalizations. In a large pond, multiple frog species often start and stop calling synchronously. It is relatively easy to understand why they all might stop simultaneously: a predator is nearby. Why should they all start simultaneously?

Steven Phelps, A. Stanley Rand, and Michael Ryan proposed that males eavesdrop on one another to assess predation risk (Phelps, Rand, & Ryan 2007). Túngara frogs frequently live sympatrically (co-occur) with *Leptodactylus labialis* frogs and together produce a mixed-species chorus at breeding sites. Can males use the vocalizations of others to assess the level of predation risk?

Phelps and colleagues captured male frogs in Panama and housed them in the laboratory for two days before testing. On the evening of testing, a male was placed in a small container. The test began with a ten-minute acclimation period of conspecific calls. Playbacks continued for the next five minutes, and the number of calls produced by the subject was recorded. At Minute 16, the playback stopped, and a small disc similar in size and shape to that of a predatory bat was passed over the subject to elicit antipredator behavior (flattening the body, freezing). From Minutes 17 to 25, one of four treatments was presented: (1) silence; (2) a repeated conspecific call; (3) a repeated call of sympatric *L. labialis*; or (4) a repeated call of an allopatric (geographically isolated) frog, *Physalaemus enesefae*, that has a similar

Figure 6.37. Túngara frog. A male calls to attract females. These frogs are common in Central America.

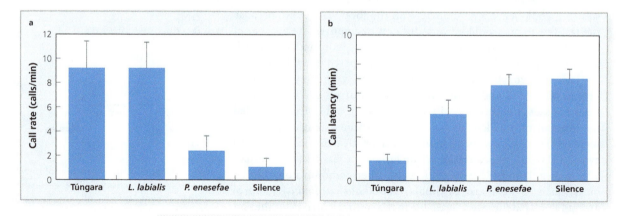

Figure 6.38. Frog calls. Mean (+ SE) calling behavior. After a simulated attack, frogs (a) produced calls at a higher rate and (b) had a lower latency to call when hearing conspecific calls and the calls of a heterospecific sympatric species (*Source:* Phelps, Rand & Ryan 2007).

call to that of the túngara frog. If túngara frogs eavesdrop on calls to assess predation risk after a simulated attack, the researchers predicted that they should respond most readily to conspecific calls, followed in order by the calls of sympatric *L. labialis*, the calls of allopatric *P. enesefae,* and silence. The researchers measured both the latency to begin calling and the call rate.

Male túngara frogs responded as predicted. They had the shortest latency to calling and the highest call rate in response to túngara frog calls, followed by the calls of *L. labialis*, the calls of *P. enesefae*, and silence (Figure 6.38). Indeed, there was no significant difference in their call rate or latency to call in response to the call of *P. enesefae* and silence, indicating that the calls of *P. enesefae* were essentially ignored. The researchers concluded that túngara frogs eavesdrop on both conspecific and sympatric heterospecifics to assess predation risk. They propose that this behavior may be one reason why many frogs participate in mixed-species choruses when courting females.

Audience effects in fighting fish

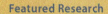

In the first two cases, we saw how receiver behavior was affected by eavesdropping within a communication network. Signalers are also affected by the presence of bystanders, as demonstrated in the audience effect we see next.

Claire Doutrelant, Peter McGregor, and Riu Oliveira investigated how a bystander affects aggressive behavior in male fighting fish (*Betta splendens*) (Doutrelant, McGregor, & Oliveira 2001). As we discussed earlier in this chapter, male fighting fish are highly aggressive and compete for territories and mates. Aggressive displays include flaring of the gill covers, as well as tail beats and bites directed at an opponent.

The researchers designed a simple experiment that examined the aggressive behaviors of a focal male as it interacted with a rival male with either a bystander present or absent. The focal male and its rival, closely matched for size and color, were separated by a removable opaque partition. A second tank was placed next to the focal tank and contained a female bystander, a male bystander, or no bystander (Figure 6.39). When the partition was raised, the focal male was allowed to interact with the rival male for a ten-minute session, after which the partition was replaced. After two hours, the partition was raised again for another ten-minute interaction session. In one session, a bystander was present, while in the second, none was present, with the order randomized across focal males. For trials with a bystander, the focal male was allowed to view the bystander for three minutes before the partition was raised.

Doutrelant and colleagues found that focal males displayed more tail beats and spent more time with erect gill covers, but directed fewer bites at their opponent when a female bystander was nearby, compared to when there was no bystander (Figure 6.40). In contrast, the presence of a male bystander did not affect male aggressive display behavior.

These results demonstrate that signaling behavior can be affected by the presence of an audience. Males used more conspicuous displays (erect gill covers, tail beats) but less overt aggressive signals (bites) in a contest with a rival male when a female was present. The researchers suggest that males were attempting to simultaneously display to the female while interacting with a rival and so used more conspicuous behavior. In addition, they speculate that females may also be less attracted to highly aggressive males. Interestingly, no such audience effect was observed when the bystander was a male, suggesting that third-party males are ignored.

Many animals live in complex social environments surrounded by both conspecifics and heterospecifics. This variation creates the opportunity for individuals to acquire information from others by intercepting their signals but also can affect the production of signals. Much work has focused on understanding how and when such audience effects influence signaling behavior.

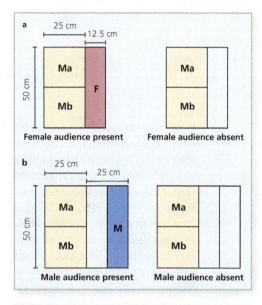

Figure 6.39. Experimental design. Ma and Mb represent the locations of a focal male and a rival (position randomized across trials). F and M represent the location of a (a) female and (b) male bystander, respectively (*Source:* Doutrelant, McGregor, & Oliveira 2001).

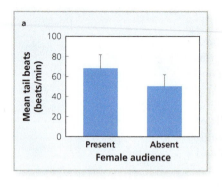
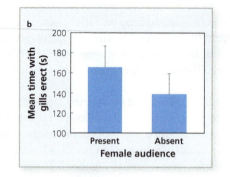
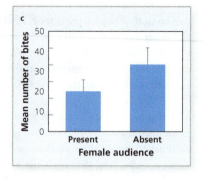

Figure 6.40. Female audience affects male behavior. Mean (+ SE) aggressive behaviors of treatment males. Male aggressive behavior was affected by the presence of a female audience (*Source:* Doutrelant, McGregor, & Oliveira 2001).

Chapter Summary and Beyond

Communication occurs when an individual uses a specialized signal to influence the behavior of another. Signals affect behavior as if they contain information about signaler phenotype or the environment and include specialized morphology, vocalizations, behavior, physiological traits, and chemical pheromones. Recent work has demonstrated that some signals affect behavior as if they conveyed information about individual identity (e.g., Tibbets & Dale 2007). Visual, auditory, and chemical signals are perceived by sensory receptors in different sensory systems. To be effective, a signal must stand out from background noise so that it can be perceived. Therefore, selection favors signal characteristics that minimize degradation and attenuation in different habitats (e.g., Peters, Hemmi, & Zeil 2007; How et al. 2008).

The fitness interests of signalers and receivers affect the evolution of signals. Signals will be accurate indicators of conditions when the fitness interests of signaler and receiver coincide, as occurs in aposematic coloration, or when signals are difficult to fake or costly, as occurs for many signals used in courtship and aggressive interactions. On the other hand, selection can favor the evolution of signals that are inaccurate indicators of conditions when the fitness interests of two individuals differ. This creates a co-evolutionary process that affects the evolution of both the signaler and receiver. Mimicry in predator-prey systems is a common example of inaccurate signaling. In Batesian mimicry, a palatable species resembles an unpalatable species and a predator is duped.

Ruxton, Sherratt, and Speed (2005) provide a comprehensive review of Batesian mimicry. In aggressive mimicry, a predator resembles a nonthreatening species and a prey species is duped, as occurs with cleaner fish and their aggressive mimics. Recent research has examined how client fish find and recognize cleaners. For example, Cheney et al. (2009) showed that the typical blue and yellow color patterns of cleaner fish contrast sharply against a background of coral reef and blue water, providing a conspicuous signal to a wide variety of client fish species. Inaccurate signaling within a species is relatively rare, because such signals must be cheap to produce and are subject to negative frequency-dependent selection. False alarm calls are one example of inaccurate signals. Individuals that produce a false alarm call can benefit from increased matings or feeding.

In communication networks, signals can be intercepted by bystanders. Such eavesdroppers learn about others and can modify their behavior accordingly. In addition, when bystanders are present, audience effects are common, as individual signalers modify their behavior. Recently, Semple, Gerald, and Suggs (2009) observed how bystanders affect mother-infant interactions in rhesus macaques (*Macaca mulatta*), and Rowell, Ellner, and Reeve (2006) examined how communication networks can affect the evolution of accurate and inaccurate signals. McGregor (2005) provides a comprehensive summary of recent work on both eavesdropping and audience effects in animal communication networks.

CHAPTER QUESTIONS

1. Fireflies that produce flash displays are rather unique among nocturnal insects in that they use a visual rather than a chemical signal (pheromones) to communicate. Provide a hypothesis to explain why visual signals are rare among nocturnal insects.

2. Male bison produce snort vocalizations during dominance contests. Vocalization frequency is affected by body size: the larger an individual, the lower the frequency of sounds it can produce. Form a hypothesis to predict how female mating behavior might be affected by male vocalization frequency.

3. In the cleaner fish example, three species were involved in the system of aggressive mimicry: the client, the cleaner, and the aggressive mimic. Predatory *Photuris* fireflies attract and eat male *Photinus* fireflies by producing a flash display that resembles the one produced by *Photinus* females. Explain how this is an example of aggressive mimicry when only two species are involved.

4. Female crayfish can realize higher fitness by mating with dominant males. Two hypotheses describe how a female can recognize a dominant individual. One is that dominant individuals produce unique pheromones after winning a contest with a rival male. The other hypothesis is that females can learn about relative fighting ability by eavesdropping on fights. Design an experiment to test these alternatives.

5. Explain how frequency-dependent selection would act on a Batesian mimic.

6. Compare the communication networks in the squirrel–blue jay system studied by Schmidt and Ostfeld (2008) and the fish system studied by Doutrelant, McGregor, and Oliveira (2009). In each case, identify the third-party eavesdropper and that individual's effects on behavior.

7. For the warbler system studied by Marchetti (1993), design an experiment to test the hypothesis that predation risk favors evolution of fewer brightly color patches for species living in bright habitats.

Chapter 7

Foraging Behavior

Figure 7.1. Mouse feeding. A mouse eats a food item that was hidden under a log.

One day we noticed quite a mess at one of our bird feeders. We had filled it full of mixed seeds, and now it was almost empty. Most of the seeds were scattered all over the ground. What had happened? After refilling the feeder, we kept careful watch and got our explanation. Carolina chickadees (*Parus carolinensis*) kept knocking smaller seeds out of the feeder until they found a larger sunflower seed, which they ate. They continued to push the small seeds out, ignoring them in their search for those larger seeds. Why were they passing up all those perfectly good seeds, spending extra time and energy to get the sunflower seeds? Animals selectively choose where, on what, and how long to forage, often passing up some food items in favor of others. This behavioral pattern is often seen in nature.

Animals eat to acquire energy and nutrients for survival and reproduction. How they feed, however, reveals a surprising degree of complexity and a large number of decisions. First, a forager needs to find food. Some animals, like many mice, feed on small, mobile prey or food hidden by sand, sediments, or bark, which can be challenging to find (Figure 7.1). After finding a potential food item, a forager is faced with a decision: should the item be eaten, or should it be ignored and the search continued? For example, some prey items, such as large nuts or hard-shelled invertebrates, may require the feeding animal to spend extensive time chewing and crushing the exterior to acquire the food inside. These items may not be worth the time and effort to consume them. And when a forager like a bee or a hummingbird finds itself in a patch of flowers, it must decide how long to stay, visiting flowers to extract nectar, before moving on to another patch (Figure 7.2).

In this chapter, we will first examine the diversity of sensory systems that animals use to locate food. Then we will focus on animals that

Figure 7.2.   Bee and flower diversity. (a) Bees have many flowers among which they can choose. (b) Each flower differs in its amount of nectar and ease of removal.

use vision to find prey and consider an adaptation that enhances their ability to find cryptic food items. Finally, we will show how foraging theory helps us understand how animals select their diet and exploit food patches.

7.1   Animals find food using a variety of sensory modalities

Before an animal can eat, it must find food. Most animals use several sensory systems or modalities when searching for food, such as vision, hearing, or olfaction. Natural selection favors modalities that most efficiently and accurately provide information about the location of food, but the efficiency of a modality can vary with environmental conditions. For instance, visually searching for food in murky water may be difficult. Given the diversity of sensory modalities that an animal possesses, what sensory systems do foragers use to find food? And are some sensory modalities favored over others? Let's examine two examples.

Featured Research

Catfish track the wake of their prey

Many fish live in habitats where visibility is reduced by turbid water or dense vegetation. In addition, many fish are nocturnal. Therefore, these animals often rely on nonvisual sensory systems to obtain information about their environment. Fish use **mechanoreceptors** both for hearing and for detecting their own body position and orientation. Mechanoreceptors also provide information about water pressure or movement that can indicate the presence of other organisms (either predators or prey). All fish have mechanoreceptors that provide hydrodynamic information, called the **lateral line system** (Coombs & Montgomery 1999). This system consists of pits or tubes that run along the side of the fish's body and head (Figure 7.3). These pits contain neuromasts, which provide information about water velocity and acceleration, as well as the direction of water movements.

mechanoreceptors Sensory receptors sensitive to changes in pressure.

lateral line system In fish, mechanoreceptors that provide hydrodynamic information.

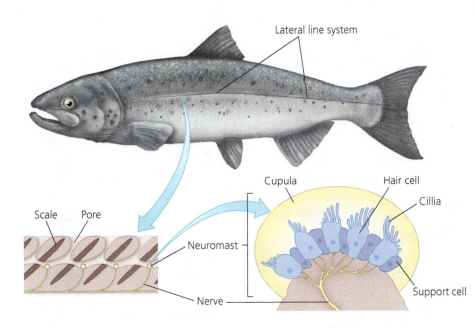

Figure 7.3. Lateral line system. The lateral line system is composed of many neuromasts that detect vibrations.

Water is an excellent medium for the detection of chemicals, so most fish also have a well-developed **chemoreception** system made up of two components. The **olfaction** (smell) component is usually localized in the nose, while the **gustation** (taste) component is usually found in the mouth but sometimes occurs throughout the whole body surface. Some fish, such as minnows, have well-developed chemoreception systems that allow them to recognize conspecific predation events and then respond accordingly with antipredator behaviors. Other fish, such as bullhead catfish (*Ameiurus* sp.), use the chemoreceptors on their barbels—the slender, whisker-like organs near the mouth—to locate stationary food.

Kirsten Pohlmann and colleagues investigated how nocturnal **piscivorous**, or fish-eating, catfish find food in the dark. They studied European catfish (*Silurus glanis*), which are 20 to 25 cm long and feed on small fish (Pohlmann, Grasso, & Breithaupt 2001). The research team used guppies (*Poecilia reticulata*) as prey fish because they create a swimming wake that has been well described. Pohlmann's team aimed to determine whether catfish can use cues provided only from the wake of individual fish to stalk and attack their prey. The researchers designed an infrared video system to track the movement paths of both the predator and the prey in an environment of complete darkness. They used four catfish and observed each one stalking up to ten individual guppies. Each catfish was allowed to acclimate to the experimental tank for one hour in the dark, after which the first guppy was added to the tank. After a guppy was consumed, another was added every 20 minutes to replace the one that had been eaten. The researchers ended a trial when ten guppies had been consumed or when no guppy had been consumed in 20 minutes. They used computer software to digitize the movement behavior sequences of both the catfish and the guppies and classified them in one of three ways: (1) path following (the catfish followed the guppy wake), (2) head-on encounters (the catfish encountered the guppy head-on without a previous encounter), or (3) attack on a stationary guppy (Figure 7.4).

Pohlman's team found that 80% of the attacks occurred on moving guppies. In the majority of all attacks, the catfish were following the same path as the guppy before the attack. The researchers concluded that the catfish appeared to be following the wake of the guppies to find and attack their prey in the dark.

However, the wake of a fish contains two cues that a predator might use to track its prey: hydrodynamic signals and chemical signals. In order to tease apart which sensory cue catfish use, Pohlmann's team conducted a second experiment using the same setup (Pohlmann, Atema, & Breithaupt 2004). In this experiment, they manipulated either the lateral line or the external gustation system of the catfish and compared

chemoreception The process by which an animal detects chemical stimuli.

olfaction The detection of airborne chemical stimuli.

gustation The detection of dissolved chemicals, often within the mouth.

piscivorous Fish-eating.

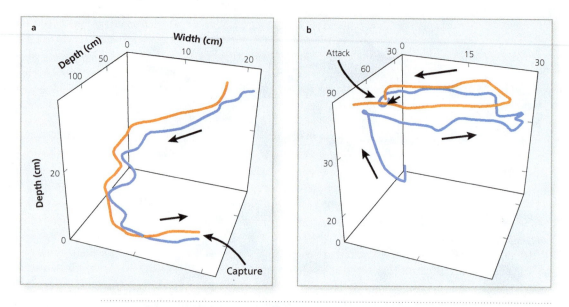

Figure 7.4. Wake following. Representation of two examples (a and b) of wake following, leading to an attack. The blue lines represent the catfish predator and the orange lines the guppy prey (*Source*: Pohlmann, Grasso, & Breithaupt 2001).

TABLE 7.1 Catfish attack mode and success. Intact catfish relied heavily on the wake-following attack mode, whereas lateral line–ablated fish most often used a head-on attack mode. Taste-ablated fish used both head-on and wake following attacks (*Source*: Pohlmann, Grasso, & Breithaupt 2001).

		PERCENT OF ATTACKS		
TREATMENT	CAPTURE SUCCESS RATE (% CAPTURE)	WAKE FOLLOWING	HEAD ON	STATIONARY
Intact	65	55	30	15
Nonfunctioning lateral line	17	6	88	6
External taste ablated	60	27	58	15

their behavior to that of controls. The lateral line provides hydrodynamic information and can be rendered temporarily nonfunctional by immersing a fish in a $CoCl_2$ (cobalt chloride) solution for six hours prior to the experiment. The external gustation system, which provides chemical information, was manipulated by surgically removing (ablating) an area of the dorsal medulla oblongata that controls this system.

Using identical procedures to record the attack behavior of the catfish, the researchers found that individuals with a nonfunctioning lateral line attacked guppies head-on 88% of the time. Most of the gustation-ablated fish also attacked guppies head-on, though less often (58% of attacks); only 27% of attacks were characterized as path-following (Table 7.1). The capture success rates also differed across treatments. Control and gustation-ablated fish exhibited similar, relatively high capture success (approximately 60%). In contrast, the success rate for individuals with a nonfunctioning lateral line system was only 17%. Pohlmann and colleagues concluded that the lateral line is a very important factor in catfish's ability to track the wake of their prey; in contrast, the external gustation system does not appear to be as important in facilitating this mode of attack behavior. The lateral line system may be key not only because it provides constant updated information about the movement of the prey but also because it provides feedback to the predator about its own movements and position. For animals that live in environments with reduced visibility, the ability to acquire such sensory information is an important adaptation that helps them capture prey.

Bees use multiple senses to enhance foraging efficiency

All animals possess multiple sensory systems. This leads to a natural question: Is foraging more efficient when multiple senses are used? Ipek Kulahci, Ana Dornhaus, and Daniel Papaj examined this question in bumblebees (*Bombus impatiens*) (Kulahci, Dornhaus, & Papaj 2008). Bees feed on nectar and pollen in flowers that can differ in color and shape (visual cues), as well as in odor (olfactory cues). Kulahci and colleagues studied the foraging behavior of individually marked bees to determine how feeding efficiency varies with the number of sensory cues that can be used to find flowers containing food.

In the laboratory, a box was attached to the colony container. The box contained a manual gate that allowed the researchers to release one individual at a time into an experimental arena. The arena contained eight artificial flowers—small circular wells that contained either 30% sucrose solution (a food reward) or water (no reward). During training, bees were exposed to two yellow flower shapes, circles or crosses, which surrounded the well (Figure 7.5). Bees were also exposed to two odors: 2 µL of either diluted peppermint or clove essential oil that was placed behind the perforated flowers.

Individual bees were exposed to one of three training treatments that varied the cues associated with reward flowers. In each treatment, half the flowers (randomly selected) contained sucrose and half contained water. In one treatment, only flower shape (a visual cue) indicated the reward (e.g., crosses indicated reward flowers and circles nonreward flowers, or vice versa). In the second treatment, only odor (an olfactory cue) indicated a reward flower (e.g., peppermint indicated reward flowers and clove nonreward flowers, or vice versa). Finally, in the last treatment, both shape and odor (visual plus olfactory cue) indicated reward flowers (e.g., clove plus circle indicated reward flowers and peppermint plus cross indicated nonreward flowers).

During training, bees visited all flowers and eventually learned which ones contained rewards. Training was deemed complete when an individual achieved 80% correct flower choices (reward flowers) over its last ten choices. Once trained, a bee was immediately tested by allowing it to enter the arena, but now all flowers were non-rewarding. The researchers recorded foraging time spent on each flower and the number of "correct" flower choices the bee made (flowers that the bee had learned contained rewards).

A total of 31 bees were tested. Bees trained on two sensory modalities had significantly higher feeding performance (correct flower visits divided by total decision time) than bees trained using only a single sensory modality (Figure 7.6); two-modality bees made more correct choices and spent less time deciding where to feed. These results suggest that the ability to use multiple sensory modes (vision and odor) facilitates efficient foraging in bumblebees. How this increased efficiency occurs requires further work.

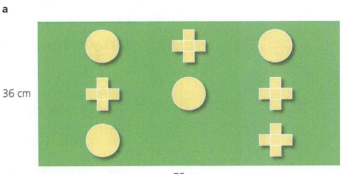

36 cm

75 cm

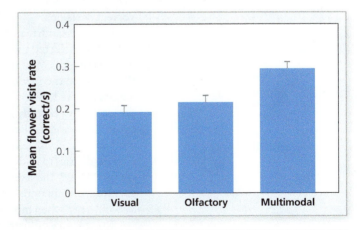

Figure 7.5. Bumblebee foraging experiment. (a) An experimental patch with eight artificial flowers. (b) A bee feeds on a "cross" flower (*Source*: Kulahci, Dornhaus, & Papaj 2008).

Figure 7.6. Flower visit rates. Mean (+ SE) flower visit rates. Bees trained to use both visual and odor cues (multimodal) had higher correct flower visit rates than those trained on just a single cue (*Source*: Kulahci, Dornhaus, & Papaj 2008).

Given that many foragers likely use multiple sensory modalities to find food like Kulahci's bumblebees, can we determine the relative importance of each modality? Recent work on a small nocturnal primate shows how we can.

Gray mouse lemurs use multiple senses to find food

Marcus Piep and colleagues (Piep et al. 2008) studied how gray mouse lemurs (*Microcebus murinus*) find their prey. Gray mouse lemurs are very small (weighing about 60 g) nocturnal primates that live in Madagascar, where they feed on a variety of arthropods and fruits. Many small nocturnal mammals, like mice, rely on olfaction to locate prey, while diurnal primates rely heavily on vision. Although gray mouse lemurs are primates, they are ecologically similar to mice because of their size and nocturnal habit. Observations of wild mouse lemurs suggest that individuals might find insect arthropod prey both visually and by hearing their rustling movements in the vegetation (Goerlitz & Siemers 2007). Other work suggests that individuals may use olfaction to find fruits (Siemers et al. 2007). Given the diversity of sensory systems that gray mouse lemurs appear to use, Piep and colleagues sought to determine the relative importance of visual, acoustic, and olfactory cues in their foraging behavior (Scientific Process 7.1).

To answer this question, they created a "choice test" experiment in an arena containing two dishes placed 10 cm apart, equidistant from the test animal's nest cage. All tests were conducted in low light that simulated nocturnal conditions. In each trial, one of the dishes contained a single mealworm (a *Tenebrio molitor* larvae) and the other was empty. The researchers manipulated the number and type of sensory cues present at the dish with the mealworm. Each dish was covered by a cone-shaped plastic lid that contained slits that allowed the passage of smells and sounds.

To manipulate auditory cues, the researchers used either live mealworms, which crawl and produce sounds that mouse lemurs can hear, or freshly killed mealworms, which produce no sound. To manipulate olfactory cues, the researchers placed the mealworms in a sealed bag (no olfactory cue) or did not seal the mealworms in a bag (olfactory cue present). To manipulate visual cues, they used a cone lid that was either black (so that the mealworm was not visible) or transparent (so that the mealworm was visible).

In the experiment, individuals were presented with different combinations of these cues so that one, two, or all three cues were present at the dish that contained the mealworm. The research team recorded the lid that the mouse lemur opened first—and thus its choice of the two dishes—and then analyzed the proportion of correct decisions.

Piep's research team found that when only olfactory or auditory cues were present, individuals selected the correct dish (the one that contained the mealworm) in 70% to 80% of the trials. When only a visual cue was present, that rate increased to 90% of the time, despite the low light level. However, when any two cues were present, mouse lemurs selected the correct dish over 90% of the time, and when all three cues were present, they correctly found the mealworm in 159 out of 160 trials (over 99% of the time). Piep and colleagues concluded that vision is the most important sensory modality for mouse lemurs. They also concluded that mouse lemurs used all three sensory modalities to find prey, and the more modalities they use, the better their hunting success. Why? The use of all three sensory modalities provides complementary sensory information under a variety of environmental conditions. Visual cues may not be available, say, when mouse lemurs forage in dense vegetation in the wild.

These studies illustrate how different species find prey using a variety of sensory modalities. This is only half the story, however. Selection should favor prey that can avoid detection. We examine the consequences of prey adaptations to minimize predation for visual foragers in the next section.

SCIENTIFIC PROCESS 7.1

Prey detection by gray mouse lemurs

RESEARCH QUESTION: *How important are vision, hearing, and smell in prey detection by nocturnally foraging gray mouse lemurs?*

Hypothesis: Lemurs will detect prey more accurately with the use of multiple sensory modalities.

Predictions: Lemur prey detection rate will be lowest with one sensory modality and highest with multiple sensory modalities.

Methods: The researchers:

- used a choice test experimental design: Food (a single mealworm) was placed under one of two dishes covered with cone lids.
- manipulated sensory cues so that one, two, or three were present at the dish with food.
- conducted experiments in low light levels similar to the nocturnal environment. Six lemurs were tested on each treatment 14 times.
- recorded the first lid opened.

Results: Individuals used all three senses to locate prey successfully. Food detection probability increased with the number of sensory modalities used.

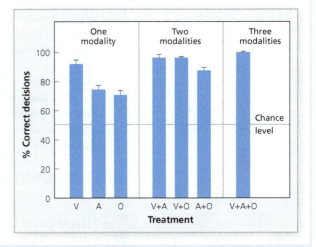

Figure 1. Correct choices. Mean (+ SE) percent correct choices for different treatments (V is visual; A is auditory; O is olfaction) (*Source:* Piep et al. 2008).

Conclusions: Gray mouse lemurs use vision, olfaction, and hearing to locate prey. Detection increased when more than one sensory modality was available. Detection was highest when a visual cue was present.

7.2

Visual predators find cryptic prey more effectively by learning a search image

Many predators, especially birds and mammals like the gray mouse lemur, rely heavily on vision to find prey. This tendency has favored the evolution of behavioral and morphological adaptations in prey species that allow them to blend into their background to avoid being detected by predators. Here, we first examine the effectiveness of such cryptic coloration. We then consider a counteradaptation found in predators to improve their ability to hunt for cryptic prey. The back-and-forth process of adaptation in one species favoring counteradaptation in another is known as an **evolutionary arms race**, as we saw in Chapter 6. It is common in interspecific interactions like the predator-prey interaction illustrated in the following study but can also occur between males and females or parents and their offspring, as we will see in Chapters 11 through 13.

evolutionary arms race The back-and-forth process of adaptation in one species favoring counteradaptation in another.

Featured Research

Cryptic coloration reduces predator efficiency in trout

Jörgen Johnsson and Karin Kjallman-Eriksson examined the ability of brown trout (*Salmo trutta*) juveniles (known as parr) to find cryptic or conspicuous prey in a simple laboratory experiment (Johnsson & Kjallman-Eriksson 2008). Brown trout are diurnal and therefore rely more heavily on visual cues for foraging than do nocturnal fish such as the catfish studied by Pohlmann's research team. These animals live in streams and are opportunistic feeders that eat a wide variety of invertebrate prey. Since aquatic invertebrates have different emergence times over the year, trout often have to learn to search for new types of prey as they become available. In addition, trout parr possess tetrachromatic color vision—that is, they have four different types of cone cells, which allow them to see in the red, green, blue, and ultraviolet spectra. This ability to distinguish colors is thought to be an adaptation that allows animals to find cryptic prey.

Johnsson and Kjallman-Eriksson asked two research questions. First, do predators find conspicuous prey more quickly than cryptic prey? Second, does hunting ability improve with experience? They predicted that trout would find conspicuous prey in less time than they would cryptic prey. They also predicted that predators would learn to find a cryptic prey item more effectively as they gained more experience searching for it (Scientific Process 7.2).

The researchers established two test aquaria, identical except for the color of the aquarium bottom: one was brown plastic covered with brown grains, while the other was green plastic covered with green aquarium grains. Before each trial, a single, small brown maggot (*Calliphoridae* sp.) prey was placed in one of six different locations on the aquarium bottom while the test fish was kept behind an opaque partition. The maggot was cryptic on the brown background and conspicuous on the green background. After the maggot was put in place, the partition was lifted and the researchers recorded the amount of time it took the test fish to find the food item.

The trout always found the conspicuous prey faster than the cryptic prey, indicating that foraging efficiency is reduced when predators search for cryptic prey. This means that background color matching can provide significant survival benefits for prey. However, search times decreased at a similar rate over the six trials both when fish searched for cryptic prey and when they searched for conspicuous prey, indicating that trout parr can improve their ability to find cryptic prey with only a moderate amount of experience.

Featured Research

Blue jays use a search image to find prey

◾ **search image** The visual distinctive features of a single prey type that, once learned, can enhance prey detection.

Why do predators have less difficulty over time finding cryptic prey? They are thought to have learned a **search image**, or the visual distinctive features of a single prey type (Tinbergen 1960). Of course, a predator may pay a price for narrowing its focus, as its probability of detecting other prey is likely to decrease. You may have experienced this effect yourself if you have ever tried to find an important school paper in your dorm room. You form a mental image of its size, shape, and unique markings. You probably do not pay attention to much else around you as you go through your room looking for the paper.

In a set of classic laboratory experiments, Alexandra Pietrewicz and Alan Kamil studied visual prey detection in blue jays (*Cyanocitta cristata*) to determine whether they use a search image to find cryptic prey (Pietrewicz & Kamil 1977; Pietrewicz & Kamil 1979). Blue jays commonly feed on moths, so Pietrewicz and Kamil trained individual birds to search for images of moths that were displayed as video images on tree trunks. The moths and tree trunks were similar in color, and so the moths were cryptic when on this background.

Pietrewicz and Kamil designed a very clever apparatus to examine whether blue jays use search images to find prey. They used two different species of bark-like *Catocala* (Noctuidae) moths, which are active during the night but then rest during the day on cryptic substrate. *Catocala relicta* rests on white birch tree trunks, whereas *Catocala retecta* rests on oak trees. The researchers took matched pictures of birch and oak tree

SCIENTIFIC PROCESS 7.2

Cryptic prey reduces predator efficiency

RESEARCH QUESTION: *How does cryptic prey coloration affect trout predator foraging efficiency?*

Hypothesis: Prey that match their background will be harder for predators to detect and predator hunting efficiency will increase with experience.

Predictions: Trout will find noncryptic prey faster than cryptic prey. Trout will find cryptic prey faster with experience.

Methods: The researchers:

- established two test aquaria, identical except for the color of the aquarium bottom: one was brown plastic covered with brown grains, and the other was green plastic covered with green grains.
- placed a single prey (a maggot) in one of six different locations on the aquarium bottom. The maggot was cryptic on the brown background and conspicuous on the green background.
- recorded the amount of time until the test fish found the food item.
- tested 42 parr, half with cryptic prey and half with conspicuous prey.

Figure 1. Background. The two experimental tanks with different colored bottoms with scattered prey (*Source*: Johnsson & Kjällman-Eriksson 2008).

Results: Individuals found noncryptic prey faster than cryptic prey. Search times to find prey decreased with experience.

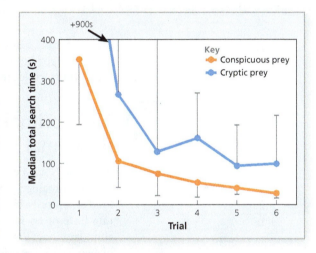

Figure 2. Search time. Median search time for conspicuous prey (orange) and cryptic prey (blue). Error bars are interquartile range, which represents one quarter of the range of the data (*Source*: Johnsson & Kjällman-Eriksson 2008).

Conclusions: Background color matching can benefit prey by reducing predator hunting efficiency. Predator search efficiency for cryptic prey can increase over time.

---

trunks with and without the moths on them. They trained jays to peck ten times at a positive image (a moth on a cryptic tree trunk) and peck one time at a negative image (no moth on a tree trunk). Both these behaviors resulted in positive reinforcement: a food reward of half a mealworm.

To examine the jays' use of search images, Pietrewicz and Kamil created two treatments. In the "run" treatment, the jays were shown eight negative images and eight positive images of one of the moth species so that they could learn a search image for that species. In the "nonrun" treatment, the jays were shown a randomly

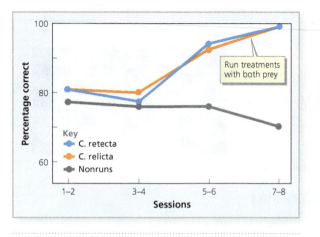

Figure 7.7. Search image results. Percent correct behavior over sessions. Jays became more accurate within runs of either species compared to nonruns in which both species were present during positive trials (*Source*: Pietrewicz & Kamil 1997).

intermixed group of four positive images of one moth species, four positive images of the other moth species, and eight negative images of no moths. In the "nonrun" treatment, the researchers assumed that jays did not have an opportunity to learn a search image for either species.

Over eight sessions, the percentage of correct responses increased in the run condition but not in the nonrun condition (Figure 7.7). In fact, the increase occurred quite rapidly. Pietrewicz and Kamil concluded that jays were using a search image of the moth species, which resulted in the increased success rate observed in the run condition. This test provided the first experimental evidence of animals using a search image. It also showed that short-term changes in the behavior of a predator can result in intense predation. In subsequent experiments, Bond and Kamil (1999) confirmed that when jays form a search image of one moth type, their ability to find other moth types is reduced.

Finding prey is just one aspect of foraging behavior. Animals must also decide, first, which prey types to include (and which not to include) in their diet and, second, how long to spend feeding in one food patch before moving on to another. To address these questions, researchers have developed models based on the assumption that animals attempt to maximize their fitness while feeding. We examine this approach next.

7.3 The optimal diet model predicts the food types an animal should include in its diet

All animals need energy and nutrients to survive and reproduce. Energy and nutrients are obtained from food, which is limited in most environments. In addition, food is typically heterogeneously distributed in the environment: most places contain little or no food, while other locations, known as food patches, may contain abundant food. For example, a dense aggregation of flowers is a food patch for a bumblebee, and a tree infested with insect larvae is a food patch for a woodpecker.

Optimal foraging theory (**OFT**) assumes that natural selection has favored feeding behaviors that maximize fitness. Many OFT models assume that fitness while feeding is a positive function of an individual's energy intake rate—the energy acquired while feeding divided by total feeding time. OFT models describe the relationship between possible behaviors and the fitness they produce. The behavior that produces the highest fitness is called the **optimal behavior** and is the behavior predicted by the model (Maynard Smith 1978; Mitchell & Valone 1990). OFT models are widely used because they produce testable predictions and can be applied to many species and behaviors. Two classic OFT models examine diet and food patch use. Let's start with the diet model, which predicts what food types an animal consumes.

The diet model

Animals encounter food types that differ in size, energy content, and **handling time**—the amount of time it takes to manipulate a food item so that it is ready to eat. Given this diversity, individuals need to decide which food items to eat (i.e., include in their diet) and which to reject. The optimal diet model (Pulliam 1974) is based on three assumptions:

1. Foragers maximize fitness by maximizing energy intake rate.
2. Food items are encountered one at a time and in proportion to their abundance.
3. All food items in the environment can be ranked according to their profitability.

optimal foraging theory (OFT) An approach to studying feeding behavior that assumes that natural selection has favored feeding behaviors that maximize fitness.

optimal behavior The behavior that maximizes fitness in an optimality model.

handling time The amount of time to manipulate a food item so that it is ready to eat.

The **profitability** of a food item is the energy it contains divided by its handling time. For example, consider a squirrel feeding on a sunflower seed and a walnut. Sunflower seeds have soft shells and small handling times but also contain relatively little food, or energy, compared to that found in a walnut. The walnut has a hard seed coat and thus a much longer handling time but contains more food, or energy. If the walnut contains 100 joules (J) of energy and takes 25 seconds to consume, its profitability is 4 J/s. If the sunflower seed contains 10 J of energy and has a handling time of one second, its profitability is 10 J/s. In this example, the sunflower seed has higher profitability and so is a higher ranked food item.

> **profitability** For a food item, the energy it contains divided by its handling time.

A graphical solution

To see how the optimal diet model works, let's first assume that all food types have the same amount of energy (E), say 10 J, but differ in handling time (h). Table 7.2 shows the handling times for each of five food types along with their profitabilities. We see that Type A has the highest profitability, whereas Type E has the lowest. Finally, let's assume that all food types are equally abundant and randomly scattered throughout the environment.

What diet maximizes the energy intake rate? Think about this question and write down your answer. You might begin by thinking that a forager should eat all food items it finds because all food items provide it with energy. A forager that eats a wide variety of food types is called a **generalist**, because it has a broad diet and consumes food items in proportion to their availability in the environment. Alternatively, you might think that a forager would maximize its energy intake rate by eating only Type A items, because they have the highest profitability. Foragers with restricted diets are called **specialists**, because they consume a small subset of potential food types. Or maybe you think that a diet somewhere in between these two extremes maximizes an animal's energy intake rate. At this point, it is not obvious which diet maximizes fitness.

> **generalist** A forager that consumes a wide variety of items in its diet.

> **specialist** A forager that has a narrow diet.

To determine the optimal diet, let's compare the profitabilities of each food type. Food Type A is the most profitable, so a forager should always include Type A in its diet. What other food types should be added to the diet to maximize energy intake rate (fitness)? You might imagine the five possible diets described in Table 7.3.

TABLE 7.2 Food item characteristics. Food types with their handling times and profitabilities.

FOOD TYPE	ENERGY (J)	HANDLING TIME (S)	PROFITABILITY (J/S)
A	10	1	10.0
B	10	3	3.3
C	10	8	1.2
D	10	24	0.4
E	10	29	0.3

TABLE 7.3 Possible forager diets. Five hypothetical diets.

DIET	EAT	REJECT	DESCRIPTION
1	Only Type A	Types B, C, D, and E	Extreme specialist
2	Types A and B	Types C, D, and E	
3	Types A, B, and C	Types D and E	
4	Types A, B, C, and D	Type E	
5	All types	Nothing	Extreme generalist

TABLE 7.4 Mean search time per item for each diet.

DIET	AVERAGE SEARCH TIME PER ITEM (S)
1	24
2	12
3	8
4	6
5	5

TABLE 7.5 Mean handling time per item for each diet.

DIET	EAT	AVERAGE HANDLING TIME PER ITEM (S)
1	Only Type A	1
2	Types A and B	2 = (1 + 3)/2
3	Types A, B, and C	4 = (1 + 3 + 8)/3
4	Types A, B, C, and D	9 = (1 + 3 + 8 + 24)/4
5	All types	13 = (1 + 3 + 8 + 24 + 29)/5

The first diet is the most specialized and the last the most general. Note that as we move down the list, each diet adds only the next most profitable food item. There are other possible diets, but it makes no sense for a forager to eat a low-ranking food item like Type E while ignoring one of higher profitability like Type A.

We can solve for the optimal diet by computing a forager's average energy intake rate per item in the diet. This is the average energy obtained per item divided by the sum of the average time it takes to find an item and the average time to handle an item in the diet:

$$\text{Average energy intake rate/item} = \frac{(\text{average energy obtained/item})}{[(\text{average search time/item} + (\text{average handling time/item})]}$$

Because we assumed that all food items contain the same energy, 10 J, the numerator is the same for all diets. We therefore need to know only how the average search time and handling time per item vary with the possible diets. The average search time per food item is the rate at which an individual encounters food in the diet. An individual with a more general diet will have a high encounter rate with food—and thus a lower search time per item—because it eats almost every food item it encounters. An individual with a narrow diet, on the other hand, will have a longer search time per item. Table 7.4 shows how search times per item eaten varies for each diet listed in Table 7.3. Note that as more food types are added to the diet, the average search time per item declines.

Average handling time per item also varies with diet breadth. From Table 7.2, we can see that if the forager only eats Type A food items, its average handling time per item is one second. If the diet includes both Type A and Type B items, the average handling time per item is two seconds (half the items are Type A, with a handling time of one second, and half are Type B, with a handling time of three seconds). Table 7.5 shows how the average handling time per item varies for the five possible diets.

We can graphically display the average search time per item and average handling time per item for the five possible diets (Figure 7.8). Our graph also shows the combination of search time *plus* handling time per item, which we need to determine the optimal diet. In our example, Diet 3 has the lowest combined search and handling time per item, and so it maximizes the energy intake rate. The model predicts that a forager should always eat Types A, B, and C when encountered and should always reject Types D and E. Why should a forager reject Types D and E? Because the cost in handling time of including them in the diet is greater than the benefit derived from reduced search time; including these food types in the diet lowers an individual's energy intake rate. The model also predicts that each food type should always be accepted or always be rejected. There is no middle ground. This prediction is called the **zero-one rule**.

Now look at the answer you wrote down. Is it the same as the one produced by the model? What logic did you use to generate

■ **zero-one rule** In the optimal diet model, the prediction that each food type should either be always eaten when found or always rejected.

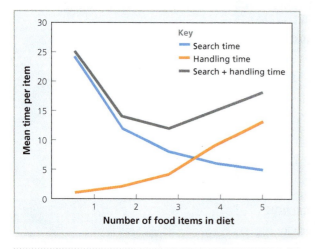

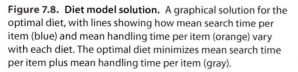

Figure 7.8. Diet model solution. A graphical solution for the optimal diet, with lines showing how mean search time per item (blue) and mean handling time per item (orange) vary with each diet. The optimal diet minimizes mean search time per item plus mean handling time per item (gray).

your prediction? This simple example illustrates the strength of the OFT approach to studying behavior. The OFT approach makes explicit assumptions and generates predictions that can be tested.

In our example, all food items have the same energy content. We can make the model more general and more realistic by allowing each food item to differ in both energy content and handling time. We can use a mathematical solution to show how this is done. In general, we find that diet breadth is affected by the abundance of the most profitable prey items and not by the abundance of the least profitable items.

The optimal diet model makes many simplifying assumptions about food types. It also assumes that there are no competitors or predators in the environment and that the forager does not need to search for mates. Can it predict the diet of real animals? Let's look at one classic field study.

Diet choice in northwestern crows

Featured Research

Howard Richardson and Nicholas Verbeek used the optimal diet model to understand the feeding behavior of northwestern crows (*Corvus caurinus*) on Mitlenatch Island off the coast of British Columbia (Richardson & Verbeek 1986) (Figure 7.9). Crows on this island eat mainly Japanese littleneck clams (*Venerupis japonica*) but often reject clams that they have found. Richardson and Verbeek used the optimal diet model to understand why.

Clams on the island range in size from 10 mm to almost 40 mm in length. Crows open clams by dropping them onto rocks repeatedly to break them and then pry open the shells with their bills. To test the model, Richardson and Verbeek needed to measure handling times, the energy content of clams, and encounter rates (search times for each item). They measured handling times for clams of different sizes in two ways. First, they observed crows feeding on clams of different sizes (ranging from 10 mm to 40 mm) and recorded the time from the first drop until the clam was eaten, which they defined as the handling time. They collected the empty shell to determine its size so they could see how handling times varied with clam size. Second, in order to increase their sample size and obtain handling time data on all sizes of clams, they dropped over 500 clams of different sizes onto rocks from heights similar to those used by crows themselves. They measured the number of drops it took (and the corresponding time) until the clam could be pried open with a fingernail.

To measure energy content, they collected and dried the soft tissue from 100 clams of different size. Previous work allowed them to determine the energy content of a clam based on the amount of dry tissue collected (James & Verbeek 1984). The researchers found that both energy content and handling time increased with clam size. Because both energy content and handling time varied with size, the researchers assumed that clams of different sizes represent different food types. They determined that the largest clam size was the most profitable and that profitability declined as size declined. Finally, they measured the availability of different-sized clams in 25 by 25 cm square plots of land called quadrats. In each of over 200 quadrats, they measured and counted all clams in the top 15 cm of substrate, which was the depth available to crows. These data provided information about the abundance of—and thus encounter rate—for each clam size.

Given these data, the model predicted that crows should eat all clams greater than 29 mm in size and reject all clams less than 29 mm in size to maximize energy intake rate. To test the model, Richardson and Verbeek collected data on the behavior of crows, noting the size of both those clams that were eaten and those that were rejected (clams that were picked up and then ignored). They found that the crows almost always ate clams that were larger than 30 mm and always rejected clams that were less than 26 mm (Figure 7.10). However, the crows did not follow the zero-one rule. Instead, they exhibited a **partial preference**: they sometimes

Figure 7.9. Northwestern crow. These birds feed on a variety of foods, including clams.

partial preference Acceptance of a food item some fraction of the time.

TOOLBOX 7.1

Mathematical solution to the optimal diet model

For simplicity, assume that there are only two food types in the environment, which we'll call Type 1 and Type 2:

1. **Type 1 contains an amount of energy E_1 and requires handling time h_1.**
2. **Type 2 contains an amount of energy E_2 and requires handling time h_2.**
3. **Type 1 is more profitable because $E_1/h_1 > E_2/h_2$.**
4. **Foragers spend time T_s searching for food.**
5. **The food types differ in their abundance in the environment. A forager encounters food Type 1 at a rate of λ_1 items per minute searching. Its encounter rate with food Type 2 is λ_2 items per minute. The inverse of the encounter rate $(1/\lambda)$ is the mean search time to find a particular food type.**

Since we have only two food types, there are only two possible diets. A forager can specialize on Type 1, the more profitable type, or it can have a general diet and eat both food types. (A third possibility, specializing on Type 2 items, makes no sense, because that is the lower ranked food type.) Which diet leads to a higher energy intake rate? To answer this question, we compare the energy intake rates of the two possible diets for a forager that searches for time T_s. We begin with the specialist diet and calculate the total energy acquired and total handling time. The total energy acquired is simply the number of Type 1 items encountered during search time T_s—or λ_1 multiplied by T_s—multiplied by the energy, E_1:

$$\text{Total energy acquired} = (\lambda_1 \times T_s)E_1$$

By the same reasoning, the total handing time of the number of Type 1 items encountered during search time T_s is the number found $(\lambda_1 \times T_s)$ multiplied by their handling time h_1:

$$\text{Total handling time} = (\lambda_1 \times T_s)h_1$$

and again we know that the total search time is T_s.

The energy intake rate R_1 for the specialist diet is the total energy acquired by feeding on Type 1 items divided by the total search time for those items plus the total time spent handling those items:

$$R_1 = \frac{(\lambda_1 \times T_s) \times E_1}{T_s + [(\lambda_1 \times T_s) \times h_1]}$$
$$= \frac{\lambda_1 \times T_s \times E_1}{T_s \times [1 + \lambda_1 \times h_1]}$$

which simplifies to

$$R_1 = \frac{\lambda_1 \times E_1}{1 + (\lambda_1 \times h_1)}$$

Now we consider the generalist diet. We go through the same steps to calculate the energy intake rate for the generalist diet, but this time we have two terms, one for each food type:

$$\text{Total energy acquired} = [(\lambda_1 \times T_s)E_1] + [(\lambda_2 \times T_s)E_2]$$
$$\text{Total handling time} = [(\lambda_1 \times T_s)h_1] + [(\lambda_2 \times T_s)h_2]$$

where the total search time is again T_s.

The energy intake rate for this diet, R_{1and2}, is again the total energy acquired divided by the total search time plus the total time spent handling those items:

$$R_{1 \text{ and } 2} = \frac{[(\lambda_1 \times T_s) \times E_1] + [(\lambda_2 \times T_s) \times E_2]}{T_s + [(\lambda_1 \times T_s) \times h_1] + [(\lambda_2 \times T_s) \times h_2]}$$
$$= \frac{T_s[(\lambda_1 \times E_1)] + [(\lambda_2 \times E_2)]}{T_s + [(\lambda_1 \times T_s) \times h_1] + [(\lambda_2 \times T_s) \times h_2]}$$

which simplifies to

$$R_{1 \text{ and } 2} = \frac{[(\lambda_1 \times E_1) + (\lambda_2 \times E_2)]}{1 + [(\lambda_1 \times h_1) + (\lambda_2 \times h_2)]}$$

A forager should generalize if the energy intake rate from eating both food types is greater than the energy intake rate from specializing on Type 1 only. That means that a forager should be a generalist if $R_{1and2} > R_1$, or

$$\frac{[(\lambda_1 \times E_1) + (\lambda_2 \times E_2)]}{1 + [(\lambda_1 \times h_1) + (\lambda_2 \times h_2)]} > \frac{\lambda_1 \times E_1}{1 + (\lambda_1 \times h_1)}$$

Rearranging produces this inequality:

$$\frac{E_2}{h_2} > \frac{E_1}{\frac{1}{\lambda_1} + h_1}$$

The model predicts that Type 2 should always either be rejected (if the inequality is false) or eaten (if the inequality is true): foragers should always either eat or reject a food type. This prediction is the zero-one rule.

The model also predicts that the decision to specialize or generalize is not affected by the encounter rate with the lower-ranked food item. The term λ_2 does not appear at all in the final inequality. Type 2 could be infinitely abundant, and the forager's diet would still not change. Why not? Look again at the inequality. Just one behavioral difference distinguishes the generalist from the specialist: when an individual finds a Type 2 food item, that item is eaten by the generalist and rejected by the specialist. The forager should eat the item (and be a generalist) if the energy intake rate for that item (the left-hand side of the inequality) is greater than the energy intake rate of rejecting that item instead of searching for a Type 1 item (the right-hand side of the inequality). Since this judgment does not involve finding the Type 2 item, its encounter rate is irrelevant. Thus, we see that diet selection is really about what *not* to eat.

ate clams of intermediate size. For example, clams that were 29 mm in size were eaten about 50% of the time. Although the diet model does not predict partial preferences, overall, the behavior of the crows was quite similar to predicted behavior. Richardson and Verbeek concluded that the observed diet yields an energy intake rate very close to that of the optimal diet.

Many experiments using a variety of animals have produced results that closely match the predictions of the optimal diet model. However, animals often exhibit partial preferences for some food types. Why? The model assumes that foragers have perfect knowledge of the abundance of food types, as well as the energy content and handling times of each type. Of course, real foragers can only estimate these parameters. According to one explanation, the uncertainty results in partial preferences. For instance, the crows probably did not know for sure when they found a clam of size 28 mm (which the model predicts should be rejected) rather than one that was 29 mm in size (which should be eaten). Given this uncertainty, it is not surprising that clams of these sizes were sometimes eaten and sometimes rejected. In addition, the diet model assumes that there are no competitors and no predation risk. These factors, too, can lead animals to modify their diet in ways that include exhibiting partial preferences (Bell et al. 1984; Berec & Křivan 2000; Lima, Mitchell, & Roth 2003).

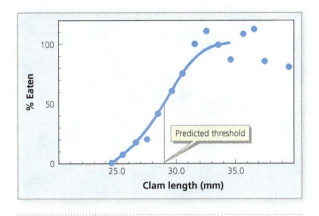

Figure 7.10. Crow diet choice. Each data point shows the percentage of clams eaten as a function of clam size (*Source*: Richardson & Verbeek 1986).

Ant foraging: the effect of nutrients

Featured Research

The optimal diet model assumes that fitness is only a function of energy intake rate. But animals require more than just energy—they also need about two-dozen elements, including carbon, nitrogen, phosphorus, and potassium (Fausto da Silva & Williams 2001). Such dietary needs can also affect feeding behavior and diet choice. For example, herbivores often face a situation of sodium limitation, because plants contain low sodium concentrations. Thus, many herbivores, ranging from butterflies to large ungulates, are attracted to natural or artificial mineral licks that contain sodium (e.g., Boggs & Dau 2004; Tracy & McNaughton 1995).

Michael Kaspari, Charlotte Chang, and Johanna Weaver tested the hypothesis that sodium limitation affects the feeding behavior of ants (Kaspari, Chang, & Weaver 2010). They examined the recruitment of ants (attraction to a food source) to two food baits: sucrose (a high-energy food) and sodium chloride (NaCl, or salt). The researchers hypothesized that to maximize their energy acquisition, ants would recruit most heavily to sucrose. If sodium is limited and essential, ants should also recruit heavily to it. To examine the relative strength of ant recruitment to the two food sources, the researchers focused on areas that differ in the availability of salt in the environment. Road salt, containing NaCl, is commonly applied to roads in winter. Residual salt can be long lasting and affect ecosystems near these roads. The researchers predicted that ants living far from salted roads (where sodium is rare) should exhibit a more pronounced preference for NaCl baits than would ants living nearer to salted roads. In other words, environmental availability of sodium should affect the diet selection of ants.

Kaspari and colleagues tested this prediction by establishing two sets of transects in Massachusetts that ran parallel to a salted road that was treated with NaCl during winter. Each set contained four transects located 1, 10, 100, and 1000 m from the road. At 1 m intervals along each transect, the researchers placed a single 2 mL vial on the ground that was half-filled with cotton soaked with either sucrose or NaCl. A total of 90 vials were used in each transect: 45 with sucrose and 45 with NaCl, randomly selected for placement at each point along the transect. One hour later, the researchers collected each vial and recorded any ants inside. At each transect, the researchers also collected soil samples to calculate soil NaCl concentration and thus estimate sodium availability at different distances from the road.

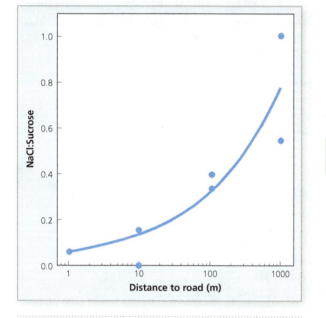

Figure 7.11. Ant NaCl preference. A greater ratio of NaCl to sucrose vials was used by ants as distance to the salted road increased (*Source*: Kaspari, Chang, & Weaver 2010). *Inset: Tapinoma sessile.*

■ **diminishing returns** A decline in instantaneous harvest rate as a food patch is depleted.

■ **marginal benefit** In foraging, the benefit obtained by feeding for one more instant.

■ **marginal value theorem** An optimal foraging model that predicts how long an individual should exploit a food patch.

NaCl concentrations were about ten times higher 1 m from the road compared to all other distances, indicating that road salt application does increase sodium availability near roads. About one-third of all ants recorded were *Tapinoma sessile*, a small omnivorous ant that consumes large amounts of vegetation. Near the road, these ants strongly recruited to vials containing sucrose: over 90% of the vials visited contained sucrose and so the NaCl vials were largely ignored. However, as distance from the road increased, ant use of NaCl vials also increased, so that at 1000 m from the road, there was almost equal recruitment to the sucrose and NaCl vials (Figure 7.11). These data demonstrate that sodium limitation affects the feeding behavior of these ants and illustrate how nutrient requirements can affect diet choice. In areas where sodium is limited, animals will often modify their behavior and diet to increase ingestion of sodium.

Diet choice is one important aspect of a forager's behavior. However, foragers also need to decide how long to exploit one food patch before departing for another. We examine this behavior next.

7.4   The optimal patch-use model predicts how long a forager should exploit a food patch

After animals exploit a patch, it can take some time before food is replenished in that patch. For example, after a bee removes nectar from a flower, it can take 24 hours for the nectar to be fully replenished. After a woodpecker removes insects from a tree trunk, it may take days to months before more insect larvae are deposited there. Therefore, when a forager enters a patch with abundant food, it initially harvests food at a high rate, but as the patch becomes depleted, the rate declines: we say that the forager experiences **diminishing returns** as it exploits a food patch. Given diminishing returns, the forager must decide how long to exploit the present (depleting) food patch before abandoning it to search for a richer patch.

To determine the optimal level of patch use, we need to examine the benefit a forager can gain by spending just a bit more time in the patch. This **marginal benefit** of feeding is the harvest rate at any point in time (or the instantaneous harvest rate). When a forager enters a patch with abundant food, the marginal benefit of feeding is high, because the instantaneous harvest rate is high. Because of diminishing returns, the marginal benefit declines as the patch becomes depleted. When the patch contains little food, the marginal benefit of feeding is very low, because it takes a long time to find the next food item.

The optimal patch-use model

Eric Charnov created an OFT model to determine how long a forager should stay in a food patch to maximize its fitness, the optimal patch-use model (Charnov 1976). Charnov's model is based on four assumptions:

1. Foragers attempt to maximize energy intake rate.
2. All patches are identical (contain the same kind and amount of food).
3. Travel time between patches is constant.
4. As a forager depletes a patch, its instantaneous harvest rate declines—that is, it experiences diminishing returns.

Charnov's model, called the **marginal value theorem**, predicts that a forager should stay in a patch until its marginal benefit of feeding declines to equal the average energy intake rate from the environment. The average energy intake rate from the environment is calculated by dividing the total energy acquired from all patches by the total time to travel to and then exploit the patches (travel times plus patch times). In Charnov's original model, the travel time between patches is held constant, and all patches are identical.

How much time should a forager spend in a patch to maximize energy intake rate? To answer this question, we first determine how much total energy the animal obtains when it stays in a food patch. Initially, there is abundant food in a patch, so the cumulative gain curve begins with a steep slope (Figure 7.12). However, the longer the time spent in a patch, the less food is left (i.e., diminishing returns), and so the slope of the curve becomes shallower as more time passes. Next, we need to include the amount of time spent traveling to a patch (T_t) in addition to time spent in the patch (T_p). The energy intake rate can now be calculated for any T_p: it is the energy accumulated, divided by T_t plus T_p. We can solve for this relationship graphically, because the energy intake rate for any T_p is the slope of the line that runs from the start of the travel time to the cumulative gain curve at each T_p. The patch time that maximizes energy intake rate is the T_p where this line is tangent to (i.e., where it just touches) the cumulate energy gain curve (Figure 7.13).

We see that the model predicts the amount of time a forager should spend in each patch for any fixed travel time. This number depends on the travel time: as travel time increases, foragers should spend longer in each patch (Figure 7.14). This prediction can be tested, as we see next.

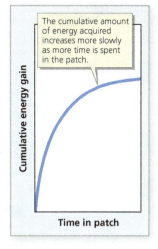

Figure 7.12. Cumulative gain curve. The graph shows how cumulative energy acquired by the forager increases with time spent in the patch.

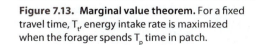

Figure 7.13. Marginal value theorem. For a fixed travel time, T_t, energy intake rate is maximized when the forager spends T_p time in patch.

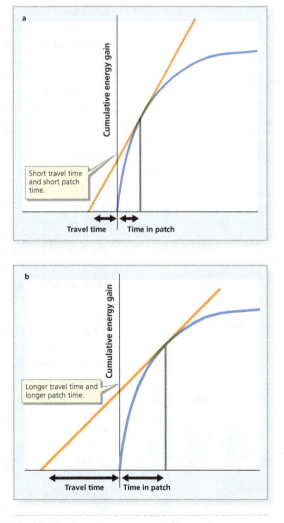

Figure 7.14. Travel time and patch time. In (a), travel time between patches is short, and so the optimal patch time is short. In (b), travel time is longer, and so optimal patch time increases.

Patch use by ruddy ducks

Michael Tome tested the predictions of the marginal value theorem with ruddy ducks (*Oxyura jamaicensis*) (Tome 1988). Ruddy ducks feed on aquatic invertebrates and vegetation on muddy lake bottoms. They repeatedly dive from the water's surface to the lake bottom to find food. A short time later, they emerge to take a breath before diving again for food.

Tome created a large concrete and glass aquarium (5 m long by 2 m wide by 2 m deep). The bottom of the aquarium contained 16 wooden trays filled with sand, and the sides had glass windows for observations. Each tray constituted a potential food patch: Tome buried wheat grains in some of the trays during his experiments and trained birds to search the sand for food. The model assumes that ducks would experience diminishing harvest rates while feeding on wheat grains in the artificial food patches. To test this assumption, Tome buried 150 wheat grains in a patch whose location was fixed, which allowed the ducks to learn which patch contained food. He then released a duck to feed from the patch for a fixed amount of time that varied from five to 200 seconds. Afterwards, he collected all the wheat grains left in the patch, which told him the amount of food eaten by the duck in that time. He plotted the number of wheat grains harvested from a patch against time spent in the patch for six ducks (Figure 7.15). He found that the harvest rate did indeed decline as ducks depleted the patch: the harvest rate curve is steep when a bird enters the patch and levels off as the patch becomes depleted. These data match the model's assumption.

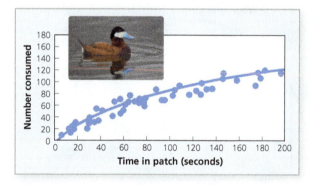

Figure 7.15. Ruddy duck cumulative gain curve. Number of food items harvested as a function of feeding time in a food patch (*Source*: Tome 1988).

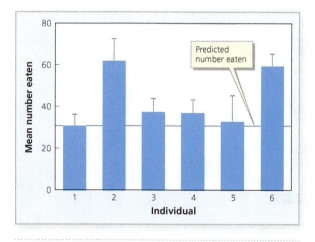

Figure 7.16. Ruddy duck patch data and prediction. Mean (+ SE) number of food items eaten by each duck. The line shows the predicted value based on the model (*Source*: Tome 1988).

Next, Tome established two identical patches on wooden trays on the aquarium bottom that each contained 150 buried wheat grains. The locations of each patch were kept constant from trial to trial, and birds were allowed to feed for seven days before the start of trials to learn their location and quality. When a bird first exploited a patch, it dove from the surface. When it abandoned the patch, it returned to the surface to breathe before diving to the second patch. Tome measured the time until a bird abandoned the first patch, as well as the travel time between patches (in other words, the time to travel to the surface to breathe and then dive back down to the second patch).

Tome already knew the relationship between time spent in a patch and the number of wheat grains eaten. From this information, he could calculate the optimal number of food items a bird should eat in a patch to maximize its energy intake rate. To test this prediction, he collected data on the number of food items left in a patch (to determine how many grains were eaten) for the same six birds, conducting ten trials for each.

Tome found that the model accurately predicted patch use for the majority of ducks (although two ducks ate more and spent longer in the patches than predicted) (Figure 7.16). He concluded that most of the ducks were maximizing energy intake rate, just as the model assumes.

Next, he tested the prediction that a forager should remain longer in a patch as travel time between patches increases. In the first experiment, the travel time between patches was small, because birds learned the location of the two patches and dove directly to them. In a second experiment, Tome increased travel time by randomly changing the location of a single food patch on the aquarium bottom in each trial. This required the ducks to search for the patch, and they needed about five search dives (rather than one) before they found it. Tome stopped the trial when a duck abandoned the patch, surfaced, and then dove again to search for other food patches. He again collected the number of food items left in the patch, allowing him to calculate the number eaten. This time, five of the six birds ate significantly more food from the food patch and thus spent significantly more time there, just as the model predicts (Figure 7.17). Tome concluded that the ducks were indeed attempting to maximize their energy intake rate, as assumed by the marginal value theorem.

The optimal patch-use model, like the diet model, makes several simplifying assumptions. It assumes that there are no foraging costs other than time spent traveling between and exploiting patches. It also assumes that a forager has perfect information about the environment: it knows that all patches are identical and that the travel time is the same between all patches. Let's see how these assumptions have been modified to better understand the feeding behavior of animals.

Optimal patch model with multiple costs

Joel Brown developed a new patch-use model that includes the energetic costs of foraging, predation risk costs, and missed opportunity costs (Brown 1988). **Energetic costs of foraging** include the energy used to exploit a patch and the metabolic costs incurred while feeding. **Predation risk costs** involve the probability of being killed while feeding. Finally, **missed opportunity costs** are the costs of forgoing other activities that might yield even higher fitness. For example, a forager experiences high missed opportunity costs if by feeding it misses opportunities to mate. Brown's model predicts that a forager will maximize its fitness when it remains in a patch until its marginal benefit of staying in the patch (again, the instantaneous harvest rate) declines to equal the marginal (instantaneous) cost of being in the patch. Since the marginal cost now includes energetic, predation, and missed opportunity costs, Brown's model is more general than the original optimal patch-use model, and it is applicable to a wider range of foraging situations.

Brown's model predicts that if two patches have the same food (identical benefits) and the same costs, they should be harvested down to the same food density. (We say they should have the same **quitting harvest rate**.) This prediction holds even if patches differ in their initial food density. Remember that the marginal value theorem model assumes that all patches are identical. Brown's patch-use model is more general because it applies even when patches differ in food amount. Let's examine one test of this prediction.

Fruit bats foraging on heterogeneous patches

A simple way to test Brown's model is to provide foragers with patches that have identical benefits and costs but differ in initial food abundance. Recall that Tome collected the amount of food left in a patch when a duck abandoned it, obtaining what is called the **giving-up density (GUD)** of food. Brown's model predicts that the patches will be harvested down to the same GUD if foragers are attempting to maximize their energy intake rate.

Francisco Sánchez tested this prediction using Egyptian fruit bats (*Rousettus aegyptiacus*) (Sánchez 2006). Fruit bats are large, nocturnal bats that live in tropical Asia and Africa. They obtain juices and pulp from a variety of fruits and also feed on nectar.

Sánchez examined the behavior of captive bats that exploited artificial feeders in a large outdoor flight cage. He wanted to determine whether bats attempt to maximize their energy intake rate while feeding on patches that differ in initial resource amount. He trained bats to feed from artificial feeders that consisted of a large cylinder with one opening that contained a mix of sugar water and protein. Brown's model, like the marginal value theorem, requires that a forager experience a decline in its harvest rate while feeding from a food patch. Bats often face diminishing returns in nature, because they often feed on fibrous fruits: juice and pulp are

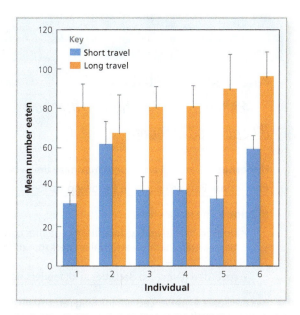

Figure 7.17. Travel time affects ruddy duck patch time. Mean (+ SE) number of items eaten increased with travel time (*Source*: Tome 1988).

energetic costs of foraging The energy used to exploit a patch and the metabolic costs incurred while feeding.

predation risk cost The fitness cost associated with being killed by a predator.

opportunity costs The fitness costs of not engaging in other activities.

Featured Research

quitting harvest rate The forager's instantaneous harvest rate when it leaves a food patch.

giving-up density (GUD) The density of food items left in a food patch after being exploited by a forager.

SCIENTIFIC PROCESS 7.3

Patch use by fruit bats

RESEARCH QUESTION: *How do fruit bats exploit food patches that differ in food amount?*

Hypothesis: Bats will attempt to maximize their energy intake rate.

Prediction: Bats will equalize the giving-up density across the patches.

Methods: The researchers:

- housed bats in an outdoor flight cage that contained feeding stations. Each station contained three artificial feeders that differed only in the amount of food they contained.
- recorded giving-up densities in all feeders each day for three days.

Results: Bats experienced diminishing returns when using the feeders. Bats equalized the giving-up densities of the three feeders at each station.

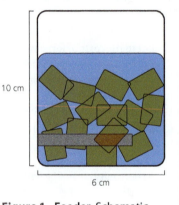

Figure 1. Feeder. Schematic drawing of artificial feeder. The cubes are the rubber tubing, and the bar represents the feeder opening (*Source*: Sánchez 2006).

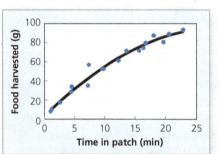

Figure 2. Food harvested. Cumulative gain curve for the feeders (*Source*: Sánchez 2006).

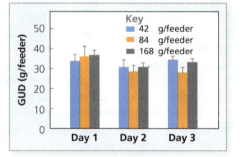

Figure 3. GUD. Giving-up densities of the three feeders (patches) on each day (*Source*: Sánchez 2006).

Conclusions: Bats maximize energy intake rate when feeding on patches that differ in initial food amount.

easy to extract at first, but extraction becomes progressively more difficult. To create diminishing returns in the artificial feeders, Sánchez placed several small pieces of rubber tubing inside the food solution. As a patch was depleted, the rubber tubing made it progressively more difficult for the bats to extract the sugar water. Before conducting his experiment, Sánchez examined the harvest rates of bats in these feeders by allowing them to feed for several nights. Each night he recorded the total time a bat spent at a particular feeder and the amount of food remaining at the end of the night—the GUD. When he plotted the corresponding harvest rate data, he found that bats did experience diminishing returns in the artificial feeders, just like the ruddy ducks in Tome's experiment.

To test the prediction that bats would equalize GUDs across patches that differed in initial resource amount, Sánchez provided bats with three identical feeders that were placed very close together to create similar energetic, predation, and missed opportunity costs. No predators were introduced into the flight cage, and the captive bats could not engage in other activities. However, the feeders contained different amounts of food: 42, 84, or 168 g of sugar per feeder in the water-protein mix. Sánchez established five feeding stations in the aviary in different locations and allowed 12 bats to feed from them all night. Each morning, he collected the amount of food remaining in each feeder to determine how much had been consumed. He repeated this procedure for three nights.

Sánchez found that the bats did equalize GUDs as predicted by the model. He concluded that the bats maximized their energy intake rate while feeding. His work and others have shown that Brown's patch-use model can explain the feeding behavior of many organisms (Brown & Kotler 2004). Brown's model can be applied to a wider range of ecological conditions because it allows for differences in patch resource amounts and foraging costs. Sánchez's work shows how the model can predict feeding behavior when the initial resource amounts of food patches differ. Next, we examine how Brown's patch-use model has been tested when patches differ in their foraging costs.

Gerbil foraging with variable predation costs

Featured Research

Burt Kotler, Joel Brown, and Oren Hasson conducted an experiment to examine how variation in predation risk cost affects patch use. Their study used gerbils, seed-eating nocturnal rodents that live in underground burrows (Kotler, Brown, & Hasson 1991). Gerbils are preyed on by owls, which are visual predators and are thought to be more effective hunters on bright nights with a full moon than they are on dark nights with a new moon. Owls often capture rodents by pouncing on them when they are in the open, so a rodent in its burrow is safe from owls. A rodent is also relatively safe when it is under a shrub, because the branches block an owl's attack.

To study gerbil patch use, Kotler and colleagues used a large outdoor aviary (18 by 23 by 5 m) with sand covering the ground. Gerbils placed in the aviary constructed burrows, and the researchers placed 16 piles of cut brush to simulate shrubs and provide additional safety from owl predators. The gerbils obtained food from 16 pairs of artificial food patches, each measuring 45 by 45 by 2.5 cm and containing 6 g of millet seed mixed into the sand. In each pair, one tray was placed under a brush-pile and another was placed 1 m away in the open, with a resulting higher predation risk cost. Previous work had demonstrated that gerbils experience diminishing returns in food trays (Kotler & Brown 1990).

The researchers studied two species, Allenby's gerbil (*Gerbillus allenbyi*) and the Egyptian sand gerbil (*Gerbillus pyramidum*), in four experimental treatments. They manipulated both the presence of barn owls (*Tyto alba*) (Figure 7.18) and illumination levels. Two owls were placed in the aviary on some nights, whereas no owls were present on other nights. Lights simulated illumination during a full moon. These lights were always on for nights near a full moon and off for nights near the new moon. The researchers examined all combinations of these treatments over the course of 48 nights.

Brown's patch-use model makes two predictions: (1) Gerbils will always feed less from the open trays compared to the brush trays, because open areas have higher predation risk; and (2) predation risk costs will be high when owls are present and on illuminated nights. In these latter treatments, the researchers predicted that gerbils would feed less and have higher GUDs in the food patches compared to treatments with no owls and no illumination.

At the beginning of the experiment, 12 *G. allenbyi* and eight *G. pyramidum* were placed in the aviary. Gerbils were allowed to feed from the trays all night, and any gerbils

Figure 7.18. Barn owl. Gerbils are frequently eaten by barn owls.

killed by owls were noted and replaced with new individuals in order to keep gerbil density constant. Each morning, Kotler's group collected all food left in the patches to calculate the GUDs and then renewed each patch with 6 g of millet. During the experiment, a total of 19 gerbils were captured by owls. Almost twice as many captures occurred on nights of high illumination (12 out of 19), and so owls were indeed more effective hunters when illumination was high.

The experimental treatments strongly affected the patch use of both species of gerbils. Both species had higher GUDs when feeding at trays in the open compared to trays under shrubs. In fact, the gerbils almost never ate at trays in the open in either treatment, supporting Prediction 1. Gerbils also had higher GUDs when owls were present than when owls were not present and in the illuminated treatment compared to the nonilluminated treatment. Both of these results support Prediction 2 (Figure 7.19).

These findings support Brown's model and show that gerbil foraging behavior is affected by predation risk costs, as well as demonstrating how a forager balances benefits and costs while feeding. Recently, conservation biologists have begun to use this technique to study patch use. They have also used GUDs to understand how human activities can negatively affect species (Applying the Concepts 7.1).

The marginal value theorem is a simple model, but it captures essential features that have shaped the evolution of feeding behavior. Brown's model is more general and has allowed a greater understanding of patch-use feeding behavior. There remains one more limiting assumption, however: individuals have perfect information about their environment.

Incomplete information and food patch estimation

Recall that the marginal value theorem assumes that an individual knows the exact number of food items in each patch. However, in nature, most animals can only estimate the number and quality of food items in each patch they encounter. This incomplete information might explain why some foragers spend slightly longer in patches than predicted, as we saw for two of the ducks in Tome's experiment. Understanding how animals estimate parameters like food patch quality also provides insight into how animals make decisions like when to leave a patch.

How do animals estimate the quality of a patch? Individuals gain **sample information** about a patch from the quantity and quality of food they find. In addition, they may have **prior knowledge** about the frequency of food patch types (their quality) in the environment from previous foraging events. Combining sample

sample information In Bayesian estimation, the information obtained by sampling an unknown parameter such as a food patch.

prior knowledge In Bayesian estimation, information about a parameter prior to sampling.

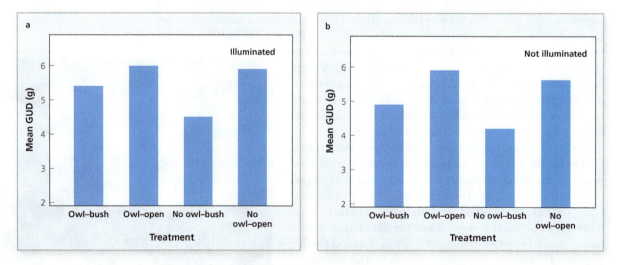

Figure 7.19. Gerbil patch use. One set of results for *Gerbillus pyramidum*. The gerbils had higher mean GUDs when the aviary was illuminated, when owls were present, and in open food patches (*Source:* Kotler, Brown, & Hasson 1991).

APPLYING THE CONCEPTS 7.1

GUDs and conservation

Conservation biologists are studying patch use and using GUDs to examine how humans affect other species. As humans expand into previously uninhabited areas, for example, they modify the environment by adding artificial lights. As seen in Kotler, Brown, and Hasson's (1991) study of gerbils, many nocturnal species perceive high light levels to be risky and therefore reduce their feeding behavior. Brittany Bird, Lyn Branch, and Deborah Miller studied how different light levels change the feeding behavior of nocturnal beach mice (*Peromyscus polionotus leucocephalus*) on Santa Rosa Island, Florida (Bird, Branch, & Miller 2004). Beach mice live in sand dune habitats and feed on seeds. At four sites, Bird and colleagues established two linear transects of food trays that contained 5 g of seed mixed into beach sand. One transect was established with a light source (two 40 W bug lights) at one end (light transect), while the other transect was placed more than 20 m away from a light source (dark transect). Each transect consisted of nine patches placed at 2 m intervals. Trials were conducted around the time of the new moon (i.e., dark nights), and GUDs were collected for two or three nights for each transect.

The research team found that mice GUDs were nearly two times higher at patches near the light source than at patches away from the light sources. In the light transects, GUDs decreased as distance from the light increased, so that at 14 to 16 m from the light source, GUDs were similar between light and dark transects. Because human dwellings often have high nighttime illumination, urbanization can negatively affect nocturnal species.

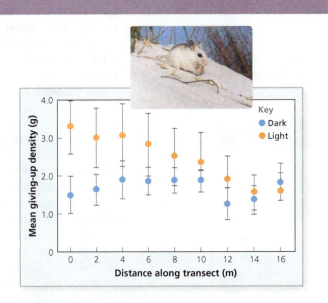

Figure 1. Mouse GUDs. Mean (± SE) giving-up density. Giving-up densities are higher on light transects (orange) compared to dark transects (blue) from zero to 12 m away from a light source (light source is at distance 0) (*Source:* Bird, Branch, & Miller 2004).

information with prior knowledge allows individuals to obtain a more accurate estimate of food patch quality. How do they do this?

You've made such estimates yourself if you have ever played blackjack. The object of this game is to acquire a set of cards that adds up to 21 without going over that value—or at least beats the dealer's hand. You can ask for any number of cards, but if your total exceeds 21, you "bust" and lose the game. How do you decide how many additional cards to obtain? You use two types of information: prior knowledge and sample information.

You already know the cards in a standard playing deck. The values range from one to 11 but are not equally represented. There are four cards of each value from two through nine, 16 cards of value ten (the tens, jacks, queens, and kings), and four cards (the aces) of a value of either one or 11. This information constitutes your prior knowledge, and it allows you to determine the probability that you will go over 21 with the next card. For instance, if your first two cards total seven, there is zero probability that the next card will result in a total that exceeds 21, so you will always take another card. On the other hand, if your first two cards total 14, there is a relatively high probability that the next card obtained will give you a total that exceeds 21.

What is that probability? You can determine this number by observing the cards that have already been played. This is the sample information, and it allows you to estimate the frequency distribution of the cards remaining in the deck. If many cards of low value (say, less than eight) have already been played, then the probability of drawing a high card is higher, and you are more likely to bust. If many cards of high value (say, cards greater than seven) have already been played, the probability that the next card will make you bust is much lower. If you did not know the cards in a standard playing deck, blackjack would be much more difficult. It is the ability to track sample information that so often separates winners from losers at the blackjack table.

Combining prior knowledge with sample information is a powerful way to estimate unknown parameters. How does this skill relate to animal behavior? Modifications of the optimal patch-use model assume that animals estimate food patches much as blackjack players estimate the likelihood that one more card will result in a set of cards totaling over 21. This process of combining sample information and prior knowledge is called **Bayesian estimation** because it is done mathematically using Bayes' theorem. Let's examine one recent test of this idea.

> **Bayesian estimation** A process of combining sample information and prior knowledge using Bayes' theorem.

Featured Research ▶

Bayesian foraging bumblebees

To answer the question of whether bumblebees estimate food patches in a Bayesian manner, Jay Biernaskie, Steven Walker, and Robert Gegear created an enclosure that allowed a single bee to search artificial flowers for sugar water (Biernaskie, Walker, & Gegear 2009). They made artificial flowers using microcentrifuge tubes that contained either sugar water (a food reward) or plain water (no food). They then created a foraging arena that contained ten food patches comprising 12 artificial flowers each (Figure 7.20). Using this design, they trained individual bees to learn different distributions of patch types. This information became the bee's prior knowledge. One set of bees learned that all patches were the same: five flowers contained rewards and seven flowers did not; hence, the mean rewards per patch were five. The location of the rewards was randomly assigned within each patch, so the bees could not memorize the exact location of rewards within patches. Such a condition is called a uniform environment, because there is no variation in the number of rewards across patches. More formally, patch variance in this distribution of patch types was zero. The researchers predicted that the patch quality estimate of bees trained in this uniform environment should decline each time the bees find a food reward, because this would indicate the presence of one less food reward in that patch. Finding rewards becomes the bees' sampling information.

This leads to our first prediction:

Prediction 1: Bees trained in the uniform environment should have a lower probability of staying in a patch after each reward is found, and they should always immediately leave after finding the fifth reward.

A second set of bees was trained in a high-variance environment. Here, half the patches contained nine rewards, and half the patches contained one reward. The mean patch in this environment contained five rewards (as in the uniform environment), but the variation in rewards among patches was much higher. For bees trained

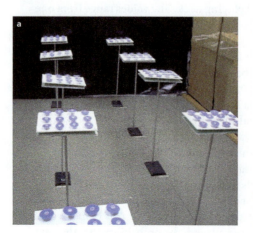

Figure 7.20. Foraging arena. (a) Food patches with 12 artificial flowers each. (b) Bee exploiting a flower.

in the high-variance environment, there was a different relationship between the number of rewards found and the expected number of rewards left in the patch. A bee using prior knowledge of the high-variance environment could now learn about the quality of a patch from sampling information. After finding one reward, it would know that there was a 50% chance the patch contained eight more rewards, and so the estimated patch quality should be relatively high. Once a bee found a second reward, it would know with 100% certainty that the patch contained seven more rewards, and so the estimate of the patch should be very high. After that, each reward found would mean that one less was left, and so the estimate of the patch quality would decline. This generates our second prediction:

Prediction 2: If bees use prior knowledge, individuals trained in the high-variance environment should have a high probability of staying in the patch after finding the second reward. The probability should decline thereafter.

The research team tested these predictions using 20 bees, half of which were trained in each environment, that fed on patches in a test environment. For each bee, the researchers recorded the number of flowers visited and how often it left a patch after finding a reward. This procedure was repeated for ten test patches for each bee.

As predicted, bees trained in the uniform environment had a lower probability of staying on the patch for each reward found. Bees in the high-variance environment, on the other hand, showed a much higher probability of remaining on the patch after finding the second reward compared to bees trained in the uniform environment. The likelihood of patch departure, however, did not decline for each reward found thereafter (Figure 7.21). These results are consistent with the idea that bees can learn different prior distributions. The data also suggest that bees combine information

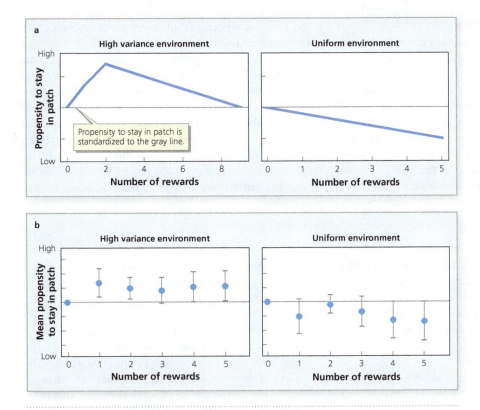

Figure 7.21. Predicted and observed patch-leaving behavior. (a) Predicted propensity to remain on a patch as a function of number of rewards found in each environment. (b) Observed (± error bar) propensity to stay in a patch as a function of number of rewards found in each environment (*Source*: Biernaskie, Walker, & Gegear 2009).

from prior distributions with sample information. It is important to note that these animals are not calculating probabilities. Rather, they display adaptive learning regarding the distribution of resources in their environment. Such estimation allows them to exploit food patches more efficiently. A wide variety of animals, ranging from insects to mammals, do the same (Valone 2006).

The original optimal foraging models are based on many simplifying assumptions. Recent work has incorporated more realistic assumptions about the environment (patches have different initial amounts of food) and the forager (multiple foraging costs and incomplete knowledge about food patch quality). Although these models may still seem somewhat simplistic, they have been remarkably successful at predicting complex foraging behaviors in a wide variety of animals.

Chapter Summary and Beyond

Animals find food using a variety of sensory modalities. Many foragers locate food visually and have trouble finding cryptic, as opposed to conspicuous, prey. Animals can increase their hunting success on cryptic prey by forming a search image that allows them to focus on a single prey type. Recent work has examined interactions among prey camouflage, predator search images, and the evolution of prey polymorphisms (Bond 2007). In addition, olfactory search images are being used to train dogs to identify explosive materials (Gazit et al. 2005).

Optimal foraging theory assumes that natural selection has favored behaviors that maximize fitness largely by maximizing energy intake rate, and it has become a powerful approach to studying the behavior of animals (Rosenzweig 2001). The optimal diet model is used to understand the diet breadth of foragers. This model predicts that animals should adopt a zero-one rule for each food type and that diet breadth should be affected by the abundance of the most profitable food items and not by the abundance of the least profitable items. Sih and Christensen (2001) provide a comprehensive review of studies that have tested this model. Their review

finds good support for the model, especially when it has been tested on foragers that consume immobile prey items.

The optimal patch-use model predicts how long a forager should remain in a patch. It predicts that foragers should spend longer in patches as travel time increases, and empirical tests show that it does often successfully predict behavior. Modification of the patch-use model includes the addition of energetic, predation risk, and missed opportunity costs of foraging. Brown and Kotler (2004) show how incorporation of these costs provides even greater understanding of feeding behavior.

While the optimal patch-use model assumes that animals have perfect knowledge of the food patches they encounter, most animals must estimate the quality of a food patch. Many appear to do so by combining prior knowledge about the distribution of patch types in the environment with current sampling information from a patch in a manner similar to Bayesian updating. Valone's (2006) review shows that many animals, including mammals, birds, fish, and insects, are capable making of such estimates.

CHAPTER QUESTIONS

1. In a review of empirical work, Sih and Christensen (2001) found that the optimal diet model better predicted the diets of animals that searched for immobile prey like seeds or clams compared to the diets of animals that hunted mobile prey. What assumptions of the model might best explain this finding?

2. Imagine the following experiments designed to understand patch-use behavior of gerbils. As in the experiment by Kotler, Brown, and Hasson (1991), you set out two seed trays 1 m apart that contain the same amount and kind of food buried in the patch.

Experiment A: One patch contains heavy soil, while the other contains light sifted sand. Which patch do you predict will have the higher GUD?

Experiment B: On a cold morning, blue jays are foraging for sunflower seeds in two nearby seed trays filled with sand. One tray is in a warm, sunny location and the other is in a cold, shady location. What prediction can you make about the relative GUDs of these two patches?

3. You record the diet selection of black-chinned humming-birds in a meadow for several weeks. Ten species of flowers are common in this field, but hummingbirds only feed on seven of them and ignore the other three. After a rainstorm, the abundance of one of the ignored flowers increases dramatically, but the abundance of all the other flowers is unaffected. Using the optimal diet model, how do you think this change will affect the diet selection of the hummingbirds?

4. Imagine an experimental arena in which a shrew (a small insectivorous mammal) must search food patches for buried invertebrates. The arena contains ten identical patches that are located 5 m apart and each contains 17 prey items. You record the patch-use behavior of the shrews and determine that, on average, they spend one minute in each patch and eat nine items before departing. What would you predict about the patch-use behavior of shrews if you increased the size of the arena so that the patches were 25 m apart?

5. Compare the experiments and results from Pohlmann, Grasso, and Breithaupt (2001) and Pohlmann, Atema, and Breithaupt's (2004) work on catfish and Piep et al.'s (2008) work on mouse lemurs. In what ways are their results similar?

6. What assumption is shared by the optimal diet and patch-use models?

Chapter 8

Antipredator Behavior

Figure 8.1. Red-tailed hawk. Red-tailed hawks are frequent predators of squirrels.

Our college campus, like many others, has a large population of gray squirrels (*Sciurus carolinensis*). They nest in trees and feed on the nuts and seeds the trees produce. Gray squirrels are typically found in dense woodlands, where many mature trees are close together, but they have also adapted well to urban environments such as our campus. In addition to their usual diet of nuts, our urban squirrels supplement their diet by scavenging food from garbage cans and people. They will visit one garbage can after another, sometimes diving to the bottom of a large can to find food. Squirrels will walk right up to a table in a large, open plaza and sit on their hind legs, a behavior that often entices students to give them cookies or French fries.

However, a couple of years ago, a pair of red-tailed hawks (*Buteo jamaicensis*) colonized our campus, made a nest in one of the light stands high above the soccer field, and raised a pair of young. Ever since, the hawks have been regulars on campus, and students and others have enjoyed watching them soar overhead and deliver food to their nestlings. The squirrels on campus have also noticed the hawks, because squirrels are one of these birds' favorite foods (Figure 8.1).

Since the hawks have been present, we have observed significant changes in the behavior of the campus squirrels. Now we rarely see squirrels looking for food far from the safety of trees, and when they do, they often dart rapidly from one place to another, rarely stopping in wide-open places to solicit food. A squirrel will often interrupt a search for food on the ground and stand up on its hind legs, presumably to look for an approaching hawk.

We have even seen a few attacks. Hawks typically attack when a squirrel is on the ground and far from a tree. Once the squirrel sees the hawk, it flees to the nearest tree, where it sometimes produces loud "chuck" calls and vigorously wags its tail back and forth. Invariably, this behavior causes other squirrels in the area to do the same, which produces quite a racket.

Have you ever wondered why a squirrel would behave this way with a predator so close? These squirrel behaviors are examples of antipredator adaptations. Many animals modify their behavior when predators are nearby to reduce the likelihood that they will be killed, or their predation risk. Animals modify their behavior by reducing overall activity and increasing their vigilance—being alert for predators. In this chapter, we will discuss the many ways that animals modify their behavior to reduce the risk of predation. Some will sacrifice one behavior in order to increase their vigilance, while others live in social groups, and still others display unusual behaviors that involve interactions with predators.

8.1   Animals modify their behavior to reduce predation risk

cryptic coloration Morphological coloration that matches the color of the environment to reduce detection by predators.

Obviously, individuals cannot be killed by a predator if predators do not detect them. One common morphological adaptation is thus to blend into the background through **cryptic coloration**—body coloration that matches the color of the environment. When that fails, individuals can modify their behavior to reduce risk. They might try to flee or simply lie low in order to be overlooked. Experiments have found evidence for each of these behavioral adaptations.

Predator avoidance by cryptic coloration in crabs

Karen Manríquez studied the effectiveness of cryptic coloration in the crab (*Paraxanthus barbiger*) (Manríquez et al. 2008). These South American crabs live and feed on the benthic substrate, or bottom, of marine coastal habitats, where they are eaten by many species of fish (Figure 8.2). The benthic habitat can consist of a uniform sandy color or a more heterogeneous mixture, thanks to small shell fragments called shell-hash. Young juveniles display color variation that ranges from brown to tan to white, with many spots and stripes. As juveniles grow, their body color changes to a more uniform light purple color. Juveniles experience much higher predation than adults, and so Manríquez and colleagues hypothesized that the complex coloration of juveniles is an adaptation to minimize detection by predators, as the color of the juveniles is more cryptic on the heterogeneous shell-hash background. The team also predicted that this coloration reduces predation.

To test these predictions, the researchers tethered individual juvenile crabs (with a mean carapace width of 11 mm) to 20 by 20 cm ceramic tiles of different colors. One set of tiles had a uniform white surface, while the other had a heterogeneous shell-hash surface. The tether was 10 cm long, which kept the crabs on the tile for the duration of the field experiment. All the plates were randomly placed in the ocean for a fixed period of time, and

Figure 8.2. Crab coloration. Juvenile crabs have a heterogeneous coloration.

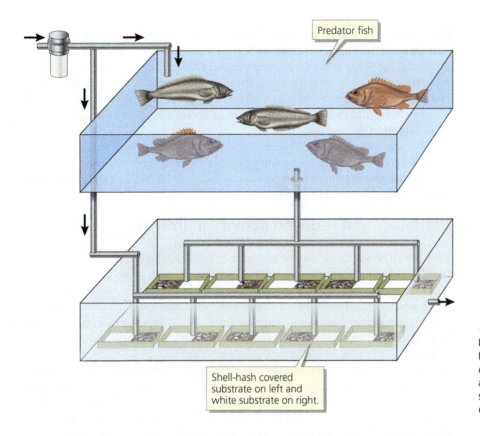

Predator fish

Shell-hash covered
substrate on left and
white substrate on right.

Figure 8.3. Crab experimental design.
Either plain water or water with fish odor
could be pumped into the experimental
arena where crabs have the choice of two
substrates: plain white or shell-hash
covered (*Source:* Manríquez et al. 2008).

the researchers recorded the number of crabs alive at the end of the experiment. As
predicted, they found very different survival rates for the crabs on the two different
surfaces. Only 30% of the crabs on the white surface tiles survived, while more than
60% of the crabs on the shell-hash-colored tiles survived. From these results, the
researchers concluded that juvenile crabs are indeed more cryptic when on the
heterogeneous shell-hash background.

The effectiveness of cryptic coloration requires that animals live in, or use, environments that match their body color. Manríquez and colleagues next conducted a
follow-up experiment to see whether crabs would select a shell-hash-colored surface
over a more uniform surface. They predicted that juvenile crabs will prefer a shell-
hash background over a uniformly colored background if given a choice, and that this
preference will be stronger when predation risk is high.

To test these predictions, they established experimental aquaria
that contained ten small plastic trays (20 by 40 by 5 cm). One half of
each tray was covered by a thin layer of shell-hash, while the other
was left empty and thus presented a uniformly white background
(Figure 8.3). This set-up gave crabs a choice of substrates to use. To
manipulate predation risk, the experimenters treated half of the
trays with predator odors by pumping seawater from a tank that
housed several predatory fish and the other half with seawater that
contained no predator odors. In each test aquarium, nine crabs were
placed in the center of each tray, and the researchers recorded the
proportion of these crabs on each substrate after three hours. To examine how crabs of different size would respond to these manipulations, they tested small crabs (less than 10 mm in length) in one set
of experiments and larger crabs (25 to 30 mm) in another.

Manríquez and colleagues found that almost all of the small
crabs preferred the shell-hash background and that even more crabs
used the shell-hash background in the predator odor treatment.

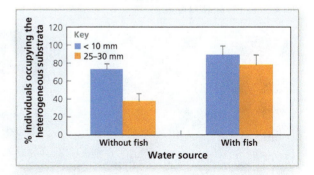

Figure 8.4. Substrate choice results. Mean (+ SE) percentage
of individuals occupying heterogeneous substrate with and
without fish predator odor. A greater percentage of both small
(blue) and large (orange) crabs occupied the substrate when
the fish odor was present (*Source:* Manríquez et al. 2008).

Larger crabs showed no preference for either background in the control tanks (perhaps because they experience lower levels of predation risk than do smaller crabs). However, the large crabs strongly preferred the shell-hash surface in the predator odor treatment (Figure 8.4). From these experiments, Manríquez and colleagues concluded that juvenile crabs can select an appropriate cryptic background when given a choice and that the complex color of juvenile crabs is an adaptation to minimize detection by predators.

This study implies that juvenile crabs can detect differences in surface coloration and select a background that allows them to be more cryptic. How animals assess their environment to match it has yet to be determined, but certainly this behavior is an effective antipredator adaptation.

Featured Research

Predators and reduced activity in lizards

All animals, even those that lack cryptic coloration, can reduce detection by predators by reducing their overall level of activity. Predators are less likely to notice stationary prey.

A simple experiment demonstrates how whiptail lizards (*Aspidoscelis uniparens*) reduce their activity level when predators are nearby. Whiptail lizards (about 7 cm long) live in deserts in southwestern North America and feed on insects and other arthropods (Figure 8.5a). While searching for food, these lizards can be killed by a variety of predators, including roadrunners (*Geococcyx californianus*), snakes, and larger lizards.

Douglas Eifler and colleagues designed an experiment to determine how whiptail lizards modify their behavior in the presence of predators (Eifler, Eifler, & Harris 2008). They established six 15 by 15 m pens in the Arizona desert. These pens had 40 cm tall walls that prevented lizards from either leaving or entering. The pens also contained natural desert vegetation. Three of the pens were designated the predator-present experimental treatment and contained two large, adult leopard lizards (*Gambelia wislizenii*), a common predator of whiptails (Figure 8.5b). The other three pens contained no predators and served as the control group. In each pen, the researchers placed six adult whiptail lizards and recorded their behavior over two weeks using focal animal sampling. Each focal observation period lasted for 15 minutes. To determine how the presence of predators affected the behavior of the whiptails, the researchers compared the average behavior of all individuals in each pen.

Eifler and colleagues noticed that the leopard lizards did hunt the whiptail lizards (two were killed) and so presented a high risk of predation. During the experiment, the lizards in the different pens exhibited significant differences in behavior. Whiptail lizards in the predator-present pens were less active: they spent less time moving and moved more slowly than did lizards in the control pens (Figure 8.6). It seems likely that this lower level of activity made them less noticeable to predators.

Figure 8.5. Lizards. (a) A marked whiptail lizard. These small insectivorous lizards are common in southwestern deserts. (b) A predatory leopard lizard.

What is the proximate mechanism for these behavioral changes? In vertebrates, exposure to a stressor, such as the sight of a predator, often causes an increase in the plasma concentration of corticosterone (e.g. Hubbs, Millar, & Wiebe 2000; Cockrem & Silverin 2002). This stress hormone affects both physiology and behavior (Sapolsky, Romero, & Munck 2000). For instance, lizards with high corticosterone levels spend more time hiding after a predator encounter than do individuals with low corticosterone levels (Thacker, Lima, & Hews 2009).

Behavioral modifications like this are common among animals, and researchers are applying their knowledge about this tendency to the challenge of minimizing crop damage from pests (Applying the Concepts 8.1).

Prey take evasive action when detected

Despite the benefits of cryptic coloration and reduced activity, predators do often find their prey. Prey must then attempt to avoid capture after being detected. One way is to take evasive action.

When a predator initiates an attack, prey often flee in an effort to escape. This predator-prey interaction can be very dramatic, as when a Canadian lynx (*Lynx canadensis*) chases a showshoe hare (*Lepus americanus*), or when a hawk chases a squirrel. Some species, like the hare, may try to simply outrun an attacker. More often, potential victims will flee to a safer location in the environment, like a squirrel that runs up a tree.

Other species cannot flee because they move slower than their predators, and in these cases, we see different kinds of antipredator behavior. For instance, in the dark of an Arizona night, a fascinating behavioral interaction occurs between big brown bat predators (*Eptesicus fuscus*) and their prey, tiger moths (*Bertholdia trigona*), which fly much more slowly (Figure 8.7). Bats hunt by sound. They emit sonic pulses and find flying insects such as moths by hearing the sonic pulses that bounce off of their prey. This sonar system is a highly effective way for bats to pinpoint the location and size of potential prey in the dark night sky. How, then, have moths and other prey evolved to avoid predation (Miller & Surlykke 2001)? First, moths' ears can also detect the sonic pulses, so they know when bats are hunting nearby. They then fly in a more erratic pattern in an attempt to evade the fast-moving bat. Second, when a bat gets close, a moth may even stop flying altogether just before it would be captured and begin a free-fall behavior. This quick, erratic drop can reduce capture probability by 40% (Radcliffe et al. 2008). Some moths, like the tiger moths in Arizona, have also evolved the ability to produce their own sonic pulses that interfere with a bat's sonar. This sonar-jamming adaptation is highly effective at reducing predation (Corcoran, Barber, & Conner 2009).

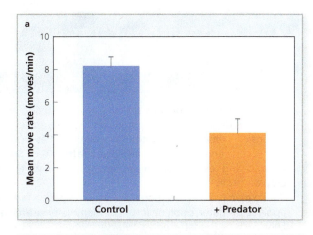

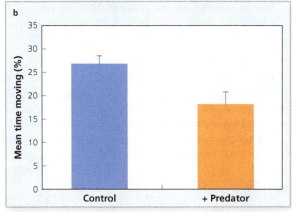

Figure 8.6. Lizard movement. Mean (+ SE) movement behavior. Lizards moved less when predators were present (*Source:* Eifler, Eifler, & Harris 2008).

Figure 8.7. Tiger moth and big brown bat. Bats use sonar to hunt moths.

8.2 Many behaviors represent adaptive trade-offs involving predation risk

A world without predators would allow animals to concentrate their behavior on activities that maximize their fitness, perhaps by increasing time searching for mates. In that light, the existence of predators represents a cost to animals: they modify their behavior to reduce the probability that they will be killed, as we saw previously. Many behavior modifications that reduce predation risk come at the cost

APPLYING THE CONCEPTS 8.1

Mitigating crop damage by manipulating predation risk

Many vertebrate herbivores are crop pests, resulting in losses of millions of dollars for farmers. Mice, rats, deer, and even kangaroos consume large quantities of grains such as wheat, soybeans, corn, and rice and can also damage fruit trees. Scientists are applying research about antipredator behavior to reduce crop damage.

To manipulate predation risk for a crop pest, a farmer can apply a predator odor near crops. Many prey use chemical cues such as predator odors to assess whether a predator is nearby and thus the level of predation risk. Prey—in this case, crop pests—often modify their behavior by decreasing the amount of time they spend foraging on crops that are near the predator odor.

Michael Parsons and colleagues conducted an experiment that applied this concept (Parsons et al. 2007). In Australia, Western gray kangaroos (*Macropus fuliginosus*) can be a source of significant crop damage (Figure 1). Parsons and colleagues tested the hypothesis that kangaroos will decrease their feeding on crops if they perceive a high level of predation risk. The researchers manipulated predation risk by applying predator urine near a food source. They used urine from a native Australian predator, the dingo (*Canis lupus dingo*); a North American predator, the coyote (*Canis latrans*); humans (*Homo sapiens*); and water as a control. They conducted their research on a population of 28 gray kangaroos that had free range on a 36 ha property on a wildlife sanctuary.

Each afternoon, when the kangaroos would normally be feeding, the researchers set up four feeding stations that each contained 400 g of grain. Next they selected one type of odor to use for the afternoon trial (one of the three urines or the water control). The four feeding stations were located at distances of 0, 6, 12, and 18 m from the odor source. Then they let the kangaroos feed from all of the stations for two hours. At the end of each trial, they collected all of the food containers and measured how much food was left to calculate how much food was eaten from each station. As we saw in Chapter 7, the amount left in each food station is known as the giving-up density, or GUD (Brown 1988), and is a useful measure of predation risk for foraging animals. If animals leave a lot of food in a food patch (a high GUD), that means they did not eat much because they perceived a high level of predation risk. On the other hand, if an animal leaves very little food (a low GUD), the animal ate a lot because it perceived a low level of predation risk.

Figure 1. Western gray kangaroo. A species that can be a crop pest in Australia.

The researchers found that when human urine or water was the odor, the kangaroos ate most of the food at all four stations. Similarly, kangaroos again ate most of the food from the stations that were 12 or 18 m away from dingo and coyote urine. However, kangaroos ate very little grain (left a very high GUD) from the feeding stations that were 0 and 6 m from the dingo and 0 m from the coyote odor. In fact, kangaroos ate almost no food from the feeding station that was right next to the dingo odor.

The researchers concluded that the urine of dingos and coyotes was effective in elevating the perceived level of predation risk for gray kangaroos because it greatly reduced the amount of food they consumed when it was nearby (less than or equal to 6 m). Use of predator urine may be an effective tool to mitigate crop damage. Research still needs to be done to determine whether kangaroos acclimate to the predator scent and thus whether its effectiveness will decline over time. Such nonlethal control of pests is one way that animal behavior research is being applied today.

---

behavioral trade-off Sacrificing one activity for another.

of behaviors that maximize fitness through feeding or mating. Sacrificing one activity for another is known as a **behavioral trade-off**. Such trade-offs are widespread, as we see next.

Increased vigilance decreases feeding time

vigilance A behavior in which an animal scans the environment for predators.

A widespread trade-off involves **vigilance behavior**, or scanning for predators (Figure 8.8). When many animals search for food, they often lower their head as they focus their attention on finding food items on or in the ground. This head-down position usually results in a reduced visual scanning range. Vegetation, such as grasses or other plants, can also increase the obstruction of an animal's scanning range. When an individual has its head down, it experiences a higher risk of predation. To counter this risk, individuals typically raise their head periodically to scan their surroundings for predators. Such vigilance behavior is ubiquitous: you can observe it by watching squirrels, birds, or deer feeding on a lawn.

Because an individual usually cannot simultaneously search the ground for food and scan effectively for predators, this represents a behavioral trade-off. In theory, doubling the amount of time spent scanning will halve an individual's time spent feeding. This results in both a decrease in the probability of being killed by a predator and an increase in the individual's risk of starvation (McNamara & Houston 1987), illustrating the trade-off. We can predict that animals will adjust their vigilance based on the level of risk in the environment: as predation risk increases, so should vigilance (Brown 1999).

Figure 8.8. Vigilance. Flamingos (*Phoenicopterus roseus*) in the foreground have their head up, engaged in vigilance behavior, while those in the background are feeding with their head down.

Vigilance and predation risk in elk

Featured Research

One test of this prediction involves elk (*Cervus elaphus*) in Yellowstone National Park. For much of the twentieth century, elk lived in an environment free of wolf (*Canis lupus*) predators, which had been killed by humans. In 1994, wolves were reintroduced into parts of the park, creating areas for elk with and without predators (Figure 8.9). John Laundré, Lucinda Hernández, and Kelly Altendorf examined how elk behavior was affected by the presence of this predator. Over a period of five years, they used focal animal sampling to quantify the time adult elk spent on feeding and vigilance behaviors (Laundré, Hernández, & Altendorf 2001). For each observation, they noted whether the individual was in an area with or without wolves, as well as its sex. Females were further classified as either with or without calves, because females with calves are often more vulnerable to predation.

The presence of wolves clearly affected elk behavior. Females, especially those with calves, spent significantly more time on vigilance behaviors and less time feeding

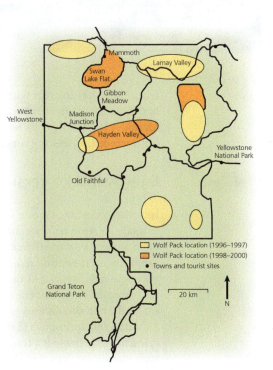

Figure 8.9. Wolves in Yellowstone. Wolf packs occurred in some parts of the park during the study (yellow and orange areas). All other areas were free of wolves (*Source:* Laundré, Hernández, & Altendorf 2001).

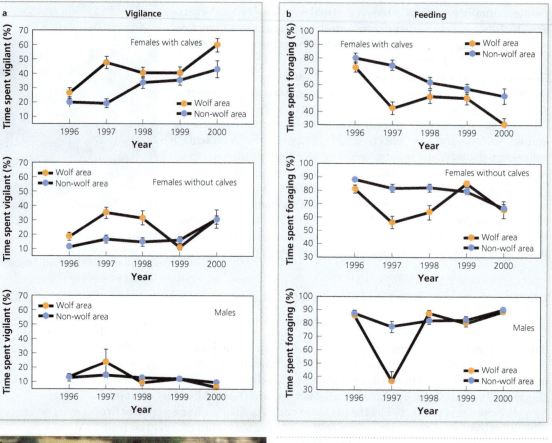

Figure 8.10. Elk behavior. Mean (± SE) vigilance (a) and feeding (b) behavior of elk. Females with calves (top) and without calves (middle) tend to exhibit higher vigilance in wolf areas (orange) than they do in non-wolf areas (blue). Males' vigilance level (bottom) does not differ in wolf and non-wolf areas. Females with calves (top) and without calves (middle) tend to spend less time feeding in wolf areas (orange) than they do in non-wolf areas (blue). Male feeding time (bottom) was not affected by wolves, except in one year. (c) Elk cow and calf (*Source:* Laundré, Hernández, & Altendorf 2001).

when they were in areas with wolves (Figure 8.10). In general, the presence of wolves had much less of an effect on males. The researchers suggest that reproductive success for males largely depends on maintaining high body mass, and so males have more to lose by reducing feeding time, even if the consequence is higher predation risk.

Energy intake versus safety in squirrels

Increasing vigilance is one way to reduce predation risk. Another strategy is to move to a safe location before eating food. You may have seen squirrels carry a nut from an open, grassy area and then eat it by or in a tree. Eating in a safe location such as a tree can decrease predation risk—but at what cost? Steve Lima observed this behavior in gray squirrels but noted that this carrying behavior could also result in increased energy costs. He posed the question: When should animals stay out in the open and eat, and when should they carry their food to safety to eat?

Lima and colleagues (Lima, Valone, & Caraco 1985) hypothesized that food carrying might represent a trade-off between feeding in safety (in a tree) and obtaining high energy intake rates. To develop this hypothesis, they created a mathematical model that includes two variables: the **energy intake rate**, or energy consumed per second, and the probability of being killed by a predator, such as a hawk or a dog. The model assumes that a squirrel's fitness increases as energy intake rate increases and the probability of being killed decreases. It also assumes that a squirrel is at risk of being killed while on the ground but safe while in a tree.

Suppose a squirrel finds a patch of ground, away from a tree, that contains many identical, large food items. In Lima's model, the squirrel has two options: it can remain at the patch and eat there, or it can carry items, one at a time, back to the nearest tree and eat there. The squirrel faces those options separately for each item. The first option is riskier, because a predator might attack, but this strategy should lead to high energy intake rates, because the squirrel doesn't spend time and energy traveling back and forth between the food patch and the tree. The second option, carrying food back to a tree, results in a lower energy intake rate but should also be safer. Clearly, a squirrel cannot simultaneously do both. How should the squirrel behave to maximize its fitness? What proportion of items should it carry back to the tree? The model made two predictions. The first prediction is that the larger the food item, the more often it should be carried to a tree. In other words, for a given distance between the food patch and the nearest tree, individuals should carry a greater proportion of food items to the tree as item size increases. This prediction relies on the assumption that larger food items take more **handling time**, or food-processing time. As handling time increases, so does the antipredator benefit in carrying a food item to safety.

The second prediction states that as the distance between the food patch and the tree increases, individuals should reduce their carrying behavior. In other words, for food items of a given size, squirrels should carry a smaller proportion to the nearest tree as the tree's distance from the food patch increases. Although the first prediction is intuitive, the second may appear a bit counterintuitive. Isn't running to the tree supposed to make the squirrel safer? However, more distance means more time spent running back and forth—and thus more time exposed to attack. And that means less antipredator benefit in carrying food to safety.

The two predictions reflect the trade-off between maximizing energy intake rates and minimizing time exposed to predators. Lima and colleagues (one of whom—one of the authors of this book—was an undergraduate then) tested these predictions in a large city park in Rochester, New York. They offered squirrels access to food patches that contained pieces of cookies, a favorite food item of urban squirrels. Each day, the researchers varied the size of the pieces (1, 2, or 3 g) and their distance from the nearest tree (3, 6, 9, or 12 m). Over the course of two months, all food size and distance combinations were offered several times.

The researchers found that at each distance, squirrels carried the largest item most frequently but rarely carried the smallest item to cover. Additionally, for a given food item size, squirrels carried fewer items to a tree as the distance between the food patch and a tree increased (Figure 8.11). Because the data matched both predictions of the model, Lima and colleagues concluded that the food-carrying behavior they observed in gray squirrels indeed represents a behavioral trade-off between maximizing energy intake rate and minimizing predation risk.

In fact, many species carry food to a safe location before eating it, and other experiments have demonstrated a similar behavioral trade-off in many species of birds (Lima 1985; Valone & Lima 1987). For example, black-capped chickadees (*Poecile atricapillus*), blue jays (*Cyanocitta cristata*), and white-throated sparrows (*Zonotrichia albicollis*) behaved just like squirrels in various tests. Other species, such as cactus wrens (*Campylorhynchus brunneicapillus*) and eastern towhees (*Pipilo erythrophthalmus*), carried more of the large than the small items but did not carry more items when close to safety. Finally, the carrying behavior of house finches (*Carpodacus mexicanus*) was completely inconsistent with both predictions of the model. Why doesn't the model work for all species? We don't know, but one possibility is that species perceive predation risk in different ways—maybe this could serve as a research project for you.

energy intake rate The energy acquired while feeding divided by total feeding time.

handling time The amount of time to manipulate a food item so that it is ready to eat.

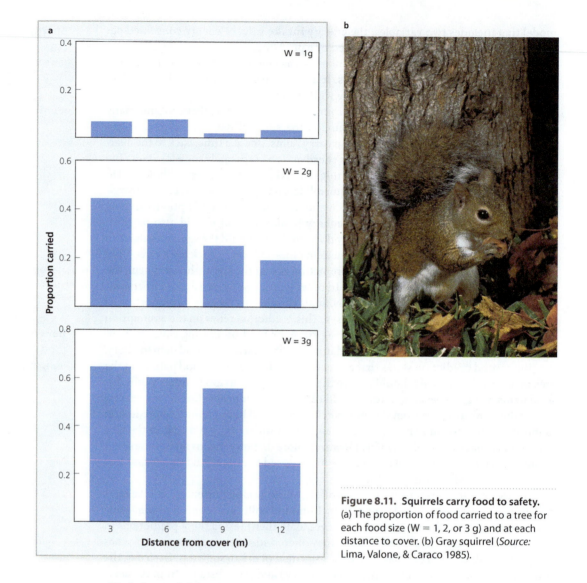

Figure 8.11. Squirrels carry food to safety.
(a) The proportion of food carried to a tree for each food size (W = 1, 2, or 3 g) and at each distance to cover. (b) Gray squirrel (*Source:* Lima, Valone, & Caraco 1985).

Rich but risky

The squirrel study demonstrates one general aspect of predation risk. In any habitat, there are risky and safe locations. In addition, in any environment, food patches are heterogeneous: some habitats or food patches contain abundant food and others contain less food. Let us call patches with abundant food *rich patches* and those with less food *poor patches*. Animals obviously should prefer to feed in rich patches because they can obtain higher energy intake rates there. However, many rich food patches also are found in places with higher numbers of predators or higher predator attack rates. That means that in order to feed in a rich patch, individuals must be willing to subject themselves to higher predation risk. We can therefore ask two research questions: How often will animals be willing to accept higher predation risk in order to feed in richer food patches, and what factors might favor animals feeding in richer but riskier food patches? The next two examples illustrate how other foragers balance predation risk and food acquisition when rich but risky places to feed are available.

Featured Research

Environmental conditions and predation risk in foraging redshanks

Redshanks (*Tringa totanus*) are medium-sized (approximately 28 cm in length) wading birds in the sandpiper family that live throughout Eurasia. They have a long, straight

SCIENTIFIC PROCESS 8.1

Feeding trade-off in redshanks

RESEARCH QUESTION: *Why do redshanks sometimes feed in the riskier salt marsh habitat instead of the safer mudflat habitat?*

Hypothesis: Energetic requirements and predation risk affect feeding behavior.

Prediction (1): The salt marsh habitat contains more food than the mudflat habitat.

Prediction (2): As the temperature drops, redshanks will feed more often in the salt marsh habitat to meet their energy needs.

Methods (1): The researchers:
- collected samples from 2 to 4 cm (depth of bill probe) of the substrate in small plots ($n = 47$) and then counted and identified all invertebrates found there.

Results (1): There were twice as many invertebrates per sample in the salt marsh habitat as there were in the mudflat habitat.

Methods (2): The researchers:
- recorded the number of redshanks feeding in the salt marsh within three hours of low and high tide and the ambient temperature over the course of 91 days.

Results (2): As ambient temperature decreased, more redshanks fed in the salt marsh habitat.

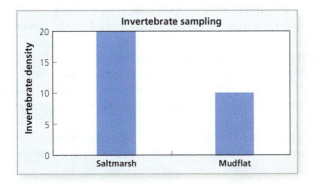

Figure 1. Food abundance. Saltmarsh habitat contains more invertebrate food than mudflat habitat (*Source*: Yasué, Quinn, & Cresswell 2003).

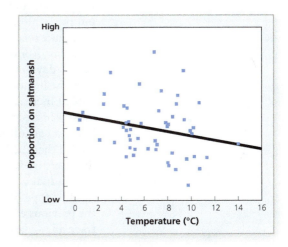

Figure 2. Redshank behavior. More redshanks fed on the saltmarsh when the temperature was low (*Source*: Yasué, Quinn, & Cresswell 2003).

Conclusions (1): There is twice as much food available for redshanks in the salt marsh habitat as there is in the mudflat habitat.

Conclusions (2): When it is cold, redshanks feed in the risky salt marsh habitat because it contains more food than the mudflat habitat. Feeding there should thus result in higher energy intake rates to meet their energetic requirements.

bill, which they use to find insects, earthworms, mollusks, and crustaceans that are buried in the soil. Redshanks can be found feeding in a variety of habitats, including salt marshes and mudflats near the coast.

Mai Yasué and colleagues (Yasué, Quinn, & Cresswell 2003) examined a trade-off between predation risk and foraging in a population of these birds (Scientific Process 8.1). In particular, they studied how weather conditions and predation risk affected daily feeding decisions during the winter in Scotland. The main predators in this area are sparrowhawks (*Accipter nisus*). The researchers first observed that predators were much more common in the salt marsh habitat than the mudflat habitat, with 20 times

endotherm An organism that generates its own body heat through metabolism to maintain its body temperature.

more predator attack rates on redshanks occurring when they fed in the salt marsh habitat than when they fed in the mudflat habitat. Yasué and colleagues concluded that the salt marsh habitat was riskier for a redshank than the mudflat habitat. Given this level of risk, one might conclude that redshanks should always feed in the mudflat habitat if they want to avoid predators. However, Yasué and colleagues observed that on some winter days, redshanks seemed to feed more often in the riskier salt marsh habitat. Why might redshanks feed more often in salt marshes during the winter?

Birds, like mammals, are **endotherms**. That means that they generate their own internal body heat, and when it is cold outside, their body temperature is much higher than the ambient temperature. To maintain their high internal body temperature, birds (and mammals) need to consume more food as the temperature drops. This characteristic led Yasué and colleagues to hypothesize that energetic requirements and predation risk affect redshank feeding behavior. That is, when the outside temperature is low, birds cannot find sufficient food by feeding only in the safer mudflat habitat. This hypothesis predicted that the salt marsh habitat contains more food than the mudflat habitat and that ambient temperature affects where redshanks will feed. On the coldest days of the year, redshanks will prefer to feed in salt marsh habitats.

To test their first prediction, the researchers measured the amount of food available in both salt marsh and mudflat habitats every two weeks over the course of the winter. They found that there were roughly twice as many invertebrates in each sample in the salt marsh habitat as there were in the mudflat habitat throughout the winter, which supported their first prediction. To test the second prediction, they recorded the number of redshanks that fed in each of the two habitats each day over the course of the winter, as well as the temperature on those observation days. As predicted, environmental temperature was strongly correlated with the behavior of the birds: the number of redshanks feeding in the salt marsh habitat increased significantly as the temperature declined. Thus, on colder days, most birds fed predominately in salt marshes, where they could obtain more food, but at a cost of suffering higher predator attack rates. From these data, they concluded that the behavior of redshanks was consistent with the hypothesis that individuals trade off higher energy intake rates in the salt marsh for higher safety in the mudflats.

This type of foraging–predation risk trade-off is widespread across taxa. While the previous examples examined vertebrates, invertebrates too behave in ways that trade off predation risk and food rewards.

Featured Research

Predation risk and patch quality in ants

Peter Nonacs has studied predation risk–foraging trade-offs in *Lasius pallitarsis*, a common ant in western North America (Figure 8.12). Much as in the study on redshanks, Nonacs asked whether ants will trade off higher predation risk for feeding in a richer food patch.

An important predator of *Lasius* ants is another ant, *Formica subnuda*, which is a large predatory species. In order to demonstrate that *Lasius* ants respond to the threat of predation by *Formica* ants, Nonacs set up a simple experiment with Larry Dill (Nonacs & Dill 1988). The researchers offered a *Lasius* colony two food patches that contained identical food. This food was a solution of sugar, proteins, vitamins, and nutrients and was a preferred ant food. The two food patches were identical distances away from a colony, but *Lasius* workers had to travel through a small arena that contained a *Formica* predator in order to obtain food from one of the patches. In the control patch, *Lasius* ants did not need to encounter a *Formica* predator to obtain food. Thus, the patches contained identical food, but the control patch was safe (no predators), while the other, the experimental treatment, was risky (*Formica* predator present). Nonacs and Dill found that after some *Lasius* workers had been killed, the rest would avoid the risky patch and obtain food only from the safe patch. From this experiment, they concluded that *Lasius* ants prefer to avoid predators.

Once they had demonstrated that *Lasius* ants prefer to avoid *Formica* predators, Nonacs and Dill designed another experiment. They wished to see if these ants would exhibit a behavioral trade-off between predation risk and food rewards. To examine this question, they offered the *Lasius* ants two food patches that differed in both predation risk and food quality (Nonacs & Dill 1990). One food patch was both risky (predator present) and rich with food, and the other was safe (no predator present) but poor in food quality. Both food patches again contained a solution of sugar, protein, vitamins, and nutrients and were placed equidistant from a colony. However, the two food patches now differed in their concentration of food: the rich food patch had a higher concentration of food than the poor food patch.

In this experiment, the *Lasius* ants always had to travel through an arena that contained a *Formica* predator if they wanted to feed from the rich patch. Thus, to feed from the richer patch, workers were subject to higher rates of predation, making it the riskier food patch. Over the course of the experiment, Nonacs and Dill varied the relative difference in quality between the two food patches. Because the food patches offered a food solution, the researchers could manipulate the quality of a food patch by altering the concentration of the food. The safe food patch always had the lowest concentration of food, while the richer food patch contained a concentration of food that was two to 16 times higher than that in the safe patch.

You can see that the greater the difference in concentration, the greater the potential rewards for feeding at the rich patch. Nonacs and Dill predicted that ants would be more willing to feed at the riskier food patch as the difference in quality between the two patches increased. In other words, the ants should be more willing to pay higher predation risk costs (and suffer higher rates of mortality) in order to obtain higher quality food.

Nonacs and Dill recorded the amount of food taken from each patch during a trial and the number of *Lasius* workers that were killed by the *Formica* predator near the richer food patch. The use of the richer patch did increase as the relative difference between patches increased, as predicted (Table 8.1). For instance, when there was only a twofold difference in patch quality between the rich and the poor patch, *Lasius* ants obtained only about 35% of their food from

Figure 8.12. ***Lasius* ant.** *Lasius pallitarsis* is a common ant in western North America that feeds on a diet high in sugar, proteins, vitamins, and nutrients.

TABLE 8.1 Ant food patch choice. As the relative difference in food concentration increases, ants feed more in the rich, risky food patch (*Source:* Nonacs & Dill 1990).

RELATIVE DIFFERENCE IN CONCENTRATION BETWEEN RICH AND POOR PATCHES	MEAN PERCENT OF FOOD TAKEN FROM RICH, RISKY PATCH
2 times richer	35
4 times richer	39
8 times richer	47
16 times richer	65

that patch. However, when the rich patch was 16 times better than the poor patch, *Lasius* ants obtained approximately 65% of their food from it. Furthermore, feeding from the rich patch was indeed risky for the ants: over 140 *Lasius* workers were killed near the rich patch by *Formica* predators during the experiment. Nonacs and Dill concluded that the ants were using information about both the quality of the food patches and the risk of predation in their foraging decisions. *Lasius* ants were willing to suffer higher predation rates to feed from richer food patches.

Featured Research

Mating near predators in water striders

As these examples show, behavioral trade-offs between predation risk and food rewards can be very sophisticated and are found in diverse taxa. However, predation risk trade-offs are not restricted only to feeding behavior. Predation risk also affects mating behaviors and results in different behavioral trade-offs.

Water striders (*Gerris remigis*) are insects in the order Hemiptera, along with cicadas and aphids. Water striders live on the surface of streams and ponds in North America and can walk or "skate" over the top of the water by trapping a layer of air between the foot and the water's surface. They feed on smaller aquatic insects by capturing individuals that are trapped on the surface of the water. During the spring and summer, males mate by attaching themselves to the backs of females. Water striders are eaten by sunfish (*Lepomis cyanellus*) and other fish, but they coexist with their predators. Water striders reduce their probability of being killed both by reducing their overall activity level and by moving to the edge of a pond or stream, where the water is too shallow for fish predators.

Andy Sih and colleagues have been studying the behavior of water striders for many years. In one classic study, they examined the effects of predators on water strider mating behavior (Sih, Krupa, & Travers 1990). The researchers asked a basic research question that had rarely been examined: Do animals reduce their mating behavior in the presence of a predator? They predicted that water striders would exhibit a predation risk trade-off and reduce mating behavior when predation risk was high but would exhibit high levels of mating behavior when predation risk was low.

To test this prediction, they created a set of four artificial wading pools in the laboratory. Water striders were individually marked and then randomly assigned to two treatments: in one, fish predators were present, and in the other, they were absent. During the study, the researchers collected data on the location of all water striders (the center or edge of a pool), the time spent active, and whether the insects were engaging in mating behavior (Scientific Process 8.2).

Sih and colleagues found that the behavior of male and female water striders was affected strongly by the presence of predators. In the pools with no fish predators, water striders exhibited high levels of activity. Individuals were commonly observed in the center of the pool and were seen mating about 20% of the time. In the pools with predatory fish, water strider activity was much lower. Individuals were rarely seen in the center of the pool and were observed mating less than 5% of the time. Based on their results, Sih and colleagues concluded that water striders do reduce their mating behavior in the presence of predators. In addition to an overall reduction of mating behavior, the researchers also found that the duration of each mating bout was shorter in the presence of predators, perhaps because a mated pair is less mobile (and so more vulnerable) if attacked.

These experiments were among the first to demonstrate a behavioral trade-off involving predation risk and the mating behavior of animals. When predation risk is low (no predators present), water striders are active and engage in much mating behavior. However, when predation risk is high (predators present), water striders reduce both their overall level of activity and their level of mating behavior.

SCIENTIFIC PROCESS 8.2

Mating behavior trade-off in water striders

RESEARCH QUESTION: *How does the presence of a predator affect the mating behavior of water striders?*

Hypothesis: Water striders modify their behavior in response to variation in predation risk to avoid being killed.

Prediction: Water striders will reduce their mating behavior in the presence of a predator.

Methods: The researchers:

- created artificial pools in the lab. Treatment pools contained predators, and the control pools did not.
- marked individual sunfish predators (three in each predator treatment pool) and water striders (four males and four females in each pool) for identification.
- observed water strider activity every 30 minutes for six hours per trial (a total of 12 observations for each treatment).

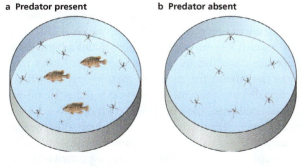

Figure 1. Water strider experiment. Predator present (a) and predator absent (b). (*Source:* Sih, Krups, & Travers 1990).

Results: In the pools with no fish (controls), water striders exhibited high levels of overall activity, with mating observed about 20% of the time. In the pools with fish (treatment), water strider activity was greatly reduced, and mating was observed only 5% of the time.

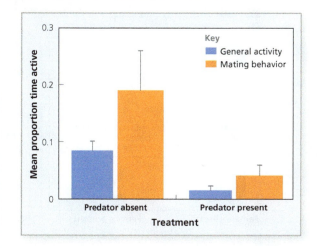

Figure 2. Water strider behavior. Mean (+ SE) activity and mating behavior were more common in the absence of predators (*Source:* Sih, Krupa, & Travers 1990).

Conclusions: Predation risk can affect the mating behavior of water striders. There is a behavioral trade-off between mating behavior and predation risk.

We now know that such trade-offs are quite common. Let's examine a recent study on fiddler crabs.

Mating and refuge use in fiddler crabs

Featured Research

Fiddler crabs (*Uca mjoebergi*) live in intertidal habitats and feed on the surface of the sand during low tide. The species gets its name from the morphology of the males, which possess a single large claw that they wave back and forth in a sexual display to attract females (Figure 8.13). Individuals construct and live in burrows in the sand, which provide a safe refuge from bird, mammal, and reptile predators. To feed or attract mates, crabs need to be active aboveground, but they will quickly retreat to their burrow in response to a predator attack.

Figure 8.13. Fiddler crabs. Males use burrows in the sand for protection from predators. They wave their large claw to attract females.

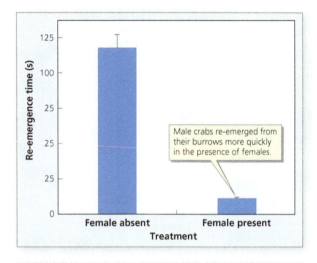

Male crabs re-emerged from their burrows more quickly in the presence of females.

Figure 8.14. Male re-emergence times. Mean (+ SE) re-emergence times of courting male fiddler crabs after a predator attack in plots with and without females present (*Source:* Reaney 2007).

Leeann Reaney examined how predators and mating opportunities affect refuge use by male fiddler crabs (Reaney 2007). The benefit of refuge use is that it provides protection from predators. The cost is that a male inside a burrow cannot court a female. Reaney hypothesized that male crabs would adjust their refuge use behavior based on the costs and benefits. Specifically, she predicted that males would emerge from their refuges more quickly after an attack (and thus expose themselves to risk) if potential mating opportunities were available aboveground.

To test this prediction, Reaney created 30 small plots (35 by 35 cm) in a coastal intertidal mudflat in Darwin, Australia. Each plot contained several males and their burrows. In 15 of these plots, she tethered a single, sexually receptive female by gluing a thin piece of string onto the carapace of the female and attaching it to a nail buried in the sand. The other 15 plots served as controls and did not contain a tethered female. Reaney first waited for all the males in a plot to become active aboveground. She then flew an artificial model of a bird predator over the plot and recorded how long the males stayed in their burrows after the "attack." She averaged the behavior of all males within each plot and compared the average values for males in each treatment.

Reaney found that all males in the control plots retreated into their burrows and stayed underground for about 120 seconds. In contrast, in the plots with a receptive female, 12% of males did not retreat into their burrow during the attack. Those males that did flee into their burrow reemerged in less than 15 seconds (Figure 8.14). From these results, Reaney concluded that males adaptively adjust their refuge use based on its benefits and costs. Males trade off the safety of the burrow for increased mating opportunities when receptive females are present. When presented with no mating opportunities, males use their burrows much more.

These examples demonstrate how individuals alter their behavior to reduce predation risk. They also show how this risk reduction comes at the cost of reduced feeding or mating activity. In the next section, we'll look at another common way for individuals to reduce predation risk: association with others.

8.3   Living in groups can reduce predation risk

social group A set of individuals that live near and associate with one another.

Many animals live in **social groups**, meaning they live near other individuals. One important benefit of this behavior for social animals is that living in groups can reduce predation risk. In Chapter 14, we examine social behavior in more detail. Here, we focus on some of the ways that social animals experience lower levels of predation risk than solitary individuals.

Featured Research

The dilution effect and killifish

dilution effect A reduction in the probability of death as a result of associating with others.

Suppose a predator attacks a group of two animals and kills one. For each prey, there is a one-half probability that it was the one killed. If instead there had been ten prey, there would be only a one-tenth probability that any one individual was killed. In a group of 100 individuals, that probability would fall to one-hundredth. The probability of any one individual's dying is diluted by the presence of others. This is called the **dilution effect** (Foster & Treherne 1981).

In the dilution effect, there is a $1/N$ probability of dying when a predator makes a single kill in a group of N individuals. As group size N increases, the probability that any single individual dies decreases. Clearly, this size increase can greatly reduce an individual's probability of dying as a result of a predator attack, especially in a group of hundreds or thousands of individuals. The only behavior required for the dilution effect to occur is to join others in a group.

D. J. Hoare and colleagues (Hoare et al. 2004) examined the dilution effect in a simple laboratory experiment using banded killifish (*Fundulus diaphanus*) (Figure 8.15). They predicted that fish should prefer to associate with larger groups rather than smaller groups, particularly when predation risk is high. Banded killifish are small (less than 10 cm standard length) freshwater fish that live in eastern North America. These fish tend to move in schools (or shoals) and can use chemical cues to perceive predation risk. The researchers compared the size of the group that killifish form in a high-predation risk treatment with the size formed in a control treatment.

Figure 8.15. Banded killifish. These small freshwater fish join groups when they perceive predation risk.

Experiments were conducted in a large tank (100 by 100 by 15 cm). To simulate high predation risk, the researchers added a diluted concentration of killifish skin extract to the tank. Many species of fish respond to the crushed skin extract of conspecifics (individuals of the same species), presumably because it indicates the presence of a predator that has made a kill nearby. In the control treatment, only water was added to the aquarium. For each trial, ten killifish were placed in the experimental tank and allowed to acclimate overnight. The next day, skin extract (treatment) or water (control) was added to the tank, and a 60-minute trial was videotaped. From the videotape footage, the researchers recorded shoal size every 30 seconds. Individuals were considered to be in the same shoal if they were within four body lengths of one another.

The results of the experiment were clear: median shoal size in the control was just two fish, with most fish swimming alone in the aquarium. However, in the high predation risk treatment, median shoal size was ten fish, the maximum group size. From these results, Hoare and colleagues concluded that banded killifish prefer to associate with more individuals when predation risk is high, as predicted by the dilution effect.

The selfish herd and vigilance behavior

A second way social groups can lower predation risk requires that individuals behave in a certain way: stick near the group's center. The **selfish herd hypothesis** (Hamilton 1971) assumes that a predator is more likely to kill a member on the outside of a group, because it will encounter outside individuals first (Figure 8.16). Individuals can therefore lower their predation risk by placing others between themselves and an approaching predator. In contrast to the dilution effect, the selfish herd hypothesis requires individuals to continually adjust their position in a group.

A third antipredator benefit of associating with others involves vigilance behavior. In social groups, more individuals can be vigilant, and hence "more eyes" can observe potential predators. Many researchers have documented a common pattern known as the **group size effect** (Pulliam 1973), in which the vigilance behavior of each individual decreases as the number of individuals in its group increases (Figure 8.17). The group size effect on vigilance behavior has been observed in a wide variety of birds and mammals and even in fish, and it is thought to result from two factors. First, as group size increases, each individual is safer because of the dilution effect, and so each individual can reduce its level of vigilance. Second, as group size increases, each individual can afford to scan less frequently because more individuals are scanning. Therefore, the collective vigilance of the entire group can remain high,

Figure 8.16. Selfish herd. Turkeys (*Meleagris gallopavo*) on the exterior of the flock are more likely to be killed by a predator than are individuals in the center.

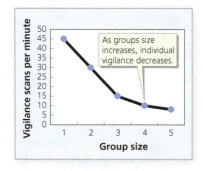

As groups size increases, individual vigilance decreases.

Figure 8.17. Group size and vigilance. The relationship between group size and individual vigilance.

selfish herd hypothesis Individuals can reduce their predation risk by moving to the center of a group.

group size effect Vigilance behavior of an individual declines as group size increases.

even though each individual scans less often. If one individual in the group sees a predator, it will quickly flee to protective cover. That motion alerts other group members to the imminent danger so that they too can respond appropriately. Next we examine a study of doves that investigated both the group size effect and the selfish herd hypothesis.

Group size effect and the selfish herd hypothesis in doves

Figure 8.18. Scaled dove. These small doves are common throughout Latin America.

Scaled doves (*Columbina squammata*) are small birds that are widely distributed throughout Latin America (Figure 8.18). Raphael Dias studied dove scanning and foraging behavior on the University of Brasília campus (Dias 2006). He investigated two research questions. First, he wondered whether scaled doves exhibited the group size effect. If so, he predicted they should exhibit reduced scanning rates as their group size increases. Second, he examined whether doves at the edge of a flock are exposed to higher predation risk, as assumed in the selfish herd hypothesis. If so, he predicted they should exhibit higher scanning rates than individuals at the center of a flock.

Dias studied wild flocks all over the campus for two months. When he found a flock of feeding birds, he recorded the group size and then selected one focal animal. He recorded the location of the focal bird in the flock (central or edge position), its scan rate (number of times it lifted its head to scan the environment over a five-minute period), scan durations, and the total amount of time it spent feeding. He also made note of any position shifts in the focal bird (i.e., any change from central to edge position or vice versa). He then moved to a different location on campus to find another flock. In total, he observed 150 flocks that varied in size from one to 19 individuals.

Dias found that individual vigilance behavior decreased significantly as group size increased, indicating that these doves did indeed exhibit the group size effect. In addition, the feeding time of individuals increased as group size increased (Figure 8.19). Furthermore, individuals in the center of a flock behaved differently than did individuals at the edge. Birds in the center exhibited lower scan rates and had higher feeding rates than individuals at the edge of a flock (Figure 8.20). And when position shifts did occur, individuals more often moved from the edge of the flock to the center—that is, they exhibited behavior consistent with the selfish herd hypothesis.

Dias concluded that doves exhibit a trade-off between scanning for predators and feeding: scan rate declined and feeding time increased strongly with group size. In addition, the data support an important assumption behind the selfish herd

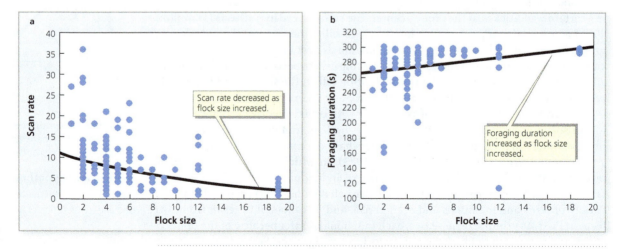

Figure 8.19. Group size effect in doves. The relationship between scan rates (a) and foraging duration (b) for different flock sizes (*Source:* Dias 2006).

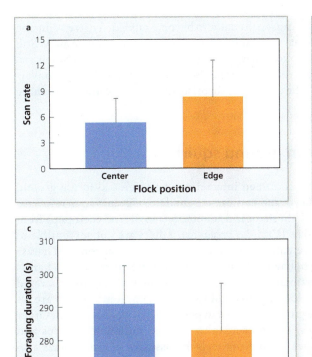

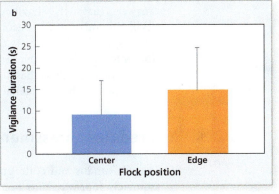

Figure 8.20. Center-edge dove data. Mean (+ SE) (a) scan rate, (b) vigilance, and (c) foraging duration for birds in the center and at the edge of flocks. Doves in the center scanned less, were less vigilant, and also foraged more than doves on the edge (*Source:* Dias 2006).

hypothesis: individuals at the edge of a group appear to experience a higher threat of predation than individuals near the center of a group.

If there are benefits of being near to the center of a group (lower predation risk, higher feeding times), though, then why don't all birds try to move to the center? Dias suggests that a dominance hierarchy maintains group position: dominant individuals most likely can maintain their central position while subordinate individuals are forced to stay at more peripheral positions (see Chapter 10).

These examples illustrate how individuals in groups can benefit from reduced predation risk. In all of these examples, individuals behave to avoid predators. In the next section, we examine how selection can favor individuals that directly interact with predators.

8.4 Some animals interact with predators to deter attack

So far, we have seen that animals modify their behavior to minimize detection by predators. However, predators cannot be entirely avoided. When an encounter occurs, how should a potential victim behave? One simple answer is that the individual should flee to a safe location. While such behavior is common, in many situations a potential victim does *not* flee. Instead, it actively interacts with a predator. For example, when a hawk or owl is observed by smaller songbirds, they will often engage in **predator harassment**, which involves rapid movement such as diving at or around a predator and is often coupled with loud vocalizations. Some species, like chimpanzees, will throw objects at a predator. Others, like ground squirrels and kangaroo rats, will kick dirt at a predator. The harassment of a predator by more than one individual is known as **mobbing behavior.**

On the surface, predator harassment behavior appears to increase the likelihood that an individual will be killed, because it means direct interaction with a predator within a short distance. So why would an individual harass a predator? One possible

predator harassment Interactions with a predator to deter attack.

mobbing behavior A behavior in which two or more individuals harass a predator.

answer is that such behavior can cause a predator to become more defensive, reducing its likelihood of attack. Second, it could simply cause the predator to move away. Still other species will engage in behaviors that appear to enhance or "advertise" their visibility to a predator. Advertisement behavior might benefit an individual by conveying to a predator that it has lost the element of surprise. Let's see how researchers have tested each of these explanations.

Predator harassment in ground squirrels

Aaron Rundus and colleagues examined interactions between California ground squirrels (*Spermophilus beecheyi*) and rattlesnake predators. Adult squirrels have evolved several behavioral and physiological adaptations to defend their pups from predation by Pacific rattlesnakes (*Crotalus oreganus*). Adults are not susceptible to rattlesnake venom, but pups are vulnerable, and so adults often confront and harass snakes to deter their hunting behavior (Rundus et al. 2007). Squirrels often harass snakes by making rapid movements, engaging in vigorous **tail flagging** (a rapid wagging of the tail), and even kicking dirt and rocks at them. Rattlesnakes are pit vipers and have heat-sensitive pit organs that can perceive infrared radiation. Other snakes, such as gopher snakes, are insensitive to infrared radiation. Rundus and colleagues investigated whether ground squirrels' harassment of rattlesnakes also had an infrared component and whether such harassment caused snakes to become more defensive (Scientific Process 8.3).

tail flagging Rapid wagging of the tail, especially when a predator is nearby.

The researchers staged encounters between 12 adult ground squirrels and different stimuli (treatments). One stimulus was an adult Pacific rattlesnake, another was an adult gopher snake, and a third was a control (another ground squirrel). Each stimulus was housed in a wire mesh cage that prevented direct contact but allowed the passage of visual, auditory, olfactory, and infrared information. All trials were videotaped with an infrared-sensitive video camera, which allowed the researchers to quantify the heat given off by the ground squirrel from different parts of its body during an encounter. The researchers measured the temperature of both the body and the tail of the ground squirrels.

During encounters with both species of snakes, all ground squirrels exhibited harassment behavior, particularly tail flagging. No harassment behavior was observed in interactions with another ground squirrel. Interestingly, analysis of the videotapes also revealed that the ground squirrel tail temperature was much higher during encounters with the rattlesnake compared to its temperature during encounters with the gopher snake. Ground squirrels appear to add an infrared component to their harassment behavior, but only when they interact with rattlesnakes that can detect such signals.

How effective is this harassment behavior? Does it indeed put predators on the defensive? To determine how the infrared signal might affect rattlesnakes, another experiment used robotic model ground squirrels, based on lifelike taxidermy mounts. The researchers could manipulate the models to exhibit tail-flagging behavior and could alter the temperature of the tail. Rundus and colleagues then trained 14 adult Pacific rattlesnakes to actively search for food in a simulated ground squirrel burrow they created in the laboratory. This set-up required the snakes to travel over 1 m to enter the burrow. After the snakes were trained, a model squirrel was placed next to the burrow entrance and a snake was introduced into the experimental arena.

The researchers exposed each snake to two treatments. In one, the model ground squirrel tail-flagged with a cold tail (thus producing no infrared signal). In the other, the model tail-flagged with a hot tail (thus producing an infrared signal in the tail). The team found that when a snake interacted with a model that gave off an infrared signal, the snake exhibited more defensive behaviors (e.g., defensive coiling) and moved less (reduced search behavior). This experiment, examining a proximate antipredator mechanism, showed that adding an infrared component to predator harassment behavior is an effective strategy for hindering a snake predator. Why do

SCIENTIFIC PROCESS 8.3

Predator harassment by California ground squirrels

RESEARCH QUESTION: *How do California ground squirrels harass rattlesnakes?*

Hypothesis: Ground squirrels use an infrared signal during rattlesnake harassment by increasing blood flow to their extremities.

Prediction: When attacked by a rattlesnake, a ground squirrel will shunt more blood to its tail than it will when being attacked by a gopher snake (*Pituophis melanoleucus*).

Methods: The researchers:

- conducted four types of randomly ordered trials using 12 adult female squirrels: (1) a baseline trial with no stimuli, (2) a control trial with a conspecific (squirrel) stimuli, (3) an experimental trial with a northern Pacific rattlesnake stimuli, and (4) an experimental trial with a gopher snake stimuli.
- placed a squirrel's home cage next to the opening of the test chamber. Trials began after the squirrel engaged the stimulus animal in the test chamber. Baseline and control trials were compared to experimental trials. Ten-minute trials were spaced two days apart.
- used infrared video to collect data on squirrel body and tail temperature.

Results: The temperature of a ground squirrel's tail was, on average, 2°C higher when it harassed a rattlesnake than when it harassed a gopher snake.

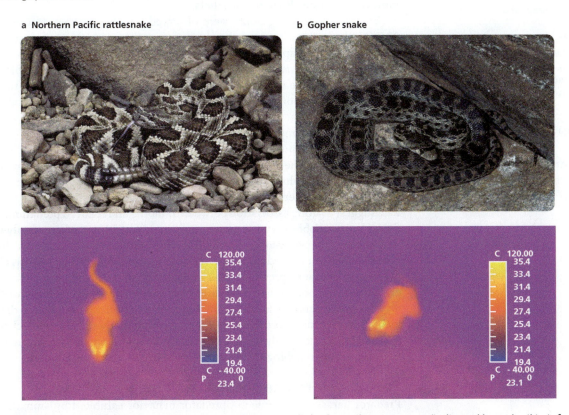

Figure 1. Ground squirrel tail temperature. Note squirrel's higher tail temperature (indicated by red tail in infrared image) when harassing a rattlesnake (a) compared with harassing a gopher snake (b) (*Source:* Rundus et al. 2007).

Conclusion: California ground squirrels include an infrared component in their harassment of rattlesnakes, presumably by increasing blood flow to the tail, which results in slowing of the snake attack.

rattlesnakes exhibit more defensive behaviors in response to the infrared signal? Squirrels often combine tail flagging with substrate throwing, and so snakes may have anticipated additional harassment. Rattlesnakes often hunt at dusk when it might be difficult to observe the tail-flagging behavior. The addition of the infrared signal by squirrels may enhance tail-flagging behavior in such low-light conditions. However, this hypothesis remains to be tested.

Mobbing owl predators

A second explanation for predator harassment behavior is that it can cause a predator to move away from an area, the **move-on hypothesis** (Curio, Ernst, & Vieth 1978). Here, the benefit of mobbing is reduced predation risk because the predator moves away from the mobbers or their offspring. Chris Pavey and Anita Smyth tested this prediction by studying the mobbing behavior of forest birds in response to the presence of powerful owls (*Ninox strenua*) (Pavey & Smyth 1998).

The powerful owl is a large (approximately 1500 g) nocturnal predator that can capture diurnal birds at dawn and dusk, as well as individuals on their nocturnal roost. The researchers studied the mobbing behavior of birds toward owls at a study site in Brisbane Forest Park, Australia, which comprised both rainforest (12% of the habitat) and open forest (88%). Pavey and Smyth noted the location of all roosting owls and all mobbing bouts by birds, as well as the response of the owl. To estimate the antipredator benefits of mobbing, the researchers compared the frequency of capture of different species of birds by analyzing owl pellets under roosts to determine the avian species in the owl diets.

Thirty-five mobbing bouts were observed, involving seven species and up to 22 individuals. Six of the seven mobbing species resided in open forest. Owls were observed roosting approximately half the time in each habitat, even though the rainforest habitat was relatively sparse. Seven bouts occurred in which the owl did not call or move in response to mobbing. Owls called or watched the mobbers during 54% of the bouts and moved to another perch during 20% of mobbing bouts (Figure 8.21).

The 39 avian prey identified in owl pellets ranged in size from 75 to 800 g. To compare the frequency of capture of mobbing and nonmobbing species, the researchers examined the frequency of capture of the six mobbing and six nonmobbing species that fell in this range. The frequency of predation on mobbing species was nearly an entire order of magnitude lower than that for nonmobbing species. These data suggest a strong association between mobbing and reduced predation. However, five of the six mobbing species were passerines (songbirds), whereas only two of six nonmobbing species were songbirds. Therefore, another explanation for this pattern is that owls prefer to feed on nonpasserines. The researchers rejected this alternative explanation because previous work indicated that owls feed on passerines almost eight times more often than nonpasserines (Pavey, Smyth, & Mathieson 1994).

The researchers concluded that mobbing of powerful owls can lead to the displacement of a predator, as predicted by the move-on hypothesis. The researchers suggest that mobbers (who mostly live in open forest) suffer lower predation than nonmobbers, perhaps because mobbing influences the owls' selection of roosting sites.

Pursuit deterrence and alarm signal hypotheses

Predator harassment is one antipredator behavior exhibited by animals that have spotted a predator. Another involves behaving in a manner that appears to make an individual more obvious to potential predators. For example, after spotting a predator, Thomson's gazelles (*Eudorcas thomsoni*) and impala (*Aepyceros melampus*) will **stot**, a behavior that resembles pogo stick–hopping as they move forward, while zebra-tailed lizards (*Callisaurus draconoides*) will raise their long black and white tail and wag it back and forth (Figure 8.22). You may have even seen such "advertisement"

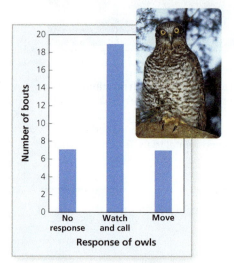

Figure 8.21. Owl response to mobbing. On 19 occasions, owls being mobbed mostly watched their mobbers and vocalized. On seven occasions, owls moved to a new perch. On seven other occasions, owls did not respond (*Source:* Pavey & Smyth 1998). *Inset:* a powerful owl.

move-on hypothesis Predator harassment functions to cause a predator to abandon an area.

stot (stotting) A high jump into the air with all four feet off the ground in which an individual moves forward.

behavior if you have ever startled a white-tailed deer (*Odocoileus virginianus*) while walking in the woods. Many times, when deer spot a predator (including humans), they lift and wag their tail, revealing the white underside as they move a short distance away.

At first, these behaviors seem to be at odds with behaving to minimize exposure to predators. Why would a prey advertise itself to a predator? One possible answer is the **pursuit-deterrence hypothesis**: advertisement behavior informs the predator that it has been detected and so pursuit is not likely to be successful. Most predators rely on surprise to successfully capture their prey, which allows them to close in rapidly before the victim has a chance to flee. If predators know that a prey has already spotted them, the likelihood of successful capture is very low, particularly when the distance between the predator and the prey is large. The predator is then better off searching for another victim.

The pursuit-deterrence hypothesis predicts that prey animals like deer that are far from a predator should more often exhibit tail-flagging behavior than individuals that are close to a predator. Why? Because when an individual is far from a predator, it is relatively safe, and so it can benefit by deterring an attack that has little probability of success. However, if the predator is nearby when spotted, the risk of capture is much higher. It is then unlikely that the predator will not initiate an attack, and so the individual should not advertise itself to the predator.

An alternative explanation for advertisement behavior is that individuals are warning nearby conspecifics, the **alarm signal hypothesis**. This hypothesis makes different predictions than the pursuit deterrence hypothesis. For example, if the function of tail flagging is to signal others, then tail-flagging behavior should occur very rarely among solitary prey, because there are no other individuals nearby to warn. In contrast, tail-flagging behavior should be commonly seen in groups of prey because of the presence of other individuals nearby. Which hypothesis best explains tail-flagging behavior? White-tailed deer have offered an interesting test case.

Tail-flagging behavior in deer

It is difficult to test the pursuit-deterrence or alarm signal hypothesis using natural predators, because few predator-prey encounters are observed. Most tests of these hypotheses therefore use humans as potential predators. Tim Caro and colleagues (Caro 1995; Caro et al. 1995) examined tail flagging in white-tailed deer in a reserve in southeastern Michigan. The deer population studied was hunted regularly and so treated humans as a threat. Caro tested the predictions of both hypotheses. The alarm signal hypothesis predicts that solitary deer should exhibit tail flagging less than deer in social groups, and the pursuit-deterrence hypothesis predicts that tail-flagging behavior should occur more often as the distance between the predator and the deer increases.

Caro and colleagues recorded deer behavior over a month in late fall. They themselves acted as predators by slowly walking through the reserve. Once they spotted deer, they recorded the number of deer nearby and whether any were aware of their presence. This part was easy, because a deer aware of an observer would stand upright and stare directly at the observer. The researchers then selected a focal animal, took note of its location, and moved slowly toward it as if stalking a prey. During the stalk, the observer would record the number of times the deer exhibited tail-flagging behavior and the distance between the observer and the deer. Once the deer fled, the observer would record its flight distance, flight time, and any tail-flagging behavior.

The researchers found no difference in the incidence of tail-flagging behavior between solitary deer and deer in social groups; both types of deer tail-flagged approximately 85% of the time. The researchers therefore rejected the alarm signal hypothesis. They also found that deer more often exhibited tail flagging as their

Figure 8.22. Advertisement behavior. (a) Impala stotting. (b) Zebra-tailed lizard tail display.

pursuit-deterrence hypothesis
Advertisement behavior informs a predator that it has been detected and so pursuit is not likely to be successful.

Featured Research

alarm signal hypothesis
Advertisement behavior functions to warn nearby conspecifics of danger.

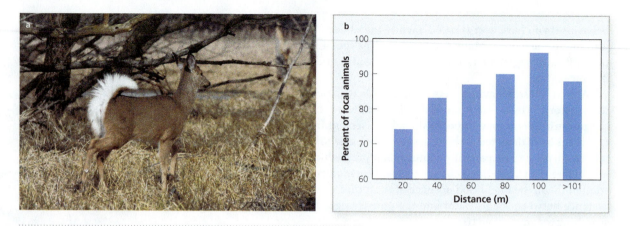

Figure 8.23. Deer tail flagging. (a) White-tailed deer. (b) Percentage of deer that tail-flag as a function of distance to predator. More deer tail-flagged as the distance to the predator increased (*Source:* Caro et al. 1995).

distance from the human increased, as predicted by the pursuit-deterrence hypothesis (Figure 8.23).

The researchers also made a new discovery. When deer fled into an open area, they continued tail flagging. However, deer that fled into dense woods would often stop tail flagging and become much more difficult to observe. These observations led Caro and colleagues to formulate a new hypothesis. Their **flash disappearance hypothesis** states that the bright white of the underside of the tail gives the predator an easily observed target. The sudden disappearance of this target against the dark background of dense woods, they predicted, means that the predator will have trouble locating its prey and will give up its attack.

At the conclusion of their study, Caro and colleagues could reject the alarm signal hypothesis. They also found support for the pursuit-deterrence hypothesis and formulated a new hypothesis for tail flagging—it functions to confuse a predator by providing a target that disappears when a deer drops its tail after running into dense vegetation. Additional tests of these hypotheses are needed using real predator-deer interactions.

flash disappearance hypothesis
The sudden disappearance of a bright target makes it difficult for the predator to locate its prey and thus encourages it to give up its attack.

Chapter Summary and Beyond

Antipredator behavioral adaptations are common because most animals live under the threat of being killed by a predator. Many animals reduce predation risk by modifying their behavior. Modifications include selecting appropriate habitats to match background color, reducing overall level of activity when predators are nearby, and adopting evasive movement behavior to avoid capture. Even more remarkable, some species, like octopuses, can rapidly change the color and shape of their body to match their background; recent research is focusing on the mechanisms responsible for these structural coloration changes (Mäthger et al. 2009).

Many behaviors that reduce predation risk also entail a cost in terms of reducing other fitness-enhancing behaviors. Such behavioral trade-offs represent a common antipredator behavioral adaptation. Several reviews show that such trade-offs affect most behaviors (Lima 1998; Caro 2005; Bednekoff 2007). Recent work has begun to examine

how animals respond to longer term exposure to predators (Ferrari, Sih, & Chivers 2009).

Many animals reduce the risk of predation by associating with others. Antipredator benefits of sociality include the dilution effect, the selfish herd hypothesis, and the effect of group size on vigilance. Recent research has examined whether large groups of individuals can benefit by confusing a predator, how individuals assess group size, and how spacing between individuals affects vigilance (Radford & Ridley 2007; Fernández-Juricic, Beauchamp, & Bastain 2007; Frommen, Hiermes, & Bakker 2009).

Finally, some animals interact directly with predators by harassing them once they have seen a predator. Such behavior can reduce predation risk by encouraging the predator to leave the area and search for another victim. Other animals appear to advertise themselves to predators, which may then deter an attack because the element of surprise is lost. Recent work has examined human-deer interactions to study additional factors regarding how deer assess the level of predation threat and how they adjust their escape behavior (Stankowich & Coss 2007).

CHAPTER QUESTIONS

1. Yasué, Quinn, and Cresswell (2003) based their conclusions on a correlation between ambient temperature and the number of redshanks feeding in the salt marsh habitat. If you could increase predation risk in the mudflat habitat by flying a trained falcon predator over that habitat, what do you predict would happen to the number of redshanks feeding in each habitat?

2. Imagine a fish that defends a territory where it attracts females for mating and that females of this species prefer males with large territories. Given what you learned about behavioral trade-offs, what predictions can you make about how the presence or absence of a predator will affect the size of the mating territory defended? Design an experiment to test your predictions.

3. Zebra-tailed lizards wave their black-and-white tails when they spot a potential predator. Could the hypotheses discussed and tested by Tim Caro (Caro 1995; Caro 2005; Caro et al. 1995) for tail flagging in deer apply to zebra-tailed lizards? How could you test these hypotheses?

4. The size of a group often increases after a predator is spotted. What antipredator benefit best explains this observation?

5. Design a study to determine whether house sparrows (*Passer domesticus*) exhibit the group size effect.

6. For each hypothesis in the Scientific Process boxes on pages 181, 185, and 191, formulate the null hypothesis.

7. We began this chapter by discussing the chuck calls and tail wagging of squirrels as a behavioral response to predators. Which of the hypotheses from this chapter could help us understand this behavior? How would you test these hypotheses?

Chapter 9

Dispersal and Migration

CONCEPTS

FEATURES

Scientific Process

Toolbox

Applying the Concepts

Figure 9.1. Jaeger. Long-tailed jaeger, a bird that lives in open ocean habitat.

One of our favorite excuses for heading outdoors is bird watching, and Arizona has a wonderful diversity of bird species—one of the greatest in the United States. One day, while in graduate school, we were birding at Las Cienegas National Conservation Area, south of Tucson. Many birders (including ourselves) keep track of the number of species seen in a day, a year, or a lifetime, and we were hoping to spot a Cassin's sparrow (*Aimophila cassinii*) to add to our list.

Suddenly, a rather large bird flew directly overhead, maybe 10 m above us. It sure looked like a jaeger, but no one in the group had ever seen one. Jaegers have a distinct tail morphology, with central tail feathers that project beyond the rest of the tail, making identification fairly easy (Figure 9.1); furthermore, these birds can migrate thousands of kilometers. However, jaegers are pelagic, or open-ocean, birds—hardly the kind found in the middle of arid Arizona. Their long, seasonal movements over the ocean take them all the way from their breeding grounds in the Arctic to their wintering grounds south of the equator. If this *was* a jaeger, it was terribly off course! No wonder other birders and experts didn't believe us.

About a week later, the curator of the bird museum at the University of Arizona called us to see what had just been brought in—a juvenile long-tailed jaeger (*Stercorarius longicaudus*). It had been found dead along a road only a few kilometers from where we had spotted our bird and had probably died from stress. Our identification was vindicated. In fact, our jaeger spotting was not the first in the state. A search of museum records revealed that several jaegers had been observed in Arizona over the past 100 years. All were likely lost, blown off course in a storm, or ill. Migrating is serious business, and not all animals make it.

In this chapter, we examine the how and why of animal movements, ranging from short-distance, one-way movements to round-trip annual migrations of thousands of kilometers. We discuss the mechanisms that allow animals to make these trips, as well as the benefits.

9.1   Dispersal reduces competition and inbreeding

dispersal A relatively short-distance, one-way movement away from a site.

competition hypothesis Dispersal functions to reduce competition for resources.

Many animals exhibit **dispersal**, a relatively short-distance, one-way movement away from a site where conditions are crowded. Competition for resources is density dependent: it increases with the number of competitors. The more individuals there are in a location, the greater the competition for food. In the **competition hypothesis**, dispersal functions to reduce competition for resources.

To test this hypothesis, researchers have manipulated both the density of individuals and the amount of food to which individuals have access. This leads to two predictions:

Prediction 1: If two sites have identical amounts of food but differ in the density of individuals, we can expect higher levels of dispersal behavior in the high-density sites.

Prediction 2: If two sites have the same density of individuals but differ in the quantity of food, we can expect higher levels of dispersal behavior in the site with less food.

Studies have tested each prediction using a variety of taxa, as we see in the next two case studies.

Featured Research

Dispersal in adult springtails

Collembolans are very small arthropods (most less than 2 mm in length) that live in high densities in the soil, feeding on fungi, algae, and detritus. They are wingless and move by crawling through mulch, leaf litter, and decaying organic matter in moist habitats. When an individual is threatened, it releases the furcula, a tail-like appendage that is held under tension underneath the body. The furcula snaps down onto the substrate, springing the individual into the air and away from danger—hence the name springtails.

Does the density of individuals affect dispersal behavior? Göran Bengtsson, Katarina Hedlund, and Sten Rundgren predicted greater dispersal behavior would occur with higher population density, because competition for resources would also be greater. To test this prediction, they established an experimental chamber in the laboratory, consisting of a series of small glass vials filled with soil, linked by rubber tubing that also contained soil (Bengtsson, Hedlund, & Rundgren 1994). Each vial constituted a habitat patch, and 60 adult springtails (*Onychiurus armatus*) were placed in one of the end vials, the home chamber. Individuals could move through the tubing to reach the second, third, fourth, and finally the fifth vial, which was 40 cm from the home vial. Although 40 cm seems like a small distance, for a springtail, it represents movement of over 400 body lengths. In order to standardize the physiology and hunger levels of each individual, the researchers maintained the population for one week without food and then for several days with excess food.

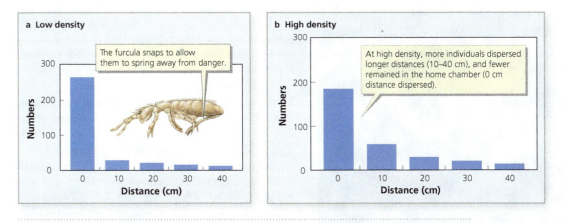

Figure 9.2. Collembola dispersal. On average, individuals in the low-density treatment (a) dispersed shorter distances than those in the high-density treatment (b) (*Source*: Bengtsson, Hedlund, & Rundgren 1994). Inset: Collembolan.

The experiment consisted of two treatments, each replicated six times. In the low-density treatment, the home chamber was 20 cm². This equates to 30,000 individuals/m², a typical density for springtails in natural soil. In the high-density treatment, the home chamber was only 7 cm², creating a density of nearly 90,000 individuals/m². A trial began when all individuals were placed in the home vial. The research team quantified dispersal behavior by counting the number of individuals in each of the vials after three days had elapsed.

Bengtsson and colleagues found that more springtails dispersed in the high-density treatment than in the low-density treatment, and those that dispersed moved a greater distance on average (Figure 9.2). For adult springtails, the density of individuals in a habitat and the resulting competition for food strongly affect dispersal behavior. Although these results support the competition hypothesis, they do rely on the common assumption that high density corresponds to intense competition, which drives dispersal behavior.

Bengtsson and colleagues examined the behavior of adults under laboratory conditions. In the next research study, we examine dispersal of juvenile birds in the field.

Natal dispersal in northern goshawks

Featured Research

Offspring of many animals aggregate after birth or hatching because they develop together in a nest or within the territory of a parent. These juveniles face two problems. First, there is competition for resources. Second, as an individual matures, it is surrounded by genetic relatives. When individuals mate with close relatives, they often suffer low reproductive success. **Natal dispersal**, a movement away from an individual's place of birth (natal location) to its first breeding location, is a one-time event that can help solve both problems.

natal dispersal A one-time movement away from an individual's place of birth.

One way to test the competition hypothesis is with food supplementation. If juveniles are provided with abundant food, they have less need to disperse, and they should exhibit lower levels of dispersal. Patricia Kennedy and Johanna Ward tested this prediction on northern goshawks (*Accipiter gentilis*), which are large (approximately 1 kg) bird-eating hawks that live in forests in the northern hemisphere (Kennedy & Ward 2003) (Figure 9.3). Goshawks are territorial, and a pair will defend an area of approximately 2 km around the nest. They produce a brood of one to two young each year. Chicks hatch in April and become independent by August, at about 80 days of age.

Figure 9.3. Goshawk. Goshawks are large, bird-eating hawks.

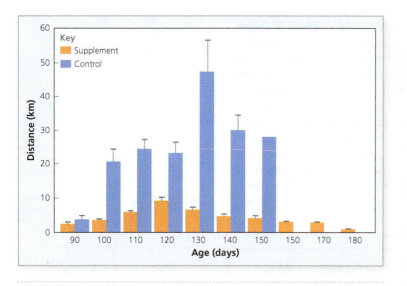

Figure 9.4. Juvenile dispersal distances. Mean (+ SE) distances traveled from the natal nest for food-supplemented juveniles (orange) and control birds (blue) at each point in time (*Source:* Kennedy & Ward 2003).

Kennedy and Ward studied a population of goshawks in the Jemez Mountains of New Mexico for a period of two years. Of the 28 broods studied, the researchers assigned 14 to a food supplement treatment and designated 14 as controls. For the food-supplemented treatment, several dead Japanese quail (*Coturnix japonica*) were provided every other day on a feeding platform 10 m from the nest. This feeding treatment began at hatching and continued through mid-October, after all juveniles had become independent. Control birds had an identical platform 10 m from the nest, but no food was provided. The researchers visited both control platforms and supplemented platforms equally to standardize human disturbance near a nest. Kennedy and Ward observed that the supplemented broods readily took food from the feeding platforms and so had access to more food than did control broods.

To track the movements of juveniles, Kennedy and Ward placed small radio transmitters on the legs of all juveniles when chicks were 21 days old. These birds were located every other day by car or airplane until mid-October. In total, Kennedy and Ward were able to collect data on the dispersal behavior of 27 juveniles, including 16 that were food supplemented and 11 controls.

Both groups became independent (defined as being away from the nest for more than a week) at essentially the same time at about 80 days of age. Dispersal behavior, however, varied greatly between the two groups. Each week following independence, control birds dispersed much greater distances from their natal site than did food-supplemented birds (Figure 9.4). By the end of the study, all control birds had moved out of the study area (more than 25 km away from their natal nest), whereas all the food-supplemented birds remained. Kennedy and Ward suggested that food-supplemented birds dispersed much shorter distances because they had a much higher estimate of food density in their natal area. That is, the birds estimated the study area to be of higher quality.

These studies demonstrate that competition for food affects dispersal. Individuals will disperse away from areas of high competition because of either high density or low food availability. What about the second proposed benefit of dispersal behavior—to avoid mating with close relatives?

Featured Research

Inbreeding avoidance in voles

inbreeding depression A reduction in fitness as a result of mating with close relatives.

inbreeding avoidance hypothesis Natal dispersal behavior minimizes the likelihood of mating with relatives.

Individuals that mate with close relatives have lower reproductive success than individuals that do not mate with close relatives—a phenomenon known as **inbreeding depression**. Dispersing away from a natal area solves this problem. The **inbreeding avoidance hypothesis** posits that one factor driving the evolution of natal dispersal behavior is that it minimizes the likelihood of inbreeding (Hamilton & May 1977; Bengtsson 1978; Pusey & Wolf 1996). This hypothesis also explains a striking pattern observed in birds and mammals: sex-biased natal dispersal. In birds, juvenile females disperse more often and to greater distances than do juvenile males; in mammals, the opposite pattern occurs: males disperse more often and farther (Greenwood 1980; Pusey 1987). When one sex disperses more often and farther than the other, there is less chance that siblings of the opposite sex will settle near each other and mate.

Eric Bollinger, Steven Harper, and Gary Barrett tested the inbreeding avoidance hypothesis using meadow voles (*Microtus pennsylvanicus*) (Bollinger, Harper, & Barrett 1993). Meadow voles are small herbivorous rodents that live in tall, dense vegetation in a variety of habitats in North America. They feed mainly on grasses and use dense vegetation to hide from predators. They exhibit male-biased natal dispersal: juvenile males are two times more likely to disperse than juvenile females (Boonstra et al. 1987). Bollinger and colleagues asked the research question: Does inbreeding avoidance affect dispersal behavior in juvenile meadow voles? If so, the inbreeding avoidance hypothesis makes the following prediction: juveniles living near their siblings should exhibit greater dispersal behavior than juveniles living near non-kin. Bollinger and colleagues tested this prediction by creating six experimental plots measuring 8 by 18 m in tall, dense grass in a natural field in Ohio: a perfect habitat for voles. Around each plot, they created a 1 m wide strip of mowed vegetation, surrounded by a 1.5 m wide strip of plowed ground with no vegetation. These two strips created a buffer area of poor-quality habitat around each plot (Figure 9.5). Voles will, however, disperse over poor-quality habitats, and so any animal found in the buffer zone was considered to be dispersing away from the plot.

The research team placed four voles into each plot: two males and two females. In one treatment, the four voles were siblings (sibling treatment), and in the control treatment, the voles were not siblings. To determine the dispersal behavior of individuals, Bollinger and colleagues placed four live traps near the center of each plot and 16 live traps around a plot in the plowed strip each day. Animals captured in the center of the plot were assumed to be not dispersing, whereas those captured in the plowed strip were assumed to be attempting to disperse.

During the ten-day experiment, Bollinger and colleagues found that, as expected for this species, males were more likely to attempt to disperse than females: more males than females were captured in the plowed strip. More important, the experimental treatment strongly affected dispersal behavior: sibling voles were much more likely to attempt to disperse than were nonsibling control voles (a mean of 2.4 voles in the sibling plots versus 1.4 in the control plots). Furthermore, sibling voles attempted to disperse several days before nonsibling voles did (4.6 days on plots for sibling voles versus 7.1 for nonsibling voles). From their results, the research team

Figure 9.5. Vole experimental plots. (a) Plots of tall grass (background) are separated by a mowed strip, seen here. (b) A captured vole.

concluded that inbreeding avoidance influenced the dispersal behavior of meadow voles. Individuals dispersed in order to move away from their close kin, as predicted by the inbreeding avoidance hypothesis.

So far, we have focused mainly on natal dispersal by juveniles. However, adults also exhibit dispersal behavior when they move from one breeding site to another. Hypotheses to explain these movements typically do not involve competition or inbreeding avoidance. Instead, they assume that individuals are attempting to move to a higher-quality breeding site, as we see next.

9.2 Reproductive success affects breeding dispersal

For adults, a major factor that affects fitness is their ability to successfully raise young. Breeding habitats differ in quality with respect to reproductive success because they differ in food abundance, predators, diseases, and so forth. Animals that settle and breed in a high-quality habitat will tend to have higher fitness than those that settle in a poor-quality habitat. However, it may be difficult for an individual to assess the quality of a habitat accurately, which could result in low reproductive success. Therefore, many animals exhibit **breeding dispersal**—movement to a new breeding site, presumably in search of a higher quality habitat.

Numerous studies have found a strong relationship between reproductive success and breeding dispersal behavior: individuals that are unsuccessful in a breeding attempt typically disperse to a new site for subsequent breeding attempts (Switzer 1993). In contrast, those that are successful typically do not disperse: they rebreed at the same site, a pattern known as **site fidelity**. This pattern holds both for individuals that breed seasonally, such as birds in temperate or boreal habitats, and those that reproduce each day, such as insects. Let's look at one example that illustrates this pattern.

breeding dispersal Abandoning one breeding site and moving to another.

site fidelity Individuals that remain at or return to a previous location to breed.

Featured Research

Breeding dispersal in dragonflies

Paul Switzer experimentally examined the breeding dispersal behavior of male eastern amberwing dragonflies (*Perithemis tenera*) on a pond in Kansas (Switzer 1997). Male eastern amberwing dragonflies defend small sites for oviposition (egg laying) that consist of vegetation or sticks rising above the surface of a pond. When a male spots a female, he leads her back to the oviposition site and courts her. If the female finds the site acceptable, she uses it to lay a clutch of eggs, which the male fertilizes. Males defend one site per day and may mate with several females in a day. Males leave the pond at night for the safety of dense vegetation and return the next day to select an oviposition site. They may exhibit site fidelity (by selecting the same site) or breeding dispersal (by moving to a new site) (Scientific Process 9.1).

To examine how a male's reproductive success affects his dispersal behavior, Switzer created artificial oviposition sites, consisting of two sticks joined together. The longer stick served as a handle to allow for manipulation of the site.

Males were captured and marked on the forewing for visual identification ($n = 10$ for both treatment and control). On the first day of an experimental trial, two males were selected while defending oviposition sticks on the pond. The treatment male was prevented from mating all day, whereas the control male was allowed to mate with females. To prevent a male from mating, Switzer simply lowered the oviposition stick under the water when the treatment male had attracted a potential mate to his territory. With the oviposition site submerged, the female would quickly abandon the male. When she left, Switzer would return the stick to its original position above the water, which allowed the male to continue to defend the site. To control for this manipulation, Switzer also submerged the control male's oviposition stick when he was on his territory, but only when females were not present. Thus, both males had their sites submerged a similar number of times. Switzer recorded

SCIENTIFIC PROCESS 9.1

Breeding dispersal in dragonflies
RESEARCH QUESTION: *What factors affect breeding dispersal in dragonflies?*

Hypothesis: Dragonflies will exhibit a win-stay lose-shift dispersal strategy based on their reproductive success.

Prediction: Dragonflies will remain on the same territory if they have high mating success and will disperse to a new territory if they have low mating success.

Methods: The researcher:

- created breeding territories on a pond using sticks.
- reduced mating success in some territories (treatment) and kept it high in others (controls).
- recorded the number of matings for each male and whether he returned to the same territory or dispersed to a different territory between Day 1 and Day 2.

Results: Control males had high mating success compared with treatment males. All control males returned to their oviposition site, whereas only 50% of treatment males returned to their oviposition site.

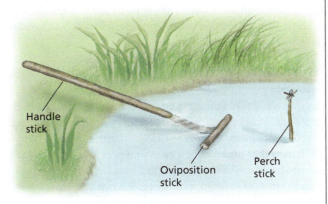

Figure 1. Experimental setup. Experimental setup showing the sticks used to manipulate the oviposition site (*Source:* Switzer 1997).

TABLE 1 Results. Pair-wise (mean +SE) comparison of experimental and control males (*Source:* Switzer 1997).

CATEGORY	EXPERIMENTAL	CONTROL
Matings	0	6.5 + 1.5
Submergences	16.9 + 4.1	19.9 + 3.6
Number of intruders chased away (by investigator)	1.8 + 0.8	2.3 + 0.9
Time of arrival on first day (min. after 0900 hours)	234.2 + 32.6	214.6 + 25.7
Total time at site during experimental day (min.)	315.9 + 37.7	374.4 + 27.3
Age (days after marking)	5.6 + 1.3	3.4 + 0.8

Conclusions: Mating success affects the breeding dispersal: individuals use a win-stay lose-shift strategy when making dispersal decisions.

the number of matings and then dispersal decisions of each male by locating them on the pond the following day.

Switzer found that control males mated an average of 6.5 times on the first day, but treatment males never mated successfully. On the following day, all control males reused the same site—that is, all exhibited site fidelity. In contrast, half of the treatment males dispersed to a new site on the following day. On average, the treatment males moved 22 m between Day 1 and Day 2, while control males did not move at all. These results clearly demonstrate that an individual's reproductive success affects its decision to disperse to a new breeding site.

The behavioral pattern exhibited by Switzer's dragonflies, and by many other species as well, is known as a **win-stay lose-shift dispersal** strategy. Here, *win* means high reproductive success, while *lose* means a reproductive failure; *stay* and *shift* refer

win-stay lose-shift dispersal When individuals return to a previous breeding location after reproductive success there, but disperse to a different location after a reproductive failure.

to the dispersal decision. We can formulate a hypothesis to explain this behavioral pattern by assuming that breeding sites vary in quality and that the quality of a site is consistent between breeding attempts. That is, we assume that (1) in poor sites, reproductive success is low, while in high-quality sites, reproduction is very successful; and (2) a site that is high quality today is likely to be a high-quality site in the future as well. Under these assumptions, the win-stay lose-shift strategy for breeding dispersal makes sense.

Public information from conspecifics affects breeding dispersal

So far, we have assumed that the win-stay lose-shift dispersal decision is based on an individual's assessment of breeding patch or site quality. This assessment, in turn, is based solely on the individual's reproductive success in the patch. Thierry Boulinier and Etienne Danchin reasoned that it might be even better for individuals to gather information about the reproductive success of conspecifics breeding in the same patch (Boulinier & Danchin 1997). Using a model, they demonstrated that information on patch-wide reproductive success of conspecifics provides more accurate information about the quality of the patch and that using such information for dispersal decisions results in higher fitness for an individual than relying only on its own experience (Boulinier & Danchin 1997; Doligez et al. 2003). Information about the quality of a resource based on the performance or behavior of others is known as **public information** (Valone 1989). Over the past decade, several studies have shown that animals use public information to improve their estimates of the quality of mates, food patches, and breeding patches (e.g., Doligez, Danchin, & Clobert 2002; Aparicio, Bonal, & Muñoz 2007; Valone 2007).

public information Information obtained from the activity or performance of others about the quality of an environmental parameter or resource.

Featured Research ▶

Kittiwakes and public information

Kittiwakes (*Rissa tridactyla*) are gulls that nest in large colonies on sea cliffs in northern Europe (Figure 9.6). The rocky habitat often creates subcolonies, or breeding patches, of just a few dozen individuals, aggregated close together and widely separated from other patches. Kittiwakes breed once a year, and previous work suggested that individuals use the win-stay lose-shift strategy when making breeding dispersal decisions from one year to the next (Danchin, Boulinier, & Massot 1998).

Figure 9.6. Kittiwakes.
(a) Kittiwakes nest in rock cliff colonies. Note the colored leg bands that researchers use to identify individuals. (b) A researcher examines an experimental bird.

Boulinier and colleagues hypothesized that patch-wide reproductive success in subcolonies might also affect breeding dispersal decisions (Boulinier et al. 2008). To test this hypothesis, the research team examined predictions regarding individuals that experience a reproductive failure. Based solely on its own experience, a failed breeder should estimate that the patch is of low quality and therefore should abandon it in favor of a new site next year—that is, it should exhibit breeding dispersal. However, if a failed breeder uses patch-wide reproductive success to assess patch quality, and if it is surrounded by many individuals that breed successfully, then it should estimate that the patch is of high quality and so should not abandon it the next year—that is, the individual should exhibit site fidelity.

To test these predictions, Boulinier and colleagues uniquely marked focal birds with color bands in each of 18 breeding patches, which each contained 40 birds on average. These patches were arranged in nine matched pairs, based on size and proximity to each other. In the first year of the experiment, one patch of each pair was the treatment and the other served as the control. In the treatment patches, all birds, including the focal individual, had their eggs removed 25 days after the first egg was laid and every sixth day after that; thus, all birds in the treatment patch experienced a reproductive failure. In the control patches, the focal bird had its eggs removed at the same time as the treatment birds, but all other birds were unmanipulated and so experienced high levels of reproductive success. The research team observed the behavior of focal birds each year to determine whether they exhibited breeding dispersal or site fidelity.

The manipulation successfully created a large difference in patch-wide reproductive success for the paired patches. No bird in the treatment patches successfully raised offspring, while more than 50% of individuals in the control patches successfully raised offspring.

Focal birds in the paired patches exhibited significantly different breeding dispersal behavior. Even though they experienced a reproductive failure, birds in the control patches exhibited a high degree of site fidelity, with over 70% returning to breed in the same patch the next year (Figure 9.7). In contrast, fewer than 50% of the focal birds in the treatment patches returned to breed in the same patch, meaning that more than half exhibited breeding dispersal. Because all focal birds experienced a breeding failure, the research team concluded that birds in the control patches, where patch-wide reproductive success was high, used public information from the reproductive success of others to estimate patch quality and to decide to remain in the same patch the next year. This study provided some of the first experimental evidence that information from conspecifics can affect breeding dispersal decisions.

So far, we have examined short distance dispersal, but what about the longer distances that animals travel? We examine this in the next section.

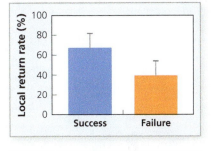

Figure 9.7. Kittiwake return rates. Mean (+ SE) return rate of unsuccessful individuals was higher on treatment plots with reproductive success of other birds compared to plots with reproductive failure of other birds (*Source:* Boulinier et al. 2008).

migration Relatively long-distance two-way movements.

9.3 Animals migrate in response to changes in the environment

Many animals undergo longer, two-way movements known as **migration**. Species in a variety of taxa migrate, including marine invertebrates, insects, amphibians, reptiles, fish, birds, and mammals (Table 9.1). One well-known example is the annual migration of more than one million wildebeest (*Connochaetes taurinus*), along with thousands of zebras (*Equus burchelli*) and Thomson's gazelles (*Eudorcas thomsonii*) (Figure 9.8). These animals travel from the southern to northern Serengeti in east Africa in the spring and return to the southern Serengeti each fall. Over the course of this migration, individuals travel over 400 km (roughly 250 mi), moving from the short-grass plains in the south to savannah habitats in the north and back again.

Figure 9.8. Wildebeest migration. Twice each year, wildebeest migrate hundreds of kilometers to more favorable feeding grounds in the Serengeti.

TABLE 9.1 Examples of migration in different taxa. (*Source:* Alerstam, Hedenström, & Åkesson 2003).

TAXA	COMMON NAME	SPECIES	TRAVEL	DISTANCE (KM)
Insect	Dragonfly	*Anax junius*	New England to Central America	2800
Crustacean	Blue crab	*Callinectes sapidus*	Chesapeake Bay	240
Fish	Bluefin tuna	*Thunnus thynnus*	Atlantic	12,000
Amphibian	Common toad	*Bufo bufo*	Hibernation site to breeding site	2–3
Reptile	Loggerhead sea turtle	*Caretta caretta*	California to Japan	11,500
Bird	Arctic tern	*Sterna paradisaea*	Greenland to Antarctica	19,000
Bird	Blackpoll warbler	*Dendroica striata*	Alaska to Bolivia	12,000
Mammal	Gray whale	*Eschrichtius robustus*	Baja California to Chukchi Sea	6000
Mammal	Elephant seal	*Mirounga leonina*	Sandwich Islands to Antarctica	3000

The proximate mechanisms involved with migration vary across taxa and are often related to a combination of internal factors, such as annual rhythms and associated physiological changes, external factors, including changes in photoperiod, and local environmental conditions. For example, in the Serengeti, mammals appear to track both the available biomass of grass and its quality. When grass growth slows and grasses dry out during the onset of drought conditions, grass nutritional value drops, and this change may serve as a cue for the onset of migration (Boone, Thirgood, & Hopcraft 2006; Holdo, Hart, & Fryxall 2009). For birds, the onset of migration is typically triggered by internal physiological changes, whereas the timing and rate of migration are often affected by changes in day length, temperature, and food availability (e.g., Berthold 1993; Marra et al. 2005).

Large-scale movements, however, come at a cost, as the jaeger at the beginning of this chapter illustrates. Not only do long journeys take a physiological toll as a result of the exertion they require, but migrating animals also often suffer risks such as increased predation. Given the costs, why do animals migrate?

Migration and changing resources

The benefits of migration include moving to an area with more favorable conditions (e.g., more food or water, less extreme temperatures). The decision to stay in one location and forgo migration comes with its own benefits and costs: permanent residents do not pay the costs of long-distance movement but must endure harsher conditions. For migration behavior to evolve in a population, the net benefit of migration must exceed the net benefit of being a permanent resident, and so examining variation in these costs and benefits can help us understand variation in the migratory behavior we see today (Pulido 2007).

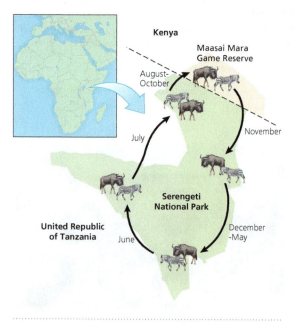

Figure 9.9. Wildebeest migration map. This map shows the circular migration route of wildebeest over the course of the year.

Bird migration and global climate change

Climate change is a phrase heard often in the news. The term *climate* refers to statistical averages about temperature, humidity, wind, and rainfall that are based on data collected over decades, whereas weather refers to these same parameters from day to day. From 1850 to 1950, the global average temperature fluctuated but exhibited no discernible trend. However, since 1950, the global average temperate has increased dramatically, demonstrating global climate change. Climate change can lead to mismatches between animals and their habitats, and it can have severe consequences for species.

Over the last 50 years, as a warmer climate has developed, particularly in the Northern Hemisphere, spring conditions have been occurring earlier than in the past (Stenseth & Mysterud 2002). For organisms that migrate to follow resources, this shift may result in a mismatch between the migration date and the availability of the resources.

David Inouye and colleagues examined the spring migration return of American robins (*Turdus migratorius*) back to high-altitude sites in the Colorado Rocky Mountains (Inouye et al. 2000). Over the past 25 years, spring temperatures have increased, but the average date of snowmelt has not changed (perhaps because of increased snowfall). The research team examined the arrival dates of robins since 1974 and the first date of bare ground. They found that robins now arrive back

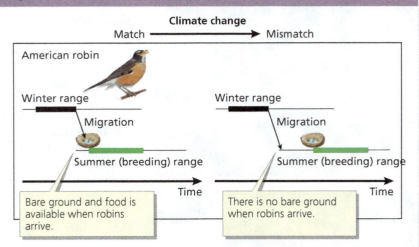

Figure 1. Bird migration and global climate change. When there is a match between the climate and migration times, robins arrive to find bare ground (green) at their summer breeding range (left). Where there is a mismatch as a result of climate change (warmer temperatures), robins arrive earlier but must face snow and no bare ground for 18 days (right) (*Source*: Stenseth & Mysterud 2002).

from their winter range 14 days earlier than they did in 1981. However, snowmelt is now rarely complete when they arrive, and bare ground typically does not appear until 18 days later (Figure 1). This is a problem, because robins need bare ground to search for buried insects. Mismatches resulting from differential responses to climate change can have negative consequences for robins in the Rocky Mountains, as well as other species.

Let's go back to the example of wildebeest, zebras, and Thomson's gazelles. These large-bodied herbivores require substantial quantities of grass each day, and grass is dependent on rainfall. Rainfall in the Serengeti is highly seasonal. These animals begin the year in the south, feeding on abundant short grasses that grow vigorously after the short fall rainy season. By spring, these grasses are depleted, and the migration begins. The herds move north and west to arrive in the northern Serengeti by July, after the onset of summer rains in this region, and stay until October, when the food supply in the north becomes scarce. At that time, herds begin to move south, back to the southern Serengeti, and arrive in December as the short grasses there begin to grow rapidly again (Holdo, Hart, & Fryxall 2009) (Figure 9.9). A similar pattern is seen in many insectivorous birds that breed in temperate regions in summer, when insects are plentiful. These birds migrate south in the autumn as insect availability declines to overwinter in warmer subtropical regions with higher food abundance, and then return north the following spring to breed. Recent changes in global climate are impacting migratory species (Applying the Concepts 9.1).

Resource variation and migration in neotropical birds

Featured Research

We can hypothesize that migration has evolved in many animal species as an important strategy that allows individuals to (1) take advantage of spatial variation in

Figure 9.10. Diet and migratory behavior. (a) Frugivorous ochre-bellied flycatcher (*Mionectes oleagineus*); (b) insectivorous royal flycatcher (*Onychorhynchus coronatus*). (c) Insectivorous species are more likely to migrate if they live in nonforested habitats (*Source:* Boyle & Conway 2007).

conditions and (2) avoid seasonal resource depression at different locations. To evaluate this hypothesis, we could evaluate the following prediction: species exposed to a high degree of fluctuation in environmental conditions and resources can benefit by migrating when conditions and resources decline. In contrast, species that live in more stable resource environments have less to gain by migrating and so should be sedentary (Levey & Stiles 1992).

W. Alice Boyle and Courtney Conway tested this hypothesis on neotropical birds in the clade Tyranni (Boyle & Conway 2007). This group includes over 300 species of tyrant flycatchers, manikins, cotingas, and becards and displays a wide range of migratory behaviors, habitats, and diets, making it ideal for testing such a hypothesis. The researchers collected previously published data from field guides and reference books on migratory behavior (migrant or sedentary), diet (insectivore or frugivore), and habitat use. They assumed that habitat types varied in degree of environmental fluctuations: tropical forest habitats, for example, tend to have relatively stable temperature and humidity compared with nonforest habitats (which include open and arid habitats, disturbed habitats, and thickets). Less was known about the relative stability of food resources, but the researchers focused on insect and fruit abundance, both of which vary seasonally.

Controlling for the effects of phylogeny, Boyle and Conway found that habitat type did correlate with migration, at least for insectivores. Insectivorous species that inhabited the forest habitats were less likely to be migratory than species inhabiting nonforest habitats (Figure 9.10). Because forest interiors have more stable temperatures and humidity levels, they may provide a more uniform environment that helps stabilize insect abundances. In the frugivores, however, habitat did not affect the propensity to migrate, perhaps because forest and nonforest areas do not differ as much in stability of fruit production. So while both habitat and diet correlated with migratory behavior, additional work is needed to more fully understand the factors that drive the evolution of migration in these birds.

These data suggest that fluctuations in resource levels and environmental conditions can affect migratory behavior. Next, we examine how migration can evolve.

The evolution of migration

How does a population change from being sedentary to migratory? Volker Salewski and Bruno Bruderer proposed a model that involves a sequence of events to explain the evolution of migration within a population (Salewski & Bruderer 2007). The model begins with a resident population of individuals that do not migrate, but do disperse when they mature. Dispersal distance varies with how far an individual must travel to find an unoccupied breeding habitat. If individuals must travel to a distant location to breed, and if resource abundance declines strongly in that location in the nonbreeding season, they and their offspring will die unless they move back toward their natal site. That is, selection will favor individuals that return to the resident population at the end of the breeding season. These individuals can again benefit by returning to breed in the same distant area when conditions improve before the following breeding season. This pattern creates a population that exhibits partial migration, or a situation in which individuals differ in their migratory behavior: some individuals are year-round residents, while others exhibit seasonal migratory behavior. Over time, if the fitness of migrants is higher than that of residents, eventually all individuals will migrate seasonally. The end result is a migratory population, in which all individuals exhibit migratory behavior. But in this model, the relative fitness benefits and costs of each behavior may vary over time and in different locations, so evolution could also proceed in the opposite direction (Figure 9.11) (e.g., Berthold 1999; Pulido 2007; Salewski and Bruderer 2007). We often see species that exhibit variation in migratory behavior, and indeed, we can find many partial migrant populations (Swingland & Greenwood 1983). The model postulates that variation in the fitness of each behavior can result in rapid changes in the proportion of individuals exhibiting migratory behavior.

Evaluation of this model requires examination of factors that might favor migration over nonmigration. Partial migrant populations have been particularly useful to understanding the relative fitness costs and benefits of migratory behavior (Chapman et al. 2011). Competition is one general factor thought to influence migratory behavior. If individuals differ in their competitive abilities, perhaps only dominant individuals will be able to secure sufficient food when it is scarce (say in winter), and so selection can favor migratory behavior by subordinates as a means to find sufficient food. This hypothesis suggests that the density of individuals should affect propensity to migrate, because as density increases, competition for food intensifies. Let's examine a recent test of this hypothesis in a migratory amphibian population.

Competition and migratory behavior of newts

Many pond-breeding amphibians like the red-spotted newt (*Notophthalmus viridescens*) breed in spring, migrate to a terrestrial habitat in fall to overwinter, and then return to the same pond to breed the next spring. Some individuals remain in the pond as year-round residents and do not migrate. Residents and migrants exhibit alternate phenotypes. Residents display the aquatic phenotype with a large dorsal fin and mucus-covered skin, while migrants change to a more terrestrial phenotype with a reduced tail fin and a skin that is more resistant to desiccation which allows them to migrate over land. These phenotypes are reversible, and migrants change between them each year. Residents face the risk of the pond drying and low oxygen levels, while migrants must pay the physiological costs associated with a change in morphology and face increased costs of travel and associated predation risk.

Kristine Grayson and Henry Wilbur designed an experiment to examine how environmental conditions affect migratory behavior in this species (Grayson & Wilbur 2009). Grayson and Wilbur manipulated the population density and sex ratio of individuals in small enclosures placed around the edge of a large pond such that each enclosure included aquatic and terrestrial substrate. In spring, the enclosures were stocked with different densities of individuals (low, medium, or high) that were either male- or female-biased in their sex ratio (2:1). Enclosures were checked daily for newts out of the water, and the researchers identified these newts as migrants by their morphology. The migrant newts were then removed from the enclosure and released in the terrestrial habitat.

Many more newts migrated out of the high-density enclosures (63%) than the low-density enclosures (39%) (Figure 9.12). Newts migrated, on average, eight days earlier from the high-density enclosures. However, the sex ratio did not significantly affect the migratory behavior of males or females. These data demonstrate that population density influences migratory behavior of newts, and that this behavior is plastic. As density increases, competition for food and space becomes more intense and favors individuals that leave.

One interesting trend was that females migrated more often (60%) than did males (37%) in both sex ratio treatments. Why did a smaller fraction of males migrate? One hypothesis is that competition among males to obtain a territory in the next breeding season favors males that do not migrate, because individuals who

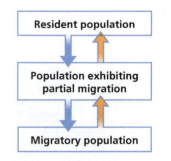

Figure 9.11. Evolution of migration. When some individuals have higher fitness when migrating, a population of residents will exhibit partial migration. If all individuals have higher fitness when migrating, the population becomes migratory. If some individuals in a migratory population have higher fitness by remaining resident, a migratory population will exhibit partial migration. If all individuals have higher fitness by forgoing migration, the population becomes resident (*Source:* Salewski & Bruderer 2007).

Featured Research

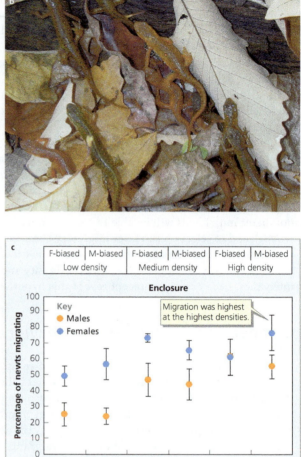

Figure 9.12. Migrating newts. (a) Experimental enclosures on the edge of the pond. (b) Migrating newts. (c) As the density of newts in enclosures increased, a greater mean (± SE) fraction of individuals migrated (*Source:* Grayson & Wilbur 2009).

arrive first are more likely to win in a territorial contest (Kokko, López-Sepulcre, & Morrell 2006). A territory can be acquired by simply staying all year—that is, by becoming a resident rather than a migrant (Kokko 1999). Of course, there are risks involved. Residents must survive the harsh winter pond conditions in order to breed the next season, but if they do, they have a higher chance of securing a breeding territory. This may explain the-sex biased migration observed in newts and likely also explains the sex-biased migration often seen in birds: males tend to migrate shorter distances than females (Cristol, Baker, & Carbone 1999).

Maintenance of polymorphism in migratory behavior

In any partial migrant population, you may wonder: How is the migratory behavior polymorphism maintained? Why do partial migrant populations exist—and are they stable? There are two possible answers to these questions. One is that migratory behavior is a fixed genetic trait of an individual, and the two genotypes (and phenotypes) are maintained by frequency-dependent selection (see Chapter 2): the fitness of each strategy declines as it becomes more common. To maintain the polymorphism, both migratory and resident behaviors must result in equal fitness at equilibrium. There is some evidence that genes influence migratory behavior (Pulido & Berthold 2010), and recent work has revealed that some variation in migratory tendency is associated with microsatellite variation in a single gene in birds (Mueller, Pulido, & Kempenaers 2011).

There is a second way that a migratory polymorphism can be maintained. All individuals in a partial migrant population might possess a genetic disposition to migrate, but the behavior may be expressed only in some, depending on conditions.

In this case, the environment or an individual's physical condition, sex, age, or dominance status influences its migratory behavior. In this scenario, each behavior need not have equal fitness; instead, individuals adopt the strategy that is best given their condition. Those in poor condition need to "make the best of a bad job" until their condition improves. Let's examine one recent test of this hypothesis.

Competition and migratory behavior of dippers

Featured Research

Elizabeth Gillis and colleagues examined migratory behavior in American dippers (*Cinclus mexicanus*) (Gillis et al. 2008). Dippers are aquatic birds that live in and around fast-flowing mountain streams. Each year, many wintering populations comprise both year-round residents and migrants, which move upstream to higher elevations in spring to breed and then return to low elevations to overwinter (Price & Bock 1983). Thus, residents and migrants share wintering grounds but breed in different locations.

The research team studied uniquely banded individuals at several sites on the Chilliwack River in British Columbia over a period of seven years. They conducted five censuses each year to locate marked individuals, along with their nests and offspring. Residents were classified as individuals that occupied a single territory year-round. Migrants were identified either as those individuals that wintered on the river and then bred elsewhere or as those that bred on tributaries upstream in spring but wintered elsewhere.

Gillis and colleagues observed 152 residents and 190 migrants. On average, migrants traveled almost 6 km to reach their breeding territory. Resident females began nesting almost two weeks earlier than migrants, although clutch size and brood sizes were similar for resident and migrant females. However, resident females were more likely than migrants to initiate a second brood if their first nesting attempt was successful and so had higher annual reproductive success (Figure 9.13). In contrast, and perhaps unexpectedly, the annual survival of migrants was slightly higher than that of residents. Still, overall, the researchers estimated that a lifelong resident is expected to fledge about 2.6 more offspring than a lifelong migrant; therefore, the fitness of nonmigratory behavior is higher than that of migratory behavior.

The researchers concluded that the persistence of alternative migratory behaviors in dippers cannot be explained by a fixed genetic dimorphism: the behaviors have different fitness, and offspring frequently exhibit a migration strategy that differs from that of their parents. Over 50 banded nestlings eventually became breeders in the population, but these birds adopted the migratory behavior of their parents less than 50% of the time. Instead, Gillis and colleagues suggest that the migratory behavior of dippers is condition dependent: individuals in the best condition (perhaps dominants) do not migrate and experience high fitness. The researchers did observe four birds that changed their migratory behavior over the course of the study, indicating that migratory behavior is flexible. The results of this study indeed suggest that less competitive individuals are "making the best of a bad job." These individuals cannot compete with residents for the limited high-quality breeding territories, but they can reproduce—albeit at lower levels than residents—by migrating to higher elevations to breed.

We have examined factors that provide ultimate explanations of migration. In the next section, we turn to several proximate factors.

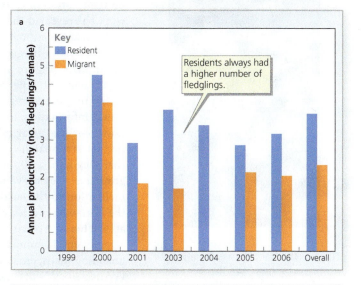

Figure 9.13. Dipper reproductive success. (a) Residents had higher mean (+ SE) reproductive success than did migrants in each year of the study. (b) Dipper. (*Source:* Gillis et al. 2008).

9.4   Animals use multiple compass systems to determine direction

orientation The determination and maintenance of a proper direction.

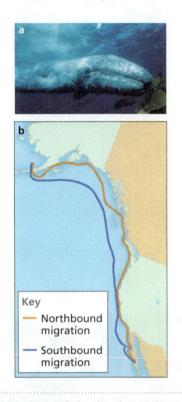

Key
— Northbound migration
— Southbound migration

Figure 9.14. Whale migration. (a) Gray whale. (b) Whale migration route along the North American coast. Blue indicates southbound migration and orange indicates northbound migration.

Figure 9.15. Star compass. Time-lapse photography shows how the stars appear to rotate around Polaris in the north sky in the Northern Hemisphere. Each streak represents a star.

Animals that migrate long distances face two issues: (1) **orientation**, or maintaining the proper direction of travel; and (2) **navigation**, or determining how to reach a particular destination. In the 1950s, Gustav Kramer suggested that birds migrate long distances by use of a "compass and map" system (Kramer 1952). He meant that birds can orient in the proper direction (using their **compass**) and can find particular locations using mental maps, much as people use physical maps (Schmidt-Koenig 1979). In the decades since, research indicates that many animals, not just birds, use a variety of compass systems to orient to a particular direction, and some appear to also possess a mental map to find particular locations. In this section, we examine orientation. In the last section of the chapter, we turn attention to navigation and mental maps.

A variety of environmental cues can provide directional information for orientation. One type of cue used by a variety of animals, including insects, fish, amphibians, birds, and mammals, is a physical landmark, such as a coastline, island, river system, or distant mountain (e.g., Collett & Graham 2004). It has long been known that some marine mammals and shorebirds migrate along continental coastlines. For instance, gray whales (*Eschrichtius robustus*) in the eastern North Pacific migrate up to 20,000 km from their summer feeding grounds in the Chukchi Sea off the coast of Alaska to their calving grounds off the coast of Baja California (Figure 9.14). During migrations north and south, whales generally remain within 10 km of the coastline (Poole 1984).

Many animals orient using a **sun compass**—the location of the sun in the sky (Åkesson & Hedenström 2007; Wehner 1998). Each day, the sun rises in the east and sets in the west, so its position in the sky relative to the time of day provides accurate directional information. Experiments reveal that invertebrates, fish, amphibians, reptiles, birds, and mammals use a solar compass (Åkesson & Hedenström 2007; Wehner 1998). Even on cloudy days, many of these animals can use the pattern of polarized skylight to determine the position of the sun, although the mechanisms of this process are not fully understood (Wehner 2001).

Nocturnal animals or those that move at night can use a **star compass,** because the constellations rotate around the celestial pole (Figure 9.15). A series of experiments by Steve Emlen demonstrated that hand-reared birds learned to orient based on the rotation of stars in the night sky (Emlen 1967a; Emlen 1967b). Subsequent work has revealed that many other taxa, including invertebrates, fish, amphibians, reptiles, and mammals, also use a star compass to orient.

The earth's magnetic field provides another source of directional information for animals (Figure 9.16). These fields flow from the south magnetic pole to the north magnetic pole. Animals can therefore use a **geomagnetic compass**, or the ability to orient using the earth's magnetic field. The magnetic field provides directional information in three ways. First, the magnetic field flows from south to north—it has polarity. Second, magnetic field lines are oriented vertically with respect to the earth's surface at magnetic north and south; at these poles, we can say that they have a 90° angle of inclination. These lines become more parallel to the ground as they move toward the magnetic equator, where they have a 0° angle of inclination to the earth's surface. Thus, their angle of inclination with respect to the horizontal varies with latitude. Third, the intensity of the field varies predictably: it tends to be strongest at the poles and weakest at the magnetic equator. Use of a geomagnetic compass has been documented in dozens of species in many taxa, including snails, crustaceans, insects, fish, amphibians, reptiles, birds, and mammals (Wiltschko & Wiltschko 2005).

Here, we examine two examples illustrating how researchers study orientation and the use of compass systems.

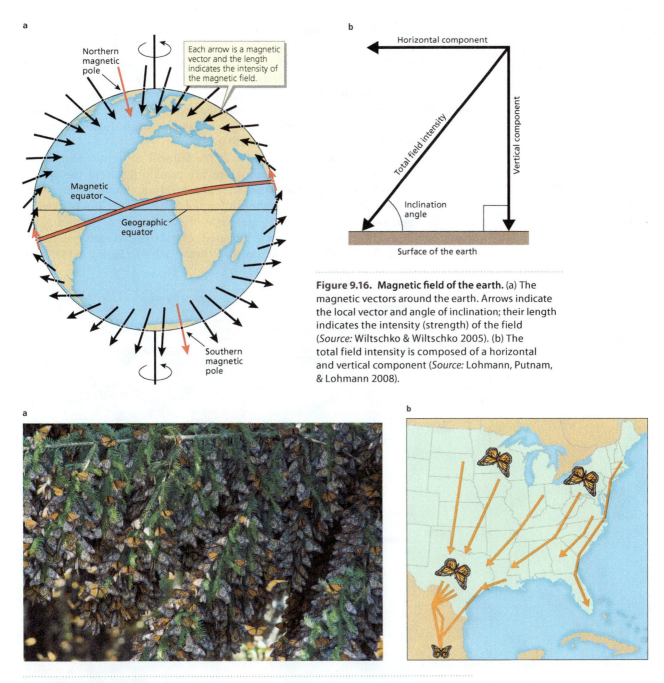

Figure 9.16. Magnetic field of the earth. (a) The magnetic vectors around the earth. Arrows indicate the local vector and angle of inclination; their length indicates the intensity (strength) of the field (*Source:* Wiltschko & Wiltschko 2005). (b) The total field intensity is composed of a horizontal and vertical component (*Source:* Lohmann, Putnam, & Lohmann 2008).

Figure 9.17. Monarch migration. (a) Monarch cluster. (b) Autumn migration routes of Eastern monarch butterflies to wintering grounds in Mexico.

The sun compass in monarch butterflies

One intensely studied animal that uses the sun compass in its amazing annual migration is the monarch butterfly. Monarch butterflies (*Danaus plexippus*) in central and eastern North America migrate to a few mountaintop sites in central Mexico, where millions of these delicate individuals gather in massive clusters to spend the winter (Davis & Rendón-Salinas 2010). Depending on the starting location, this journey can take several weeks and cover over 1000 km (Figure 9.17).

Monarch butterflies have a life cycle comprising several stages. Like many insects, adults lay eggs on plants that develop into larvae, or caterpillars. After growing for

Featured Research

navigation The process of determining a particular location and moving toward it.

sun compass The use of the sun for orientation.

star compass The use of stars or constellations to orient.

geomagnetic compass The ability to orient using the earth's magnetic field.

intergenerational migration Migration that occurs over more than one generation.

several weeks, each larva forms a pupa, or chrysalis, that is characterized by a jade color with gold markings. The chrysalis becomes clear just before the emergence of the adult monarch butterfly. Each adult usually lives only a few weeks before mating and laying the eggs of the next generation, and this cycle takes place several times during the spring, summer, and early fall. However, as the temperatures drop at the end of the fall, the last generation of the year does not reproduce. Instead, it consumes large quantities of nectar to store energy, so that it can begin its long migration to the warmer climate of Mexico, where it will overwinter.

The generation that migrates to Mexico in the fall will not return to its natal area in the spring. Instead, many generations are produced during the spring migration—a cycle known as **intergenerational migration**. The first generation stops in northern Mexico and the southern United States, where the butterflies mate, lay eggs, and die. The emerging generation heads farther north and again mates, lays eggs, and dies. This pattern is repeated several times as each generation moves farther north (Holland, Wikelski, & Wilcove 2006) (Figure 9.18). This

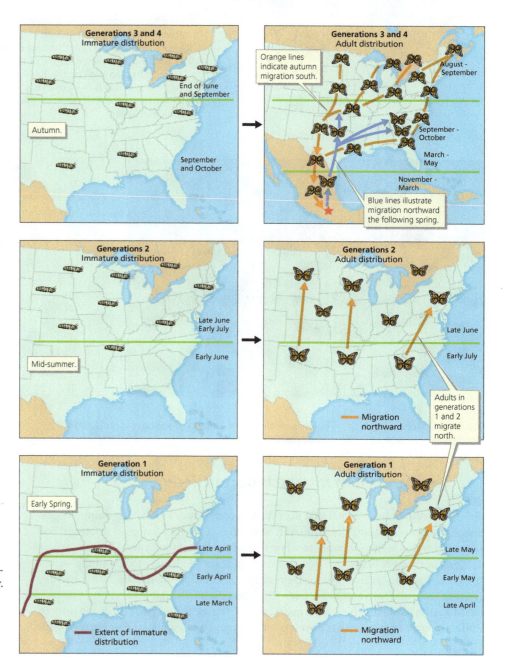

Figure 9.18. Intergenerational migration of monarch butterflies. Monarchs go through several generations in their migration. Panels indicate the presence of larvae (left) and the subsequent movement of those adults (right) over the year. Generation 1 starts at the bottom of the figure. Green lines represent northern range distribution in spring and southern range distribution in autumn (*Source:* Monarch Watch).

raises a critical question: How do butterflies that have never been to their wintering site get there?

Several experiments have demonstrated that monarchs use a sun compass to orient properly for their journey. Animals, including monarchs, have internal timekeeping mechanisms that regulate such daily activities as their active and resting cycles. The internal clock is itself regulated by daily light and dark cycles. After a 12-hour dark cycle (and at the beginning of a light cycle), the position of the sun provides directional information, because the sun rises in the eastern sky. In the autumn, after six hours of daylight, the sun has moved to the southern sky (in the Northern Hemisphere), and at the end of a 12-hour light cycle, the position of the sun provides information about the direction west.

Sandra Perez, Orley Taylor, and Rudolf Jander demonstrated that monarchs use a sun compass by bringing two groups into the laboratory (Perez, Taylor, & Jander 1997). One group was "clock-shifted" six hours over the course of several weeks (Figure 9.19). That is, these individuals still experienced a light/dark cycle of 12 hours each, but their lights were turned on at a later time each day until eventually the lights came on at noon, when the sun outside was in the southern sky in the autumn. The control group was kept on the naturally occurring light/dark cycle, so their lights were turned on at 6:00 A.M. Both groups were then released outside in September and followed. The control group migrated south-southwest, as is typical in Kansas, where the work was conducted. In contrast, the clock-shifted animals moved in a westward direction. Why? Their internal clock had been reset so that their "daybreak" occurred at noon. The sun was in the southern sky, but they interpreted this position as east—the normal direction of the sun at daybreak—and so migrated in the wrong direction (Figure 9.20). While you probably do not use the sun as a compass very often, you may have experienced similar clock-shifting problems if you've ever traveled across many time zones and had trouble sleeping, it is because your internal clock does not match the local time.

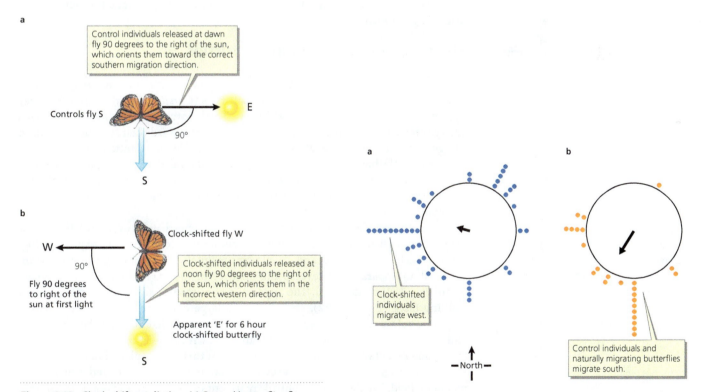

Figure 9.19. Clock-shift prediction. (a) Control butterflies fly south, 90° clockwise to the rising sun. (b) Butterflies clock-shifted six hours see the sun in the southern sky at their "dawn" and are predicted to travel 90° to its right and so will fly west instead of south (*Source:* Perez, Taylor, & Jander 1997).

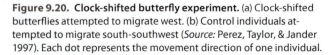

Figure 9.20. Clock-shifted butterfly experiment. (a) Clock-shifted butterflies attempted to migrate west. (b) Control individuals attempted to migrate south-southwest (*Source:* Perez, Taylor, & Jander 1997). Each dot represents the movement direction of one individual.

Antennae and the sun compass system in monarchs

Christine Merlin, Robert Gegear, and Steven Reppert investigated whether antennae play a role in the monarch's sun compass (Merlin, Gegear, & Reppert 2009) (Scientific Process 9.2). They examined migration orientation by tethering individuals in an outdoor flight simulator during fall migration. They first compared the flight direction movement behavior of controls, whose antennae were intact, and individuals that had had their antennae removed. Individuals without antennae could fly, but they failed to orient to any particular direction—just as many tried to move north as tried to move south. In contrast, the flight direction of control butterflies was to the southwest, the proper direction for migration in early October.

To determine whether these results were due to differences in light reception in the antennae, the researchers conducted a second set of experiments a few weeks later, from mid-October to mid-November. This time, they painted antennae either an enamel-based black or clear. The black paint blocked perception of light, while the clear paint allowed light to pass through. They again examined the flight direction of tethered individuals in the outdoor flight simulator. The movement behavior of individuals whose antennae were painted with clear paint was to the south (the proper direction at this time of the year), whereas individuals whose antennae were painted black attempted to move to the north. The lack of proper movement in individuals with either no antennae or black-painted antennae that blocked light indicates that antennae photoreceptors are crucial to this species' sun compass.

These studies demonstrate how migrating butterflies orient in the proper direction to reach their wintering grounds in Mexico. However, given that it takes multiple generations to make the journey, much work still needs to be done to understand how they reach their specific destination. Part of this effort involves the use of citizen scientists all over North America who collect data each year on migrating butterflies (Applying the Concepts 9.2).

The magnetic compass in sea turtles

Many species migrate between breeding sites and feeding sites that may be thousands of kilometers apart. Loggerhead sea turtles (*Caretta caretta*) (Figure 9.21) lay eggs in underground nests on sandy beaches. Turtle hatchlings from southern Florida emerge and, after traveling to the water, must orient toward their feeding grounds in a large, circular system of ocean currents called the North Atlantic gyre. For loggerhead sea turtle hatchlings, this journey takes several days. Once at the feeding grounds, individuals may spend several years in the ocean, until sexual maturity, when they migrate back to their natal beaches for nesting (Lohmann, Putnam, & Lohmann 2008).

Kenneth and Catherine Lohmann and colleagues have studied sea turtles for over 20 years. In an early study, Kenneth Lohmann tested the ability of juvenile loggerhead sea turtles to detect and orient using magnetic fields, or **magnetoreception**. He used an apparatus to reverse the polarity of the magnetic field experienced by turtles (Lohmann 1991). The apparatus consisted of an open chamber with a walking table (Figure 9.22), encircled by a large coil that could create a strong magnetic field. Individuals were attached to a harness that allowed them to orient and swim in any direction, while the apparatus recorded their individual movements and orientation during a trial. One set of turtles was placed in the chamber while the magnetic field was reversed; a control set experienced the normal magnetic field. Turtles in the control treatment oriented and moved to the northeast, the direction of the North Atlantic gyre, and those that experienced a reversed magnetic field oriented and moved in the opposite direction (Figure 9.23). Thus, both groups oriented and moved toward a magnetic north, demonstrating the use of a geomagnetic compass.

magnetoreception The ability to detect magnetic fields and orient using them.

SCIENTIFIC PROCESS 9.2

The role of the antennae in the monarch butterfly sun compass
RESEARCH QUESTION: *How do monarch butterflies use a sun compass?*

Hypothesis (1): The sun compass is associated with the antennae.

Prediction (1): Individuals without antennae will not be able to orient properly.

Methods (1): The researchers:

- removed antennae from ten butterflies (antennae-less treatment) and left the antennae of ten control butterflies intact.
- tethered individuals and recorded flight behavior direction.

Results (1):
Control butterflies oriented southwest, while antennae-less butterflies did not orient any particular direction.

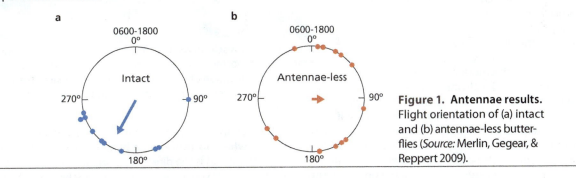

Figure 1. Antennae results. Flight orientation of (a) intact and (b) antennae-less butterflies (*Source:* Merlin, Gegear, & Reppert 2009).

Hypothesis (2): Antennae contain photoreceptors.

Prediction (2): Butterflies with clear paint or no paint on antennae will be able to orient properly, whereas butterflies with black paint on antennae will not be able to orient properly.

Methods (2): The researchers:

- painted antennae with opaque black paint, clear paint (control 1), or no paint (control 2).
- tethered individuals and recorded flight behavior direction.

Results (2):
Butterflies with no paint or clear paint on their antennae oriented south. Butterflies with black-painted antennae oriented north.

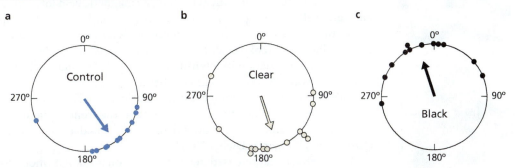

Figure 2. Paint results. Flight orientation of (a) control, (b) clear painted antennae, and (c) black painted antennnae (*Source:* Merling, Gegear, & Reppert 2009).

Conclusions: Antennae and functioning photoreceptors are required for sun compass orientation in monarch butterflies.

Citizen scientists track fall migration flyways of monarch butterflies

Not only researchers are tracking the mysteries of butterfly migration and their habitat. You, too, can participate in this research.

Monarch butterflies are of concern to conservation biologists because of the loss of their required habitat in Mexico, the United States, and Canada. To better monitor monarch butterfly populations, citizen scientists are helping to collect data on different life stages of the butterflies. Citizen scientists are individuals trained by biologists to participate in field projects across the country. The monarch butterfly project includes a Monarch Larva Monitoring Project (http://www.mlmp.org/), in which individuals at nature centers throughout the United States learn to conduct weekly surveys of monarch eggs and larvae on plants. Monarch Watch (http://www.monarchwatch.org/), based at the Kansas Biological Survey at the University of Kansas, oversees numerous citizen scientist programs that involve tagging individual butterflies and creating monarch waystations.

One monarch watch program, Journey North (http://www.learner.org/jnorth/) collects citizen scientist reports of monarch sightings during the fall migration. Elizabeth Howard and Andrew Davis used three years of these data to examine the migration flyway of monarchs in eastern North America (Howard & Davis 2009). Their analysis of thousands of sightings indicates that there is one major flyway south through the central portion of the United States (Figure 1).

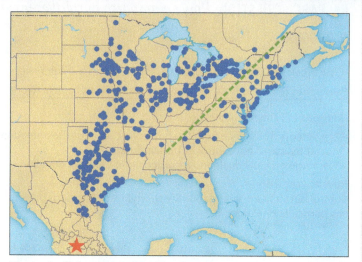

Figure 1. Citizen scientists. The dots represent roost sightings of monarchs reported by citizen scientists in 2005–2007. The star represents the location of the overwinter site in Mexico. The green line separates the central and eastern flyways (*Source:* Howard & Davis 2009).

This is important information, because it allows scientists to know where to concentrate habitat conservation efforts. Such information is the direct result of data collection by citizen scientists. If you want to be a citizen scientist, see http://www.fs.fed.us/monarchbutterfly/citizenscience/index.shtml.

Figure 9.21. Loggerhead sea turtle.
(a) An adult turtle migrating in the ocean.
(b) A hatchling.

Magnetoreception

It is easy to understand that animals can use the sun or stars as a compass. But how do they use the earth's magnetic field to orient themselves? Different systems appear to exist across taxa, but the most widespread involves the substance magnetite (Gould 2008). Magnetite is a small crystal of iron oxide whose magnetic properties depend on its size and shape. It has been found in many taxa, including honeybees, fish, amphibians, reptiles, birds, and mammals (and, originally, in bacteria). Because crystal magnetite is a permanent magnet, its magnetic field interacts with the earth's magnetic field, creating a force (pressure) that can be detected. Experiments have shown that exposure to external magnetic fields temporarily disrupts animals' orientation ability (Wiltschko et al. 1998). More recent work has demonstrated that birds can detect both variation in intensity of an external magnetic field and its direction (Stapput et al. 2008; Wu & Dickman 2012).

So where is the magnetoreception system located? The answer is still incomplete, but in vertebrates, magnetite is often found in the front of the head, in connective tissue near the sinus cavity, orbital (eye) and nasal cavities, and the inner ear (Kirschvink, Walker, & Diebel 2001).

However, magnetite-based magnetoreception is not the whole story. Some birds and amphibians cannot orient properly in total darkness, which suggests that light may also play a crucial role in one type of geomagnetic compass system (e.g., Wiltschko & Wiltschko 1981; Deutschlander, Phillips, & Borland 1999). This finding has led to the idea that magnetoreception might also be accomplished when a receptor

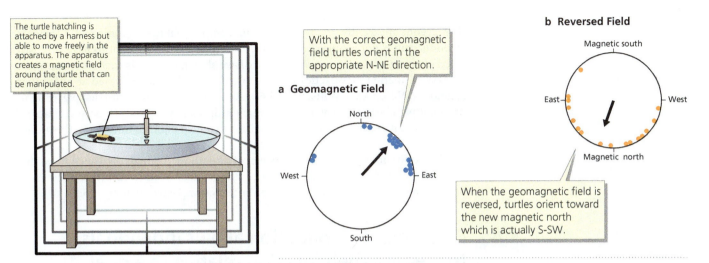

The turtle hatchling is attached by a harness but able to move freely in the apparatus. The apparatus creates a magnetic field around the turtle that can be manipulated.

With the correct geomagnetic field turtles orient in the appropriate N-NE direction.

a Geomagnetic Field

North

West   East

South

b Reversed Field

Magnetic south

East   West

Magnetic north

When the geomagnetic field is reversed, turtles orient toward the new magnetic north which is actually S-SW.

Figure 9.22. Turtle orientation apparatus. Diagram of the apparatus used to characterize orientation direction of sea turtles. The magnetic field inside the apparatus can be changed (*Source:* Lohmann 1991).

Figure 9.23. Turtle orientation. (a) Control turtles attempted to move north-northeast. (b) When the magnetic field was reversed, turtles attempted to move toward the new position of magnetic north-northeast. Each dot represents data for one turtle, and the mean direction is indicated by the arrow (*Source:* Lohmann 1991).

molecule absorbs photons (light particles). The energy absorbed temporarily changes the chemical property of the molecule in a manner that makes it responsive to external magnetic fields (Ritz, Adem, & Schulten 2000). Photopigments known as crypto-chromes possess such chemical properties and so are considered to be possible receptor molecules (Gegear et al. 2010). There is still much to learn about the biophysical basis of magnetoreception in animals, and this area of research remains active (Wiltschko & Wiltschko 2005; Gould 2010).

Multimodal orientation

The previously described research studies illustrate how research-ers have demonstrated the use of different compass systems. But one particular aspect is important to note: most, if not all, animals use multiple cues to orient. For instance, much research has examined the homing behavior of rock doves (*Columba livia*). This domesticated species has remarkable ability to find its way back to a home roost after being displaced up to 1800 km. How do they do this? Most studies of pigeon homing allow birds to live in a home loft for several weeks. Birds are then displaced to a new location and the flight direction or the entire flight of released birds is recorded using radio tags or flight recorders.

What mechanism are these birds using to orient? Birds that are temporarily blinded can orient properly and return home, so vision is not required to orient properly (Walcott 2005). However, birds that are clock-shifted six hours fly in the wrong direction—they make a 90° error in their orientation (Schmidt-Koenig 1990), just as we saw for monarch butterflies. This finding demonstrated that pigeons use a sun compass to orient. However, this error only occurs if birds can see the sun; on overcast days, clock-shifted birds orient the proper direction (Keeton 1974), indicating that pigeons use another cue to orient when the sun is hidden behind clouds. To investigate the possibility that pigeons use a magnetic compass, researchers placed magnets on their back (Figure 9.24) or magnetic coils on their head to alter the magnetic field around each bird. Such birds failed to orient properly, but only when they could not see the sun (Walcott 1996). These experiments reveal

A magnet glued to the body of a pigeon exerts an external magnetic force that disrupts the ability to orient properly.

Figure 9.24. Pigeon magnetoreception. When small magnets are placed on their body, pigeons have trouble orienting properly.

that pigeons apparently first rely on the sun compass but use a geomagnetic compass when the sun is not visible. And, while most birds are thought not to have sophisticated olfactory capabilities, some studies have shown that birds that are made anosmic (lack functioning olfaction) fail to orient properly when released at unfamiliar locations (Wiltschko & Wiltschko 1989). These studies demonstrate that pigeons can use multiple cues to orient properly. Such redundancy is likely widespread across taxa because it provides backup mechanisms that help animals to travel in the proper direction.

The ability to orient is just one aspect of migration. Animals also need to be able to move to specific locations, like the sea turtles that need to return to their natal beach to nest. Next, we examine how animals accomplish that navigation.

9.5   Bicoordinate navigation allows individuals to identify their location relative to a goal

bicoordinate navigation The ability to identify a geographic location using two varying environmental gradients.

Some animals are capable of **bicoordinate navigation**: they can identify their geographic location using two varying environmental gradients (Figure 9.25), much as humans use latitude and longitude. As an animal migrates, its geographic location changes. As long as it knows its present location relative to a target location or goal, such as its starting point or its final destination, it can adjust its direction to arrive at the destination. An important feature of bicoordinate navigation is that it can be used even when an individual is in an unfamiliar location, as might occur when an individual is blown off course during migration. Because bicoordinate navigation provides positional information, it is often referred to as a mental map.

Here we briefly examine three examples of bicoordinate navigation. We begin by revisiting the work of the Lohmanns and colleagues to see how they determined that sea turtles use bicoordinate navigation and a mental map.

Featured Research

Bicoordinate navigation and magnetic maps in sea turtles

Recall that the angle of inclination of the earth's magnetic field changes with latitude (Figure 9.16). In addition, the intensity of the magnetic field also varies across the globe (Figure 9.26). Because these aspects provide two environmental gradients, they can be used for bicoordinate navigation.

To determine whether sea turtles use bicoordinate navigation, the Lohmanns took advantage of an important fact: turtles need to orient themselves in order to remain in the North Atlantic gyre. For example, near Portugal, a northward current springs off the gyre, toward the colder northern waters of England. If sea turtles end up in these colder waters, they will die, and so turtles near Portugal need to move south. But turtles in the South Atlantic need to move north to remain in warm water.

The Lohmanns and colleagues (Lohmann et al. 2001) tested the ability of loggerhead sea turtle hatchlings to orient in the gyre. The experiment, conducted in eastern Florida, again used a computerized coil system to control the magnetic field. Turtles were exposed to one of three different magnetic fields, characteristic of three different geographic locations in the North Atlantic gyre, thanks to variations in the angle of inclination and field intensity. The hatchlings all oriented to stay within the gyre at each of the perceived locations (Figure 9.27). This finding indicates that migrating sea turtles can use the earth's magnetic field to orient in different (and appropriate) directions. In addition, because hatchlings that had no prior experience could navigate properly, this ability must be inherited and not learned.

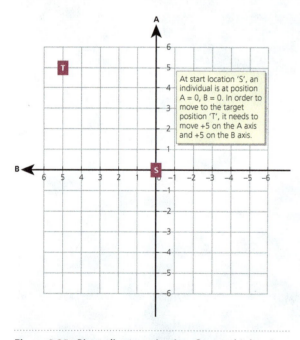

At start location 'S', an individual is at position A = 0, B = 0. In order to move to the target position 'T', it needs to move +5 on the A axis and +5 on the B axis.

Figure 9.25. Bicoordinate navigation. Geographic location can be identified with respect to two environmental gradients (A and B). Knowledge of start position S and target location T in bicoordinate space A and B allows an individual to reach its target goal (*Source:* Thorup & Holland 2009).

To further investigate bicoordinate navigation in sea turtles, the Lohmanns and colleagues examined whether green sea turtles (*Chelonia mydas*) use variation in the earth's magnetic field to navigate to specific locations (Lohmann et al. 2004). They captured juveniles near Melbourne, Florida, at their coastal feeding grounds and transported them to an orientation apparatus at a nearby test site. The turtles were divided into two groups. One group was exposed to a magnetic field that simulated the magnetic conditions 337 km north of the capture location. The other group of turtles was exposed to a magnetic field that simulated the magnetic conditions 337 km south of the capture location. Turtles were allowed to orient and move freely within the apparatus, and the directions in which they swam were recorded.

The research team found that the turtles in the two test groups behaved in different ways. Those experiencing simulated conditions of an area 337 km north of the capture site oriented and tried to move south, while those experiencing simulated conditions of an area 337 km south of the capture site oriented and tried to move north (Figure 9.28). These results show that sea turtles can use variation in the earth's magnetic field as a map to guide navigation north and south to a specific location—in this case, back toward their capture site. But how do individuals return to a specific location? Do they use landmarks, or do turtles use both the inclination and intensity of the magnetic field to identify a specific place on earth? Additional work is required to determine exactly how the map sense is used to find a specific location, such as a natal beach.

Turtles are not the only animals that can use bicoordinate navigation. Birds, too, can use bicoordinate navigation, as we see next.

Bicoordinate navigation in birds

Many birds migrate thousands of kilometers, and use a geomagnetic compass, a sun compass, or a star compass (for a review, see Wiltschko & Wilschko 2009). But do birds display bicoordinate navigation? To answer this question, Nikita Chernetsov, Dmitry Kishkinev, and Henrik Mouritsen examined migration movements in Eurasian reed warblers (*Acrocephalus scirpaceus*), which spend the winter in sub-Saharan Africa and then migrate to Eurasia to breed.

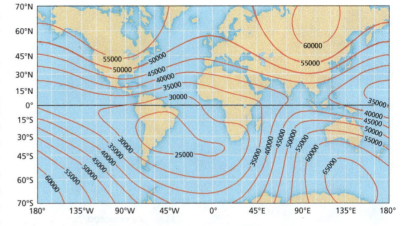

Figure 9.26. Intensity of the earth's magnetic field. Lines indicate variation in the intensity of the magnetic field, measured in nanoteslas (nT). Intensity ranges from about 25,000 to 65,000 nT.

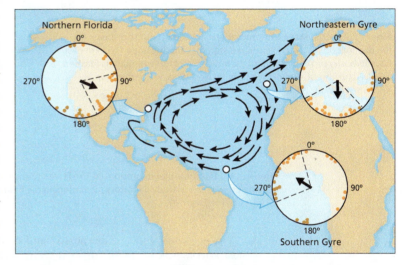

Figure 9.27. Sea turtle gyre results. The North Atlantic gyre is represented by the circular arrows. Each circle represents orientation data from turtles at the geomagnetic fields of three geographic locations. The mean is indicated by each arrow (*Source:* Lohmann et al. 2001).

Featured Research

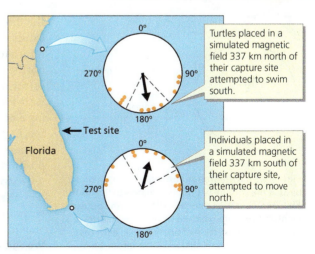

Turtles placed in a simulated magnetic field 337 km north of their capture site attempted to swim south.

Individuals placed in a simulated magnetic field 337 km south of their capture site, attempted to move north.

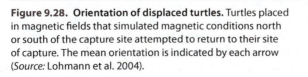

Figure 9.28. Orientation of displaced turtles. Turtles placed in magnetic fields that simulated magnetic conditions north or south of the capture site attempted to return to their site of capture. The mean orientation is indicated by each arrow (*Source:* Lohmann et al. 2004).

Figure 9.29. Displacement predictions. Birds migrating northeast (to the destination indicated by the red star) that are displaced 1000 km east from Rybachy to Zyenigorod might travel one of three directions (dashed lines): (1) northeast, (2) northwest, or (3) west (*Source:* Chernetsov, Kishkinev, & Mouritsen 2008).

Figure 9.30. Reed warbler migration. (a) The Emlen funnels used by Chernetsov to study migratory movement in the field. (b) Eurasian reed warbler.

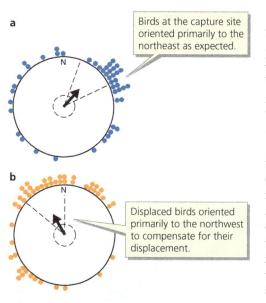

a

Birds at the capture site oriented primarily to the northeast as expected.

b

Displaced birds oriented primarily to the northwest to compensate for their displacement.

Figure 9.31. Orientation results. (a) Orientation of birds at the capture site, predominately to the northeast. (b) Orientation direction of birds at the displacement site, predominately to the northwest. Arrows indicate mean direction of orientation in each case (*Source:* Chernetsov, Kishkinev, & Mouritsen 2008).

Chernetsov and colleagues conducted an experiment during the warblers' spring migration (Chernetsov, Kishkinev, & Mouritsen 2008). Over a period of three years, they captured individuals near Rybachy, Russia. Half were flown 100 km east to Zvenigorod, while the other half were kept in Rybachy (Figure 9.29). The birds were placed in small Emlen funnels, an apparatus developed by Stephen Emlen to study use of the star compass in birds. The test birds had a clear view of the sky and so could use a variety of compass cues (Figure 9.30).

The Eurasian reed warblers tested at the capture site oriented northeast, the direction of their destination. The birds displaced 1000 km east could exhibit one of three behaviors: (1) they might orient due west, trying to return to the capture site, as we saw in green sea turtles; (2) they might orient northeast, the direction they were originally heading; or (3) they might orient northwest, trying to navigate to their final destination. The research team found that displaced birds oriented to the northwest and so were orienting a new direction to compensate for being displaced, suggesting that they, too, have the capacity for bicoordinate navigation (Figure 9.31). These results suggest that Eurasian reed warblers are able to determine their geographic position and adjust their migration route after being displaced, much as we use maps to find our way home after getting lost—although it is still unclear how they do this. Obviously, this ability can be of adaptive significance if birds are blown off course or need to modify their migratory route.

One final set of well-known migrants include salmon and their relatives. We end by examining how these fish accomplish their navigational feats.

Homing migration in salmon

Salmonids (salmon, char, and trout) are well known for their ability to return to spawn in their natal stream. Many salmon migrate downstream to the ocean, where they spend up to several years feeding prior to their long journey home to reproduce (Figure 9.32). This journey can be up to thousands of kilometers long and can occur more than ten years after leaving the natal stream. How is this feat accomplished?

Two mechanisms are thought to be used: olfaction and geomagnetic reception. Both rely on imprinting, which is thought to occur when salmon undergo a physiological change that allows them to adapt to both freshwater and saltwater—a process known as smoltification. Smoltification is associated with increases in several hormones, including thyroxine, which is thought to be associated with olfactory imprinting of the natal stream (Dittman & Quinn 1996). In essence, fish are thought to learn the odor and the magnetic properties of their natal stream in order to return to it.

The olfaction imprinting hypothesis was first proposed by Arthur Haesler and Warren Wisby and assumes that streams possess a characteristic and persistent odor, that fish can detect such odors, and that fish retain a memory of the odor they experience early in life (Haesler & Wisby 1951). Haesler and Wisby demonstrated the importance of olfaction in a simple experiment in which they captured salmon migrating upstream from two forks of a Y-shaped stream near Seattle (Wisby & Haesler 1954). Each fish was uniquely tagged to identify the fork in which it was captured. Half the fish had their olfactory pits blocked to impair their olfactory sense, and the half (controls) were left untreated. All individuals were then released 1.6 km below the fork so that they could resume their migration upstream. Of the fish recaptured, 89% of controls but only 40% of treatment fish were captured in the same branch that it was originally captured in. The researchers concluded that olfactory-impaired fish

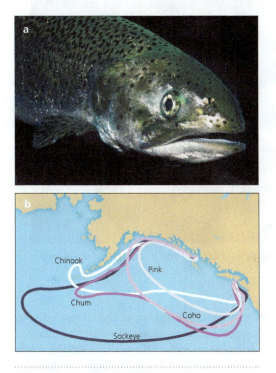

Figure 9.32. Salmon migration. (a) Salmon. (b) Migration regions for five species of Pacific salmon: Chinook (*Oncorhynchus tshawytscha*), chum (*O. keta*), coho (*O. kisutch*), pink (*O. gorbuscha*), and sockeye (*O. nerka*). All breed in freshwater streams in North America.

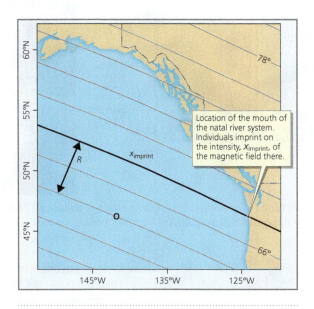

Figure 9.33. Salmon geomagnetic navigation. Salmon can imprint on the intensity of the earth's magnetic field ($X_{imprint}$) at the mouth of their natal river system (bold line). To return to their natal site from the open ocean (O), they need to travel north (or south) toward this intensity (R arrow) and then turn east and remain along this intensity until they reach the coast (*Source:* Bracis & Anderson 2012).

TOOLBOX 9.1

Emlen funnels

Early observations about bird migration indicated that caged birds increased their activity levels each spring and fall, a behavior known as migratory restlessness, or Zugunruhe. Stephen Emlen studied this behavior in a series of classic experiments using indigo buntings (*Passerina cyanea*). Indigo buntings are beautiful blue birds that breed each spring in the eastern United States and then migrate and spend the winter in southern Mexico, Central America, and the Bahamas. Interestingly, these birds are normally diurnal, except during the migration season, when they become highly active in the evening and throughout the night. These observations suggested that perhaps these birds were using celestial cues such as stars to orient.

Emlen designed a simple yet elegant apparatus to take advantage of the buntings' migratory restlessness behavior (Emlen & Emlen 1966) (Figure 1). An Emlen funnel consists of a small funnel with a wire mesh top that allows a bird to see out. The bottom has an inkpad that transfers ink to a bird's feet. The sides of the funnel are lined with paper. A bird's movements in hopping forward as it exhibits migratory restlessness are recorded by the ink tracks left by its feet on the paper.

Emlen conducted experiments in a planetarium, where he could manipulate the night sky to examine how it affected the birds' orientation and movements (Emlen 1967a; Emlen 1967b). Stars can provide directional information in the Northern Hemisphere, because all stars rotate around the celestial pole. In a series of experiments, he placed birds in Emlen funnels and manipulated the location of stars and their rotation. Emlen convincingly showed that indigo buntings use celestial cues to orient. In particular, he determined that maturing birds learn to identify the direction around which all stars rotate and use this fixed location as an indicator of north. In the Northern Hemisphere, the celestial pole is centered on the North Star, Polaris. Emlen demonstrated that he could train young birds to use a different star (Betelgeuse, in the constellation Orion) as an incorrect indicator of north after he allowed birds to view all stars in the planetarium rotating around it (Emlen 1970). More recent research indicates that indigo buntings may also use the earth's magnetic field to orient the proper direction during migration (Sandberg et al. 2000).

Figure 1. Emlen funnel. (a) Side view of an Emlen funnel, showing a bird standing on an ink pad and its footprints are then recorded on the paper sides (*Source:* Emlen & Emlen 1966). (b) Indigo bunting.

randomly selected a fork, whereas those with functioning olfaction used that sense to return to their natal stream. What exactly provides the unique odors of natal streams is less clear. Some work suggests amino acids (Yamamoto, Hino, & Ueda 2012), while other work suggests that salmon detect the odor of related, younger fish moving downstream (Nordeng 1977).

For adult salmon in the open ocean, they first need to find the mouth of their natal river system. It is difficult to do this using odor alone, and so a second mechanism is needed: this may be geomagnetic imprinting. Salmon are hypothesized to imprint on the specific properties of the earth's magnetic field at the mouth of their natal river system after undergoing smoltification (Lohmann, Putnam, & Lohmann 2008).

In particular, they are thought to learn the inclination angle and/or intensity of the earth's magnetic field at this location, both of which vary with latitude (Figure 9.33). Studies have shown that salmon can respond to different magnetic field intensities, and juvenile salmon will change their orientation in tanks in response to a change in the external magnetic field (DeBose & Nevitt 2008). We expect further evaluation of this hypothesis in the coming years.

These research studies illustrate how researchers study bicoordinate navigation and reveal that a variety of animals possess such capability. Little is known about the navigational abilities of most species, which is an active area of research.

Chapter Summary and Beyond

Animals move from one location to another during dispersal and when migrating. Juveniles embarking on natal dispersal leave their place of birth to find a site for reproduction. Natal dispersal can benefit individuals by reducing both competition with siblings and the likelihood of settling nearby and mating with close kin. Two recent reviews on dispersal behavior suggest that to more completely understand these movements, we need to focus on how the environmental context of an animal (e.g., competition, habitat quality, inbreeding risk) and the internal state of an individual (e.g., body condition, physiology) interact (Handley & Perrin 2007; Clobert et al. 2009). Breeding dispersal by adults is commonly observed after a reproductive failure. Recent work has focused on how individuals combine and use information about their own reproductive success, the success of others nearby, and the density of individuals when deciding whether to disperse to a new site (e.g., Citta and Lindberg 2007). For

a comprehensive review on the evolution of dispersal, see Ronce (2007).

Longer, two-way movements are known as migration. Animals use a variety of cues to orient to the proper direction, including the position of the sun and stars and the earth's geomagnetic field. Ongoing work continues to reveal that animals use multiple cues for orientation (Bingman & Chen 2005; Muheim, Moore, & Phillips 2006; Murray, Estepp, & Cain 2006). Mouritsen and Hore (2012) provide additional details about magnetoreception in birds. The study of migration is being revolutionized by the development of small geolocators that can be attached to individuals to record long-distance movements (Reynolds & Riley 2002; Stutchbury et al. 2009). For a comprehensive review of mechanisms that regulate migration, see Ramenofsky and Wingfield (2007), and for a review of physiological adaptations involved in migration, see Åkesson and Hedenström (2007).

CHAPTER QUESTIONS

1. Identify two ways that dispersal differs from migration behavior. In what way are they alike?

2. Competition for resources and inbreeding avoidance are two explanations for natal dispersal. Discuss each of these hypotheses in the context of proximate and ultimate explanations for natal dispersal.

3. A researcher examined natal dispersal distances in a population of chickadees over the course of two years. The population size and number of juveniles dispersing was the same in both years. However, in Year 1, the average dispersal distance was 10 km, but in Year 2, the average dispersal distance was only 5 km. Propose an explanation for this difference.

4. What assumption is required to explain the win-stay lose-shift behavioral strategy often seen in breeding dispersal?

5. Flycatchers, like the Eastern wood pewee (*Contopus virens*), eat flying insects, including bees, flies, butterflies, and wasps. Woodpeckers, such as the downy woodpecker (*Picoides pubescens*), eat beetle larvae buried under tree bark, as well as berries, seeds, and sap. In temperate climates like midwestern North America, flycatchers migrate each fall to the tropics, whereas woodpeckers do not. In this chapter, we discussed the relationship between resource fluctuations and migration. How would you explain the migratory difference between flycatchers and woodpeckers? To do this, you need more information. Gather information on the life cycle or life history of these birds' main food items. You can find this information in either the primary or secondary literature. Using the information you collect, describe the relative stability of food sources of these birds over the course of the year.

6. In a northern population of sparrows, you have determined that 80% of individuals migrate each fall to warmer locations, while 20% remain residents year-round and overwinter on the breeding grounds. Climate scientists predict that at this location, winters will become milder over the next 50 years. Explain how this shift might affect the proportion of adults that migrate in the future.

7. How does geomagnetic orientation differ from bicoordinate navigation that relies on the earth's magnetic field?

Chapter 10

Habitat Selection, Territoriality, and Aggression

Figure 10.1. Mockingbird and mirror. A mockingbird attacks its mirror image, a perceived rival, to defend its territory.

10.1

10.2

10.3

10.4

On a recent outing with students to our university's field station where students take courses in field biology, we left our vehicle parked near a trail as we hiked. When we returned, a northern mockingbird, *Mimus polyglottos*, was vigorously pecking at the car's sideview mirror (Figure 10.1). We all watched for a while, trying to understand this seemingly odd behavior. We realized this bird probably perceived its reflection in the mirror as a rival threat that required active defense. Once we placed a shirt over the mirror, the behavior stopped. Our observation was not that unusual, and in fact, researchers often use mirror reflections to study aggressive behavior (e.g., Kusayama, Bischof, & Watanabe 2000).

Many animals, like the mockingbird we observed, defend territories, particularly during the breeding season. Before this season, however, they must select a habitat in which to settle. In this chapter, we'll see that conspecifics play a major role in habitat selection. The presence of another individual means that resources must be shared, and so competition for these areas can be high. Yet the presence of a rival may also indicate that a habitat is of high quality. Once in a habitat, territorial defense can provide exclusive access to food or mates, but it is also costly. We explore how animals fight over resources such as territories and conclude with an examination of how aggressive behavior is regulated.

10.1 Resource availability and the presence of others influence habitat selection

Recall that in Chapter 1, we examined a hypothetical situation in which there was variation in the number of robins in yards: some yards had many robins, while others had few. In nature, such variation in the density of individuals in different habitat locations is common. This naturally leads to two questions: Why does this variation exist, and how do animals decide where to live? Two important factors may come to mind: the amount of resources and the number of other individuals in a habitat. Resource availability can be an important predictor of habitat quality, and the number of individuals in a habitat can indicate the level of competition for those resources. Let's see how researchers have examined the importance of these factors to explain the distribution of animals.

The ideal free distribution model

■ **ideal free distribution (IFD) model**
A model that explains how animals distribute themselves among habitats or food patches.

Steve Fretwell and Henry Lucas developed the **ideal free distribution (IFD) model** to explain how animals distribute themselves among habitats (Fretwell & Lucas 1969). The model's name comes from its assumptions that animals are "ideal" in that they select habitats that maximize survival and reproduction and "free" to enter any habitat.

The model contains five assumptions:

1. Individuals attempt to maximize their fitness when settling in a habitat.
2. Habitat locations differ in the resources they contain.
3. The fitness of individuals in a habitat decreases as more individuals settle there because of increased competition for resources. In other words, fitness is negatively density dependent.
4. Individuals have equal competitive ability and can accurately assess the fitness payoffs of each habitat.
5. Individuals are free to move between habitats at no cost.

The model then makes the following predictions:
Prediction 1: Individuals will settle in habitats based on the relative fitness payoffs: the number of individuals in each habitat will be proportional to habitat quality, resulting in more animals settling in higher quality habitats.
Prediction 2: All individuals will have the same fitness no matter where they settle.

We can understand these predictions using three hypothetical habitats, or habitat patches, of high, medium, and low quality (Figure 10.2). Imagine a situation in

Figure 10.2. Ideal free distribution model. The model predicts the abundance of individuals in hypothetical habitats A, B, and C, at population densities X, Y, and Z. (a) Habitat A is the highest quality because it yields the highest fitness for any fixed population size. Habitat C is the lowest quality because it yields the lowest fitness. At low population density, X, all individuals select habitat A and have high fitness. At moderate density, Y, most individuals select habitat A, but some choose habitat B, and all have moderate fitness. At high density, Z, every habitat is used: most individuals settle in habitat A, the fewest settle in habitat C, and all have low fitness. (b) The number of individuals in each habitat at population densities X, Y, and Z.

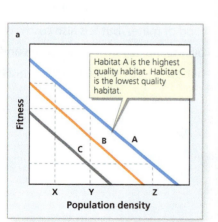

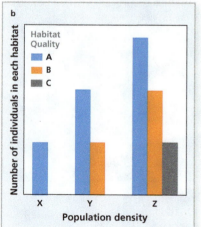

which individuals settle sequentially in these habitats. The first few individuals should settle in the high-quality habitat because this yields the highest fitness. However, as more and more individuals enter the high-quality habitat, the fitness of each individual will decline because of increasing competition for resources; this explains the negative relationship between number of individuals in a habitat and their fitness. At some point, the fitness payoff is higher for an individual settling in the medium-quality habitat, so a new settler should go there, and additional settlers should then select both of these habitats equally. As even more individuals settle in both habitats, eventually the best choice for a new settler will be the low-quality habitat. In this process, we see that as more individuals settle, more habitats of increasingly lower quality are used.

This model also allows us to make predictions about the relative number of individuals in each habitat. If habitat quality is determined by resources, the model predicts that the number of individuals will match the resources available. For just two habitats, if one contains twice the resources of the other, then twice as many individuals should settle there. If the resources differ by a factor of ten, so will the number of individuals. In this way, each individual obtains the same fitness no matter which habitat it settles in. During the settling process, individuals are always free to move to a higher quality habitat if one is available. However, once all individuals settle, none can do better by moving to another habitat. Let's look at two tests of these predictions.

The ideal free distribution model and guppies

Featured Research

The IFD model is easiest to test at a small spatial scale rather than in large habitats; therefore, numerous tests have used small food patches. At this smaller scale, patch quality (i.e., habitat quality) can be manipulated by changing food density or provisioning rate. Mark Abrahams did this when he tested the model using guppies, *Poecilia reticulata* (Abrahams 1989). Abrahams created two food patches using flasks filled with water that delivered food (fly eggs) to two feeding bars, or patches, on opposite sides of an aquarium. He modified the quality of each patch by changing the number of eggs in each flask. In this way, the relative patch quality varied from equal (1:1 ratio) to highly skewed (1:9 ratio). He placed ten individuals (either all males or all females) in the aquarium and fed them at three distinct times each day using the same relative patch qualities. In each trial, he recorded the number of fish at each patch (Scientific Process 10.1).

The IFD model predicts that the fish should distribute themselves among the patches in the same ratio as the food delivery ratio. Abrahams found that the model accurately predicted the distribution of fish late in the day, but not the distribution in the first trials of the day. Why? Apparently, it took the fish some time each day to learn the relative qualities of the patches. Abrahams also quantified individual feeding rates for a smaller sample of fish by counting the number of eggs eaten by each individual at a patch. He found that they all had the same feeding intake rate and saw no evidence of unequal competitive ability—a finding that would have violated an assumption of the model. Abrahams's results support the predictions of the IFD model, demonstrating how resources and competition affect the distribution of individuals among patches.

The IFD model has been very effective in predicting and explaining variation in the number of individuals in food patches. Can the model also explain habitat selection over larger geographic scales and over many years (a longer time scale)? Our next study provides the answer.

SCIENTIFIC PROCESS 10.1

Ideal free guppies

RESEARCH QUESTION: *How do guppies distribute themselves among food patches?*

Research Hypothesis: Food delivery rate and number of competitors will affect the distribution of individuals among food patches, as outlined by the IFD model.

Prediction: Individuals will distribute themselves so that all obtain equal food intake rates. Their relative abundance will match the relative rate of food delivery.

Methods: The researchers:

- conducted trials using groups of ten males or ten females in 90-l aquaria. Feeders (Erlenmeyer flasks with a stir rod to keep fly eggs in the water solution) were placed on two sides of the aquarium. A piece of tubing that ran from the flask into the aquarium created a horizontal feeding bar. The presence of different amounts of food in each flask created patches of different quality but kept the overall total food the same across treatments.
- conducted three trials per day (one in the morning, one at noon, and one in the evening) using the same relative patch qualities.
- counted the position of guppies (left feeder, right feeder, or not feeding) every 30 seconds for 24 minutes per trial.

Figure 1. Experimental tank. Feeders on each side create two feeding patches. A trough under the feeding bar collects extra food (*Source:* Abrahams 1989).

Results:

- The distribution of male and female fish was strongly affected by the amount of food in a patch.
- In the morning trials, fewer fish than predicted used the patch with more food.
- In the noon and evening trials, the distributions matched the IFD prediction.
- Within a patch, the feeding intake rate of each of the six males was similar, as the model assumed.

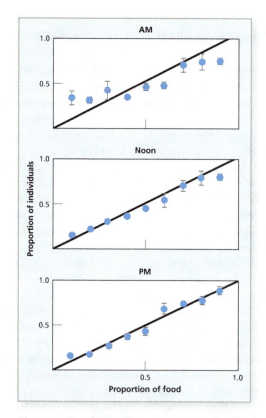

Figure 2. Patch use. The proportion of individuals at the left feeder versus the proportion of food available at that feeder for the morning, noon, and afternoon trials (*Source:* Abrahams 1989).

Conclusion: Guppies distributed themselves according to the predictions of the IFD model once they had time to learn the distribution of food.

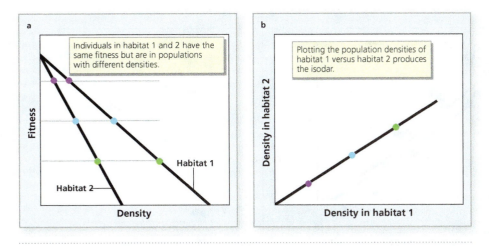

Figure 10.3. Isodar. (a) The ideal free distribution model shows population sizes in two habitats in each of three years (denoted by different colors). Note that in all three years, the fitness is equal for individuals in the two habitats. (b) Data from both densities of each year are plotted against one another to yield a single point. Data from multiple years (different colors) form a linear isodar with a positive slope (*Source:* Morris 2006).

The ideal free distribution model and pike

Featured Research

Douglas Morris examined the IFD model using a different approach. He examined the distribution of individuals among habitat patches over a number of years to see if the number of individuals in each habitat changed over time (Morris 1988). Recall that the model predicts that individuals in all patches will have equal fitness. However, if total population size changes each year, so too will the predicted number of individuals in each patch. Morris showed that if you plot the predicted number of individuals in one habitat against the predicted number in the other habitat over time, the model predicts a straight line with a positive slope, which Morris called an **isodar** (*iso* means same, and *dar* comes from *Darwinian fitness*) (Figure 10.3).

isodar The number of individuals in one habitat plotted against the number of individuals in another habitat at several points in time.

Thrond Haugen and colleagues tested the isodar line prediction of the IFD model for a population of pike (*Esox lucius*) in Lake Windermere in England (Haugen et al. 2006). Lake Windermere is a long, narrow lake with a shallow portion in the middle, which creates two deep basins, or habitat patches: one in the north and one in the south. Pike are top predators that often exceed 1 m in length (Figure 10.4). The midpoints of the basins are only 7 km apart, and pike regularly move over 2 km per day, so the cost of moving between basins should be relatively small. Researchers have conducted mark-recapture studies to estimate the density of pike in these basins for 50 years, and repeated captures have revealed that individuals often move back and forth between the basins, as assumed by the model.

Recall that both resource availability and the presence of others affect habitat selection. In this study, the researchers did not manipulate habitat quality, but rather took advantage of manipulations of the number of fish in each habitat. At Lake Windermere, high-intensity fishing was conducted in the south basin for three years and was then followed by high-intensity fishing in the north basin for three years. These manipulations reduced pike populations by 15%–28% within each basin in a given year.

Figure 10.4. Pike. Pike are common top predators in freshwater systems.

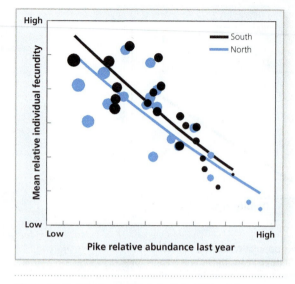

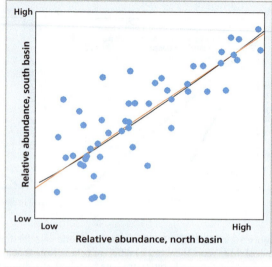

Figure 10.5. Pike fecundity. In both habitat patches, mean fecundity declined as density of pike increased, and so pike show negative density-dependence in fecundity in each basin, as assumed by the IFD model. Circle size reflects estimated body mass of females (*Source:* Haugen et al. 2006).

Figure 10.6. Pike isodar. The observed isodar (red line) for the pike data matches the predicted isodar from the IFD model (black line). Each point represents one year of data (*Source:* Haugen et al. 2006).

Haugen and colleagues first examined the assumption that the fitness of individuals in each basin was negatively density dependent. To determine this, they used the long-term mark-recapture data to plot survivorship and reproductive rates of individuals in years of different density. Within each basin, they found that in years of low pike abundance, both survivorship and reproduction were high, while in years of high abundance, survivorship and reproduction were low. This was especially evident in years when density manipulation occurred (Figure 10.5). Survivorship and fecundity declined as population size increased, just as the IFD model assumes. Figure 10.5 also indicates that the south basin is the higher quality habitat because the fecundity (fitness) of individuals is higher there than in the north basin for all population levels.

Over the 50-year period, there was tremendous variation in total population size—and thus the number of pike in each basin. This provided sufficient variation in densities in the basins to plot the isodar, which turned out to be a straight line with a positive slope, just as the IFD model predicts (Figure 10.6).

These examples illustrate how variation in patch quality and the presence of others affect the distribution of individuals in both small-scale food patches and larger scale habitats in a manner consistent with Fretwell and Lucas's model. It is important to note that although the model assumes that individuals know the relative quality of their choices, the guppy experiment reveals that it can take time for individuals to assess the relative quality of food patches. How do animals assess the relative quality of larger scale habitats? Recent research provides one answer.

Featured Research

Cuckoos assess habitat quality

Animals may assess the relative quality of different habitats by estimating the level of food resources they contain. Nicholas Barber and colleagues used a novel approach to examine how some birds might do this. Black-billed (*Coccyzus erythropthalmus*) and yellow-billed (*C. americanus*) cuckoos migrate from Central and South America to breed in woodland habitats in eastern North America each spring, returning to their nonreproductive habitat in autumn. They eat a variety of insects but consume large quantities of hairy caterpillars, including exotic gypsy moths (*Lymantria dispar*) (Figure 10.7). Gypsy moths, native to Europe and Asia, were accidentally introduced into northeastern North America in the late 1860s and have expanded west and south ever since, making them a major forest pest. Females lay egg masses on tree branches in the fall, and larvae (caterpillars) emerge in spring when trees leaf out. At this stage, larvae can disperse on silken threads to reach new trees. The larvae develop through several molting stages over a period of several weeks and consume leaves as they mature.

Gypsy moth populations fluctuate over space and time. When population densities reach unusually high levels in outbreak years, trees become defoliated as a result of the caterpillars' high leaf consumption (Figure 10.8). Such defoliation is easy to see in aerial photographs. Do cuckoos use defoliation to locate gypsy moth outbreaks and thus habitats with abundant food? To test this idea, Barber and colleagues obtained digitized maps of gypsy moth outbreaks for each year from 1975 to 2003 (Barber, Marquis, & Tori 2008). They also used Breeding Bird Survey (BBS) data to estimate the abundance of cuckoos over a period of 16 years—five years before an outbreak year, the outbreak year, and ten years following the outbreak. BBS data record the abundance of all birds seen or heard each spring along fixed 39 km census routes. The researchers used BBS routes that came into contact with defoliation areas as focal routes. For each route, they calculated average cuckoo abundance over the 16-year period.

The research team found that cuckoo abundance correlated with caterpillar outbreaks. The abundance of cuckoos peaked in the year of defoliation, but in the five years preceding defoliation, abundances were at or below average (Figure 10.9). Black-billed cuckoo densities remained above average in the year following defoliation but returned to average in subsequent years; yellow-billed cuckoo densities were at or below average in all years following an outbreak. This pattern suggests that cuckoos can zero in on local areas in each year that gypsy moths are abundant and defoliating trees.

Were the birds actually tracking the outbreaks? To answer this question, the researchers attempted to characterize patterns of bird movement. Barber and colleagues examined all BBS routes within 500 km of a gypsy moth outbreak. For each route, they determined the distance and compass direction to the area of defoliation. They then compared the abundance on a route in the outbreak year to its ten-year average. They found that routes within 40 km of an outbreak had substantially higher-than-average abundances in outbreak years, whereas routes that were 45 to 140 km away from an outbreak had lower-than-average abundances in outbreak years, suggesting that birds 45 to 140 km away from an outbreak moved toward the outbreak—and did so from all directions.

Barber and colleagues concluded that gypsy moth outbreaks strongly affect the local abundance of cuckoos. They suggest that cuckoos will move fairly long distances (up to 140 km) to find high-quality habitats. Gypsy moth outbreaks provide just such high-quality habitats. Large defoliated areas may provide the cues, but this hypothesis has yet to be directly tested.

Up until now, we have assumed that individuals assess the relative quality of their options and that rivals reduce fitness of a habitat. So if two habitats of equal quality exist and one is already occupied, a new settler should opt for the unoccupied habitat. Sometimes, however, individuals prefer to settle in the same habitat that a competitor already occupies, even when a high-quality, unoccupied habitat is nearby. This requires a new explanation, as we see next.

Conspecific attraction

The IFD model assumes that competition for resources results in a decline in individual fitness as the number of individuals in a habitat increases. Because each additional individual reduces fitness for all, owing to competition, new settlers should

Figure 10.7. Cuckoo and moth. (a) Yellow-billed cuckoo. (b) Gypsy moth larva.

Figure 10.8. Gypsy moth defoliation. A patch of trees defoliated by gypsy moth larvae.

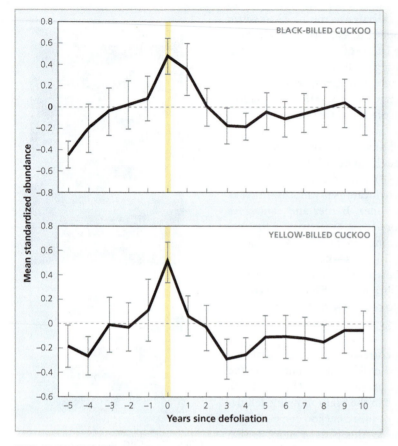

Figure 10.9. Cuckoo abundance over time. For both species, mean (± SE) cuckoo abundance on a route was highest in the year of gypsy moth defoliation, indicated by the yellow line (zero years since defoliation). The horizontal dashed line is the 16-year average abundance on the route. Abundance was standardized to account for the population declines in both species over the entire study period (*Source:* Barber, Marquis, & Tori 2008).

avoid rivals if a better option exists. Sometimes, however, individuals exhibit the opposite pattern. Instead of avoiding others, they settle near them, a pattern known as **conspecific attraction**.

Why might individuals exhibit conspecific attraction? Two hypotheses exist. First, for very low population sizes, fitness may increase as more individuals settle, a phenomenon known as an **Allee effect** (Allee 1931). For example, a solitary individual might be at high risk of predation, whereas two or more individuals might experience lower risk. Similarly, a group might be more attractive to potential mates than a single individual.

Second, it may be difficult for individuals to assess the relative quality of some habitats—an ability assumed by the IFD model. Habitat quality is a function of many factors, such as resource levels, the availability of nesting locations, and the number of predators, some of which may be difficult to assess. Instead, a settler can use the presence of another individual as an indicator of habitat quality, a phenomenon called **conspecific cueing** (Stamps 1988). In many species, experienced breeders will resettle in a habitat if they successfully reproduced there the preceding year, as we saw in Chapter 9. If a high-quality habitat last year is likely to be a high-quality habitat this year, then the location of experienced breeders provides a valuable cue. This hypothesis predicts that inexperienced individuals, whether first-time breeders or new immigrants, are most likely to exhibit conspecific cueing. The next two studies test these hypotheses.

Conspecific attraction and Allee effects in grasshoppers

conspecific attraction A phenomenon in which individuals are attracted to others of their species, particularly during habitat selection.

Allee effect A situation in which the fitness of individuals increases with increased density.

conspecific cueing A hypothesized mechanism to explain attraction to members of the same species. A settler uses the presence of another individual as a cue to habitat quality.

The creosote bush grasshopper, or desert clicker (*Ligurotettix coquilletti*), is abundant in the deserts of North America, where it feeds on creosote bush (*Larrea tridentata*) leaves. For a male desert clicker, a bush represents a breeding habitat. Each spring, males set up mating territories on creosote bushes. Males tend to be faithful to a site and remain on the territory throughout the spring and summer, when they make a loud clicking sound to attract females (Greenfield & Shelly 1985). Earlier studies indicated that males seem to aggregate on some bushes and not use others at all, even though the bushes seemed to be similar in quality (Greenfield, Shelly, & Downum 1987). Why do they congregate? Something other than resource quality appears to affect their habitat choice.

Katherine Muller wondered whether male desert clickers were using the presence and vocalizations of others as a cue for habitat selection (Muller 1998). At a site in the Mojave Desert, she selected pairs of creosote bushes that were 3.5 m apart and similar in both size and shape. She used playback calls from speakers placed at the base of each bush to standardize the vocalizations. In the first experiment, she gave males the choice of one bush with a speaker playing conspecific male calls or another bush that was either unoccupied or had a speaker playing heterospecific male calls. She played the calls at one bush at a time while a focal male was held in a cage equidistant from each bush. She then lifted the lid of the cage and recorded the movement behavior of the focal male (Scientific Process 10.2).

SCIENTIFIC PROCESS 10.2

Conspecific attraction in grasshoppers
RESEARCH QUESTION: *How do desert clickers select habitat patches?*

Hypothesis: Males use the presence of another male as an indicator of patch quality.

Prediction: Males will prefer creosote bush patches from which conspecifics call.

Methods (1): The researchers:

- chose pairs of creosote bushes (3.5 m apart) that were similar in size and shape.
- played conspecific call from one bush and either heterospecific call or silent control from the other bush.
- placed a caged focal male 2 m from a pair of habitat patches (creosote bushes) and measured which bush the male selected.

Focal male

Figure 1. Design for experiments. Males can select one of two habitat bushes.

Results (1): Seventeen out of 20 males flew to the conspecific call bush versus the silent bush. Sixteen out of 20 males flew to the conspecific rather than the heterospecific call bush.

Methods (2): The researchers:

- used a similar design as Experiment 1 but placed a resident male on one bush and left the other empty.
- recorded the presence of the focal males on the bushes at 12 and 36 hours after release.

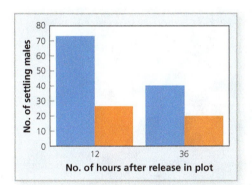

Figure 2. Habitat choice. Males prefer bushes with a conspecific (blue bars) over an empty bush (orange bars) (*Source:* Muller 1998).

Results (2): Most focal males selected the bush with a conspecific at 12 hours. More males were counted on the bush that contained a conspecific at 36 hours.

Conclusion: Males prefer habitat patches that contain other males, an example of conspecific attraction.

In both experiments, at least 80% of males flew to the conspecific call bush. Clearly, males preferred the habitat with an apparent conspecific.

To determine what happens after a male arrives on a bush that also contains a rival, Muller conducted another experiment using similar protocols. This time, one of the bushes was occupied with a calling conspecific male that had been a resident of the bush for three days and the other bush was empty. After the focal male was released, Muller counted the number of crickets on both bushes 12 and 36 hours later. Of the males released, 72 of 98 were found on occupied bushes at the end of 12 hours, again displaying conspecific attraction. At the end of 36 hours, some of these males had abandoned the bushes, and a higher proportion had left the occupied bush than had left the empty bush. Muller observed aggressive territorial interactions among males on the bushes and assumed that this behavior explained the decline in males on occupied bushes over time.

Why did focal males prefer the occupied bushes? One explanation is female behavior. In previous work, Muller found that females exhibit a strong preference for bushes with multiple calling males (Muller 1995); Greenfield and Shelly (1985) found that aggregated males have higher mating success than singletons. In other words, female preference may create an Allee effect: males that aggregate on a bush with rivals can obtain more mating partners than can solitary males. Thus, even though a second male may experience increased aggression from an established male, its reproductive success can still be higher than if it settled alone. This example illustrates an Allee effect that can explain conspecific attraction in grasshoppers and their habitat selection behavior. Next, we turn to an example of conspecific cueing to explain conspecific attraction in birds.

Featured Research

Conspecific cueing in American redstarts

American redstarts (*Setophaga ruticilla*) are migratory warblers that breed in deciduous forests in eastern North America (Figure 10.10). Each spring, migrants arrive on their breeding grounds in May and establish territories. Adult males return first, followed by yearling males that will breed for the first time, and then females. The breeding season is relatively short, and so males need to find a territory quickly to breed successfully.

Beth Hahn and Emily Silverman studied conspecific cueing and habitat selection in these birds at a breeding site in Michigan (Hahn & Silverman 2006). Three types of males settle on a territory. The first are returning males, or experienced birds that bred the preceding year at the study site. The second are new adult immigrants, or males that bred in a different location the preceding year. The third are first-time breeders, yearling males. These yearlings can be distinguished from older adults by plumage differences: they have only scattered black feathers on their head, breast, and back, while adults are solid black (Rohwer, Klein, & Heard 1983).

Both immigrant adults and yearlings lack experience in the habitat. Hahn and Silverman predicted that they would use the presence of a singing male as a cue to the location of a high-quality habitat. In contrast, they predicted that returning males would not be affected by this singing.

The researchers captured and uniquely color-banded all birds on 12 habitat plots measuring 300 m by 300 m. Six plots were designated as treatment plots and six as controls. In each treatment plot, the researchers played songs of male American redstarts every day in May, starting before any males arrived and continuing until all males had settled. The next year, the treatment plots were switched: treatment plots became controls, and controls became treatment plots. The researchers placed speakers 2 m up in trees and connected them to a compact disc player. They standardized the redstart playback amplitude so that songs could be heard 100 m away. In order to avoid habituation to the songs, they used a total of 75 individuals to make the discs. Hahn and Silverman surveyed each plot four times per week, recording the arrival of males along with their location, age, and identity. Any new, unmarked adult male was captured and uniquely color-banded.

As predicted, new immigrant males strongly preferred to settle on plots that played a conspecific song, while returning adults did not, indicating that conspecific songs do attract new immigrant males to plots through conspecific cueing. However, first-time breeders did not exhibit conspecific attraction after all: they settled equally on treatment and control plots (Figure 10.11). Why the unexpected result? First-time breeders did not arrive at the site until six to 10 days *after* the experienced

Figure 10.10. Calling American redstart. A common bird in North American woodlands.

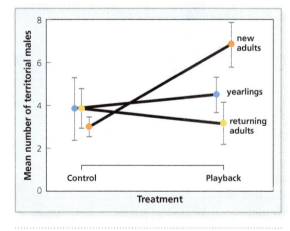

Figure 10.11. Male American redstart settlement on plots. Mean (± SE) number of territorial males on control and treatment plots. New immigrant adults settled preferentially on plots with playback songs of conspecifics, exhibiting conspecific attraction. Yearlings and returning adults were not affected by the playback treatment (*Source:* Hahn & Silverman 2006).

adults. By then, the adults had already settled and begun to sing, so song production was high on all plots, effectively swamping out the playback treatment. Nonetheless, these results indicate that conspecific attraction is an important mechanism for habitat selection for new males. We now know that many birds respond positively to song production when searching for habitat and breeding territories (Ahlering & Faaborg 2006). Conservation biologists have used this finding to help restore populations of birds threatened with extinction (Applying the Concepts 10.1).

In this section, we have seen that habitat selection is a complex process that is influenced by resource distribution, the number of competitors, habitat patch assessment, and Allee effects. Selection of a habitat patch is only the first step for many species prior to reproduction. Individuals from many species need to establish a territory in order to successful reproduce. We examine this behavior next.

10.2 Individual condition and environmental factors affect territoriality

When animals settle in an area, they can establish a **territory**, an area defended to obtain the exclusive use of the resources it contains, such as food, nesting sites, or potential mates. Territorial defense provides exclusive use of resources but also requires that an animal expend effort to defend it from rivals. Territories differ from **home ranges**, which are areas of repeated use that are not defended. The home ranges of many individuals can overlap while territories do not. Territory defense requires physical effort: it takes both time and energy to chase off intruders. Can all males defend a territory?

territory An area defended to obtain the exclusive use of the resources it contains.

home range An area of repeated use by an individual that is not defended from others.

Body condition affects territoriality in damselflies

Featured Research

Aggressive behaviors such as fighting require substantial effort, and so only individuals in good body condition may be able to acquire and defend a territory. Body condition can be measured by body mass, body size, or measures of fat, which indicates extra energy storage that could be used in territory defense. Jorge Contreras-Garduño and colleagues studied how the physiological status of damselflies affects their ability to acquire a territory.

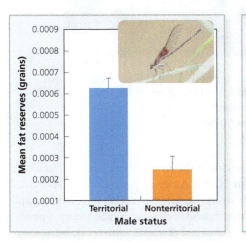

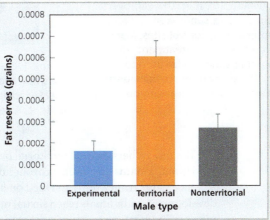

Figure 10.12. Fat reserves in male American rubyspot damselflies. Mean (+ SE) fat reserves in grains (1 grain = 0.06 g). Territorial males had higher fat reserves than nonterritorial males (*Source:* Contreras-Garduño, Canales-Lazcano, & Córdoba-Aguilar 2006). *Inset:* territorial male.

Figure 10.13. Male damselfly fat reserves after experimental challenge. Experimentally challenged males forced to defend their territory had lower mean (+ SE) fat reserves than did nonchallenged territorial males or nonterritorial males (units are grains) (*Source:* Contreras-Garduño, Canales-Lazcano, & Córdoba-Aguilar 2006).

APPLYING THE CONCEPTS 10.1

Conspecific attraction and conservation

Many bird populations are declining because of human exploitation and habitat change. Natural habitats are often converted for the purposes of urban development or agriculture. After conversion, they may no longer be suitable for particular species or may contain novel competitors or predators. In response, conservation biologists are working to restore degraded places and even create new habitat areas. Once a population has left a location, how do you attract it back to the new or restored habitat?

One method increasingly used to restore bird populations takes advantage of conspecific attraction. For instance, populations of colonial nesting seabirds such as Atlantic puffins (*Fratercula arctica*) have been restored to former breeding grounds by placing models and dummy nests at former breeding islands to attract individuals (e.g., Kress & Nettleship 1988; Parker et al. 2007). Song playback has also been used to attract species back to areas.

Michael Ward and Scott Schlossberg used conspecific calls to attract endangered black-capped vireos (*Vireo atricapilla*) to restored habitat sites in Texas (Ward & Schlossberg 2004). Black-capped vireos (Figure 1a) are small, territorial songbirds whose populations have declined steeply over the past few decades, largely because of habitat change and an increase in brown-headed cowbirds (*Molothrus ater*), a species that lays eggs in the nests of other birds (hosts) and so decreases hosts' reproductive success.

Ward and Schlossberg used song playbacks to attract birds to settle in restored habitats at Fort Hood, Texas, from which cowbirds had been removed. They selected seven large, restored sites (15–71 ha) from which vireos had been absent for at least two years. Over two breeding seasons, they played black-capped vireo songs on some sites and left others silent as controls, switching certain treatments and controls between years (Figure 1b). In each year, over 70 vireos were observed at sites where songs were played, while fewer than five were observed on control sites. Many of the birds reproduced successfully on treatment sites (Figure 1c). These data indicate that song can be used to attract new settlers to a habitat. This gives resource managers and conservation biologists a new tool for increasing populations of endangered birds.

Figure 1. Black-capped vireo. (a) Black-capped vireos are small, endangered songbirds in Texas. (b) The speaker setup used to play songs on experimental plots. (c) A nesting female.

The researchers studied territorial behavior in the American rubyspot, *Hetaerina americana*, a damselfly found throughout North America near sunny riverbanks. Adults emerge from a pupa and feed on mosquitoes. After several days of feeding and development, individuals reach sexual maturity and then fight for a territory, a perching site. Territory ownership is determined in aerial contests, which can be as short as three seconds or as long as two hours (Córdoba-Aguilar & Cordero-Rivera 2005). Males that cannot gain a territory simply search for females in their home range, but this tactic results in very limited mating success. What determines whether a male gains a territory? Territory disputes can be energetically expensive, and so the research team examined whether or not territorial males had higher fat reserves than nonterritorial males. Fat serves as fuel for energetically costly behaviors such as fighting, and so larger reserves of fat could allow individuals to better defend a territory.

The researchers first identified and then collected 30 territorial and 30 nonterritorial males of similar age from the Xochitepec River in Mexico (Contreras-Garduño, Canales-Lazcano, & Córdoba-Aguilar 2006). In order to measure body fat content, they weighed the body both before and after extracting the fat. The difference yielded the fat mass of a male.

They found that territorial males had much higher fat reserves than nonterritorial males (Figure 10.12), indicating that territorial males did indeed have higher potential energy reserves. Are these fat reserves used to maintain a territory? In a second experiment, Contreras-Garduño and colleagues staged territorial fights. They created "intruders" by capturing sexually mature males and tying them to a wooden stick, with a thread attached to their thorax, that still allowed them to fly. These males were used to intrude upon a territorial male in the field. The intruders were brought close to the territorial male repeatedly for 20 minutes to simulate a long territorial contest. After the experiment, the fat reserves of the experimental males were measured and compared to those of nonmanipulated territorial males and those of a set of nonterritorial males.

Territorial males who interacted with intruders had the lowest fat reserves of all three groups (Figure 10.13). The nonmanipulated territorial males had fat reserves that were almost four times higher than those of the experimental males. The researchers concluded that the long staged contests reduced a territorial male's fat reserves to very low levels, indicating that the aerial displays required for defending a territory are indeed energetically very costly. The study suggests that only males with high energy reserves can hold a territory.

Environmental factors and territory size in kites

Featured Research

If an individual is capable of defending a territory, it must decide the size of the territory to defend. Larger territories likely contain more resources but are accompanied by a higher cost of defense. Several factors affect the benefits and costs of territory defense (Brown 1964). Two of the most important are the densities of both resources and competitors. First, as the density of resources increases, so do the benefits of defending a territory. As resources become more abundant per unit area, animals can defend a smaller territory and still have exclusive access to abundant resources. Second, as the number of competitors increases, the cost of defense also increases, because there are more potential intruders. This cost can be reduced by defending smaller territories, because they require that less area be defended. These factors produce two predictions, which are not mutually exclusive:

Prediction 1: Territory size should be negatively correlated with resource density.

Prediction 2: Territory size should be negatively correlated with competitor density.

Jeffery Dunk and Robert Cooper tested these predictions by examining the relationship between territory size and resource and competitor density in white-tailed kites (*Elanus leucurus*) (Dunk & Cooper 1994). White-tailed kites are birds of prey that feed primarily on rodents and defend feeding territories in grasslands with scattered trees (Figure 10.14). To find food, they hover, or kite, while searching the ground for prey. This behavior allows researchers to readily determine the boundaries of a feeding territory.

Dunk and Cooper studied a population in northern California in which kites fed primarily on California voles (*Microtus californicus*). Individual kites were captured and uniquely marked with colored leg bands. Over an 18-month period, the researchers determined the feeding territories of 26 kites. All locations of hovering, perching, and

Figure 10.14. White-tailed kite and vole. Kites (a) defend feeding territories that vary in size with environmental conditions. They feed on rodents, such as voles (b).

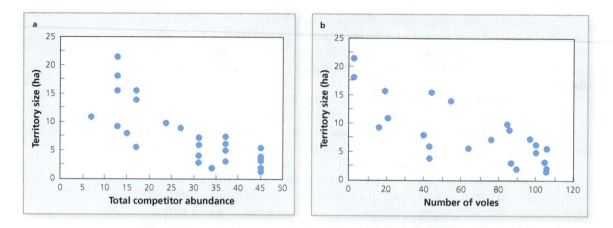

Figure 10.15. Kite territory sizes. Territory size correlates negatively with (a) the number of competitors and (b) the availability of food (number of voles per ha) (*Source:* Dunk & Cooper 1994).

interactions with other birds were recorded for several days and plotted on an aerial photograph to map territory sizes. To estimate food abundance in the area, the researchers trapped and counted voles each month. To estimate competitors, they surveyed hawks, kites, falcons, and owls at the study site each month. These counts of rodent-eating birds were conducted within one hour of sunset, the time of highest competitor activity. For each kite territory mapped, the researchers correlated its size with both rodent and competitor abundance.

Territory size ranged from about 1 to 21 hectares. The researchers observed an average 1,400 individual voles per territory, while competitor abundance varied from about five to 45 birds. Kite territory size was negatively correlated with both competitor abundance and vole density, examples of negative density-dependence (Figure 10.15). Statistical analyses revealed that competitor abundance affected kite territory size more strongly than did food density. This finding was supported by the observation that prey abundance was correlated with competitor abundance, and this in turn affected kite territory size. As competitor abundance increased, kites defended smaller territories, presumably to reduce the costs of defense and because more food was available. The researchers noted that while their data support the predicted relationship between food availability and territory size, other factors such as the availability of perches or prey selection by competitors may also be important in determining territory size. Further work, including experimental manipulations, can help to elucidate the importance of different factors in affecting territory size.

These examples demonstrate that individual condition and environmental factors affect territoriality. The acquisition and defense of a territory often require aggressive interactions with rivals. Next, we examine factors that affect the duration and intensity of such interactions.

10.3   The decisions of opponents and resource value affect fighting behavior

Territory defense and other contests over resources involve interactions with others. These contests can be quite complex because species often exhibit a variety of behaviors that differ in intensity. The costs of fighting are affected by the behavior of opponents. If the opponent decides to flee, then fighting has a low cost. If the opponent decides to fight intensely and is a superior fighter, then the decision to fight intensely could be costly.

For example, male fruit flies commonly fight over food. Fights can be as short as a few seconds or longer than four minutes (Chen et al. 2002) and can include chasing, pushing, tussling (in which opponents tumble over one another), and even boxing—a fight in which opponents rear up on their hind legs and strike each other with their forelimbs (Figure 10.16). The longer the contest lasts, the more kinds of behaviors are exhibited (Chen et al. 2002), but rarely is an individual harmed or killed. Why is there so much variation in contest behavior? Why don't more animals become injured while fighting?

We can answer these questions using game theory, a cost-benefit approach used to understand the evolution of behavior when an individual's fitness is affected by how other individuals behave (Chapter 2). Let's see how this works for a game theory model that examines contests between two animals: the hawk-dove model.

Figure 10.16. Flies boxing. Fruit flies engage in contests over food.

The hawk-dove model

John Maynard Smith and George Price (1973) introduced the **hawk-dove model** to examine the behavior of two individuals engaged in a contest over a resource, such as a territory. The winner of the contest obtains a benefit B that enhances fitness. The loser receives no benefit. In this simple game, both individuals have the same fighting ability and adopt one of two behavioral strategies: hawk or dove. Hawks are always willing to escalate an interaction into an intense, prolonged fight in order to win. In fact, they fight until they are injured (when they give up) or until they win. Losers of these intense fights are injured and pay a cost C. The cost of losing is assumed to be larger than the benefit of winning ($C > B$). Doves, on the other hand, back down immediately if their opponent escalates their behavior and so never actually fight or pay the cost C. When two doves fight, they engage in a short, no-cost display, and each has a 50% chance of winning.

The model predicts that a population will contain both hawks and doves and that the proportion of each will be determined by the fitness benefit (B) of winning and the fitness cost (C) of being injured while fighting. At the **evolutionary stable strategy** (**ESS**; see Chapter 2), the frequency of hawks works out to B/C. By definition, at the ESS, both hawks and doves have the same fitness.

Such a population will exhibit tremendous variation in fighting behavior. When two hawks meet, fights will be long and intense. When a hawk interacts with a dove, the contest will be very short. Contests between two doves will also be short but will involve only low-cost displays, and so any one individual will win some fights and lose others. This variation is similar to that observed in the real-life contests between animals that Maynard Smith and Price wanted to explain.

The game theory approach is a powerful way to understand contests over resources. However, the hawk-dove model assumes that individuals are identical in terms of fighting ability. In real populations, individuals can differ in age, sex, experience, size, and condition. In addition, contests often occur between the owner of a resource such as a territory and an intruder. Owners may value the resource more because they have greater knowledge about its true value (e.g., the resources it contains). These differences can affect an individual's probability of winning a contest, or the **resource holding power**, of an individual. For instance, large owners of a resource such as a territory often have a higher probability of winning a contest against a small intruder (e.g., Kemp & Wiklund 2004). Differences in resource holding power are incorporated into game theory models known as **asymmetric games**, in which the strategies or payoffs for each player differ. This aspect has been incorporated into the model we examine next.

The sequential assessment model: fiddler crab contests

Featured Research

In an asymmetric game involving a contest over a resource, individuals assess their opponents' resource holding power relative to their own in order to determine how much effort they need to exert. For example, in the **sequential assessment model**, individuals can exhibit a variety of strategies and may differ in fighting ability: some may be superior fighters, while others may not (Enquist & Leimar 1983). The ESS is

hawk-dove model A game theory model that examines the behavior of two individuals engaged in a contest over a resource.

evolutionary stable strategy (ESS) A strategy that, if adopted by a population, cannot be trumped by another strategy because it yields the highest fitness.

resource holding power The ability to win an aggressive encounter.

asymmetric game In game theory, a situation in which the strategies or payoffs for each player differ.

sequential assessment model An asymmetric game involving a contest over a resource.

The hawk-dove model

John Maynard Smith and George Price introduced the application of game theory to evolutionary biology (Maynard Smith & Price 1973). They wanted to understand why conflicts between individuals of the same species rarely resulted in serious injury, even when individuals possessed effective weapons such as horns or fangs. Their model, the hawk-dove model, examines the outcomes of fights between individuals who can display only two behavioral strategies: hawk or dove. The hawk fights until it wins or is injured. The dove never fights. Maynard Smith and Price assigned benefits and costs for individuals playing these strategies against one another. There are three possible interactions:

Hawk versus hawk:

Each player has a probability of 0.5 of winning or losing
Winners get a benefit B
Losers receive no benefits, are temporarily injured, and pay a cost of C

When hawks play hawks, the average payoff is $0.5\,B - 0.5\,C = 0.5\,(B - C)$. In other words, half the time an individual wins and gets a payoff B, while half the time the individual loses and pays a cost C.

Dove versus dove:

Each player has a probability of 0.5 of winning
Winners obtain a benefit B
There are no costs, because the players never escalate the conflict and so are never injured
When doves play doves, the average payoff is $0.5\,B$.

Hawk versus dove:

Hawks always win and receive a benefit B
Hawks pay no costs, because the dove always retreats
Doves always lose and get no benefits, but they also pay no costs
When a hawk plays a dove, the payoff to the hawk is B and the payoff to the dove is 0.

Let's look at the benefits and costs of a focal hawk or dove in a matrix. The payoffs in the matrix are read left to right and are based on the opponent's behavior:

	Hawk opponent	Dove opponent
Focal hawk payoff	$0.5\,(B - C)$	B
Focal dove payoff	0	$0.5\,B$

Maynard Smith and Price assumed that in this game, the cost of losing a contest was greater than the benefit of winning the resource. This means that $C > B$.

Given this game, what strategy is the ESS? Let's evaluate some possibilities.

A POPULATION OF ALL DOVES

This is known as a pure strategy, because all individuals adopt the same strategy in the game. If all individuals are doves, all interactions are dove versus dove, and so all individuals obtain a fitness $0.5B$. However, if a hawk individual enters this population, or if one of the doves changes its behavior and adopts the hawk strategy, it will receive a reward B during all encounters with doves and so will have a higher fitness, because $B > 0.5B$. *In a population of pure doves, hawks can successfully invade, and so pure dove is* not *an ESS.*

A POPULATION OF ALL HAWKS

Each individual receives a payoff of $0.5\,(B - C)$. Because we assumed that $C > B$, here, individuals actually receive a negative payoff! Remember that they all receive the same negative payoff, because all interactions are hawk versus hawk. However, if a dove enters this population, it will lose all contests with hawks but never get injured. The dove's payoff in these encounters will be 0, which is greater than $0.5\,(B - C)$ when $C > B$. That means that in a population of pure hawks, a dove will have higher fitness—even though the dove can never win a contest. *Doves will therefore increase in the population, and so pure hawk is also* not *an ESS.*

Given that pure dove and pure hawk are not ESSs, is there an ESS for this game? As you might have guessed, the ESS must be a mixture of the two strategies. In a mixed evolutionary stable strategy, a fraction of individuals play each strategy, or else each individual plays a particular strategy in some fraction of the encounters. To solve for the mixed ESS, we need to find the frequency of hawks and doves in the population such that each strategy receives the same fitness. We can solve this mathematically by letting

p = the frequency of hawks in the population

q = the frequency of doves in the population

By definition, $p + q = 1$, because all individuals are either a hawk or a dove. Next, we calculate the payoffs for a hawk and a dove based on their frequency in the population. Hawks will encounter hawks a fraction p of the time and doves a fraction q of the time. There are p hawks in the population, and each time a hawk encounters another hawk, it receives a payoff of $0.5\,(B - C)$. There are q doves, and each time a hawk encounters a dove, the hawk receives a payoff of B.

The overall payoff to a hawk will then be

$$p\,[0.5\,(B - C)] + qB$$

By the same reasoning, the overall payoff to a dove will be

$$p\,(0) + q\,(0.5B)$$

Now all we need to do is to set the payoff to a hawk equal to the payoff to a dove:

$$p\,[0.5\,(B - C)] + qB = p\,(0) + q\,(0.5B)$$

A little algebra gives

$$p = B/C$$

Therefore, in a population at the ESS, the frequency of hawks will be B/C.

The hawk-dove model introduced game theory to animal behavior and has also been used extensively in economics, political science, and even cancer research (e.g., Hanauske et al. 2009; McEvoy 2009).

to fight as long as the potential benefits of winning exceed the costs of continuing to fight. If an individual determines that its opponent is superior, the best choice will be to back down and abandon the fight. However, if the individual determines that it is the superior fighter, the best choice will be to continue the contest, because it will have the better chance of winning (Applying the Concepts 10.2).

At the beginning of a contest, an individual may not know its opponent's fighting ability. How do individuals assess each other's ability? One way they can do so is by varying fighting behavior during the contest. The model predicts that individuals should begin with low-cost behaviors, such as simple displays or pushing. These behavioral interactions may provide sufficient information—but only if individuals differ greatly in ability. If these low-cost behaviors do not settle the fight, then individuals should use more costly behaviors, as can be seen with boxing fruit flies. This pattern leads to two predictions:

Prediction 1: Low-cost behaviors should be used early in fights, followed by more intense, high-cost behaviors.

Prediction 2: The relative difference in fighting ability of opponents will affect contest duration: fights between evenly matched opponents will be longer than fights between unequally matched opponents.

Ann Pratt, Kelly McLain, and Grace Lathrop tested these predictions of the sequential assessment model in male sand fiddler crabs (*Uca pugilator*) (Pratt, McLain, & Lathrop 2003). A female fiddler crab selects males by the quality of their burrow, particularly its width (Reaney & Backwell 2007), because she uses it for gestation of her eggs. Males fight intensely over burrows. Contests between males include a variety of behaviors that range from low-cost claw-waving or pushing to higher cost behaviors such as interlacing claws, pinching, and even flipping an opponent (Table 10.1). Winners, who tend to be larger males, take or keep possession of the burrow, while losers leave the area.

To test the model, the researchers observed natural contests between males on a beach in South Carolina. They assigned males to one of three size groups (small, medium, large) based on the width of their carapace, the hard dorsal surface of the exoskeleton. For each contest, they recorded the sequence of behaviors observed and their durations, the duration of the entire contest, and the winner (Scientific Process 10.3).

TABLE 10.1 Crab ethogram. Crabs use a variety of agonistic and nonagonistic behaviors in contests (*Source:* Pratt, McLain, & Lathrop 2003).

BEHAVIORAL ELEMENTS	DESCRIPTION
Agonistic	
Extend	Claw is swept toward opponent; no contact
Jump	One opponent lunges at the other; no contact
Manus Align	Opponents face each other with manus of one claw held adjacent to the other; no shoving
Manus Push	Opponents face each other; each opponent pushes the claw of the other with his claw held level to the substrate
Dactyl Slide	Pollex and dactyl of the claw of each opponent are intercrossed and slide back and forth near their distal ends
Heel and Ridge	Intercrossed pollex and dactyl slide proximately to the manus; some shoving occurs
Tap	Rapping of dactyl of pollex during Heel and Ridge
Downpush	After a Burrow Retreat by one, the other opponent reaches in with its claw; claws often interlock with pinching
Interlace	Intercrossed pollex and dactyl are clamped tight on opponent's claw; vigorous shoving and pinching occurs
Flip	With interlaced claws, one opponent is lifted from the substrate and tossed
Nonagonistic	
Burrow Retreat	One opponent retreats into the burrow, often with at least part of the claw visible
Motionless	One opponent freezes, with the claw held aloft
Leave	One opponent walks away

SCIENTIFIC PROCESS 10.3

Sequential assessment in crab fights
RESEARCH QUESTION: *How do crabs fight?*

Hypothesis: Individuals assess their relative resource holding power to determine how much effort to put into a contest.

Prediction 1: Low-cost behaviors should be used early in fights, followed by more intense, high-cost behaviors.

Prediction 2: Fights between evenly matched opponents will be longer than fights between unequally matched opponents.

Methods: The researchers:

- observed 152 natural contests over burrows between males. Each male was assigned to one of three sizes: less than 14 mm, 14 mm to 16 mm, or greater than 16 mm.
- recorded the sequence of behaviors, their durations, total contest duration, and the winner.

Results:

- Low-intensity behaviors were most often used early in contests; higher intensity behaviors were used later in contests.
- Contests between evenly matched opponents lasted significantly longer than contests between a small opponent and a large opponent.

Figure 1. Fighting behavior. Low-intensity fighting behaviors (on the left) are used early in fights (blue bars) while higher intensity behaviors (on the right) are used later (orange bars) (*Source:* Pratt, McLain & Lathrop 2003).

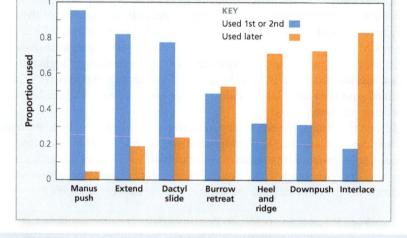

Conclusion: Males appear to assess relative resource holding power in contests over burrows.

Pratt and colleagues found that contests usually began with a low-intensity behavior and later escalated. Longer contests contained a greater diversity of behaviors, and the intensity increased with contest length. Contest duration was negatively correlated with size difference: fights that involved individuals of the same size class lasted longest, while fights between small and large individuals ended quickly. These observations matched the model predictions, leading the researchers to conclude that fiddler crabs appear to sequentially assess their opponent to determine the appropriate strategy, which results in complex fighting behavior. This example—and the sequential assessment model—helps to explain why so few contests result in death: once an individual determines that it has a low probability of winning, it usually withdraws to avoid injury and death.

In the examples so far, a contested resource has had a fixed value. In reality, the value of a resource might differ over time or in different situations. Game theory models predict that fighting intensity should increase with the value of the contested resource (Maynard Smith & Parker 1976; Enquist & Leimar 1987). Let's examine two tests of this prediction.

Resource value and fighting behavior in penguins

What happens when the value of a resource changes over time? How do animals adjust their behavior? Magellanic penguins (*Spheniscus magellanicus*) (Figure 10.17) are medium-sized birds, roughly 70 cm tall, that breed along the coasts of southern South America from September through February. Males establish and defend nest sites under bushes or in burrows. A male will use the same location year after year but must defend it from rivals. Contests over nesting sites occur during two periods. The first, from mid-September to mid-October, is when males first arrive at the breeding grounds to set up and defend nests prior to egg laying. By mid-October, males have established nest sites, mated, and left the breeding grounds to feed for two weeks. By then, females have laid their eggs and begun egg incubation, and aggression is low. In early November, males arrive back at the breeding sites and agonistic interactions occur again, but mostly at the sites of failed nests—nests where eggs have been destroyed by predators. By November, it is too late to successfully nest again, and so these contests are presumably over ownership of a site for the next breeding season.

Figure 10.17. Magellanic penguin. These birds commonly use burrows as nests, a limited resource that males fight to obtain.

Daniel Renison and colleagues examined fighting behavior among these birds. The researchers reasoned that the value of a nest site differs before and after mid-October: it is highest before egg laying and substantially lower later in the season (Renison, Boersma, & Martella 2002). Renison and colleagues used October 21 as the cutoff date between these two periods, because egg laying had not yet occurred at that time, and predicted that fighting behavior should be more intense before this date. Nests also vary in terms of quality, determined by the amount of cover they provide. Nests in deep burrows with small entrances provide better protection for eggs and chicks and so are of higher value than nests that are only depressions in the soil and lack cover.

Magellanic penguins exhibit two distinct fighting behaviors: bill duels and overt fights. In a bill duel, individuals stand face to face and knock their bills together. A duel represents a relatively low level of aggression, because the only physical contact is between bills and individuals are not injured. Overt fights are more intense: they involve sharp pecks and typically result in cuts to one or both opponents.

The research team tested the prediction from game theory models that variation in the resource value of a nest should correlate with the frequency and type of agonistic interaction. Fights over high-quality nests (with high-quality being determined both by time of season and the amount of cover over the nest) should involve higher levels of fighting intensity than fights over less valuable nests.

The researchers recorded the relative frequency of bill duels and overt fights between males for nests in a single large colony (2,000 breeding pairs) in Punta Tombo, Argentina, over a four-year period (Renison et al. 2006). Each day, they observed the colony for 13 hours and recorded all bill duels and overt fights. They examined all fights that occurred before egg laying (prior to October 21) and fights over failed nests after egg laying. In order to assess the severity of a fight, they measured its duration and the total length of cuts on the contestants that resulted from the fight. The researchers did this by capturing penguins after each fight and measuring the length of cuts on the bare skin of the face around the base of the bill, the area most commonly cut. They added these cut lengths to calculate a summed length of cuts. They also assigned each nest a nest-cover score based on the amount of cover the nest provided the eggs and parents, ranging from 1 (least cover) to 5 (most cover).

Renison and colleagues found that the type and relative frequency of fights varied before and after the egg-laying period. Before

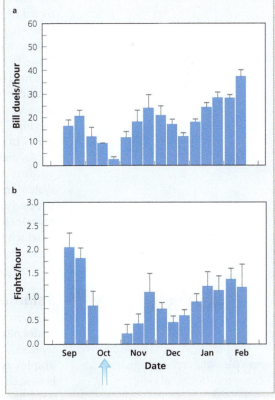

Figure 10.18. Penguin fighting behavior. (a) Bill duels are more common late in the season while (b) overt fights are more common early in the breeding season. The arrow indicates the cutoff date between early and late breeding dates (*Source*: Renison et al. 2006).

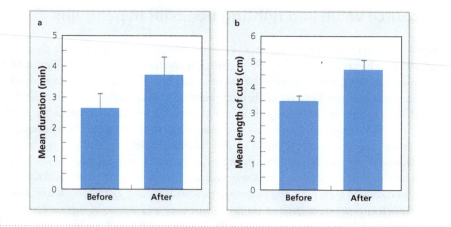

Figure 10.19. Duration of penguin fights and injuries. (a) The mean (+ SE) duration of fights and (b) mean (+ SE) total length of cuts during fights were lower before compared with after egg laying (*Source:* Renison et al. 2006).

egg laying, there were more frequent, intense fights, while after egg laying, there were more frequent bill duels (Figure 10.18). However, both the duration of fights and the summed length of cuts were shorter prior to egg laying (Figure 10.19). The quality of the nest also affected fighting behavior. More fights than duels occurred over higher quality nests. Furthermore, males with higher quality nests fought longer and received more cuts than males with lower quality nests.

The researchers concluded that the variation in fighting behavior observed was affected by the value of the contested resource, as predicted by game theory. When the contested resource had high value—before egg laying and when nests had high amounts of cover—individuals fought intensely. When the value of a nest site was lower, fighting behavior was less intense. But why were the duration of fighting and the summed length of cuts shorter before the egg-laying period? This is a time when nests should be at their highest value, and so we would predict higher levels of fighting. The researchers speculated that fight duration may have been short because the intensity was high. Alternatively, early in the season, before egg laying, there is a larger size asymmetry in contestants, and so individuals may be able to rapidly determine the better fighter and end contests quickly.

This study examined natural variation in fighting behavior in an unmanipulated system in the wild. As such, many uncontrolled environmental variables may have affected the fighting behavior of the penguins. In the next example, researchers examined variation in fighting behavior in staged contests conducted in captivity, in which the value of a resource was manipulated. This approach allowed environmental factors to be more controlled.

Featured Research

Salamander fights for mates

Males often engage in contests with other males for access to a female (or a group of females). This is particularly common in amphibians. Reproduction often involves **amplexus** behavior, in which a male grasps a female and holds her for up to several hours. During amplexus, rival males will attempt to displace an amplexed male by wrestling with him. How much effort should an intruding male invest in these displacement bouts?

Paul Verrell examined wrestling behavior in red-spotted newts, *Notophthalmus viridescens* (Verrell 1986). These small salamanders (less than 125 mm in length) are common in eastern North America (Figure 10.20). In amphibians, female fecundity, as measured by number of oocytes, typically increases with body size, indicating

amplexus When a male grasps and holds a female prior to copulation.

APPLYING THE CONCEPTS 10.2

Reducing duration and intensity of piglet fights

In the agricultural production of domestic pigs (*Sus scrofa domesticus*), many litters are born simultaneously. The piglets in a litter are housed together with their mother but are kept physically and visually isolated from other litters. After several weeks, piglets from different litters are mixed together, which results in aggressive fights between piglets. These fights are often intense and can result in injury, making them a major welfare and production issue. Farmers have tried to reduce fighting with drugs and masking odors, but they have had little success.

Per Jensen and Jenny Yngvesson applied the sequential assessment model to piglet fights (Jensen & Yngvsson 1998). Fights often begin with a low-intensity nosing phase, in which two individuals walk together and exchange nose contact. After a period of time, fights can escalate to a biting phase, in which individuals push and bite each other until the loser retreats. Jensen and Yngvsson reasoned that if opponents rely on fights to learn each other's relative fighting ability, as assumed by the sequential assessment model, then allowing interactions of individuals before a contest should produce shorter and lower intensity fights.

To test this prediction, they staged contests between piglets. Some individuals (the treatment group) were allowed to see each other in adjacent pens separated by a wire mesh. Control animals were housed in adjacent pens separated by a solid wooden wall that prevented visual contact. Contests were conducted between piglets that were either matched for size or differed in weight by about 20%. Jensen and Yngvsson found that fight intensity and duration were affected by both weight difference and prior visual contact. Fights between opponents of similar size lasted longer than fights with a large weight asymmetry, as predicted by the model. In addition, fight duration and intensity were lower in fights between opponents that had had prefight visual contact. The researchers suggested that farmers and animals could benefit from husbandry that allows visual contact between piglets of different litters.

---

that larger females should be more valuable to a male. Verrell first verified that larger females had a higher resource value. He allowed 13 females of varying sizes, measured by length from snout to vent, to mate with a male. He then counted the total number of eggs produced as a measure of female fecundity. Larger females laid significantly more eggs than smaller females, indicating that larger females are a higher value resource.

With this information, Verrell predicted that the fighting intensity of males should increase with the size of the amplexed female. To test this prediction, he staged encounters between size-matched males in a large aquarium. In each trial, one female, whose body size ranged from 43 mm to 51 mm, was placed in the aquarium along with two males. One of the males quickly initiated amplexus, and in 33 trials, the rival male began a wrestling contest with the amplexed male. The wrestling bout ended either when the rival left the amplexed pair or when he successfully detached the amplexed male from the female.

The total duration of the wrestling contest increased with the size of the female, as predicted (Figure 10.21), and the intruder did displace the amplexed male in four of the trials. These data illustrate how the value of a contested resource affects the intensity of aggressive behavior. Males spent more time wrestling over large females than they did over small females, as one might expect, given that larger females lay more eggs.

In this section, we've seen how game theory models can provide unique insight into and help explain the fighting behavior of many species. These models are being increasingly used in agricultural settings to manage animal aggression. In the final section, we turn our attention to proximate explanations of fighting behavior as we examine how aggression is regulated.

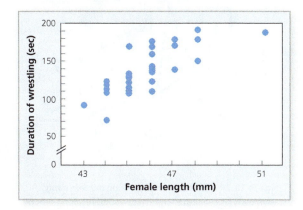

Figure 10.20. Red-spotted newts. Males wrestle with rivals for access to mates.

Figure 10.21. Duration of wrestling in male red-spotted newts. The duration of wrestling increased with the size of the female that was amplexed (*Source:* Verrell 1986).

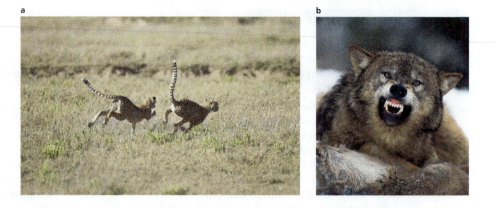

Figure 10.22. Aggression. (a) A cheetah chases a rival to defend its territory. (b) Wolves bare their teeth during threat displays.

10.4 Hormones influence aggression

As we've seen throughout this chapter, contests over resources and territorial defense often involve aggression: behaviors such as chasing and fighting, as well as those that indicate threat of attack, such as particular postures and behaviors (Figure 10.22). Some species defend territories year round and thus may aggressively defend their territories regularly, while others defend a territory only during the breeding season. For example, many birds defend territories during spring and summer (the breeding season) but live in large social groups during autumn and winter (the nonbreeding season). Therefore, the aggression levels in these species will vary greatly over the course of a year, with high levels at the start of the breeding season and low levels outside the breeding season.

testosterone A steroid hormone produced in the gonads and regulated by the hypothalamus and pituitary gland.

challenge hypothesis Male-male interactions increase plasma testosterone and thus sustain subsequent aggressive behavior.

What regulates aggression? In many species, variation in aggression is coincident with variation in plasma androgen hormone levels such as **testosterone**. This steroid hormone is produced in the gonads and is regulated by the hypothalamus and pituitary gland. Testosterone has widespread effects on both physiology and behavior. Increased testosterone levels are associated not only with increases in aggression, sexual behavior, spermatogenesis, and energy costs but also with suppression of parental care behavior, the immune system, and fat stores, as well as increased mortality (Wingfield, Lynn, & Soma 2001). Thus, testosterone levels affect physiological and behavioral traits that strongly affect fitness in ways that require its regulation.

In many vertebrates, testosterone secretion levels coincide with changes in photoperiod. There is increased production in spring, as days grow longer, and decreased production in autumn, as days become shorter. John Wingfield and colleagues documented changes in plasma testosterone levels in male song sparrows, *Melospiza melodia* (Wingfield et al. 1990; Wingfield et al. 2000; Wingfield, Lynn, & Soma 2001). Free-living males displayed very high spikes in testosterone levels in early spring during territory acquisition, as well as a few weeks later when mating, but testosterone levels declined greatly after that (Figure 10.23). Socially isolated captive birds displayed a similar seasonal pattern, but with much lower spikes (Wingfield, Lynn, & Soma 2001). High levels of testosterone may promote acquisition of territories early in spring, with a reduction corresponding to the need for males to provide increased levels of parental care once eggs hatch (Chapter 13).

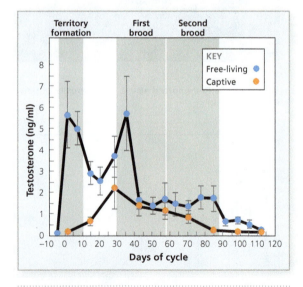

Figure 10.23. Testosterone variation in male song sparrows. Free-living males (blue circles) have high mean (± SE) plasma testosterone levels during territory formation and reproduction of the first brood. Captive males (orange circles) show similar patterns of variation over the spring, but with lower levels (*Source:* Wingfield, Lynn, & Soma 2001).

Why might free-living birds display higher levels of testosterone compared to isolated captives? One possibility is that social interactions play a role in testosterone secretion. Wingfield's **challenge hypothesis** states that male-male interactions (over territories, dominance status, or mates) increase plasma testosterone and thus sustain subsequent aggressive behavior (Wingfield, Lynn, & Soma 2001). When males fight over territories and mates, success in a contest can strongly enhance fitness. Enhanced testosterone levels are predicted to increase aggressive behavior and therefore the probability of winning a contest. This can also explain a phenomenon known as the **winner effect**: for many animals, winning an aggressive interaction often enhances the likelihood of winning a subsequent interaction (Hsu, Earley, & Wolf 2006). A combination of the winner effect and the challenge hypothesis, the winner-challenge effect, states that winning a challenge increases plasma testosterone levels that enhance aggressive behavior and thus the likelihood of winning subsequent interactions (Oyegbile & Marler 2005).

> **winner effect** A phenomenon in which winning an aggressive interaction enhances the likelihood of winning a subsequent interaction.

Winner-challenge effect in the California mouse

Featured Research

Temitayo Oyegbile and Catherine Marler examined the winner-challenge effect by exploring the links between testosterone levels, aggressive behavior, and the outcome of contests in the California mouse, *Peromyscus californicus* (Oyegbile & Marler 2005). California mice aggressively defend their territory year-round, and, like many other species, larger individuals typically win aggressive interactions (Huntingford & Turner 1987).

The research team randomly assigned focal males to one of four training treatments, which differed in the number of times (from zero to three) that a male won an interaction with a rival. In a fifth no-interaction control treatment group, males never interacted with a rival. Prior to training, all males were paired with a female for ten days. On Day 11, the pair was placed in a large observation cage that contained a small nest box that allowed the mice to establish residency of the cage. Training interactions were performed on Days 13, 15, and 17 and consisted of either a winning encounter with a rival or simple handling by the experimenters. To create winning encounters, the female was removed and a smaller, mildly sedated male intruder was introduced to the observation cage for ten minutes. On Day 19, the testing phase began. The female was removed, and an unfamiliar and slightly larger male that had previously won an encounter was introduced to the cage. Oyegbile and Marler recorded the resident male's latency to attack, the number of attacks, and the outcome of the contest (winning behaviors included chasing and wrestling; losing behaviors included jumping away, freezing, and retreating). Baseline blood samples were taken from all males before training began and again 45 minutes after the test encounter.

The number of winning encounters a male had in the training phase correlated with an increased likelihood of winning the test encounter. However, the only statistically significant differences were between males with no prior experience and males with three prior winning experiences. Males that won more training-phase encounters also had significantly shorter latency to attack the rival (Figure 10.24). Testosterone levels were higher in mice that had previous winning experiences compared to controls that had none (Figure 10.25). These results support the challenge hypothesis, because variation in plasma testosterone levels correlated with higher levels of aggression and contest outcome. This finding also provides a proximate explanation for the winner effect: winning fights results in elevated plasma testosterone levels, which makes males more aggressive and enhances their likelihood of winning subsequent fights. Together, these results illustrate the winner-challenge effect in mice.

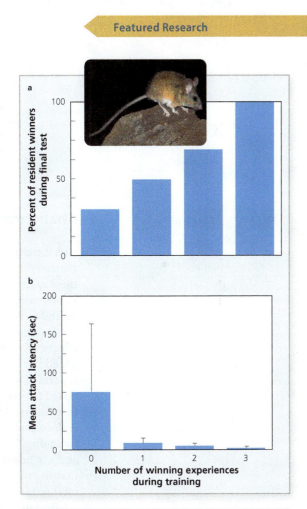

Figure 10.24. Mouse contest results. Mice that won more previous contests in training had (a) a higher winning percentage against a larger rival in the final test and (b) a lower mean (+ SE) latency to attack (*Source:* Oyegbile & Marler 2005). *Inset:* California mouse.

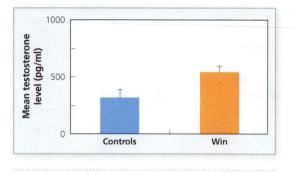

Figure 10.25. Testosterone levels in mice. Mean (+ SE) testosterone levels were higher in males that won fights compared to controls (*Source:* Oyegbile & Marler 2005).

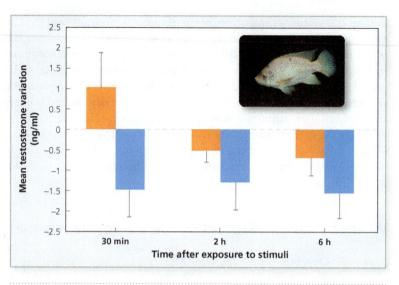

Figure 10.26. Testosterone variation in fish. Mean (+ SE) changes in testosterone levels for male bystanders that observed an aggressive interaction (orange) compared with changes in levels for controls (blue). Testosterone levels increased in bystanders that had observed a fight 30 minutes earlier compared to controls (*Source:* Oliveira, Lopos, & Carneiro 2001). *Inset:* Mozambique tilapia.

Featured Research

Challenge hypothesis and bystanders in fish

Can the challenge hypothesis be extended to bystanders? Does simply observing aggressive interactions affect hormone levels? Rui Oliviera and colleagues examined these questions using the Mozambique tilapia, *Oreochromis mossambicus* (Oliveira, Lopos, & Carneiro 2001). These cichlid fish are native to east Africa, and males defend territories during the breeding season.

The researchers first kept all fish in isolation for seven days to minimize any effects of previous interactions. Treatment focal male bystanders were then allowed to view two conspecifics interacting. The conspecifics were initially visually separated by an opaque partition and so could not interact. The partition was then raised for one hour, during which the two rivals engaged in a fight for dominance that the focal bystander could view through a one-way mirror. The research team collected urine from the bystander to assess testosterone levels at different times: two hours before viewing the males fight and then again 30 minutes, two hours, and six hours after the interaction. Control fish were treated similarly, except that the opaque partition was never raised and so they did not view an aggressive interaction.

Bystanders that observed an aggressive interaction had higher testosterone levels 30 minutes after the interaction than bystanders that did not observe an aggressive interaction (Figure 10.26). This experiment demonstrates that social observations alone can result in increased testosterone levels, thus extending the challenge hypothesis. Individuals do not need to engage in male-male aggressive encounters directly, but rather can simply view aggression resulting in an increase in their plasma testosterone levels. The researchers suggested that this kind of viewing experience can prepare a bystander for future aggressive interactions: a rapid elevation of testosterone levels can increase aggressive behavior and the likelihood of winning a contest. The challenge hypothesis has even recently been extended to people viewing sporting events and provides a proximate explanation for variation in aggression (Applying the Concepts 10.3).

APPLYING THE CONCEPTS 10.3

Sports, aggression, and testosterone

Psychologists have long been interested in links between plasma testosterone levels and human aggression (e.g., Persky, Smith, & Basu 1971). Evidence to date suggests a weak association between testosterone levels and aggression, although debate over this relationship continues (Book, Starzyk, & Quinsey 2001; Archer, Graham-Kevan, & Davies 2005). John Archer suggested that the weak association can be explained by the challenge hypothesis: human testosterone levels in males are affected by recent challenges involving competition with men. Archer's examination of existing data provided general support for this hypothesis (Archer 2006). For example, a few studies report that testosterone levels are higher in winners compared to losers after a sporting competition (e.g., Booth et al. 1989).

Justin Carré and Susan Putnam extended the application of the challenge hypothesis to human spectators of a competitive interaction (Carré & Putnam 2010). They collected plasma testosterone samples from the saliva of elite male college ice hockey players before and after these athletes watched videos.

In the first experiment, players first watched a video of their team winning a prior contest and then two weeks later watched a video of their team losing a prior contest. In the second experiment, players first watched a video of their team winning a prior contest and then two weeks later watched a neutral video—a documentary film. The researchers collected saliva samples from each player ten minutes before and ten minutes after he watched a video.

In both experiments, there was a significant increase in testosterone concentrations in males' saliva after they had watched a prior winning contest, but not after they had watched a prior losing contest or the documentary (Figure 1). These results suggest that simply viewing an aggressive contest that led to a victory elevates testosterone levels in men. In fact, the increase in testosterone concentration occurred even in male team members who did not play in that particular game. The researchers suggested that although more study is needed, their results may have ramifications for athletic competition.

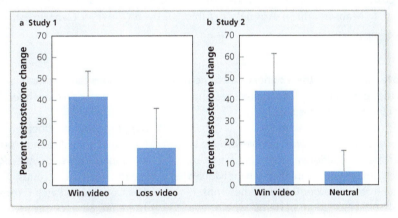

Figure 1. Testosterone levels in male hockey players. Mean (+ SE) testosterone levels increased more in male hockey players after they watched a video of their team winning rather than watching either a video of their team (a) losing a prior contest or (b) a neutral video (*Source:* Carré & Putnam 2010).

Chapter Summary and Beyond

Animals select habitats in which to live and reproduce based on the resources those habitats contain and the presence of others. When habitats exhibit negative density-dependent fitness and movement is cost-free, individuals distribute themselves among habitat patches in an ideal free distribution. In this pattern, the number of individuals in a habitat is proportional to habitat quality, and all individuals obtain the same fitness. Additional work has examined how distributions change when individuals are not free to select a habitat because of the behavior of dominant individuals (for recent examples, see Murray, Mane, & Pusey 2007; Purchase & Hutchings 2008). In some cases, fitness does not decline as density increases, and we then observe conspecific attraction, either because of Allee effects or because of conspecific cueing. Much work has been devoted to understanding the relative importance of these mechanisms (Donahue 2006).

Once a habitat is selected, many individuals defend a territory to secure the resources it contains. How territory holders assess the costs of defense is an active area of research (e.g., Müller & Manser 2007). Animals exhibit

much variation in fighting behavior that can be understood by game theory models, which explain how behavior evolves when fitness depends on how others behave. Game theory models have successfully been used to understand the fighting behavior of many taxa (Arnott & Elwood 2008), predicting that fighting behavior will be affected by the relative difference in opponents' fighting ability and the value of the contested resource. Much recent work has been devoted to understanding how individuals assess their relative fighting ability during a contest (Elias et al. 2008; Hsu et al. 2008). Territory defense and contests over resources require aggressive interactions that are regulated by androgen hormones such as testosterone. Androgen hormone levels are often elevated by social stimulation, as articulated in the challenge hypothesis. Hirschenhauser and Oliveira (2006) have reviewed empirical studies of the challenge hypothesis in relation to vertebrates, while Scott (2006) has extended the hypothesis to insects. Other work has examined how these hormones mediate winner effects (Gleason et al. 2009; Oliveira, Silva, & Canário 2009).

CHAPTER QUESTIONS

1. In the hawk-dove model, the fitness benefit is assumed to be less than the cost of injury C. Change that assumption so that now $B > C$. What is the ESS?

2. Thick-billed parrots (*Rhynchopsitta pachyrhyncha*) in groups of fewer than ten suffer very high predation by falcons. However, groups with more than ten individuals are relatively safe from predators. Based on these observations, what habitat selection patterns would you expect to find in this species?

3. Imagine that you stage a contest between two male crickets that differ in mass by 20%. The contest lasts for 30 seconds and contains one fighting behavior. Next you stage a contest between two males that differ in size by 5%. What

predictions can you make about fighting behavior in this contest based on the sequential assessment model?

4. Suppose you conduct an experiment with mallards (ducks) at a small pond. You create three food patches 20 m apart from each other. Patch A receives one piece of bread every six seconds, Patch B receives one every 12 seconds, and Patch C receives one every 20 seconds. There are 36 ducks in the pond. What is the predicted number of ducks in each patch according to the IFD model?

5. In the experiment in Question 4, the three patches were placed 20 m apart. How might moving the patches 100 m apart change the results of the experiment? Why?

6. The IFD model is fairly simple to test with a number of species. You can test the predictions of the model by throwing pieces of bread to house sparrows on a lawn or going to a local park with a pond and throwing bread to ducks or koi fish. You can change the quality of each patch by varying the rate of food delivery or food size. You can also watch human behavior. Just as animals choose food patches, humans choose lines at the bank, grocery store, or tollbooths on a highway. In each situation, the object is to get through the line as quickly as possible. Describe how human behavior in these situations would follow the predictions of the IFD model.

7. What prediction can you make about the androgen levels of the winner as compared to the loser in a contest between red-spotted newts, following the interaction in Verrell's experiment?

Chapter 11

Mating Behavior

Figure 11.1. Tree frog. Male tree frogs vocalize in spring to attract females for mating.

Our yard in St. Louis is quiet all winter long—but each spring, this situation changes dramatically. The first sign of spring is the sound of male birds singing each morning. As it gets still warmer, we hear field crickets, and then hundreds of male tree frogs calling in a loud chorus on warm, humid evenings (Figure 11.1). Why the change? Sex. The silence of winter gives way to the sounds of animals attracting mates.

Some years are especially noisy. Every 13 or 17 years, a brood of periodical cicadas emerge from underground, where they have been feeding on the roots of plants, to mate. Cicadas are hemipterans (true bugs) and North America is home to seven periodical species. In 2011, two 13-year species emerged (*Magicicada tredecim* and *M. neotredecim*), and our region became awash with cicadas (Figure 11.2). Millions of males were calling females each day. It was so loud that we couldn't spend much time outside, because the sound level frequently exceeded 90 decibels.

In this chapter, we start by discussing the differences between males and females and how this affects their mating behavior. We then see why many individuals are choosy when selecting mates. Next, we examine how selection can favor the evolution of multiple mating tactics when males cannot successfully compete for mates. Finally, we'll see how mating behavior can be affected by the social environment.

11.1 # Sexual selection favors characteristics that enhance reproductive success

primary sexual characteristics
The genitalia and organs of reproduction.

secondary sexual characteristics
Morphological differences between the sexes that are not directly involved in reproduction.

sexual selection A form of natural selection that acts on heritable traits that affect reproduction.

mate competition Selection in which one sex competes with other members of the same sex for access to the other sex for reproduction.

Figure 11.2. **Cicadas.** Periodical cicada species emerge at high densities only in certain years and create a very loud chorus.

As we saw in Chapter 2, Charles Darwin proposed natural selection as a mechanism to explain the evolution of adaptive traits in species (Darwin 1859). Darwin was a devoted naturalist and made copious notes on the morphological differences between male and female genitalia (organs of reproduction), or the **primary sexual characteristics**, in many different species. He also noted many other differences between males and females in traits not directly involved in reproduction, called **secondary sexual characteristics**. For example, in birds such as widowbirds, only males have colorful plumage and very long tail feathers (Figure 11.3). Many male mammals and insects possess large antlers, horns, or horn-like projections that are lacking in females. Darwin observed that such elaborate and exaggerated traits, in conjunction with complex behavioral displays and vocalizations, are often involved in conspecific interactions during the mating season.

Secondary sexual traits puzzled Darwin because they did not seem to fit into his theory of natural selection. Exaggerated morphological and behavioral traits should be energetically expensive to produce and maintain, and they can make individuals more obvious to predators and so reduce their survivorship. How could a peacock's tail increase survival? And if it did, why was the exaggerated trait found only in males? In *On the Origin of Species*, Darwin first proposed that exaggerated male traits might be advantageous for reproduction rather than for survival. In a subsequent book, *The Descent of Man and Selection in Relation to Sex*, Darwin in fact hypothesized that these traits might arise from a different form of selection (Darwin 1871). **Sexual selection** represents "the advantage certain individuals have over others of the same sex and species solely in respect of reproduction" (Darwin 1871, 210). As such, sexual selection is a subset of natural selection.

How does sexual selection explain the evolution of exaggerated male traits? First, Darwin proposed, in many species there is intense competition within one sex (often males) for mating opportunities with the other sex (often females). For example, males may engage in direct physical combat, in which only the winners mate with females. Today, we describe this process as **mate competition** (or intrasexual selection)—members of one sex compete with one another for mating opportunities

Figure 11.3. **Exaggerated male traits.** (a) Male long-tailed widowbirds (*Euplectes progne*) possess long tail feathers that (b) females lack. (c) Male stag beetles (*Lucanus cervus*) possess large horn-like projections that females lack.

(Figure 11.4). While such competition is an important aspect of sexual selection, Darwin recognized another aspect as well: females may be choosy. Darwin thus proposed that males may also compete among themselves to increase their attractiveness to females. In this way, he envisioned that females often play an active role in reproductive decisions in **mate choice** (or intersexual selection)—members of one sex exhibit distinct mating preferences. Together, mate competition and mate choice create sexual selection, a process that favors characteristics in one sex that allow the trait bearers to be more successful reproductively. Sexual selection is one of the most active research areas in animal behavior today.

Why two sexes?

Obviously, we could not have sexual selection without two sexes, but how did the two sexes evolve? One fundamental difference between males and females is the size of their gametes (Figure 11.5). Males tend to produce many small, motile gametes (sperm), while females tend to produce much larger, nutrient-rich, and nonmotile gametes (eggs), a phenomenon called **anisogamy**. In many algae, fungi, and unicellular protozoans, however, all individuals produce similar-sized gametes; this **isogamy** appears to be the ancestral form (Bell 1978; Bulmer & Parker 2002). How did anisogamy evolve from isogamy?

Geoff Parker and colleagues developed a model to answer this question (Parker, Baker, & Smith 1972; Bulmer & Parker 2002). This model is based on the following assumptions:

1. In the ancestral marine environment, individuals in a population produce different-sized gametes.
2. Each parent has a fixed amount of energy to allocate to gamete production, resulting in a size-number trade-off: as the number of gametes produced increases, their size will decrease.
3. Zygote viability is related to its size. Larger zygotes have higher viability because they contain more resources for survival.

Parker considered the fitness of small, large, and intermediate-sized gametes. Small gametes have a numerical advantage: they will create the most zygotes. On the other hand, large gametes always produce zygotes with the highest survival. Intermediate zygotes have neither advantage and so have the lowest fitness. Thus, there is disruptive selection against intermediate-sized gametes. The result is high fitness for either "proto-males," which produce many small gametes, or "proto-females," which produce fewer large gametes.

We now can understand why one sex (males) produces many small gametes that contain only genetic material, while the other sex (females) produces a smaller number of large, nutrition-rich gametes. The evolution of two different sexes leads to different reproductive strategies and sets the stage for sexual selection, as we see next.

Bateman's hypothesis and parental investment

Angus Bateman was one of the first researchers to examine sexual selection in males and females using fruit flies (*Drosophila melanogaster*). In a series of experiments, Bateman placed equal numbers of virgin females and males in milk bottles for three or four days and allowed them to mate (Bateman 1948). Adults varied in age from one to six days old. Females can take up to four days to become sexually mature, while males are sexually mature within 24 hours of eclosion (emergence of the adult from the pupal case). Each adult was heterozygous for a different dominant mutation, and so Bateman could determine the parents of three-fourths of the offspring by their phenotype. Across all experiments, 96% of the females produced progeny (as evidenced by the presence of their mutation in some progeny). In contrast, only 79% of the males successfully produced progeny. From these data, Bateman concluded that males had a higher variation in reproductive success than females. Bateman inferred that the

Figure 11.4. Mate competition. Male northern elephant seals (*Mirounga angustirostris*) compete for females.

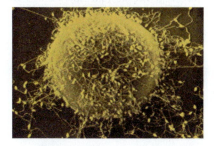

Figure 11.5. Gametes. A single sea urchin egg cell surrounded by hundreds of sperm cells.

mate choice Selection by one sex for members of the other sex for reproduction.

anisogamy The existence of different sized gametes (small and large) in the different sexes.

isogamy The production of gametes of the same size by all individuals.

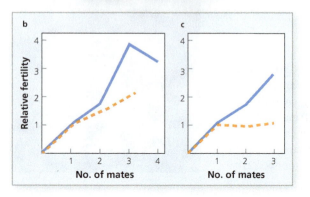

Figure 11.6. Bateman's results. (a) Fruit flies mating. (b) Relative reproductive success for males (solid line) and females (dash) in Bateman's experiments 1–4 combined. (c) Same as (b) except data are from experiments 5–6 combined (*Source:* Bateman 1948).

▪ **Bateman's hypothesis** Female reproductive success is most strongly limited by the number and success of eggs that she can produce, while male reproductive success is limited by the number of mates he obtains.

▪ **parental investment theory** The sex that has a greater investment in offspring production should be choosier when it comes to mates.

▪ **weapons** Exaggerated morphological traits used in male-male competition.

▪ **ornaments** Exaggerated morphological traits used to attract females.

intensity of sexual selection, as measured by the variation in reproductive success, was, in general, higher on males than females. He also inferred that this increased intensity of sexual selection was due to male-male competition.

Furthermore, Bateman's experiments suggested that a male's reproductive success increases more strongly with the number of mates obtained than does female reproductive success (Figure 11.6). Bateman concluded that reproductive success for a female is primarily limited by egg production, because a female can obtain sufficient sperm to fertilize all her eggs from a single mate. In contrast, sperm production should not limit male reproductive success. Instead, male reproductive success is primarily limited by the number of matings a male obtains. Because females produce fewer gametes than males, there will be intense competition among males for fertilization of female gametes (Kokko & Jennions 2003). This situation sets the stage for greater variation in male reproductive success—and thus more intense sexual selection on males. In sum, the evolution of male-male competition and female choice follows from competition for access to gametes of the opposite sex (Kokko, Jennions, & Brooks 2006).

In **Bateman's hypothesis**, female reproductive success is most strongly limited by the number and success of eggs that she can produce, whereas male reproductive success is limited by the number of mates. This difference between the sexes has become a foundation of sexual selection theory. Following Bateman, Robert Trivers expanded on this concept by identifying all forms of parental investment as a key difference between the sexes, in addition to gamete size. As defined by Trivers, parental investment includes "any investment by the parent in an individual offspring that increases the offspring's chance of surviving (and hence reproductive success) at the cost of the parent's ability to invest in other offspring" (Trivers 1972). Parental investment can include gestation, incubation, defense, and food provisioning and these often differ between males and females. Trivers's **parental investment theory** predicts that the sex that pays the higher cost of parental investment should be choosier when it comes to mates. The other sex will then experience more intense sexual selection. Because females in most species invest more in offspring, we can expect to see choosy females and a higher variation in male reproductive success.

As a result of sexual selection, we expect that males will often exhibit exaggerated traits used in competition for females. These traits may be morphological or involve behaviors such as vocalizations and courtship rituals. Exaggerated morphological traits used in male-male competition are often called **weapons**, and those used to attract females are known as **ornaments**. Exaggerated male traits are widespread; mammals and arthropods are well known for their elaborate weapons (Emlen 2008), and many male birds possess ornaments, as we see next.

Featured Research

Antlers as weapons in red deer

The antlers of deer provide a classic example of an exaggerated morphological trait used as a weapon in mate competition. Tim Clutton-Brock, Loeske Kruuk, and colleagues have studied sexual selection in a population of red deer (*Cervus elaphas*) on the Isle of Rum in Scotland (Kruuk et al. 2002). Within this population, individual males will defend and mate with several females each year. Antlers are present only in males and are used in aggressive contests to defend females from rival males (Suttie 1979; Clutton-Brock & Albon 1989).

This population has been intensively observed (40 times/year) since 1971. Individuals are marked or known by unique natural markings. Researchers take blood samples from calves and adults for genetic analysis to determine individuals' reproductive success. They also collect and weigh the antlers shed by males each year

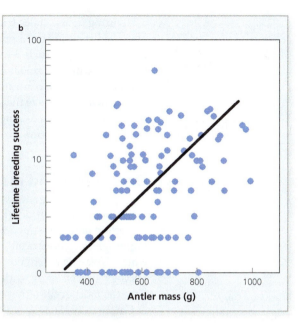

Figure 11.7. Lifetime breeding success and antler size. (a) Red deer males possess large antlers lacking in females. (b) The average antler mass over an individual's lifetime was positively correlated with his total lifetime breeding success. Each dot represents one male (*Source:* Kruuk et al. 2002).

to determine their mass (size). When individuals die, post mortem analyses provide information on body size (hind leg length). These data have allowed researchers to examine the effects of both male body size and antler size on reproductive success.

The researchers determined that males with larger antlers had higher breeding success in any given year, as well as over their lifetime (Figure 11.7). Although larger, older males had heavier antlers, the effect of antler size on reproductive success was still significant after accounting for differences in body size. In fact, male body size per se had little effect on breeding success.

These data indicate that antler size is significantly correlated with male breeding success in this population: males with larger antlers sire more offspring because they obtain more mates. Given that antlers are used in contests with rivals, the data suggest that sexual selection has influenced this exaggerated male trait.

In red deer, we see a positive correlation between weapon size and reproductive success. Next we examine an experimental test of this relationship.

Weapon size and mating success in dung beetles

Featured Research

In many beetles, males possess a horn-like projection that females lack (Figure 11.3c). Joanne Pomfret and Robert Knell studied the role of these horns in competition for mates in male dung beetles (*Euoniticellus intermedius*) (Pomfret & Knell 2006). Males fight with one another to defend dung piles, while females dig subterranean tunnels to mate and raise their offspring. Pomfret and Knell observed much variation in body size and horn size among males and conducted an experiment to determine whether large horns provided an advantage in these fights.

They staged fights between males in artificial nesting arenas (similar to ant farms) consisting of two panes of glass placed 5 mm apart, filled with soil and dung. The researchers collected adult males and females from a field and placed a single female in the arena, which quickly dug a tunnel. Two males, closely matched for body size but not horn length, were then introduced into the arena with the female. Male body length was measured with calipers, and horn length was measured from photographs of the head. The researchers recorded the outcome of subsequent interactions. The winner was deemed to be the male that successfully entered and remained in the tunnel after 24 hours, and most winners subsequently mated with the female. The loser was excluded from the tunnel and thus mating opportunities.

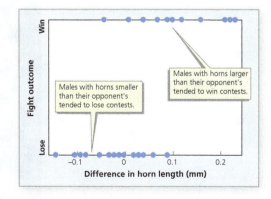

Figure 11.8. Beetle contest outcomes. Focal males with a horn shorter than that of their opponent (negative difference in horn length) won few contests. Focal males with horns larger than that of their opponent (positive difference in horn length) won most contests (*Source:* Pomfret & Knell 2006).

During fights, males tried to hook their horns under their opponent's head. They also used their horns to push an opponent out of the tunnel. For small beetles, the larger male or the male with the longer horn won the most contests. In contests between large males, large horn size alone was the most important predictor of success (Figure 11.8). In all contests, the greater the difference in horn length, the more likely was the winner to have the longer horn. These findings suggest that the maintenance of long horns in male dung beetles can be explained by sexual selection driven by male-male competition.

Featured Research

Ornaments and mate choice in peafowl

Ever since Darwin, peacocks (*Pavo cristatus*) have been the quintessential example of the action of sexual selection. Males possess extremely large tails whose feathers contain brightly colored ocelli, or eyespots (Figure 11.9). Because males fan out their tail when displaying to females, it seemed likely that female mate choice involves tail morphology. Indeed, early work demonstrated that females prefer to mate with males that have the most ocelli—that is, males with the most ornamented tails (Petrie, Halliday, & Carolyne 1991).

lek A location where males aggregate and display to females.

However, before peahens select a peacock mate, males must compete to establish and defend a display site. Many males aggregate on a **lek**, a location that lacks resources, where they display to females using behaviors such as vocalizations and fanning of the tail. Males fight for status and preferred positions on the lek.

Adeline Loyau, Michel Saint Jalme, and Gabriele Sorci examined mating behavior in a population of peacocks at Parc Zoologique de Clères in France (Loyau, Saint Jalme, & Sorci 2005). The researchers captured and color-banded males and females in early spring. They measured the length of the tail and male body size by measuring their tarsus (a bone in the lower leg that provides an index of body size) and photographed each male's tail when open to count the number of ocelli. The researchers then recorded male-male interactions during the breeding season, along with the number and duration of male tail displays to females, the number of vocalizations, and the number of copulations.

The research team found intense competition for display sites, as only 45 of 61 males successfully defended a site. They also found significant positive correlations between the likelihood of defending a display site and both body size and tail

Figure 11.9. Peacock tail. Females (peahens) are gray and brown, while males (peacocks) have large, colorful, and elaborate tails.

length; larger males and those that had longer tails were more successful in acquiring a site (Figure 11.10). Females were very selective in their choice of mates. Only 12 males obtained copulations, and over one-third of all copulations were obtained by a single male. In general, males that had high display rates and the most ocelli obtained the most copulations, but these were not necessarily the largest males (Figure 11.11).

The research team concluded that both mate competition and mate choice has played a role in the evolution of the peacock tail. First, only a subset of males successfully defended display sites, and all of these males were large individuals with long tails. Males frequently used their tail in aggressive interactions to chase nonterritorial males from display sites, a behavior that could also favor larger tails. Second, females strongly preferred males with high display rates and tails with many ocelli. Together, male competition and female choice have favored the evolution of males with long, elaborately ornamented tails and high display rates.

These three studies illustrate how both mate competition and mate choice are important in the evolution of elaborate and extravagant traits used in sexual selection. Each study examined the current function of male exaggerated traits. How did these traits arise? We look at that question next.

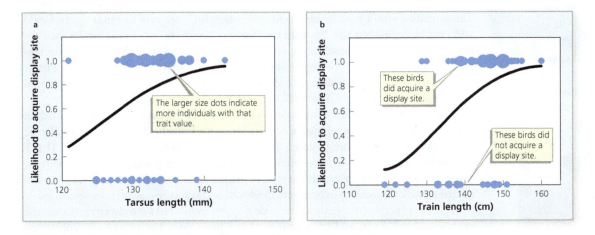

Figure 11.10. Probability of acquiring a display site. (a) Larger males (longer tarsus length) and (b) those with longer trains (tails) were more likely to acquire a display site. Dot size indicates number of data points (*Source:* Loyau, Saint Jalme, & Sorci 2005).

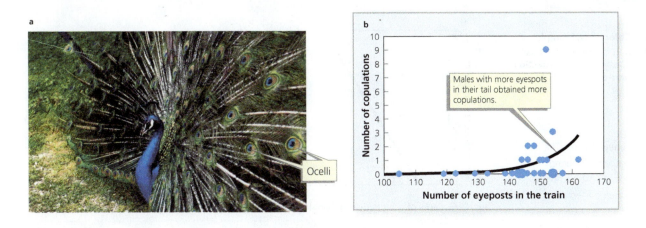

Figure 11.11. Number of tail eyespots and copulations. (a) Ocelli. (b) Males who had tails with a large number of ocelli obtained more copulations. Dot size indicates number of data points (*Source:* Loyau, Saint Jalme, & Sorci 2005).

The origin of sexually selected traits: the sensory bias hypothesis in guppies

At this point, you may be wondering how a male trait becomes a focus of female mate preference. One explanation is the **sensory bias hypothesis**, which states that female mating preferences are a byproduct of preexisting biases in a female's sensory system. These biases are presumed to have evolved in a nonmating context, with males evolving traits to match those biases (West-Eberhard 1984; Basolo 1990; Ryan & Rand 1990). One test of this hypothesis involves guppies (*Poecilia reticulata*). Females prefer to mate with males that display the greatest amount of orange coloration, which is derived from carotenoids obtained in the diet (Figure 11.12). Carotenoids are an important nutrient in many species, and they play a key role in immune system function (e.g., Bendich 1989). Colorful guppies are also easier for predators to spot and so suffer higher predation risk (Endler 1980). Prior work revealed that both natural and sexual selection has affected the strength of female preference and male coloration across populations. In stream populations where predation risk is high, female preference for orange is reduced, and males display less orange body color than do males in other populations with low predation risk (Houde & Endler 1990).

Helen Rodd and colleagues noticed that guppies in the wild are attracted to and voraciously consume orange-colored fruits that fall into the water (Rodd et al. 2002). These fruits contain high levels of carotenoids, which are rare in other foods. Rodd and colleagues tested the hypothesis that guppies have an innate preference for orange food items because they contain carotenoids and that this preexisting bias favors males that display orange color on their body. This hypothesis makes a basic prediction: both male and female guppies should be attracted to orange-colored objects because their color is associated with the presence of carotenoids.

The researchers tested their prediction in several natural populations in Trinidad. In one set of studies, they placed small discs on a leaf in the water held in place by a small rock. The discs could be red, orange, green, purple, blue, white, yellow, or black (Figure 11.13). For five minutes, the researchers noted all guppy approaches, pecks at the object, and the sex of the individual. To standardize environmental conditions, the researchers also conducted a similar laboratory experiment in which the discs were placed in an aquarium.

In both experiments, male and female guppies exhibited a strong attraction to the orange and red discs, and the strength of the preference for orange was correlated with female mating preference for the color orange. In populations with the strongest female preference for orange males, both sexes displayed the strongest attraction to orange discs (Figure 11.14).

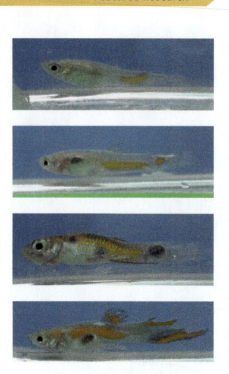

Figure 11.12. Male guppies. Males vary in the amount of orange pigment on their body.

Figure 11.13. Color discs. Examples of color discs used by Helen Rodd and colleagues (2002) to study guppy attraction to different colored objects.

sensory bias hypothesis The hypothesis that female mating preferences are a byproduct of preexisting biases in a female's sensory system.

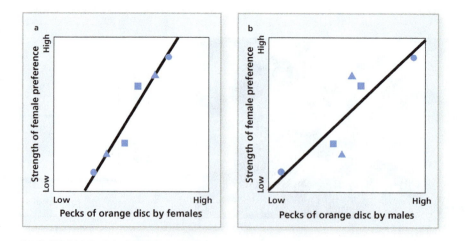

Figure 11.14. Mating and food color preferences. In populations where females most strongly preferred orange males (y-axis), both (a) females and (b) males pecked the orange-colored disc most frequently. Each symbol represents a different population (*Source:* Rodd et al. 2002).

These experiments indicate that both male and female guppies display an innate preference for orange objects. This finding provides support for the hypothesis that female mating preference is linked to a preexisting preference for orange food objects. In Section 11.3, we'll examine additional hypotheses to explain female trait preferences.

Male mate choice in pipefish

Featured Research

While female mate choice and male exaggerated traits are common, they are by no means universal. Depending on circumstances, males can also be selective, and this choosiness has favored the evolution of female secondary sexual traits in some species (Clutton-Brock 2007). Pipefish are an example of a **sex-role reversed species**, in which females compete for males, who invest heavily in parental care. Females transfer their eggs to the males' egg-brooding structure. The male then fertilizes the eggs and cares for them as they develop—that is, males become pregnant. Because males invest more heavily in offspring than females as a result of their high levels of parental care, sexual selection theory predicts that males should be the choosy sex and that female-female competition for mates may be intense. We might also expect females to display secondary sexual ornaments.

sex-role reversed species A species in which females compete for males, who invest heavily in parental care.

Anders Berglund and Gunilla Rosenqvist have been studying deep-snouted pipefish (*Sygnathus typhle*) for many years. This small (less than 40 cm long), slender species, a close relative of seahorses, lives in shallow seagrass habitats in coastal Europe (Berglund, Rosenqvist, & Svensson 1986; Berglund & Rosenqvist 1993). Females possess a secondary sexual ornament: a temporary striped pattern on the side of the body that is displayed only during competition between females and when courting males, a behavior known as a courtship dance. Normal pipefish body coloration is quite cryptic, but the ornament makes the female stand out and so is likely to increase predation risk, making it costly (Scientific Process 11.1).

Berglund and Rosenqvist tested whether this ornament affects male mate choice (Berglund & Rosenqvist 2001). Using a simultaneous choice experimental design with aquaria divided into three sections, they conducted an experiment that allowed a male to choose between two females on consecutive days. Females were placed in two sections that were separated by an opaque barrier, preventing them from viewing each other. The male was in the third section and could view both females. On Day 1, the three fish were placed in the aquarium and their behavior recorded. On Day 2, the same fish were placed in the same aquarium, but now all dividers were removed so individuals could interact. On both days, fish behavior was filmed for ten hours.

On Day 1, males spent more time near the female that displayed her ornament more (the more ornamented female), and spent more time in courtship dancing with her. On Day 2, the male again spent more time courtship dancing with the more ornamented female and copulated with her more often.

These results show that males prefer females that are more ornamented. Berglund and Rosenqvist suggest that female ornamentation reflects dominance status and overall vigor and may make it easier for males to assess female body size. Body size affects egg production, and so males should prefer larger mates.

In all these cases, we see that sexual selection can explain the evolution of certain traits because they enhance reproductive success. For traits used in mate competition, the selective advantage is easy to see: larger weapons can provide an advantage in direct competition with rivals. Furthermore, it is straightforward to understand how a trait favored by one sex in their choice of a mate will evolve in the other sex, because those that possess the most extreme form of the trait will have high reproductive success.

At the beginning of the chapter we noted that in most species, females are often the choosy sex. What benefits can females gain by such choosiness? We examine that question next.

SCIENTIFIC PROCESS 11.1

Male mate choice in pipefish
RESEARCH QUESTION: *How do male pipefish select a mate?*

Hypothesis: Males prefer females that have the highest expression of a sexually selected trait.

Prediction: Males should spend more time and mate more often with the female that displays her temporary striped pattern (ornament) the most.

Methods: The researchers:

Day 1: (No Female-Female Interaction)

- conducted a simultaneous choice test with a single male and two enclosed females that were matched for size.
- recorded the amount of time the male spent in front of each female, how often a female displayed her ornament, and how often the male engaged in a courtship dance with each female.

Day 2: (Female-Female Interaction)

- conducted the same experiment, but now all dividers were removed so that all individuals could interact.
- recorded total time dancing, latency to dance, and copulations with each female. All eggs were subsequently removed from the male's brood pouch and counted.

Results:

- Males spent more total time, more time dancing, and had a shorter latency to dancing with the female that displayed her ornament more.
- The more ornamented female obtained more copulations.

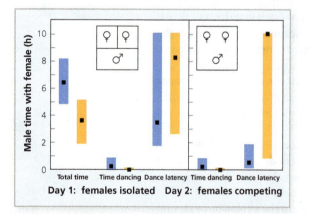

Figure 1. Male behavior. Time spent with the more (blue bars) and less (orange bars) ornamented female. Black square is the median time with female. Each box is the middle 50% of data (*Source:* Berglund & Rosenqvist 2001).

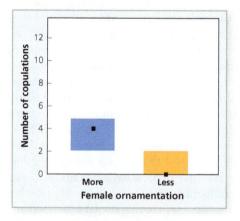

Figure 2. Copulations. Number of copulations with the more (blue bar) and less (orange) ornamented female. Black square is the median time with female. Each box is the middle 50% of data (*Source:* Berglund & Rosenqvist 2001).

Conclusion: Females with greater expression of the ornament have higher mating success.

11.2 Females select males to obtain direct material benefits

Parental investment theory predicts that the sex that invests the most time and energy in offspring can benefit by being selective in its choice of mates. As we saw, this usually means that females are choosy. What are the benefits of being selective? Two possibilities exist (Andersson & Simmons 2006). First, females can benefit by choosing males that provide **direct material benefits**, such as food gifts, access to territories with abundant food, or enhanced parental care (Møller & Jennions 2001). Second, females can benefit by mating with males based on indirect genetic benefits that enhance the fitness of offspring. In this section, we examine how females can benefit directly from selective mate choice.

direct material benefits Material resources obtained by a female from mating with a particular male.

Female choice and nuptial gifts in fireflies

In many arthropods, males provide a **nuptial gift** to a female prior to mating that provides nutrition and so can increase her reproductive success (Figure 11.15). These gifts can be prey, carrion, plant material, glandular secretions, or regurgitated food (Vahed 2007). Sara Lewis, Jennifer Rooney, and Christopher Cratsley studied how nuptial gifts affect female firefly preference for males and their resulting fitness (Rooney & Lewis 2002; Cratsley & Lewis 2003; Lewis, Cratsley, & Rooney 2004).

nuptial gift A physical resource such as a food item that a male provides to a female to enhance his mating success.

Male *Photinus ignitus* fireflies produce a spermatophore as a nuptial gift—sperm packaged within a protein-rich structure produced by male accessory glands (Lewis, Cratsley, & Rooney 2004). Once inside a female, the spermatophore disintegrates, releasing its nutrients, which the female allocates directly to her developing oocytes (Rooney & Lewis 1999). Because adults do not feed, the spermatophore represents an important resource. Does it affect fitness?

Firefly courtship involves species-specific bioluminescent flash patterns; males fly and flash to females sitting on nearby vegetation, and females flash in response to attract a male to her location (see Chapter 6). Cratsley and Lewis (2003) noted variation within a species in the duration of male flashes and examined how this might affect female choice and fitness.

To investigate this question, they first examined whether there was a relationship between male flash duration and spermatophore size. They collected 36 males and brought them into the laboratory, where each male was weighed and video-imaged to analyze its flash pattern. Most had flash durations that ranged from 56 to 89 ms and

Figure 11.15. Nuptial gift. (a) A male hanging fly (*Bittacus* sp.) (right) presents a nuptial gift (dead fly) to a female (left). (b) The pair copulate. (c) The female takes the gift to eat.

were highly repeatable among males (Figure 11.16). To quantify spermatophore mass, the researchers allowed each male to mate and then sacrificed the female to collect and weigh the spermatophore that was transferred. They found a significant positive correlation between male flash duration and spermatophore mass (Figure 11.17). The mechanism behind this correlation is unclear, but the association does mean that females could use variation in flash duration to choose males with the largest spermatophores, a possibility that requires further evaluation.

Next, Cratsley and Lewis examined the effect of flash patterns on female mating behavior. Because they wanted to manipulate the duration of the flashes, the research team created simulated male flashes to which females responded by flashing. To examine the effect of the duration of the male flash signal on female behavior, the researchers created flashes of different durations (55, 63, 71, 79, or 87 ms) and presented them to 25 females. They found that females responded most frequently to the longest flashes (Figure 11.18).

Together, these results suggest that females prefer to mate with males that have the longest flash duration and thus likely have large spermatophores. Does this preference affect female fitness? South and Lewis (2011) have recently examined this question from a comparative perspective. Across a broad range of arthropod species, they found a positive relationship between spermataphore nuptial gift size and female fecundity, an important component of fitness. Future experiments will help clarify this relationship in fireflies.

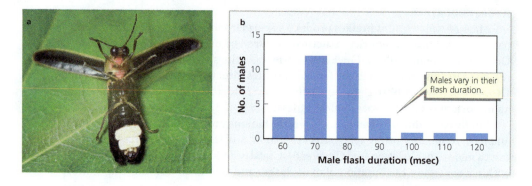

Figure 11.16. Male flash durations. (a) A male producing a bioluminescent flash. (b) Variation in the duration of male flashes in a population (*Source:* Cratsley & Lewis 2003).

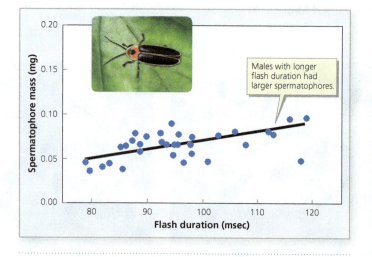

Figure 11.17. Spermatophore mass. There is a significant positive correlation between male flash duration and spermatophore mass (*Source:* Cratsley & Lewis 2003).

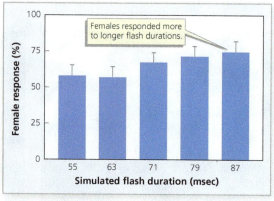

Figure 11.18. Female response to flashes. Mean (+ SE) female response to simulated flashes. Females respond more to longer flash durations (*Source:* Cratsley & Lewis 2003).

Female choice and territory quality in lizards

Another direct benefit of mate choice by females is access to high-quality resources defended by a territorial male. Ryan Calsbeek and Barry Sinervo examined how territory quality affects female choice and fitness in side-blotched lizards (*Uta stansburiana*) (Calsbeek & Sinervo 2002) (Figure 11.19). Males defend territories where they display to females from rock perches. The quality of a territory is based on its rockiness: rocks provide perches not only to display to females but also to spot predators. Rocks also increase the range of "microclimates" (both hot and cool locations) available for thermoregulation, an important physiological factor for ectotherms like lizards. Previous work demonstrated that offspring growth and survival are greater when females lay eggs within male territories that have many rocks (Calsbeek et al. 2002). In this system, males compete for territories with abundant rocks, and females prefer to mate with males who control the rockiest territories—that is, the best males defend high-quality territories. However, that preference makes it difficult to know whether female mate choice is based on male quality or the quality of his territory.

To find out, the researchers manipulated territory quality after males had established their territories but before females became receptive and selected mates. They first mapped male territories and observed that large, dominant males settled on those territories with the most rocks. Then the researchers moved ten to 40 rocks from the territories of large, dominant males to the territories of small neighboring males. The added rock piles increased territory quality for small males while reducing territory quality for large males, because males remained on their territory after the manipulation. This design allowed the researchers to successfully separate high-quality males from high-quality territories. At the end of the breeding season, after females had selected mates, females were captured and brought into the laboratory to lay their eggs, which were incubated under standard conditions. The researchers measured the egg-laying date and egg mass of each female.

They found that females strongly preferred the improved territories, even though they were occupied by small males. Of the 51 females in the population, 37 moved on to the improved plots. Females on improved territories laid eggs sooner and produced larger egg masses (Figure 11.20), demonstrating a direct fitness benefit for females that select high-quality territories. In this system, females appear to select mating partners based on male territory quality and derive a significant fitness benefit by doing so.

We have seen that females can benefit by selecting males that provide access to a material benefit: resources. In many species, however, males do not provide material benefits, and still females are selective in their choice of a mate. Why? This is where indirect benefits enter the equation, as we are about to see.

Figure 11.19. Side-blotched lizard. Males defend territories to attract females.

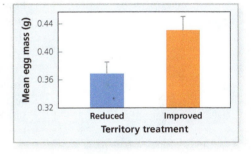

Figure 11.20. Female egg mass. Mean (+ SE) egg mass of females on treatment territories. Females on improved territories produced larger egg masses (*Source:* Calsbeek & Sinervo 2002).

Female mate choice can evolve via indirect benefits to offspring

In many species, females do not benefit from direct material benefits but rather from indirect benefits. Females can obtain **indirect genetic benefits** when potential mates differ in genetic quality that affects their and their offspring's fitness. Females can benefit by mating with such males to obtain those alleles for their offspring. How do females assess the genetic quality of potential mates if they cannot directly observe their genes? One way is by selecting males based on their secondary sexual traits. This idea, that female choice can affect the evolution of male traits, has stimulated much debate and research, starting as early as the nineteenth century.

indirect genetic benefits Genetic benefits females can obtain for their offspring by mating with males that have high genetic quality.

Fisherian runaway and good genes

Perhaps no area in animal behavior has generated as much controversy as sexual selection, starting with Darwin and his contemporary Alfred Russell Wallace, who independently proposed a theory of evolution based on natural selection. Recall that Darwin proposed two aspects of sexual selection: male-male (mate) competition and female (mate) choice. Wallace disagreed vehemently and considered *only* male-male competition to be a selective force (Wallace 1878). Their argument about the importance of female choice persisted because there was no clear mechanism to explain how mate choice could drive the evolution of secondary sexual traits.

In 1930, Ronald Fisher wrote a milestone paper proposing a genetic explanation for the evolution of exaggerated secondary sexual traits (Fisher 1930). In his verbal model, Fisher assumed that females selected mates based on a particular trait that varies among males. Such a trait could evolve because of (1) its fitness advantage, independent of female choice; and (2) female preference for it. Thus, initially, the trait indicated male quality, because males possessing it had higher fitness and so were more attractive to females. If female preference for a trait and the trait itself have a genetic basis, female offspring will prefer the trait, and male offspring will express it. The intensity of female preference for the trait will increase as long as male offspring have a mating advantage by possessing it.

In sum, the male trait should co-evolve with female preference and become increasingly exaggerated. Fisher called this a **runaway process**, one that would continue until the benefit it provided through sexual selection was outweighed by the disadvantage it entailed through natural selection (e.g., increased predation risk). Russell Lande formalized Fisher's idea into the notion of runaway selection. He showed mathematically how this process could occur through linkage disequilibrium when the genes for the trait and for female preference for the trait are genetically linked (Lande 1981), even if the trait was unrelated to male fitness. Linkage disequilibrium occurs when the genotype at one locus is not independent of the genotype at another locus.

Amotz Zahavi proposed an alternative hypothesis to explain how female choice could affect the evolution of male traits. He suggested that exaggerated secondary sexual characteristics allow the choosing sex to assess mate quality (Zahavi 1975). For Zahavi, the exaggerated trait is "a test imposed on the individual" by natural selection. Only high-quality males should thus be able to display the most exaggerated form of the trait—an idea he called a "handicap." Zahavi's **handicap principle** hypothesis states that well-developed secondary sexual characteristics are costly, because they handicap a male's survival. Females should prefer such males because they must have excellent genotypes to overcome the handicap.

Subsequent theory expanded on Zahavi's work to demonstrate that exaggerated secondary sexual traits must be costly to produce in order for them to be a reliable indicator of male genetic quality, or **good genes**. These genes may be associated with an enhanced immune system, greater fighting ability, or increased vigor and viability (Moller and Alatalo 1999). The cost of the trait need not be a handicap to survival; however, such traits simply need to be costly to produce or maintain (Kodric-Brown & Brown 1984; Grafen 1990). If the trait were not costly, then all males, regardless of their quality, would be able to express the trait, and it would no longer indicate quality. Thus, a long, physically demanding vocal or behavioral mating display, or physiologically costly morphologies and chemical pheromones can be reliable indicators of male genetic quality if only the healthiest and most vigorous males (i.e., those with the best genes) can produce them (see Chapter 6).

At first, researchers debated the merits of Fisher's runaway and Zahavi's handicap principle hypotheses by focusing on their different assumptions. More recent work has shown that these models share important similarities (e.g., Ryan 1998; Kokko et al. 2002; Radwan 2002). In both models, the male trait initially preferred by females indicates a high degree of male vigor and survivorship. In good genes models, a genetic correlation is assumed between the female preference for the male trait and the male "good genes" that provide enhanced physical vigor and are indicated by the trait. In Fisherian runaway selection, there is a presumed genetic correlation between the female preference for the male trait and the male trait itself.

runaway process An evolutionary process in which a male trait co-evolves with a female preference for it and becomes increasingly exaggerated.

handicap principle Well-developed secondary sexual characteristics are costly to survival but reliable signals of fitness.

good genes The alleles of a high-quality individual.

Today, much research on sexual selection focuses on understanding (1) the traits used by females in mate choice, (2) how sexually selected traits indicate male genetic quality, and (3) the fitness benefits of mate choice. Let's see how these areas are addressed.

Mate choice for good genes in tree frogs

Sexual selection theory predicts that females should base their mating decisions on the presence of costly male traits to obtain indirect genetic benefits. Many species produce vocalizations to attract mates; for example, among frogs that aggregate in ponds, many males call to attract females in a chorus. Calling is costly, because it is energetically expensive and can make an individual more susceptible to predators (e.g., Bucher, Ryan, & Bartholomew 1982; Ryan, Tuttle, & Ryan 1982). Do females assess male quality on the basis of their vocalizations?

Julie Jaquiéry and colleagues tested the hypothesis that female European tree frogs (*Hyla arborea*) select males for genetic benefits based on their vocalizations (Jaquiéry et al. 2009) (Figure 11.21). Females initiate all matings, and males do not interrupt a pair that is mating (Friedl & Klump 2005), so male-male competition is assumed to be weak. The species is thus ideal for examining the genetic benefits of female choice. Furthermore, the species is threatened in parts of its range, so understanding its mating behavior may help with conservation (Arens et al. 2006).

Each night of the 22-night breeding season, the research team visited four neighboring ponds in western Switzerland to determine which males were calling in the chorus. A total of 15 calling males were captured and identified from a photographic data bank (individuals' black lateral line has unique color patterns), and a buccal (or cheek) sample was taken for genetic analysis (Figure 11.22).

To determine male mating success and the fitness of their offspring, the team visited the ponds every four days in daylight to collect egg masses. All clutches were reared in the laboratory until hatching. Ten offspring from each clutch were reared with unlimited food, and each tadpole was weighed twice to record its growth rate. Tadpole survival was determined at 28 days, and all individuals were then genotyped. The resulting genetic analysis allowed determination of the mating success of adults. From these data, the researchers could determine the attractiveness of each male, defined as the number of females with whom he mated divided by the number of nights he called during the breeding season.

Only ten of the 15 males sired any clutches, indicating that males varied in their attractiveness to females and that females were selective. Offspring growth rate was positively correlated with the father's attractiveness; males that sired more offspring produced tadpoles with higher growth rates (Figure 11.23). In amphibians, offspring growth rate is

Figure 11.21. European treefrog.
A calling male.

Figure 11.22. Buccal swab. Collecting a buccal swab for genetic analysis.

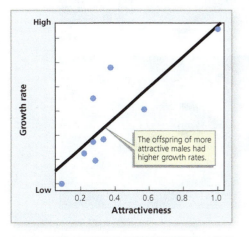

Figure 11.23. Tadpole growth rate. The offspring of more attractive males had higher growth rates (*Source:* Jaquiéry et al. 2009).

positively related to fitness, and so the research team concluded that females do indeed select males for genetic benefits. Additional work is required to determine the features of male vocalizations that attract females. The researchers speculate that these features will be accurate indicators of male quality, because calling is so costly (see Chapter 5).

This example illustrates how female mate choice for a sexually selected, costly trait can enhance her fitness. Next, we consider a different way that females can benefit by selecting males that have the greatest expression of a costly secondary sexual trait.

Featured Research

Good genes and immune system function in birds

William Hamilton and Marlene Zuk proposed that parasites and pathogens play an important role in sexual selection when secondary sexual traits are costly and condition dependent (Hamilton & Zuk 1982). For example, males may vary in their immune response: some males will have alleles that confer greater resistance to pathogens than will others. If pathogens reduce male vigor, then only males with superior disease resistance will be able to display exaggerated secondary sexual traits. Females that mate with such males will benefit by passing these alleles to their offspring. This is the **Hamilton-Zuk hypothesis**, which makes two predictions:

> **Hamilton-Zuk hypothesis** A hypothesis that parasites and pathogens play an important role in sexual selection when secondary sexual traits are costly and condition dependent.

1. Females should prefer to mate with males that have the greatest expression of secondary sexual traits
2. High parasite loads will reduce that expression in males

Marlene Zuk and colleagues tested these predictions in a captive population of red jungle fowl (*Gallus gallus*) (Zuk et al. 1990).

Jungle fowl are native to southeast Asia. Like domestic chickens, males possess a variety of secondary sexual traits that involve the plumage and a red comb on their head (Figure 11.24). Zuk's team designed a simple experiment. Half the males were infected with intestinal nematodes at one week of age and kept isolated from control males that were not infected. The researchers measured the size of the comb each week as the birds matured, and at ten months of age (when males were sexually mature), they measured the color intensity of the comb by comparing it to a reference set of Munsell color chips. Next, they conducted mate choice tests in which an adult female could select either a parasitized or a control male. Males were visually isolated from one another and restrained to prevent them from interacting. The researchers recorded the male that the female solicited or with whom she copulated.

Parasites affected the development of both comb length and color. Parasitized males had smaller, duller combs than controls (Figure 11.25).

Figure 11.24. Male jungle fowl. (a) A male jungle fowl with a dull red comb. (b) A male with a brighter red comb that females prefer.

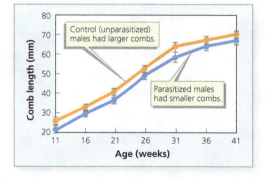

Figure 11.25. Male jungle fowl comb length over time. Mean (± SE) comb length of treatment and control males. As males matured, the combs of parasitized males (blue) were significantly smaller than the combs of unparasitized controls (orange) (*Source:* Zuk et al. 1990).

In the choice tests, females exhibited a strong mating preference for control males. The results support the predictions of the Hamilton-Zuk hypothesis and illustrate one way in which females can obtain genetic benefits from mate choice. Growing evidence demonstrates a link between genetic variation and immune system function and so illustrates how females can select males based on this aspect of their good genes (e.g., Olsson et al. 2003).

Mate choice fitness benefits in spiders

Featured Research

Another approach to understanding the indirect fitness benefits of mate choice is to manipulate the ability of females to make a choice. In many invertebrates, chemical signals, or pheromones, are commonly used in species recognition, but less is known about their role in mate choice (Johansson & Jones 2007). However, recent work suggests that pheromone production is costly and so may be a sexually selected trait that advertises quality (Johansson, Jones, & Widemo 2005).

Teck Hui Koh and colleagues examined the importance of pheromones as signals of male quality in mate choice in spitting spiders (*Scytodes* sp.) (Koh et al. 2009), which expel a sticky spit when catching prey (Figure 11.26). In this species, males provide no resources to females and so are ideal for investigating the indirect fitness benefits of mate choice.

The research team collected juveniles from two sites in Singapore and raised them individually until they matured. To ensure that females did not have experience with males that might affect their mate choice, the researchers used virgins. They first conducted a two-phase mate choice experiment using a Y-shaped apparatus to test female choice for both live males (Phase 1) and male pheromones alone (Phase 2).

In the first phase, two randomly selected males were placed behind wire mesh screens, one in each long arm, while a female was placed in the end of the stem (Figure 11.27). The screen prevented visual contact but allowed chemical communication. To prevent tactile communication, 3 cm thick foam was placed under the entire apparatus to prohibit any vibrations. Females were given one hour to select a mate; selection was determined by the female making contact with a screen for one minute.

In the second phase, the researchers collected male pheromones. They placed each male from the first phase in a sterile Petri dish with moistened filter paper for 48 hours, during which the male deposited silk draglines and chemicals. The paper was collected and pairs randomly placed in the Y-apparatus with the same female, and she again was allowed to make a choice over a period of one hour.

Of the 44 females that selected a male in the first phase, 42 selected the arm with the same male's filter paper in the second phase. The other two failed to make a choice. These results demonstrate that females can use male pheromones to select a mate.

To examine how mate choice affects female fitness, the 42 females from Phase 2 were divided into two treatments, such that the mean weight of females in the groups was similar. One set was allowed to mate with their preferred male, while the others were restricted to mating with the unpreferred male.

Females mated to their preferred male produced significantly larger egg sacs, with more and heavier eggs and recorded a higher hatch rate, which should correlate with

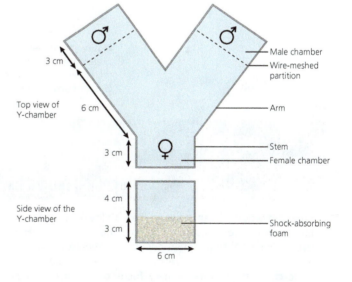

Figure 11.26. Spitting spider. These spiders expel a sticky substance to aid in prey capture.

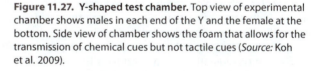

Figure 11.27. Y-shaped test chamber. Top view of experimental chamber shows males in each end of the Y and the female at the bottom. Side view of chamber shows the foam that allows for the transmission of chemical cues but not tactile cues (*Source:* Koh et al. 2009).

higher fitness (Figure 11.28). In this species, males provide no resources, and so females presumably benefit by obtaining indirect fitness benefits that result in enhanced offspring fitness. The researchers suggest that male pheromones may function as if they convey information about male size, nutritional status, immune function, fertility, parasite load, genetic heterozygosity, or other attributes that affect offspring. Additional work is needed to identify the pheromone(s) involved and how it varies with male phenotype. Fitness benefits of mate choice like these are starting to influence the management of captive breeding programs (Applying the Concepts 11.1).

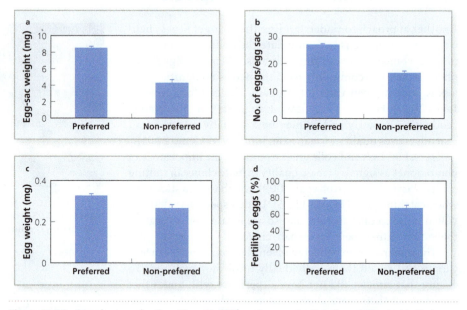

Figure 11.28. Female reproduction. Mean (+ SE) female reproduction. Females mated to their preferred male had (a) larger egg sacs, (b) more eggs per egg sac, (c) higher egg weight, and (d) higher fertility of eggs compared to those of females mated with their nonpreferred male (*Source:* Koh et al. 2009).

APPLYING THE CONCEPTS 11.1

Mate choice in conservation breeding programs

Breeding programs can maintain the viability of captive populations and serve as a source for reintroduction efforts. However, it is crucial to maintain genetic variation and minimize inbreeding. Because captive populations are small, two genetic concerns are the primary focus of researchers. First, genetic diversity can decline rapidly as a result of genetic drift, and second, inbreeding can lead to reduced viability of offspring. Both issues can create problems for program success, especially given small captive populations.

Historically, managers have selected breeding pairs based on genetic considerations alone and have hoped that viable offspring would result. Many pairings fail, however, and one factor might be the lack of mate choice. Cheryl Asa, Kathy Traylor-Holzer, and Robert Lacy (2011) have noted that litter sizes and offspring survival rates are higher when individuals select mates compared to the outcomes produced by forced pairings (see, e.g., Drickamer, Gowaty, & Holmes 2000; Anderson, Kim, & Gowaty 2007).

For instance, the captive breeding program for the Mauritius kestrel (*Falco punctatus*) has attempted to increase the population of birds by hand-rearing eggs. Some eggs are harvested from wild pairs of birds who then re-nest, whereas others are produced by captive birds. All eggs are hand-reared in captivity but differ dramatically in outcome. Over 80% of eggs harvested from the wild hatch, and 94% of those are successfully reared. In contrast, 65% of eggs produced in captivity hatch, and only 85% of those produce viable offspring (Cade & Jones 1993). Eggs harvested from the wild presumably result from females that exhibited mate choice, while eggs produced in captivity come from females that cannot exhibit choice because their mate was assigned by researchers.

Asa and colleagues suggest that the success of captive breeding programs can be enhanced by incorporating mate choice while taking into account other genetic concerns (Asa, Traylor-Holzer, & Lacy 2011). For instance, a focal female might be given the option of mating with several genetically acceptable mates.

In spitting spiders, female egg production and hatch rate was higher when the female mated with a preferred male. One explanation for this outcome is differences in male genetic quality. Another is that the female reduced her reproductive investment based on the quality of her mate. We examine such sexual selection that occurs after mating next.

11.4 Sexual selection can also occur after mating

Up to now, we have considered aspects of sexual selection prior to copulation. This is only part of the story. Both mate competition and mate choice can occur after gamete transfer. Postcopulatory male competition can occur when females mate with more than one male and the sperm of different males compete to fertilize the eggs (Parker 1970). In addition, postcopulatory female choice can occur after gamete transfer when a female influences the fertilization success of sperm from one male over that of others. Both phenomena require that females mate with multiple males, and we address that aspect of behavior in Chapter 12.

Mate guarding in warblers

One behavior that males can adopt to enhance their paternity is **mate guarding**: before and after copulation, a male follows his mate to prevent her from mating with rivals. However, this behavior has obvious costs. The time and energy spent guarding a mate can preclude a male from mating with other females or from acquiring food and other resources.

Helen Chuang-Dobbs and colleagues studied the effectiveness of mate guarding in black-throated blue warblers (*Dendroica caerulescens*) at the Hubbard Brook Experimental Forest in New Hampshire (Chuang-Dobbs, Webster, & Holmes 2001). This small (10 g) passerine bird is common in the forests of eastern North America (Figure 11.29). A single pair mate and raise offspring together, but females will often mate with other males, and so males risk raising another male's offspring, called **extra-pair young** (Chuang, Webster, & Holmes 1999). The researchers examined the effectiveness of mate guarding through a combination of observations and an experiment.

Figure 11.29. Mate guarding. Male black-throated blue warblers remain close to their mate.

mate guarding A behavior in which a male follows his mate to prevent her from mating with rivals.

extra-pair young Offspring of a pair-bonded female produced outside the pair bond by a third-party male.

The researchers monitored the breeding behavior of all adults by capturing and color-banding them for identification. In the observational aspect of their study, the research team followed focal males and their mates for up to an hour on days just before egg laying, the period of high fertility for females. Every two minutes, they recorded the distance between individuals and noted how often males followed females or vice versa each time the birds moved to new locations. Evidence of effective mate guarding would be short inter-individual distances and a preponderance of male rather than female follows. They then collected a blood sample from all nestlings for genetic analysis to determine parentage.

In the experimental part of their study, the team used a different study plot and removed some males for one hour during the female's fertile period. To do this, they attracted males away from their nests by playing conspecific songs from a speaker located near a territory and placed mist nets next to the speakers to capture the birds. Five males were captured and held in captivity, while three others spent most of the hour interacting with the speaker and so spent no time with their mate during the song playback. These eight treatment birds were presumed to exhibit a lower level of mate guarding than the eight control males that were not manipulated. The paternity of treatment and control birds was compared to determine the effect of reduced mate guarding.

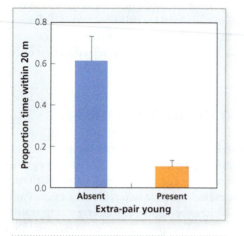

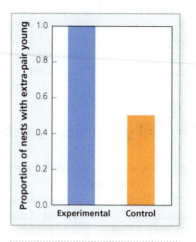

Figure 11.30. Intensity of male mate guarding. Mean (+ SE) proportion of time males spent close to their mate. Males with no ("absent") extra-pair young spent a greater proportion of their time within 20 m of their mate (mate guarding) than males with extra-pair young present in their nest (*Source:* Chuang-Dobbs, Webster, & Holmes 2001).

Figure 11.31. Proportion of nests with extra-pair young. Males that were experimentally removed and spent less time mate guarding had a higher proportion of nests with at least one extra-pair young than did controls (*Source:* Chuang-Dobbs, Webster, & Holmes 2001).

Observed males varied in their intensity of mate guarding, and this behavior was associated with variation in the paternity of offspring. Males who spent more time mate guarding had fewer extra-pair young than those who spent less time mate guarding (as measured by more male follows and greater time spent within 20 m of the female) (Figure 11.30). Similarly, the nests of all experimental males who had been away from their mate for one hour contained extra-pair young, while only 50% of control nests contained such young (Figure 11.31). These findings indicate that mate guarding can be an effective strategy for increasing paternity assurance and thus fitness for males. However, given that control nests did contain extra-pair young, mate guarding does not prevent females from mating with other males. Furthermore, remember that mate guarding comes at a cost: males that spend more time mate guarding have less time to spend seeking matings with other females (see Chapter 1).

Featured Research

Sperm competition in tree swallows

sperm competition The competition between sperm of different males to fertilize eggs.

cryptic female choice A situation in which a female influences the fertilization success of sperm from one male over others.

inbreeding depression A reduction in fitness as a result of mating with close relatives.

Not all males can continuously guard their mates. If males cannot prevent a female from mating with other males, then the sperm of different males may compete to fertilize the eggs, a phenomenon known as **sperm competition**. One way to deal with this is to swamp a rival male's sperm by mating frequently with a female. Susan Crowe and colleagues tested the hypothesis that paternity assurance can be enhanced through frequent copulations in tree swallows (*Tachycineta bicolor*) (Crowe et al. 2009). In this species, extra-pair fertilization is common. Males also spend significant time on parental care, so selection should favor behaviors that minimize loss of paternity. Crowe's team predicted that frequent copulation rate would increase paternity for a male.

They tested this prediction on a population of tree swallows at Queen's University Biological Station in Ontario. Adults and offspring were captured and uniquely marked, and blood samples were taken for genetic analysis. In tree swallows, copulations occur mainly at the nest site and so are relatively easy to observe: males hover over females and make cloacal contact (Figure 11.32). Previous work has shown that most fertilization results from copulations that occur during the three days before egg laying (Lifjeld & Robertson 1992; Lifjeld, Slagsvold, & Ellegren 1997). Consequently, beginning nine days prior to egg laying, each of 43 focal pairs were observed for three hours a day for over two weeks, starting at dawn, when most copulations occur.

The copulation rate varied among pairs and peaked in the three days prior to egg laying, averaging about five copulations per hour. Paternity analysis revealed that 52% of young were sired through extra-pair copulations. However, an increase in the frequency of copulations by a male increased the percentage of young he sired, as predicted (Figure 11.33). Thus, males can reduce the likelihood of paternity loss from sperm competition by increasing their copulation frequency.

High copulation frequency could increase paternity through sperm competition or could act as a type of mate guarding (because frequent copulations prevent other males from mating). Both of these behaviors can result in higher paternity for a male. Females, however, can also exert mate choice *after* copulation, as we see next.

Cryptic female choice

As we've seen throughout this chapter, female mate choice prior to copulation is usually obvious: females either do or do not mate with a particular male. Research has also found evidence of female choice after copulation. In species with internal fertilization, a female mates with multiple males and then influences the fertilization success of sperm from one male over that of others, which is known as **cryptic female choice** (Thornhill 1983). The usual result is that one male's sperm fertilizes a disproportionate frequency of her eggs. Why does cryptic female choice exist? How would you differentiate it from sperm competition?

Cryptic female choice will benefit a female who mates with several males that differ in quality. She can enhance the fitness of her offspring by biasing fertilization success in favor of the higher quality male. One way to test this idea is to examine the fertilization success of males that differ in degree of relatedness to a female. A close relative is a low quality mate, because such matings often result in low fitness for offspring, a process known as **inbreeding depression**. In comparison, an unrelated male is a higher quality mate. Can females alter the fertilization probability of sperm from males that differ in degree of relatedness?

Figure 11.32. Swallow copulation. Males stand on or hover over the backs of females to copulate.

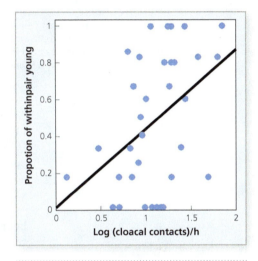

Figure 11.33. Copulation frequency and paternity assurance. There is a significant positive relationship between the frequency of copulations (cloacal contacts per hour) and paternity assurance (proportion of within-pair offspring) (*Source:* Crowe et al. 2009).

Inbreeding avoidance via cryptic female choice in spiders

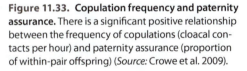

Featured Research

Klaas Welke and Jutta Schneider examined whether female orb spiders (*Argiope lobata*) avoid inbreeding depression by cryptic female choice (Welke & Schneider 2009). Females occupy webs that are visited by roving males and commonly mate with two males (Figure 11.34). Each female contains two independent sperm storage organs, and these spermathecae, one on each side of the body, are filled by two separate matings, either by the same male or by different males. Males inseminate females using their paired pedipalps, which are filled with sperm, but use only one pedipalp per mating attempt. In this species, the right pedipalp of a male is always inserted into the right spermathecae of a female, and the same is true for the left side.

Welke and Schneider took advantage of this reproductive morphology to conduct double mating trials. Females were mated with either (1) two sibling males (SS), (2) two nonsibling males (NN), or (3) a sibling and a nonsibling male (switching which one was first, SN or NS). Each male was matched for size and age but had one pedipalp removed, so the pair of males had complementary pedipalps (one only had the right, and the other only had the left).

Figure 11.34. Orb-web spider. Males visit females' webs to mate with them.

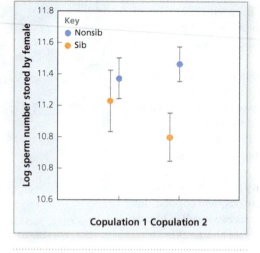

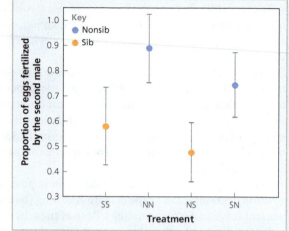

Figure 11.35.   Sperm storage. Mean (± SE) sperm number stored. Females stored more sperm during the second copulation from nonsibling matings (blue) than they did from sibling matings (orange) (*Source:* Welke & Schneider 2009).

Figure 11.36. Fertilization success of second males. The mean (± SE) proportion of eggs fertilized by the second male is lower when he is a sibling than when he is a nonsiblings. S = sibling; N = nonsibling (*Source:* Welke & Schneider 2009).

Subjects were offspring of wild-caught females who had laid egg sacs in the laboratory. Hatchlings from each egg sac were kept in separate groups, so that siblings (offspring from the same egg sac) and nonsiblings (offspring from different egg sacs) could be identified. Male and female hatchlings were separated until they reached adulthood. Females were placed in individual containers, in which they built a web, while males were kept in separate containers.

Males were introduced to the female's container one at a time, allowed up to one hour to copulate, and then removed. After both males copulated, the researchers estimated the number of sperm transferred to each spermathecae from the first and second mating by sacrificing a subset of females, dissecting the spermathecae, and counting spermatozoa under a microscope. To determine paternity for each male in each treatment, the researchers conducted a second experiment, in which they sterilized one of the males by irradiating him to make his sperm inviable. They could then calculate the proportion of eggs fertilized by each of the two males, because they knew which one was sterilized and the eggs his sperm fertilized did not hatch.

Females mated readily with both siblings and nonsiblings, with no signs of inbreeding avoidance prior to copulation. The number of sperm in the spermatheca from the first male mating did not differ based on his relatedness to the female. However, for the second male, females stored more sperm from the nonsibling than they did from the sibling (Figure 11.35). With respect to paternity, the researchers focused on the proportion of eggs fertilized by the second male. They found that the proportion of eggs fertilized by the nonsibling second male was higher as well (75% versus 48%) (Figure 11.36).

These data demonstrate cryptic female choice by orb spiders. When females mate with two males that differ in relatedness, they store more sperm from the nonsibling male and thus bias paternity toward those males. Both results should reduce inbreeding within a brood and illustrate how sexual selection can operate after copulation.

11.5   Mate choice by females favors alternative reproductive tactics in males

Throughout the chapter, we have discussed individuals that all use one reproductive strategy to maximize success, such as winning contests over mates, possessing a high-quality territory, or displaying an exaggerated secondary sexual trait. However, as we have seen, not all males are able to win contests, defend high-quality territories, or display exaggerated traits. What do they do? When males cannot effectively compete, selection will favor the evolution of alternative strategies that enhance reproduction (Gross 1996). The presence of multiple behavioral mating phenotypes in a population are known as **alternative mating tactics** (Taborsky, Oliveira, & Brockmann 2008).

Alternative mating tactics exist in many species and occur as two basic tactics. In the bourgeois tactic, competitive males defend a territory to monopolize its resources for females or else possess traits attractive to females. Typically, the bourgeois males are large, older, and in the best physiological condition. The parasitic tactic, on the other hand, is used by less competitive males to usurp matings from bourgeois males. Several types of parasitic behaviors exist. **Satellite males**, for example, associate with bourgeois males by remaining near them to intercept females that are attracted to the bourgeois individual. In contrast, **sneaker males** attempt to avoid detection so that they can quickly enter a bourgeois territory to fertilize eggs being deposited in a nest.

The evolution of alternative reproductive tactics

Two hypotheses attempt to explain the evolution of alternative reproductive tactics. One hypothesis views such tactics as a **conditional strategy**: individuals of high condition adopt the bourgeois tactic, while those in poor condition adopt the parasitic tactic. In this hypothesis, bourgeois males have higher fitness than parasitic males, and individuals can adopt either tactic depending on their condition. Essentially, parasitic males are "making the best of a bad situation" (Kodric-Brown 1986).

The second hypothesis postulates that the tactics coexist in an **evolutionary stable strategy (ESS)** that is maintained by frequency-dependent selection (see Chapter 6). Here, the fitness of a strategy increases as it becomes increasingly rare. This idea makes sense when we think about the parasitic tactic. These males do not pay the costs associated with the bourgeois tactic of attracting a mate or defending a resource. They can thus parasitize more individuals and obtain high fitness when there are many more bourgeois males in the population than parasites. If most males in a population exhibit the parasitic tactic, they will have low fitness, because there will then be too few bourgeois males to parasitize. This hypothesis predicts equal fitness for the tactics if they coexist as an ESS (Gross 1996). Let's see how these hypotheses have been tested.

Conditional satellite males in tree frogs

Featured Research

Male frogs produce advertisement vocalizations near ponds and streams to attract females. Females prefer to mate with those males that produce the most energetic calls (the loudest and at the highest rate) and calls of the lowest frequency. Typically, only large males can produce the lowest frequency sounds (Gerhardt 1987).

Sarah Humfeld examined male reproductive behavior in green tree frogs (*Hyla cinerea*) in southeastern Missouri (Humfeld 2008). In this population, males exhibit both a bourgeois calling tactic and a parasitic satellite tactic. In the first, they call to attract females. In the second, a satellite male takes up residence near a calling male but remains silent, presumably to intercept females attracted to the caller. Humfeld noticed that some males use both tactics. Could this represent a conditional reproductive strategy? Humfeld's hypothesis made two predictions: (1) males differ in attractiveness to females, and those that produce unattractive calls are more likely to be satellites; and (2) satellite males should exhibit the same call preference as females—that is, both should be attracted to males that exhibit the lowest frequency calls.

To test these predictions, Humfeld characterized the advertisement calls of males in the wild. Each male in the population was assigned a status of either bourgeois calling male or silent satellite. She recorded 20 consecutive calls and then captured, measured, and marked each male. To obtain calls from satellites, she removed the nearest caller, which induced the satellite to begin calling, and then she recorded his calls.

alternative mating tactics Multiple behavioral mating phenotypes in a population.

satellite male An alternative, parasitic mating tactic in which a male remains near a bourgeois male to intercept females that are attracted to the bourgeois male.

sneaker male An alternative male reproductive tactic in which a male attempts to avoid detection so that he can quickly enter a bourgeois territory to fertilize eggs being deposited in a nest.

conditional strategy The use of a particular strategy based on an individual's condition.

evolutionary stable strategy (ESS) A strategy that, if adopted by a population, cannot be trumped by another strategy because it yields the highest fitness.

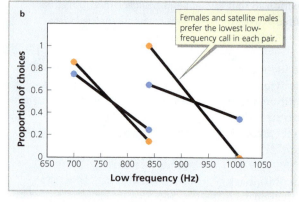

Females and satellite males prefer the lowest low-frequency call in each pair.

Figure 11.37. Female and satellite male choices. (a) Green treefrog. (b) Both females (orange) and satellite males (blue) preferred the call with the lowest low-frequency peak in two experiments that differed in the calls presented. Lines connect the two calls of an experiment (*Source:* Humfeld 2008).

Next, she had to determine female vocal preference. She created pairs of synthesized calls that were identical except for the lowest frequency they reached. One pair had lowest frequencies at 700 Hz and 840 Hz. The other pair had lowest frequencies at 840 Hz and 1120 Hz. Humfeld captured females and placed them individually in a choice arena, covered with a dark cloth and away from all natural populations of calling males. She placed two speakers on either side of the arena and recorded the movement of the female toward one of the speakers, an index of her vocal preference. Finally, to determine how different male calls affect satellite males, Humfeld repeated the first experiment, but this time with individual satellites in the arena.

Humfeld found that bourgeois calling males were larger, in better condition, and produced lower frequency calls than satellites, and so these data support the first prediction: satellites produce less attractive calls. She also found that females and satellites both strongly preferred those male calls with the lowest frequency calls, as predicted (Figure 11.37).

Humfeld concluded that male green tree frogs adopt a conditional strategy. Females prefer large males in good condition, and only these can produce low-frequency calls. Small, perhaps younger males call when away from an attractive male. But because of their size and condition, they cannot produce attractive calls, and so when an attractive male is nearby, they adopt a satellite strategy and remain silent but stay close to a male producing a very attractive call. This behavior should increase their encounters with females until they attain a larger size. It is unclear how satellite males obtain fertilizations, and so the relative fitness of the alternative strategy is unclear. However, the data suggest that if satellite males attempted to attract females by calling alone, they would be unsuccessful.

ESS and sunfish sneaker males

As we just saw, testing hypotheses about the evolution of alternative reproductive tactics is challenging, because it is often difficult to quantify fitness for different mating tactics. For example, some male sunfish, called parental (i.e., a bourgeois tactic), defend nests in breeding territories and provide parental care for the eggs there. Females approach these males and spawn at a nest site within the territory. Parasitic, sneaker males do not defend territories but instead intrude while a parental male is spawning with a female in an attempt to fertilize eggs as they are laid. The advent of molecular techniques can help researchers understand the fitness consequences of these tactics, because they can be used to establish paternity.

Oscar Rios-Cardenas and Michael Webster used molecular techniques to examine the mating behavior and reproductive success of parental and sneaker male pumpkinseed sunfish (*Lepomis gibbosus*) in a small (4 ha) pond at the Huyck Preserve in upstate New York (Rios-Cardenas & Webster 2008). Pumpkinseed sunfish are medium-sized (15 to 20 cm in length) freshwater fish that are common in ponds and streams throughout North America (Figure 11.38). Over a period of three years, the researchers mapped the location of all active nests. When a spawning event occurred, the researchers captured nesting males, females, any sneakers involved in the spawning attempt, and all subsequent developing eggs. Tissue samples were taken from the adults and eggs for genetic analysis of paternity. Males were weighed, and their age was determined based on scale morphology.

The researchers observed 435 active nests, 60 spawning events, and 26 sneaker intrusions. They found that sneaker males were younger and smaller than parentals; most sneaker males were about two years old, while parental males were at least four years of age (Figure 11.39). On average, parental males

Figure 11.38. Pumpkinseed sunfish. These fish are found in all the Great Lakes and are popular with anglers.

sired about 85% of the offspring in their nest, and sneakers sired 15% (Figure 11.40). At first glance, parental males might seem to have higher fitness than sneakers, but sneakers are much rarer in the population than parentals. In fact, the researchers estimated that sneaker males represent only about 15% of the population. Because the reproductive success of sneakers is in proportion to their abundance in the population, the two strategies actually have equal reproductive fitness.

The researchers concluded that in this population, the two strategies are in an ESS. About 15% of individuals mature rapidly at two years of age and adopt the sneaker tactic. The remainder continue to grow and mature at four years of age, when they begin to compete for territories.

Additional work is required to understand how this reproductive decision occurs, but it seems likely that physiology plays a role. Rosemary Knapp and Bryan Neff studied closely related bluegill sunfish (*Lepomis macrochirus*) and found that parental and parasitic males exhibit very different hormone profiles (Knapp & Neff 2007). Parasitic males had higher testosterone and cortisol and lower 11-ketotestosterone levels than parental males (Figure 11.41). This finding suggests that hormones may control the development of reproductive tactics in sunfish.

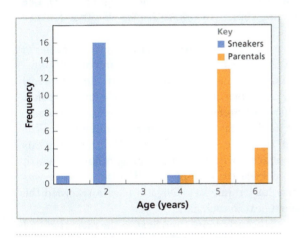

Figure 11.39. Male ages. Relative frequency of sneaker and parental males by age. Sneaker males are younger and smaller than parental males (*Source:* Rios-Cardenas & Webster 2008).

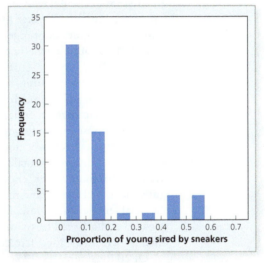

Figure 11.40. Proportion of young sired by sneakers. Frequency of reproductive success by sneakers. Most sneaker males sired less than 10% of eggs in a nest (*Source:* Rios-Cardenas & Webster 2008).

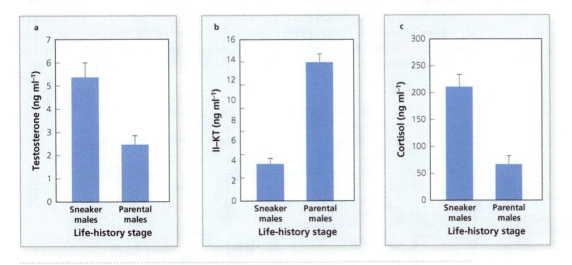

Figure 11.41. Hormone profiles. Mean (+ SE) hormone levels of males. Sneaker and parental males exhibit different levels of (a) testosterone, (b) 11-ketotestosterone, and (c) cortisol (*Source:* Knapp & Neff 2007).

Clearly, mating behavior is complex and can involve a variety of tactics. In the final section, we examine how the decisions of others can also influence mating behavior.

 11.6

Mate choice is affected by the mating decisions of others

mate choice copying A situation in which one individual observes and copies the mating decisions of another individual.

Throughout the chapter, we have assumed that the mate choice decisions of individuals are independent. Yet mating for many species takes place in a social environment, where individuals can observe and copy the mating decisions of others, a behavior known as **mate choice copying**.

Featured Research

Mate copying in guppies

The first experimental evidence of mate choice copying comes from guppies (*Poecilia reticulata*), small freshwater fish native to Trinidad. Recall that males are brightly colored, with spots of orange, as well as black and iridescence (Figure 11.12). Lee Dugatkin examined whether females copied the mate choice of other females (Dugatkin 1992).

To answer this question, he placed a focal female in a clear Plexiglas container in the center of an aquarium. He then placed males, matched for size and overall coloration, into chambers on either side of the aquarium. Dugatkin first allowed a focal female to choose one of the males. In guppies, the amount of time a female spends near a male is a good indicator of her mate preference. Next, he arranged the aquarium so that a different female (the model) was now near the nonchosen male. This was done to simulate her "choosing" this male as a mate (Figure 11.42). During this time, the model female and male interacted with typical courtship behaviors, which the focal female could view. The model was then removed, and the focal female was released and allowed to swim freely for ten minutes. Dugatkin again recorded the amount of time the focal female spent next to each male.

Of 20 focal females, 17 spent significantly more time next to the male that had been near the model female. These females switched their mate choice preference to copy the mate choice of the model female. Why might females copy the mate choice of another female? One possibility is that because guppies are schooling fish, a female might simply prefer to be in the area where she recently observed more fish (the side where she observed the male and model female). Dugatkin tested this explanation by conducting the same experiment but using only females in the end chambers. In this experiment, only ten out of 20 females spent more time at the side that had more fish, which eliminated this "schooling" hypothesis. To further examine mate-copying behavior, Dugatkin reasoned that if females are

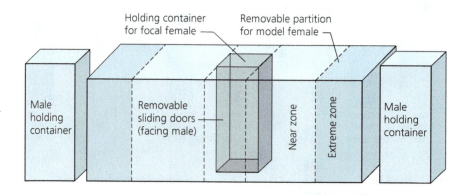

Figure 11.42. Choice test design. The test apparatus had separate containers for males and moveable doors to adjust what the female viewed (*Source:* LaFleur, Lozano, & Sclafani 1997).

really copying the mate choice of another female, they should choose that male even after he has been moved to a different location. So in another experiment with different females, Dugatkin switched the position of the males after the model had been observed next to one of them. In this case, 16 of the 20 females spent more time near the male who had been seen near the model female, even though he was now in a different location. Dugatkin conducted even more experiments to eliminate additional possible explanations (e.g., Dugatkin & Godin 1992) and concluded that female guppies do indeed copy the mating decisions of other females. In this species, mate choice is not always independent; instead, females may copy the mate choice of others they observe.

Mate copying in fruit flies

Featured Research

Mate choice copying has now been documented in many species, but most often in vertebrates (Hoppitt & Laland 2008). Recently, Frédéric Mery and colleagues examined this behavior in fruit flies (*Drosophila melanogaster*) (Mery et al. 2009) (Scientific Process 11.2). The research team gave females a choice of males that were made to differ in quality by raising some males on a rich-food medium that contained 100% of standard nutrients (high-quality males) and others on a poor-food medium that contained only 25% of standard nutrients (low-quality males). They then conducted a three-phase experiment in a small Plexiglas box. In the Pre-test, on Day 1, females were allowed to choose between a high-quality and a low-quality male, and they spent more time with the high-quality male. Next, half the females saw a model female with the high-quality male, and the others saw a model female with the low-quality male. In the Posttest, females were again allowed to choose between the high-quality and low-quality males. Now the behavior of the treatment females differed. Those that saw a model with the low-quality male now spent significantly more time with him—that is, they exhibited mate choice copying perhaps to gather more information about the relative qualities of the two males. No such change was observed in females that saw a model with the high-quality male.

These data reveal the effect of social environment on the mating behavior of fruit flies. Female mating behavior can be affected by observing another female associate with a male.

The benefit of mate copying

Why should nonindependent mate choice evolve? One answer comes from a simple model (Nordell & Valone 1998). If a female has the option of mating with two males that differ in quality, she increases her fitness by mating with the higher quality option. However, it may be difficult for a female to determine which male is of higher quality if the males are similar or if she has little experience discriminating between males. In the absence of other information, she must select one at random and so will have only a 50% chance of making the best decision. Alternatively, she can observe another female and copy *her* mate choice. If the observed female is a better discriminator, the copying female will have a better chance of selecting the better male. If not, she is no worse off.

This model predicts that mate copying should be observed only when discrimination is difficult—that is, when the males are matched for quality. In guppies, for instance, females prefer males with the most orange coloration, and females mate copy when males are matched closely for color. When males are very different in color, female copying is rarely observed (Dugatkin & Godin 1993). In addition, if discrimination ability increases with age and experience, then the model predicts that young females should mate copy more often than older females, a tendency that has been demonstrated in guppies (Dugatkin & Godin 1993).

SCIENTIFIC PROCESS 11.2

Mate copying in fruit flies
RESEARCH QUESTION: *What factors affect the mating behavior of female fruit flies?*

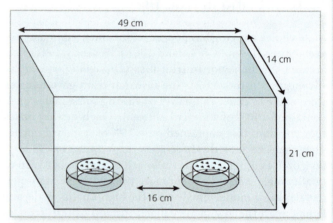

Hypothesis: Male condition will affect the mate choice copying of female fruit flies.

Prediction: Female fruit flies will spend more time with an unpreferred male after observing another female associate with him.

Methods: The researchers:

- created groups of high-quality males (fed nutrient-rich food) and low-quality males (fed nutrient-poor food).
- conducted a female mate choice trial with one high- and one low-quality male (Pretest).
- allowed half the females to then observe a model female next to the high-quality male and the other observed a model female next to the low-quality male.
- repeated the mate choice trials (Posttest).

Figure 1. Experimental arena. Males were restrained under lids that allowed females visual and olfactory information (*Source:* Mery et al. 2009).

Results:

- In the Pretest females spent significantly more time near the high-quality male.
- In the Posttest, females that observed a model female with the low-quality male now spent more time near the low-quality male.

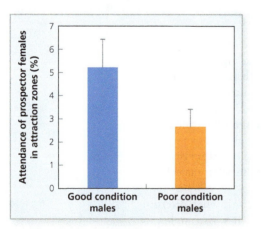

Figure 2. Pretest female behavior. Mean (+ SE) attendance of prospector females. In the pretest, females preferred the high-quality male (*Source:* Mery et al. 2009).

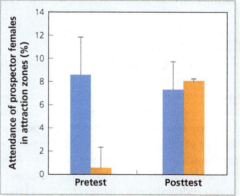

Figure 3. Posttest female behavior. Mean (+ SE) attendance of prospector females. Blue bars: high-quality males; orange bars: low quality males. In the posttest, the attractiveness of the high-quality male did not change. However, the attractiveness of the low-quality male increased significantly (*Source:* Mery et al. 2009).

Conclusion: Female fruit fly mating behavior is affected by observing the mating decisions of other females.

Nonindependent mate choice by male mosquitofish

Now we know that female mate choice is affected by the mating decisions of other females. Is the same true for males? Recall that males compete with one another to fertilize eggs, often by sperm competition inside a female's reproductive tract. We might therefore predict that males should try to avoid mating with females they have recently seen mate.

Bob Wong and Miranda McCarthy asked: Does the risk of sperm competition affect mate choice in male eastern mosquitofish (*Gambusia holbrooki*)? They used a standard two-choice test with three aquaria (Wong & McCarthy 2009) (Figure 11.43). First, Wong and McCarthy examined the preference of focal males for two randomly selected females. In the second stage, the focal male was constrained, and a rival male was placed next to each female. Here, the researchers could control whether the focal male observed these other males. In a final stage, the rivals were removed and the focal male was again allowed to associate with both females.

In Experiment 1, the focal male could not see either rival. In Experiment 2, the focal male saw a rival near the female he preferred in the first phase of the experiment. And in Experiment 3, the focal male saw a rival next to only the female he did not prefer in the first phase.

In Experiments 1 and 3, males preferred the larger female, and their preference was consistent between the two stages (Figure 11.44). However, in Experiment 2, male behavior was affected by observing a rival near the preferred female. When tested again in the final stage, the males now spent less time near their initially preferred female (Figure 11.45). These

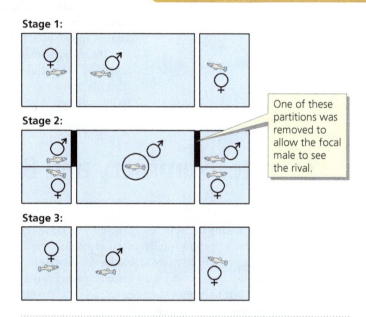

One of these partitions was removed to allow the focal male to see the rival.

Figure 11.43. Male preference experimental setup. Each experiment consisted of three stages. Stage 1 measured male preference; in Stage 2, males saw a rival next to a female; Stage 3 measured male preference again (*Source:* Wong & McCarthy 2009).

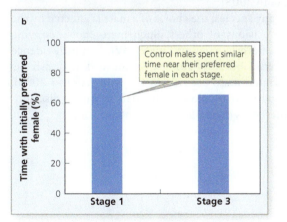

Control males spent similar time near their preferred female in each stage.

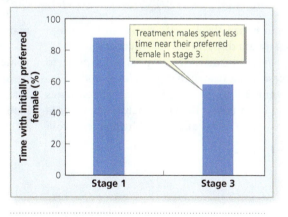

Treatment males spent less time near their preferred female in stage 3.

Figure 11.44. Time in association for control male mosquitofish. (a) Mosquitofish. (b) Median percentage of time males spent in association with an initially preferred female. Males display high repeatability in their initial mate preference (*Source:* Wong & McCarthy 2009).

Figure 11.45. Time in association for treatment males. Median percentage of time males spent in association with an initially preferred female. Males spent less time with their initially preferred female in Stage 3 after observing her with another male (*Source:* Wong & McCarthy 2009).

results show that males have a mating preference for larger females and that their choice of mates is influenced by observing the choice of other males. They will spend less time near a preferred female if they just saw her associate with another male. This change in behavior is more likely to minimize the risk of sperm competition.

These experiments demonstrate that the mate choice of individuals is often affected by the social environment and observations of the mating behavior of others. This phenomenon appears to be widespread, and its implications are just beginning to be explored (e.g., Dubois, Drullion, & Witte 2012).

Chapter Summary and Beyond

The study of mating behavior is organized around the concept of sexual selection—differential reproduction as a result of competition to mate and selective mate choice. Sexual selection predicts that, in general, male fitness increases most strongly with the number of sexual partners, whereas female fitness increases more by selecting mates that provide resources or genetic benefits. Therefore, males will often compete for access to females, and females will be choosy in their selection of mates. It also explains the evolution of elaborate secondary sexual characteristics and courtship displays, and both are still the focus of much research (Jones & Ratterman 2009). Barry and Kokko (2010) summarize why male mate choice is rare in most species, and a recent investigation into Bateman's research prompted significant theoretical critiques of his work, but many of his basic conclusions remain valid (Tang-Martínez & Ryder 2005). Other work focuses on how genes regulate the development of traits in one sex but not the other (Williams et al. 2008) and on the neural activation involved in sexual selection (Hoke, Ryan, & Wilczynski 2010). Although sexual selection is typically used to explain traits within a species, recent work has examined how it can promote speciation as well (Gray & Cade 2000).

Females can benefit from mate choice in two ways. First, they can obtain direct material benefits by selecting males that provide nutrient-rich nuptial gifts prior to mating or access to a territory that provides abundant resources (Gwynne 2008). Second, they can select mates that provide indirect fitness benefits, such as good genes that will be passed on to her offspring. Much recent work has focused on understanding the indirect benefits of mate choice, including female choice based on immune system function and heterozygosity (e.g., Milinski 2006; García-Navas, Ortega, & Sanz 2009). One example is a focus on the major histocompatibility complex (MHC), genes that encode for proteins involved in the immune system response to pathogens (Tregenza & Wedell 2000). Mate choice based on cognitive traits is also attracting attention (Boogert, Fawcett, & Lefebvre 2011).

Sexual selection also occurs after mating. Sperm from different males often compete to fertilize eggs within the female reproductive tract. Female physiology can bias the outcome of such competition, a phenomenon known as cryptic female choice, which can reduce fitness costs associated with inbreeding depression. Recent work has begun to investigate the mechanisms of cryptic female choice (e.g., Luck et al. 2007). When males cannot successfully compete for females, selection will favor alternative mating tactics. Common examples in many taxa are satellites and sneakers, and recent work has attempted to determine whether such tactics yield the same fitness as those used by competitive males (Taborsky, Oliveira, & Brockmann 2008).

In many species, the choice of a mate is affected by the mating decisions of others. Such mate choice copying occurs in both females and males. While most evidence comes from laboratory studies, recent work with fish has demonstrated mate copying in the wild (Alonso 2008). Mate copying may allow individuals to better discriminate differences in quality among potential mates.

CHAPTER QUESTIONS

1. Under what circumstances would you expect sexual selection to be more intense on females than on males?

2. Bateman's (1948) experiment used flies that varied in age from one to six days old. Given the difference in age of maturation of males and females, how might this design have affected his results?

3. In a polygynous mating system, a few dominant males mate with many females and many males do not obtain a mate. In a monogamous mating system, individual males and females often form pair bonds, mate, and raise young together. Which mating system would you predict to have the highest intensity of sexual selection on males, and why?

4. Chuang-Dobbs, Webster, and Holmes (2001) used both observation and experiments to examine the hypothesis that mate guarding facilitates paternity assurance for males. Explain why both methods are important.

5. Crowe et al. (2009) studied sperm competition in tree swallows and found a correlation between copulation frequency and paternity. Design an experiment to test the hypothesis that more frequent copulation helps ensure paternity.

6. You discover a species with alternative reproductive tactics. Large males defend territories to attract females. Small males do not defend territories but instead use a roving tactic: they attempt to copulate with females they encounter while searching for them. In the population studied, half the adult males defend territories and half do not. A genetic analysis of paternity indicates that territorial males sire 80% of all offspring. Is the roving tactic a conditional strategy, or are the two behaviors in an ESS in this population?

7. Imagine two populations of the same species that both have identical numbers of adult males and females. In one population, females frequently exhibit mate choice copying, while in the other, this behavior does not occur. How might this difference in female behavior affect the variance in male reproductive success in these populations?

Chapter 12

Mating Systems

Figure 12.1. Squirrel interaction. A gray squirrel interaction during the mating season.

We have a long walk between the parking lot and the biology building on our campus, and a fair number of animal species live along the way. In spring, animals are occupied with the business of reproduction. It is hard to walk any distance in our region and not see the brilliant red of male northern cardinals (*Cardinalis cardinalis*). A male and female pair up, defend a territory, and raise their offspring together. Along the banks of campus ponds, male red-winged blackbirds (*Agelaius phoeniceus*) display their red epaulets to rivals to defend their territories and their one or two mates. And gray squirrels (*Sciurus carolinensis*) engage in long chases, or mating bouts; males have agonistic interactions with each other and chase females as well (Figure 12.1). Some of these males will mate successfully with several females, and others won't mate at all. All female gray squirrels will mate, often with several different males.

Cardinals, blackbirds, and squirrels differ in the kinds of social associations and number of mates they have during a breeding season, and they represent a few of the mating systems that exist. In this chapter, we review mating system diversity and examine theories to explain such variation. We also discuss how social mating systems that are based on observed behavior differ from genetic mating systems, which are based on reproduction.

12.1 Sexual conflict and environmental conditions affect the evolution of mating systems

▧ **mating system** A description of the social associations and number of sexual partners an individual has during one breeding season.

Animals exhibit **mating systems**—the social associations and number of sexual partners an individual has during one breeding season (Table 12.1). In some mating systems, two individuals live in close association with one another during the breeding season and are said to form a pair bond. However, pair bonds are far from universal, and an individual's number of mates and their types of social association range widely across taxa.

Over the past few decades, our understanding of animal mating systems has changed dramatically. In the past, we assumed that each species had a characteristic mating system. Today, we understand that not just species but also populations and even individuals can exhibit tremendous variation in mating systems. For instance, Sharon Pochron and Patricia Wright studied the reproductive behavior of sifakas (*Propithecus diadema edwardsi*) (Figure 12.2), social primates that lives in Madagascar (Pochron & Wright 2003). Over a period of 15 years, they found roughly equal occurrences of four different mating systems. Although this is an unusual amount of variation within one species, we do often find more than one type of mating system within a species. To understand why, we need to consider the factors that influence the evolution of mating systems. As we see next, these factors involve the reproductive decisions of males and females. Such decisions are affected by their fitness benefits and costs, which in turn vary with environmental conditions.

The evolution of social mating systems

Stephen Emlen and Lewis Oring argued that mating systems can be understood by examining two factors. One is evolutionary: **sexual conflict**, or the differential selection on males and females to maximize their fitness. The other factor is ecological: the ways in which resource limitation and distribution affect the fitness benefits and costs for each sex (Emlen & Oring 1977). We can examine these factors together to understand the variability in mating systems.

▧ **sexual conflict** The differential selection on males and females to maximize their fitness.

Emlen and Oring's model assumes that for a female, fitness will often be more strongly limited by the resources she can obtain to invest in offspring than by the number of her sexual partners (Bateman 1948; Emlen & Oring 1977). This idea follows from the assumption that females typically invest more energy in offspring than males do. First, eggs, because of their relatively large size, should require more energy to produce than sperm. Female birds and reptiles may invest up to 50% of their body mass into production of a single clutch of eggs (e.g., Vitt & Price 1982; Monaghan & Nager 1997). Second, in many species, females expend more energy on parental care (e.g., incubation, feeding, and defense), as we will see in Chapter 13. In addition, most females can obtain sufficient sperm from a single male to fertilize all eggs for

Figure 12.2. Sifaka. This small primate from Madagascar displays a variety of mating systems.

TABLE 12.1 Mating systems. Mating systems differ in the number of mating partners for each sex.

MATING SYSTEM	NUMBER OF FEMALES	NUMBER OF MALES
Monogomy	1	1
Polygyny	More than 1	1
Polyandry	1	More than 1
Polygynandry or plural breeding (includes social associations)	More than 1	More than 1
Promiscuity (no social associations)	More than 1	More than 1

one reproductive event. In contrast, for a male, fitness is most strongly affected by the number of sexual partners he obtains. The more partners, the higher his fitness. This difference sets the stage for sexual conflict.

Emlen and Oring suggest that for a female, selection will favor a mating system that provides the greatest access to resources. This system might be monogamy—one male with one female—if her sexual partner provides high levels of resources to the young, such as high-quality territory and a large quantity of parental care. It might be polyandry—one female with multiple males—if multiple males all provide care to her offspring. In contrast, for males, selection will favor polygyny—one male with multiple females—when males that mate with multiple partners have higher fitness than those that mate with a single female. If sexual conflict does not exist because care from both parents is required to successfully raise offspring, selection will favor social monogamy and biparental care. Emlen and Oring suggest that this system might often be favored in resource-poor environments or when resources are difficult to obtain. Biparental care and monogamy might also be favored when predation on unattended young is high.

When biparental care is not required, each sex benefits when the other provides most of the parental care. How is the sexual conflict resolved? When parental care is female biased, selection will favor the evolution of polygyny. Polygyny is predicted to evolve when environmental conditions lead to the aggregations of females, because such aggregations are more easily defended from rivals.

Two factors can promote female aggregations. First, females may aggregate for reasons other than reproduction. For instance, aggregations of females may experience lower predation risk than solitary females, or a group of females may more successfully avoid harassment from males (e.g., Dada, Pilastro, & Bisazza 2005). In such cases, **female defense polygyny** occurs when males monopolize aggregations of females directly (Figure 12.3). The second factor that can lead to aggregations of females is the distribution of resources in the environment. If the resources required by females are clumped, males can monopolize females by defending territories with large amounts of such resources. **Resource defense polygyny** occurs when males defend territories rich in the resources that are used by and attract multiple females. Males then mate with all females that remain near the resources defended.

Sometimes resources and females may be more uniformly distributed in an environment, rather than clumped. Alternatively, resources may be too unpredictable in space or time, making them costly to defend. If it is too costly for males to successfully defend resources or females, potential mates must seek each other out, which can involve high energetic and predation costs. One way to reduce these costs is for males to settle in fixed locations, called **leks**, and then display to females there (Figure 12.4). Leks have evolved in many taxa, and males typically form dominance hierarchies within the lek, with dominant males occupying the most preferred, central locations. Such males typically mate with several females, a mating system that Emlen and Oring called **male dominance polygyny**. Males on leks can benefit as a result of reduced predation risk, and such aggregations may be more attractive to females (Alatalo et al. 1992), which will increase encounters with them. Females, too, can benefit from the existence of leks, because they reduce the time required to search for a mate: females can quickly assess the relative quality of different males on a lek to find the best option.

In polygynous mating systems, there will be a large amount of variation in male mating success. Males that can successfully defend females or resources will obtain many matings, and males that cannot will obtain few. In contrast, there should be much less variation in female mating success, because most females will mate. Such knowledge about mating systems can provide important information for conservation programs (Applying the Concepts 12.1).

When might polyandry evolve? Emlen and Oring suggest that polyandry can evolve when it is advantageous to both sexes that females be freed from providing

Figure 12.3. Female defense polygyny. A male (a) impala (*Aepyceros melampus*) and (b) northern fur seal (*Callorhinus ursinus*) defend their harem of females.

Figure 12.4. Lek. A male sage grouse (*Centrocercus urophasianus*) displays on a lek. Note another displaying male in the background.

female defense polygyny A single male monopolizes and mates with two or more females.

resource defense polygyny A male defends resources and mates with multiple females attracted to the resources.

lek A location where males aggregate and display to females.

male dominance polygyny In a lek system, only a few males mate with most females.

APPLYING THE CONCEPTS 12.1

Mating systems and conservation translocation programs

Many species have suffered great population declines and now face an increased probability of extinction. One management strategy to reduce extinction risk is to add individuals to a small population or create new populations by translocating individuals from one place to another. Animals are moved from either large captive or wild populations, but translocations are expensive, and many have failed as a result of lack of knowledge about a species' biology or ecology.

A study by Dominique Sigg, Anne Goldizen, and Anthony Pople illustrates the importance of understanding mating systems in carrying out a successful translocation program (Sigg, Godizen, & Pople 2005). The researchers studied the bridled nailtail wallaby (*Onychogalea fraenata*), a marsupial whose population and geographic range have declined dramatically over the past century. Over a period of three years, a total of 133 uniquely marked animals were translocated to establish a new population. Both captive- and wild-born animals were moved, and an equal number of males and females were released. The researchers collected tissue samples from these adults. Four times each year, animals in the new population were captured, identified, and weighed, and tissue samples were collected from offspring to determine their parents.

Bridled nailtail wallabies are polygynous, and males compete with each other for access to receptive females. Not surprisingly, only about 25% of males bred successfully, whereas almost all of the females did. Male mating success was not affected by origin (captivity or the wild), but body size did strongly affect success. Males larger than 5800 g were six times more likely to sire offspring than were smaller males (Figure 1). The researchers suggest that in polygynous species such as this, it is inefficient to introduce equal numbers of males and females into a new environment, as is often done. Instead, they recommend introducing a greater number of females, because most males will not breed successfully. Furthermore, they suggest that only large males should be translocated, because small males have little chance to mate. These changes could both reduce the costs of translocation and increase the success of a program.

Figure 1. Wallaby reproductive success. (a) Male wallaby. (b) Males over 5800 g had the highest reproductive success (*Source:* Sigg, Goldizen, & Pople 2005).

parental care, making parental care male biased. One possible situation that might produce a polyandrous system is when there is very high predation on offspring. If many offspring are lost to predators, both sexes can benefit if a female can quickly reproduce again. However, recall that egg production is costly. If a male provides more care, a female can provide less care and thus feed more to quickly replenish her energy stores for reproduction. Once males provide high levels of care and females are able to quickly lay multiple clutches, selection can favor females that mate with multiple males and lay a clutch of eggs that is then tended by each one.

Polygynandry and promiscuity both involve multiple mating partners for each sex but differ in regards to social associations. There are social associations in polygynandry, but not in promiscuity. Polygynandry, or plural breeding, is rare but does occur in some social species, and particularly in mammals that defend a territory from other such groups. For example, African lions (*Pathera leo*) live in social groups, which facilitate hunting success (Figure 12.5). A pride consists of two or more adult

Figure 12.5. African lions. Multiple male lions defend and mate with multiple females.

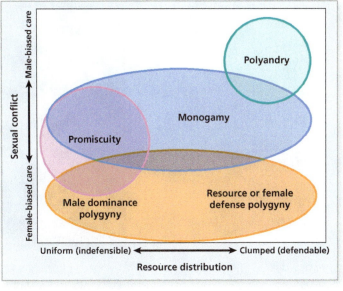

Figure 12.6. Emlen and Oring model. Mating systems are the result of the interaction between sexual conflict and resource distribution.

males, often relatives, and up to nine adult females. Males defend their females and multiple males mate with multiple females (Packer et al. 1991).

Promiscuity, on the other hand, should evolve when the benefits of social living are low. Both females and males are then solitary, and pair-bond formation provides no fitness benefits to either sex. Conditions also favor its evolution when the defense of mates or resources is uneconomical, as might occur when population density is high or when it is too costly for males to aggregate on leks, as in high predation risk conditions. Such a system might also occur when individuals cannot afford to forgo feeding to spend time on a lek.

In sum, mating system variation can be understood by examining (1) the competing interests of the sexes in attempting to maximize their fitness and (2) how environmental conditions affect the benefits and costs of resource defense and of different mating behaviors for each sex (Figure 12.6).

Clearly, mating systems and parental care behavior have co-evolved. In this chapter, we'll see how the model proposed by Emlen and Oring helps to explain the evolution of, and variation in, social mating systems. We cover parental care in the next chapter.

Mating systems in reed warblers

Featured Research

One way to test the Emlen and Oring model is to use the comparative approach by examining a group of related species that exhibit different mating systems. The model allows us to predict, for example, how variation in resource abundance should affect a mating system. In resource-poor environments, biparental care will often be required to successfully raise offspring, and so monogamy should be favored. When resources are more abundant, biparental care will be less essential, and so polygyny or perhaps promiscuity should be favored.

Bernd Leisler, Hans Winkler, and Michael Wink examined these predictions for the evolution of mating systems in 17 species of Acrocephaline reed warblers (Leisler, Winkler, & Wink 2002). These insectivorous birds are widely distributed throughout the Old World in a variety of marshy and shrubby habitats. All female reed warblers provide more parental care than males through behaviors such as egg incubation and food delivery to young, but the level of male care varies across species. Most species are monogamous, but several are polygynous, and one exhibits promiscuity (Figure 12.7). Leisler and colleagues used the comparative method to address two research questions. First, how does habitat quality—defined as the amount of food resources—correlate with mating systems? Second, how are habitat quality and mating system related to the level of male care?

Figure 12.7. Reed warbler phylogeny and mating system. (a) *Acrocephalus schoenobaenus* (polygynous). (b) *A. palustris* (monogamous). (c) Phylogeny and mating system. Only four species display polygyny (triangles), indicating that it is probably not the ancestral condition. Because polygyny is found in two separate groups, it appears to have evolved independently at least twice. The majority of species are monogamous (circles), indicating that it is likely the ancestral condition. The square indicates a single case of promiscuity. Gray: medium habitat; pink: good habitat (*Source:* Leisler, Winkler, & Wink 2002).

The researchers needed several types of information to answer these questions: the quality of habitat in which each species occurs, each species' level of male care, and a phylogeny of all the species. The research team used published data to estimate habitat quality for each species. They used data on the food biomass and prey size available to categorize habitat quality as poor (only small prey size), medium (larger prey size), and good (many large prey). Poor habitats tended to have dense stands of reeds, bushes, or trees that prohibited light penetration. Medium habitats had more light, and good habitats had abundant light and were highly productive, containing many large prey. The researchers used previously published studies and

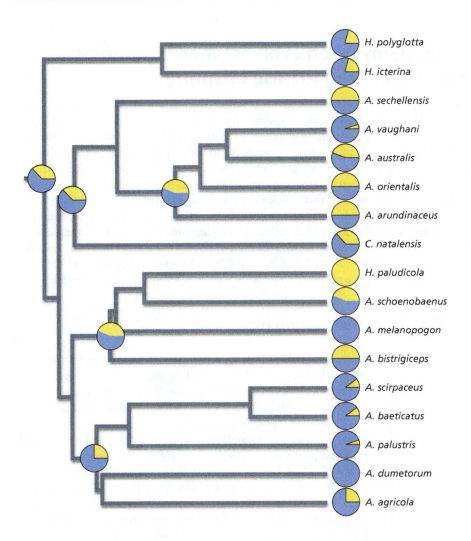

Figure 12.8. Reed warbler phylogeny and paternal care. High levels of paternal care (amount of blue in circle) appear to be ancestral, with reduced male care (less blue in circle) having evolved several times (*Source:* Leisler, Winkler, & Wink 2002).

characterized the relative level of paternal care as full (equal to that of the female), reduced, or none. Finally, they created a molecular phylogeny using previously published sequences of the cytochrome-b gene.

They found a strong association between habitat quality and mating system across the 17 reed warbler species they studied. Monogamy and high levels of male care were predominant in poor-quality habitats, whereas both polygyny and promiscuity and reduced levels of male care were associated with medium- and good-quality habitats. When behavior was mapped onto the phylogeny, two patterns emerged. First, polygyny appears to have evolved independently in this group at least twice. Second, the phylogeny suggests that the ancestral species was monogamous, with a high level of male care (Figures 12.7 and 12.8). Leisler and colleagues' research suggests that in this group of species, male care co-evolved with mating system and was reduced or absent in habitats of higher quality that contained more abundant resources. This analysis supports the ideas of Emlen and Oring. In their model, habitats with limited resources should favor social monogamy and high levels of male care. Changes to higher quality habitats are associated with reduced levels of parental care and with polygyny and promiscuity. Leisler, Winkler, and Wink's work represents one approach to understanding the evolution of mating systems. An alternative is to examine the mating systems of individual species in light of the model proposed by Emlen and Oring. Such work often involves experimental manipulations and tests of different predictions for different mating systems. We now examine each mating system in turn.

Monogamy often evolves when biparental care is required to raise offspring

Monogamy is rare in mammals, an observation consistent with the Emlen and Oring model. That is because for most species, female lactation—the secretion of highly nutritious milk from mammary glands—alone can provide all the food needed to raise offspring successfully. Monogamy does occur in some species, such as the California mouse (*Peromyscus californicus*), which is found throughout central and southern California and northern Mexico (Figure 10.24). Adults form exclusive pair bonds, and genetic data indicate that this species does not mate outside the pair bond (Ribble 1991). The pair defends a territory, and males assist with all aspects of parental care except lactation. This assistance includes huddling (maintaining close body contact with pups in order to warm them), grooming the young, and carrying them from one location to another (Gubernick & Alberts 1987). The Emlen and Oring model predicts that monogamy will be observed in situations when biparental care is essential.

Featured Research

California mouse monogamy

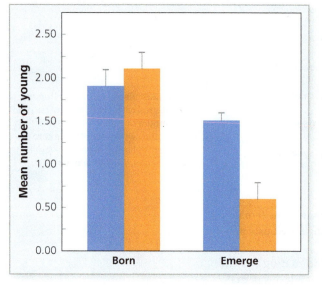

Figure 12.9. Offspring survival. Mean (+ SE) number of young. Families with fathers (blue) had more offspring survive (i.e., emerge from the nest) than families without fathers (orange) (*Source:* Gubernick & Teferi 2000).

David Gubernick and Taye Teferi were curious about how male parental care affected reproductive success in California mice (Gubernick & Teferi 2000). They conducted an experiment at the Hastings Natural History Reservation in California. All adults were captured regularly with live traps and uniquely marked with numbered ear tags. Females were checked for reproductive condition (pregnant or lactating) and weighed. The researchers compared the number of young that survived and emerged from "male-absent" families with those from control "male-present" families. In male-absent families, males were removed within three days of the birth of their first litter of the season. In the control treatment, males were trapped and handled but then released and allowed to return to their families. The researchers estimated the number of young born to each female by comparing her weight loss between the last days of pregnancy and the time of first capture after giving birth. A standard formula was used to convert body mass loss to offspring number. After giving birth, each female was trapped and dusted with one of six uniquely colored fluorescent pigment powders that were transferred to her young and mate in the nest. All individuals subsequently captured were checked with ultraviolet light for the presence of fluorescent pigment to determine the pair-bonded male and the female's offspring (Ribble 1991).

There was no difference in the number of offspring born to females that had their male removed and the number of offspring born to those that did not. However, significantly more young survived from the male-present families than did from the male-absent families (Figure 12.9). The absence of the father and his parental care significantly reduced offspring survival. Although males cannot feed offspring, they can assist with thermoregulation and protection. The researchers concluded that the need for biparental care has favored the evolution of monogamy in this species. It is not clear why this rodent is unusual in its requirement for biparental care.

Featured Research

Monogamy and biparental care in poison frogs

A second example of the role biparental care plays in the evolution of monogamy comes from tropical frogs. Jason Brown, Victor Morales, and Kyle Summers examined

SCIENTIFIC PROCESS 12.1

Biparental care and monogamy in poison frogs

RESEARCH QUESTION: *Why does* R. imitator *have a monogamous mating system while* R. variabilis *is promiscuous?*

Hypothesis: When resources are limited, monogamy will evolve because biparental care is required to successfully raise offspring.

Prediction: Tadpoles in resource-limited small pools will exhibit high survivorship only with biparental care.

Figure 1. Poison frogs with different mating systems.
(a) *R. imitator* is monogamous.
(b) *R. variabilis* is promiscuous.

Methods: The researchers:

- manipulated phytotelma pool size. Individual tadpoles of one species were placed in either small or large pools and did not receive female feeding. In the control, *R. imitator* individuals were placed in a small pool and received parental care in the form of female feeding.
- visited each pool weekly to record size (growth rate) and mortality of individuals.

Results:

- For both species, individuals in large pools had high growth rates and high survivorship.
- In small pools with no parental care, individuals of both species had low growth rates and low survivorship.
- *R. imitator* controls in small pools with biparental care had high growth rates and high survivorship.

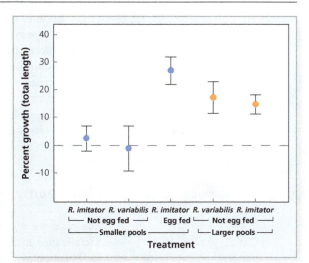

Figure 2. Growth rate. Mean (± SE) percent growth. In the small pool treatments (blue), only eggs that were fed exhibit high growth. All eggs in large pools (orange) exhibit high growth rates even though they were not fed (*Source:* Brown, Morales, & Summers 2010).

Conclusion: The need for biparental care in *R. imitator* likely provided strong selection that favors monogamy in this species.

the mating system of two closely related species of poison frogs, *Ranitomeya imitator* and *R. variabilis*, in Peru (Brown, Morales, & Summers 2010) (Scientific Process 12.1). Both of these species rear their young in a phytotelma, a small pool of water that collects at the base of leaves or petals in a plant. The small size of these pools reduces predation, but it also limits the quantity of nutrients and other resources available for the developing offspring. *R. imitator* rear their tadpoles in very small pools, averaging 24 mL in volume, while *R. variabilis* prefers larger pools, averaging 112 mL in volume.

Previous work suggested that *R. imitator* was monogamous with biparental care (Brown et al. 2008). Females lay from one to three eggs in dense foliage. Males fertilize the eggs, carry each hatchling to a phytotelma for development, and provide protection. Females deposit trophic (unfertilized) eggs in the phytotelma to nourish the young.

In contrast, *R. variabilis* are promiscuous and raise their young with uniparental male care. Females lay eggs fertilized by a male, who then carries the young to a large phytotelma. Males provide parental care, but this care involves only defense; offspring are not fed by the female (Brown et al. 2008).

Brown and colleagues examined whether an ecological factor, the size of the breeding pool, is associated with the evolution of parental care. Do larger pools provide enough resources for uniparental care, whereas smaller pools require biparental care? To examine the importance of parental care for these species, the researchers conducted a reciprocal transplant experiment. They placed tadpoles of either species into either a large or a small natural phytotelma, with just one species per pool of water. Large pools contained more than twice the water volume of small pools, and the researchers used filtered rainwater to maintain water levels throughout the experiment. They placed a mesh screen over the top of the phytotelma to prohibit the addition of resources. None of the experimental tadpoles received parental feeding. In the control, *R. imitator* tadpoles were placed in small pools and did receive parental feeding. The researchers surveyed the pools each week and measured and weighed each tadpole at the start and end of the experiment (after 21 days).

Tadpole growth rate differed significantly across treatments. Individuals of both species in large pools had higher growth rates than those in small pools that were not fed. However, control *R. imitator* tadpoles in small pools that were fed also had high growth rates. The only mortality observed occurred in small pools where the tadpoles received no feeding: four of the eight *R. variabilis* and three of the eight *R. imitator* in this treatment died during the experiment.

In this species, offspring growth rate strongly affects survivorship, which demonstrates the benefit of biparental care for *R. imitator* offspring raised in a small phytotelma. Because they rear offspring in small pools with limited resources, *R. imitator* adults must provide biparental care and feeding to their offspring in order for them to develop successfully. The researchers concluded that this need has favored the evolution of monogamy, a mating system that is very rare in amphibians. But why don't *R. imitator* lay their eggs in large phytotelma, where there would be no need for biparental care? One possibility is that *R. imitator* is a poor competitor, and competition for large phytotelma may be intense. By laying eggs in a small phytotelma, *R. imitator* faces less competition from other species, but doing so requires increased parental investment in the form of biparental care.

Featured Research

Monogamy without biparental care: snapping shrimp

The need for biparental care is thought to explain the overwhelming majority of socially monogamous species. Yet social monogamy does occur in some species that lack biparental care, including coral reef fishes, some ungulates and rodents, and several species of crustaceans (Mathews 2002a). Two hypotheses, which are not mutually exclusive, have been proposed to explain social monogamy in such species. The **territorial cooperation hypothesis** states that because two individuals can better defend a critical resource, such as a safe refuge, selection may favor pair formation and shared defense. The **mate guarding hypothesis** states that selection may favor males that guard females by remaining in close association over the course of one or more reproductive cycles. This behavior allows a male to monopolize a single female, but doing so also prevents him from seeking other mates. Both hypotheses make assumptions about the environment. In the first, competition for limited territories is assumed to be intense or predation high outside a safe refuge within the territory. In the second, a male's encounter rate with females is assumed to be low, perhaps because females are rare or difficult to locate.

Lauren Mathews tested both hypotheses in the snapping shrimp (*Alpheus angulatus*) in a series of studies (Mathews 2002a; Mathews 2002b; Mathews 2003). Snapping shrimp are small crustaceans (less than 5 cm in length) that live in burrows in intertidal rubble habitats. Most individuals live in male-female pairs within a single burrow (Mathews 2002a). However, the female provides all parental care: she broods the eggs on specialized abdominal appendages and then carries, cleans, and aerates the eggs during their development. Both males and females are territorial and will

territorial cooperation hypothesis Two individuals (one of each sex) can better defend a critical resource, such as a safe refuge, and this can select for pair formation and shared defense.

mate guarding hypothesis Selection favors males that mate with and guard one female over one or more reproductive cycles by remaining in close association with her.

construct and defend a burrow that provides protection from predators. Females are sexually receptive only for a short period of time after molting. Mate guarding by a male can thus enhance his fitness, particularly if a male can accurately assess time to molting. After all, a female close to molting should be of high quality.

Mathews tested a prediction of each hypothesis. The territorial defense hypothesis predicts that pairs of shrimp should defend a burrow more successfully than a solitary individual, and the mate guarding hypothesis predicts that males should tend to guard females that are closer to sexual receptivity.

To test the first prediction, she collected shrimp from a rubble intertidal habitat in Florida and housed them in the lab in aquaria (Mathews 2002b). She created four treatments with uniquely marked individuals identified as residents or intruders. Residents (either a single or male-female pair) were placed into aquaria for 24 hours to construct a burrow. Then a single intruder (either female or male) was placed into the tank, and all agonistic interactions were recorded for 30 minutes. After 24 hours, she determined the location of all shrimp: any shrimp inside the burrow was deemed a "winner," while those outside were "losers."

Females in pairs were winners in 82% of trials, whereas single females won only 57% of trials (Table 12.2). Males in pairs were winners more often than single males, but the difference was not statistically significant. These results indicate that pairs are more effective at territory defense than are single individuals.

Next, Mathews tested the prediction of the second hypothesis, on mate guarding and sexual receptivity. She used the same basic experimental design, introducing a single male intruder to a male-female resident pair in which the female was either close to sexual receptivity (pre-molt, a high-value female) or far from sexual receptivity (post-molt, a low-value female).

Mathews observed that males paired with a high-value female won 72% of their contests, whereas males paired with a low-value female won only 34% of their contests. This finding suggests that males can assess female condition and modify their territorial defense behavior according to female molting state, in line with the mate guarding hypothesis.

To more fully examine this hypothesis, Mathews tested whether males could determine female sexual receptivity (Mathews 2003) (Figure 12.10). In this experiment, a water source was attached to each arm of a Y-maze. One water source (the treatment) contained a shrimp, and the other did not (untreated water). She used seven different treatments, varying the sex of the shrimp and its molting state. She tested water with pre-molt, inter-molt, and post-molt males and females and a control (untreated water). Mathews recorded the behavior of test males as water flowed through the Y-maze. She found that males only moved toward the pre-molt female water, indicating that they are able to determine, and are attracted to, females close to sexually receptivity.

In a follow-up experiment, Mathews placed two females in a test aquarium and allowed them to construct a burrow. One was a pre-molt female and the other

TABLE 12.2 Resident effects. Females in pairs won significantly more fights than single females. A similar but weaker pattern was observed for males (*Source: Mathews 2002b*).

RESIDENT	INTRUDER	PERCENT WON BY RESIDENT
Female in pair	Single female	82
Single female	Single female	57
Male in pair	Single male	53
Single male	Single male	38

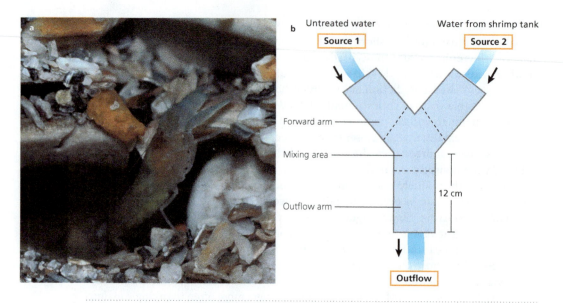

Figure 12.10. Y-maze design. (a) Snapping shrimp. (b) Y-maze used to study snapping shrimp behavior. One source tank contained water from a shrimp and the other contained untreated water. A male was placed in the outflow arm and its movement was recorded to determine if it exhibited a preference for one forward arm or the other. Solid lines show direction of water flow (*Source:* Mathews 2003).

a post-molt female brooding eggs. She then placed a single male in the aquarium. After 24 hours, over twice as many males had paired with the pre-molt female than had paired with the post-molt female.

Mathews concluded that both territorial cooperation and mate guarding have favored social monogamy in snapping shrimp. Paired individuals, especially females, can more successfully defend territories. Males can assess female condition and thus tend to guard females close to sexual receptivity. Mathews does point out that because of the experimental design, she is not able to rule out the possibility that female behavior could influence male pairing decisions as well as the effect of chemical cues, and that more work is needed to investigate these possibilities.

The previous three examples reveal conditions that favor the evolution of monogamy. Next, we turn our attention to polygamy.

12.3 Polygyny and polyandry evolve when one sex can defend multiple mates or the resources they seek

Because females tend to invest more in parental care than males, most examples of polygamy involve polygyny. Much work has centered on determining how males defend and mate with multiple females. According to the Emlen and Oring model, they can do so either directly, by defending these females, or indirectly, by defending resources. Let's see how researchers have distinguished these alternatives.

Featured Research

Female defense polygyny in horses

Feral horses (*Equus caballas*) are found throughout the world in grassland habitats. They live year-round in social units called bands. A typical band consists of one male and up to a dozen adult females. If there are multiple males, a single dominant male obtains most or all of the matings within the band, and so horses exhibit social

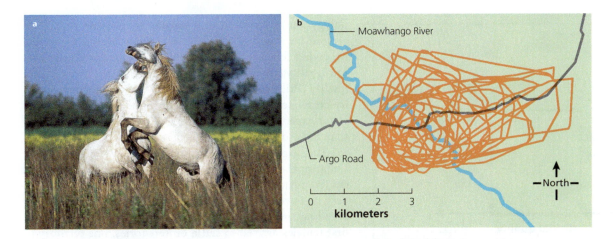

Figure 12.11. Horse bands. (a) Feral horse aggressive interaction. (b) Bands use similar areas. Here, the area used by 28 bands is plotted as separate polygons, which overlap greatly (*Source:* Linklater et al. 2000).

polygyny. Adult males that do not associate with females in bands live in bachelor groups. How can we distinguish between female defense and resource defense polygyny? To do so, we need to determine whether males defend females per se, or whether they simply defend the resources they use.

Wayne Linklater and colleagues studied a large population of horses in New Zealand over a three-year period (Linklater et al. 1999; Linklater et al. 2000). Each month, the researchers traveled along regular routes throughout the study site to record the location of bands. Bands were identified by natural markings or by freeze brands previously applied to their coat. The research team plotted the home range of each band as a convex polygon using these sightings. In addition, they recorded social behaviors, particularly aggressive interactions between males from different bands.

They found that males aggressively defended their band from other males whenever two bands met or when bachelor males were nearby. During the study, bands overlapped greatly in their home range (Figure 12.11). In fact, all bands essentially used the same area, although at different times, and there was no indication of exclusive defense of territories. This finding suggests that males actively defend females and not resources. In other words, they display female defense polygyny.

Resource defense polygyny in blackbirds

Featured Research

In horses, males defend aggregations of females directly. In other species, males defend an area of resources used by females and so can mate with multiple females. If access to resources affects female reproductive success, the **polygyny threshold model** (Verner 1964; Verner & Wilson 1966; Orians 1969) predicts that females should mate polygynously only when the benefits of doing so—access to greater resources—outweigh the cost of sharing resources with other females. Of course, males differ in their ability to defend resources. Some will defend territories rich enough in resources to meet the requirements of multiple females. Others will only be able to defend territories with lower levels of resources—enough, for example, to fulfill the requirements of just a single female. This variability means that females can mate monogamously on a resource-poor territory or can mate polygynously on a resource-rich territory. Females should select the option that leads to higher fitness (Figure 12.12). The model predicts that only males defending resource-rich territories will mate polygynously.

Stanislav Pribil and William Searcy tested the model using red-winged blackbirds (*Agelaius phoeniceus*), a species known to mate both monogamously and

polygyny threshold model
A model that predicts the occurrence of polygyny based on the amount of resources available to females in male territories.

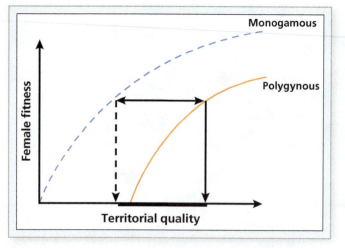

Figure 12.12. Polygyny threshold model. The fitness of a female is plotted as a function of territory quality for two options: mating monogamously or mating polygynously. To benefit by mating polygynously, territory quality must be substantially higher (thick line on x axis).

Figure 12.13. Nesting blackbird. Females nest in dense vegetation around ponds.

polygynously. Males defend territories in dense vegetation around lakes and ponds. The researchers tested the model by giving females a choice of mating monogamously in a low-quality territory or polygynously in a territory of higher quality (Pribil & Searcy 2001). In this species, territory quality correlates with nest placement: nests in vegetation over water experience less predation and so are of higher quality than those on land.

The experiment was conducted at three marshes in Ontario, Canada. At each marsh, the researchers identified similar adjacent territories, or territory dyads, at the beginning of the reproductive season, when males migrated back to breeding sites and started defending territories. When the first females arrived to settle on territories, one territory in each dyad was randomly chosen to be high quality, and the other was chosen to be low quality. Territory quality was manipulated by adding nesting platforms with cattail (*Typha latifolia*) shoots, which created nesting sites for the blackbirds. On the high-quality territory, the platforms were placed over open water, whereas in the low-quality territory, they were placed on land. All females on the low-quality territory were removed, but one female was allowed to remain on the high-quality territory. This set-up created mating options on adjacent territories for newly arriving females. Because the high-quality territory already contained both a resident male and a female, any newly arriving female would mate polygynously if she settled there. In contrast, the low-quality territory contained only a resident male, which ensured that an arriving female would mate monogamously if she chose to settle there. Twice each day, the experimenters recorded the presence of newly settled females (Figure 12.13) and the behavior of the residents.

In 12 of the 14 dyads, newly arriving females settled first on the high-quality territory. All of these females began building a nest in the platforms, and ten eventually laid a clutch of eggs. The two females that settled first on the low-quality territory also began nest building. However, one subsequently abandoned her nest and the other laid a clutch that was lost to predators, demonstrating the territory's low quality. Territory quality clearly affected the mating decisions of females. Females preferred to mate polygynously on high-quality territories over monogamously on low-quality territories. These results support the polygyny threshold model, because only males defending high-quality territories mated polygynously.

Previous work in this system allowed the research team to calculate the fitness benefits and costs for females settling on the experimental plots (Pribil 2000). Nesting over water rather than land had the benefit of reduced predation—a benefit estimated at approximately one additional offspring raised. Mating polygynously also meant less food provisioning by the male, which reduced the number of offspring that a female can successfully raise by about 0.6 (Pribil 2000). Thus, the net benefit of mating polygynously on the experimental plots was about 0.4 offspring. For females, the benefits of polygyny outweighed the costs, and the newly arriving females were indeed selecting the best option.

These findings offer strong support for the polygyny threshold model. They show how resource distribution—in this case, the availability of high-quality nesting sites—and the ability of males to defend these resources can affect mating systems. This model has been applied to many taxa, including invertebrates, as we see next.

Resource defense polygyny in carrion beetles

For carrion beetles, the resource defended by a male is very apparent: a vertebrate carcass, or carrion. Carrion beetles are large insects (2 to 4 cm in length) that live in a variety of habitats around the world. They all use small vertebrate carcasses as a food source for their larvae, but these carcasses are rare and represent a highly clumped resource. Females lay eggs on or next to the carcass, and after hatching, the larvae feed on the carcass (Figure 12.14). Previous research documented male-male aggression but not female-female aggression over carrion (Suzuki, Nagano, & Trumbo 2005). Does this variation in aggression among the sexes indicate that these beetles exhibit resource defense polygyny?

Seizi Suzuki and colleagues studied *Ptomaucopus morio*, a carrion beetle found in Japan (Suzuki, Nagano, & Kobayashi 2006). They placed four virgins (two males and two females) in an arena with a piece of meat. Within each sex, individuals differed in size. Large males were dominant to small males, won fights, and actively defended the resource. Females moved throughout the arena, were observed mating with both males, and oviposited eggs on the meat. After a few weeks, the developing larvae crawled away from the meat to pupate. Subsequently, the researchers used DNA fingerprinting to determine the parent of each offspring. Large and small females had approximately equal reproductive success, but large males sired three times more offspring than small males (Figure 12.15).

The researchers concluded that *P. morio* exhibit a resource defense polygyny mating system. Large, dominant males defend carcasses to keep smaller males at bay. Females do not exhibit aggression and share access to a carcass. By staying near the carcass, the dominant males obtain multiple matings and sire most of the offspring of visiting females, just as predicted by the polygyny threshold model.

Male dominance polygyny: the evolution of leks—hotspots or hotshots?

In horses, blackbirds, and carrion beetles, polygyny results from the defense of either females or resources that are spatially aggregated. However, in many circumstances, resources and females are more uniformly distributed in an environment. It is then too costly for males to defend females or resources, because both are so widely dispersed. In these cases, the Emlen and Oring model predicts male dominance polygyny. Males will compete with one another to establish a dominance hierarchy, often on a lek, and females will mate with high-ranking males. A few high-ranking individuals then obtain most of the matings, while many others have low reproductive success. So why do low-ranking males decide to aggregate with more attractive males? In other words, how do low-ranking males benefit by lekking?

Two hypotheses explain the evolution of male aggregation into leks. The **hotspot hypothesis** predicts that all males should aggregate where they are likely to encounter many females (Emlen & Oring 1977; Bradbury, Gibson, & Tsai 1986), such as near food resources. Settling in such locations increases encounters with females for all males, even those of low rank. In contrast, the **hotshot hypothesis** states that low-ranking males should aggregate around high-ranking males, because females are more likely to visit attractive males (Beehler & Foster 1988). In both cases,

Figure 12.14. Carrion beetle and carcass. Carrion beetles use vertebrate carcasses for reproduction.

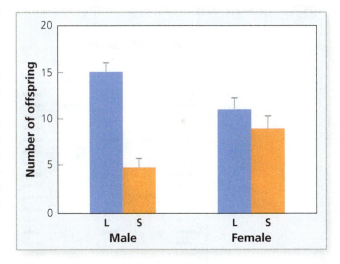

Figure 12.15. Carrion beetle reproductive success. Mean (+ SE) number of offspring. Large males (L) who defended the carcass had three times more reproductive success than did small males (S). In contrast, both the large and small female had similar reproductive success (*Source:* Suzuki, Nagano, & Kobayashi 2006).

hotspot hypothesis All males can benefit by aggregating in a location where they are likely to encounter many females.

hotshot hypothesis Low-ranking males can benefit by aggregating around high-ranking males because females are more likely to visit attractive males.

TOOLBOX 12.1

DNA fingerprinting

DNA fingerprinting is the use of molecular techniques to assess relatedness (Queller, Strassman, & Hughes 1993). One common method for determining paternity uses microsatellites—regions of noncoding DNA that consist of multiple repeats of very short nucleotide sequences. For example, one microsatellite locus might consist of the sequence adenine-cytosine-cytosine repeated multiple times, which we indicate as $(ACC)_n$. Here, n represents the number of times the sequence is repeated. For most microsatellites, individuals are often heterozygous, and there is tremendous variation in the number of repeats in a population. For instance, for a socially monogamous pair, the male's genotype might be $(ACC)_8$ and $(ACC)_{28}$, while his mate might be $(ACC)_{12}$ and $(ACC)_{20}$.

Microsatellite loci often exist adjacent to known DNA sequences, and standard techniques exist to identify both the microsatellite locus and the adjacent DNA. The extracted DNA can then be amplified using the polymerase chain reaction (PCR) to generate large quantities of the DNA sequence for analysis from a small sample of tissue. The amplified DNA is placed in a polyacrylamide gel that is exposed to an electric current. DNA fragments of different lengths move at varying speeds in the gel, depending on their size, and can be identified by their final location or band pattern. Assessing paternity requires examining the microsatellite fragment lengths from both offspring and parents. In practice, researchers often examine many microsatellite loci for greater certainty about parentage. At each locus, individuals will contain one allele from the mother and one from the father.

In our example, an offspring of the parents noted previously might have a genotype that is $(ACC)_8$ and $(ACC)_{12}$. To determine when an offspring is the result of an extra-pair partner, we simply identify when the social parents could not produce its genotype. For instance, an offspring of the previous parents whose genotype is $(ACC)_4$ and $(ACC)_{20}$ is the result of an extra-pair mating (Figure 1), because neither social parent possesses the microsatellite locus $(ACC)_4$.

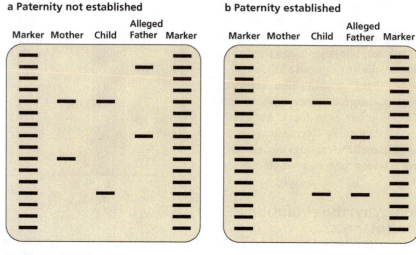

Figure 1. DNA fingerprinting. This technique involves comparing microsatellite bands from offspring and alleged parents. (a) Lack of band sharing between offspring and social father indicates an extra-pair mating. (b) Band sharing establishes paternity between social father and offspring.

low-ranking males can benefit by encountering more females at a lower cost than by searching for females on their own.

Distinguishing these hypotheses is challenging because they are not mutually exclusive: a hotshot could be at a hotspot. One approach is to remove males from a lek, because the two hypotheses then make contrasting predictions about how males and females will respond. If males settle near high-quality (hotshot) males, the removal of hotshots should decrease the lek's attractiveness to both females and remaining males. Removal of lower quality males near a hotshot should affect only the remaining males, who should then move to the vacated territories. In contrast, if males settle in areas of high female activity (a hotspot), removal of some males should not influence the attractiveness of a display location. New males will settle on the vacated sites, which will still be attractive to females. Which hypothesis makes the correct prediction? Two experiments have reached contrasting conclusions.

Lekking behavior in the great snipe

Featured Research

Jacob Höglund and Jeremy Robertson investigated lekking behavior in the great snipe (*Gallinago media*), a small shorebird that breeds in moist terrestrial habitats in Europe (Höglund & Robertson 1990) (Figure 12.16). Males display on leks for several hours each evening for two weeks. The researchers studied three leks that contained seven to 20 males. On each lek, they first observed male interactions to determine dominance relationships. Each lek contained one or two dominant individuals that won most interactions. A few subordinate snipes lost most interactions, and many subdominant males were of intermediate dominance level. Dominant individuals had the highest display rates and mating success on each lek and mated polygynously.

Höglund and Robertson removed individual males for one night and recorded the behavior of the remaining males. When dominant snipes were removed, their territory was left unoccupied, and neighbors tended to move away. In contrast, when subdominant or subordinate males were removed, neighbors quickly added the vacant space into their own territory. These results support the hotshot hypothesis: experimental removal of a dominant male had negative effects on his neighbors, but removal of subdominant and subordinate males resulted in no major changes on the lek. Other subdominants and subordinates quickly usurped the vacated territories, presumably to move closer to the territory of a dominant male, the hotshot.

Figure 12.16. Great snipe. During the breeding season, males display to females on leks.

Peafowl leks

Featured Research

Höglund and Robertson were among the first experimenters to remove males on leks, and their work was groundbreaking. However, they did not record how the manipulations affected female behavior. Remember that the hotshot hypothesis predicts that if the hotshot is removed, then both males and females should be affected, whereas the hotspot hypothesis predicts that they should not be affected. Adeline Loyau, Michel Saint Jalme, and Gabriel Sorci examined male and female behavior in a peafowl (*Pavo cristatus*) lek (Loyau, Saint Jalme, & Sorci 2007).

The research team studied a large lek at the Parc Zoologique de Clères in northern France. The population of the park was around 100 individuals. About 75% of males aggregated and defended sites on the lek, while 25% were "floaters" that did not defend lek territories. The research team recorded male-male interactions, male display rates, female visitation rates, and the mating success of 29 males. The lek was clustered near a feeding location, where birds were fed twice each day (Figure 12.17). Males with sites near the feeding location had the highest display rates, attracted the most females, and had a high mating success rate; they also mated polygynously (Figure 12.18). Males farther from the feeding location did obtain some matings, but their success rate was much lower.

To examine the two lek hypotheses, the research team removed each of the 29 males, a few at a time, for about two weeks. The hotshot hypothesis predicts that the removal of the best males should affect female behavior. In contrast, the hotspot hypothesis predicts that such a removal should have no effect on females, because of the lek's location near a resource valued by females—the feeding site.

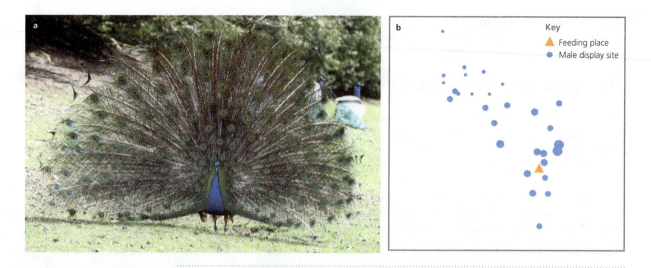

Figure 12.17. Peafowl lek. (a) Peacock display. (b) Diagram of a peafowl lek, showing the location of male display sites (blue) and the location of the feeding place (orange triangle). Size of display dots represents intensity of display by the male (*Source:* Loyau, Saint Jalme, & Sorci 2007).

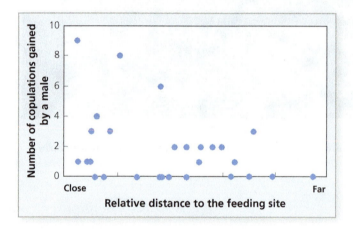

Figure 12.18. Copulations. Males closest to the feeding place obtained the most copulations (*Source:* Loyau, Saint Jalme, & Sorci 2007).

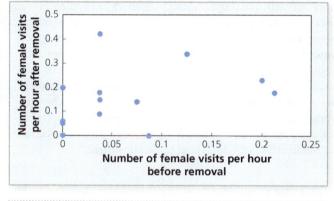

Figure 12.19. Female visits. Female visitation rates to males on individual display sites (dots) before and after the original male was removed (*Source:* Loyau, Saint Jalme, & Sorci 2007).

After male removal, floaters quickly settled on the vacated sites and began displaying to females. Display rates of newly settled males were similar to those of the prior site owner. Based on their visitation rates, females found these males just as attractive (Figure 12.19). In this system, females most often visited and mated with males whose territory was closest to the feeding location, regardless of male identity. These results support the hotspot hypothesis.

These studies show that the hotspot and hotshot hypotheses can both explain the evolution of leks in different species. Additional studies are needed to understand the relative importance of these hypotheses in more species. To this point, we have focused on polygyny because it is more common than polyandry. However, polyandry does occur in some birds and fish, as we see next.

Polyandry and sex-role reversal

In social polyandry, a single female associates and mates with two or more males. This rare mating system occurs in a few species of pipefish and in some families of birds (Figure 12.20). It is particularly common in shorebirds. Jaçanas, for example, are medium-sized birds with very long toes that allow them to walk on floating vegetation. Females in some species aggressively defend large territories from other females; two or more males settle and defend smaller territories within a female territory. The female mates with each male and lays multiple clutches, one for each male, who then provides most of the parental care. These females exhibit high levels of territorial defense and aggression, while males provide most of the parental care. Because these reproductive behaviors usually occur in the opposite sex, the birds are known as a **sex-role reversed species**.

Social polyandry has been a challenge to explain because it is difficult to determine how it benefits both males and females. Malte Andersson proposed a three-step hypothesis (Andersson 2005). First, we start with a species in which parental care is male biased. This behavior might be favored in resource-poor habitats, where females need to spend extensive time feeding to replenish their energy stores for egg production to complete a clutch. Alternatively, as suggested by Emlen and Oring, if predation on clutches is high, selection can favor male-biased care. Again, increased male parental investment gives females time to acquire sufficient energy to lay a replacement clutch.

The second phase in the development of social polyandry is the evolution of high female fecundity. In particular, a female may be able to lay more eggs than a single male can successfully raise. Female fitness is then constrained by the number of mates she can find, and selection will favor females that mate with multiple males. Here, males are a limiting resource for females, which naturally leads to the final phase—intense competition among females for mates. Increased competition will favor the evolution of increased fighting ability (e.g., large body size) and aggression, both of which are common in polyandrous, sex-role reversed females.

Andersson's hypothesis has yet to be tested experimentally, but multiple lines of evidence are consistent with it. First, in many polyandrous species, predation on clutches is very high. For example, 80% of the clutches of the polyandrous comb-crested jacana (*Irediparra gallinacea*) are lost before hatching, presumably to predation (Mace 2000). Second, social polyandry is frequently observed in species in which male care limits female reproduction. For instance, shorebirds produce relatively large eggs, and so a male can properly incubate a clutch of only four or five eggs (e.g., Székely & Cuthill 2000), but females can produce eight. Two clutches of four eggs each are thus typically tended by two males. In pipefish, females deposit eggs on or into a male that fertilizes and broods them (Wilson et al. 2003). Females can produce more eggs than can be brooded by a single male, and so females typically mate with multiple males, with each male receiving eggs from only one female (Berglund, Rosenqvist, & Svensson 1989; Jones, Walker, & Avise 2001). Third, many polyandrous birds live in resource-rich environments but appear to have evolved from species that lived in resource-poor habitats (Andersson 1995). Thus, the ancestral habitat probably favored male-biased care, but the current habitat allows high female fecundity, because there is abundant energy for egg production.

Polygyny and polyandry involve multiple mating by one sex. Next, we examine mating systems that involve multiple matings by both sexes.

Figure 12.20. Pipefish and jacana. (a) Pipefishes and several shorebirds like (b) this jacana exhibit social polyandry.

sex role-reversed species
A species in which females compete for males, who invest heavily in parental care.

12.4 The presence of social associations distinguishes polygynandry from promiscuity

Emlen and Oring suggest that polygynandry, or plural breeding, will evolve when group defense of a territory is more effective than defense by a single individual. In these cases, a breeding group will consist of several males and females that defend a large territory and so associate closely with one another. Plural breeding occurs in social carnivores such as African lions (*Panthero leo*), banded mongooses (*Mungos mungo*), and spotted hyenas (*Crocuta crocuta*) (Figure 12.21), as well as several species of primates,

Figure 12.21. Plural breeding. The banded mongoose exhibits plural breeding.

including chimpanzees (*Pan troglodytes*) and bonobos (*Pan paniscus*).

Such groups are often challenged by rivals, and large groups are typically more successful than smaller ones in a conflict over a territory. For example, encounters and territorial challenges often occur between lion prides, resulting in intense chases. The larger pride wins over 80% of such interactions (Packer, Scheel, & Pusey 1990), demonstrating the advantage of group defense. In many plural breeding species, the number of males in a social group often correlates positively with the number of females, because higher male numbers are required to defend all females successfully as group size increases (Clutton-Brock 1989). The males are often relatives, and most sire some proportion of offspring (e.g., Cant 2000; Engh et al. 2002).

Polygynandry in European badgers

Figure 12.22. European badgers. A species that lives in social groups in Great Britain and exhibits a plural mating system.

One striking example of plural breeding is found in the European badger (*Meles meles*) (Figure 12.22). This large carnivore ranges throughout Europe and Asia and exhibits much variation in sociality. In Great Britain, badgers live in social groups of up to two dozen individuals. Such groups contain multiple reproductively active males and females, suggesting the possibility of polygynandry.

Hannah Dugdale and colleagues investigated the mating system in a population of badgers in Wytham Woods, Oxford (Dugdale et al. 2007). Over a period of 17 years, adults and cubs from different social groups were captured and uniquely marked with a tattoo. The researchers took blood samples for DNA extraction to determine the parentage of over 300 cubs. By determining the parentage of the cubs, they could also determine which individuals were breeding in each group.

On average, social groups contained 12 adults, six of each sex, and approximately 27% of all adult males and 31% of adult females bred successfully each year. Within social groups, an average of 1.6 females per group gave birth each year and 1.5 males sired offspring, but as many as five males and five females reproduced in a social group (Figure 12.23). Because multiple males and females reproduced within social groups, these data provide clear evidence of a plural breeding system. Why do multiple males, rather than just one, mate with many females? Mating occurs year-round in

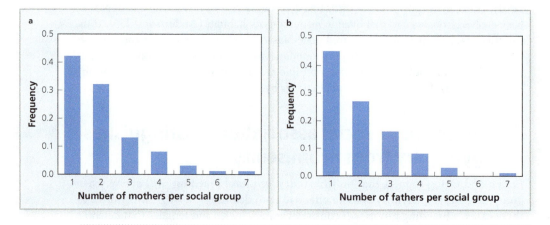

Figure 12.23. Badger reproduction. The frequency of social groups with different numbers of adult (a) females and (b) males that reproduced within the group. In most groups, multiple males and females produced offspring (*Source:* Dugdale et al. 2007).

England, which reduces the likelihood of male mate guarding, because such a behavior would likely be too costly for a single male attempting to successfully guard all the females (Yamaguchi, Dugdale, & MacDonald 2006). In addition, females could benefit by mating with multiple males to assure fertilization, reduce harassment from males, or increase genetic diversity of offspring, as we will see in Section 12.5. Males from outside a group do sire offspring in some groups, and perhaps multiple males can better defend a group of females from such outsiders. Further work is required to better understand the fitness benefits that such a mating system provides to males.

Promiscuity and scramble competition: seaweed flies and red squirrels

Featured Research

Emlen and Oring suggest that promiscuity will evolve when costs exceed the benefits in the defense of mates or resources and when there is no need for biparental care. In a promiscuous mating system, individual males and females mate with multiple partners in the absence of social associations. There is either uniparental care, typically provided by the female, or no parental care.

Promiscuous mating is characterized by **scramble competition**, as individuals compete indirectly with each other to find and secure copulations with multiple mates. This system occurs in many invertebrates. For example, seaweed flies (*Coelopa frigida*) live and breed in decaying seaweed on beaches. They engage in no courtship behavior, and males attempt to copulate with females as soon as they encounter them. A recent study estimated that in one high-density population on St. Mary's Island in England, males mount a female every eight minutes during the day and a female may mate with hundreds of males during her three-week life span (Blyth & Gilburn 2006).

A less extreme example of promiscuity is found in red squirrels (*Tamiasciurus hudsonicus*). These North American diurnal rodents live in coniferous forests, and males and females defend separate, permanent territories. Jeffrey Lane and colleagues studied the mating system of a population of red squirrels near Kluane National Park in southwest Yukon, Canada (Lane et al. 2008). They captured and marked individuals with numbered ear tags and collected tissue samples for DNA analysis. Each ear tag also contained two uniquely colored wires that facilitated identification from a distance. Females were fitted with radio-transmitters and monitored daily to assess reproductive behavior. Females are receptive for a single day, when they relax territorial defense; on these days, multiple males visit the territory, often simultaneously, and engage in "mating chases." Some chases result in copulations (Figure 12.24). On these days, the researchers followed the female for an average of 9.5 hours and recorded all chases and copulations. After females gave birth, researchers visited each nest and collected tissue samples from all offspring for DNA analysis to determine paternity.

Females copulated with an average of 5.8 males. For females that raised more than one young, 82.5% of litters were sired by multiple males, with a mean of 2.3 sires per litter. These data demonstrate that the mating system of the red squirrel is promiscuity with scramble competition for mates. Multiple males search for receptive females and then attempt to mate with them. Many males are successful, and so both males and females mate with multiple partners in this species.

In this system, males and females each have large territories. It is unlikely that a single male could successfully defend a territory of twice than normal size so as to secure sufficient food for himself

scramble competition
A competition in which individuals compete indirectly with each other to find and secure copulations with multiple mates.

Figure 12.24. Red squirrels. Each individual red squirrel defends a separate territory from others. Here, two marked individuals mate.

and his mate. In addition, a female's ability to move about freely and her running behavior make defense of her also unlikely. These two factors therefore favor the evolution of promiscuity in this species. You may be wondering why females mate with multiple males. What benefit does this behavior entail? We address these important questions next.

12.5 Social and genetic mating systems differ when extra-pair mating occurs

social mating system The social associations and presumed mating behavior of individuals based on those associations.

genetic mating system The actual number of sexual partners of each sex in a social mating system that contribute to a set of offspring.

extra-pair copulations Copulations of a pair-bonded individual with a third individual outside the pair bond.

Mating system classifications were originally developed from observations of individuals and their pair-bond associations, and are known as **social mating systems**. With the advent of DNA fingerprinting techniques, we have learned that social mating systems often differ from **genetic mating systems**, which describe the actual number of sexual partners that contribute to a set of offspring, because both males and females mate outside of existing pair bonds and social associations. For instance, roughly 90% of socially monogamous bird species engage in **extra-pair copulations** (**EPCs**), which result in extra-pair offspring (Griffith, Owens, & Thuman 2002). This behavior is widespread across taxa (e.g., Sefc et al. 2008; Cohas & Allainé 2009). Together, these two observations have led to the definition of extra-pair mating systems, which occur when social and genetic mating systems differ because of extra-pair paternity (e.g., Stutchbury & Morton 1995). Why are these social and genetic mating systems so different?

To understand extra-pair matings, consider a socially monogamous, pair-bonded species. From an evolutionary perspective, it is easy to understand why males seek sexual partners outside the pair bond: doing so can increase their reproductive success. The risk to a male of seeking extra-pair partners is that other males are behaving similarly, so while he is away from his partner, he may lose paternity to other males. On the other hand, a female that seeks extra-pair partners may risk losing parental care from her mate if his paternity is not certain. In general, females cannot increase the number of offspring they produce by mating with multiple males. Given the potential cost and apparent lack of benefits, why do socially monogamous females mate with multiple partners?

Several hypotheses explain multi-male mating by socially monogamous females. In some cases, females can gain additional care for their offspring if each sexual partner provides some parental care (e.g., Blomqvist et al. 2005). Alternatively, in a few social species with multiple males in the group, dominant males may kill unrelated offspring (infanticide). However, if a female mates with multiple males—especially dominant ones—she may be able to protect her offspring from infanticide because paternity will be uncertain (Wolff & MacDonald 2004). However, the most general explanation for multi-male mating is that females can increase the genetic quality of their offspring.

How can multi-male mating result in increased genetic quality of offspring? In social monogamy, females are limited to mating with one male. Only one female can pair bond with the highest quality male, and so many females will pair bond with lower ranking males. Mating with a low-quality male can affect not only the survival of her offspring but also her own condition: socially monogamous females, in fact, experience higher levels of stress when they mate with low-quality mates (Griffith, Pryke, & Buttemer 2011). From this perspective, it is easier to identify the potential benefits of extra-pair matings for females. First, a female can benefit by mating with a male of higher quality than her social partner, as we saw previously in the good genes hypothesis (see Chapter 11). Second, high levels of heterozygosity commonly increase fitness, because it reduces the likelihood that recessive deleterious alleles will be expressed (Brown 1997a). A female can benefit by mating with an extra-pair mate who is genetically more dissimilar to herself than her social mate, because this will increase the level of heterozygosity of her offspring. These hypotheses are not mutually exclusive. Taken together, they constitute the **genetic quality hypothesis**, which states that females that engage in multi-male matings can improve the fitness of their offspring via genetic mechanisms. Let's look at two empirical tests of this hypothesis.

Extra-pair mating in juncos

Featured Research

genetic quality hypothesis The hypothesis that females that engage in multi-male matings can improve the fitness of their offspring via genetic mechanisms.

The main prediction of the genetic quality hypothesis is that extra-pair offspring will have higher lifetime fitness than within-pair offspring. This prediction has proven challenging to test, in particular because it requires a direct measure of fitness. Nicole Gerlach and colleagues tested the hypothesis by analyzing a long-term dataset of dark-eyed juncos (*Junco hyemalis*) from a population at the Mountain Lake Biological Station in Virginia (Gerlach et al. 2012). This population had been studied by Ellen Ketterson and her students since 1983, so detailed behavioral and demographic information, as well as data on mating behavior, were available. DNA analyses were started in 1990, making it possible to determine paternity for thousands of birds. Gerlach and colleagues examined microsatellite data on over 2200 nestlings.

They found no difference in the fledgling success (F1 generation) between extra-pair and within-pair offspring. However, both male and female F1 individuals from extra-pair matings produced more offspring than those from within-pair matings (Figure 12.26). The researchers found that over a lifetime, females that engaged in extra-pair matings had almost twice as many grand offspring as females who remained truly monogamous. In this case, the benefit of extra-pair mating is seen in the "grandchildren" generation and supports the genetic quality hypothesis. The availability of a long-term dataset allowed the researchers to identify the fitness difference. Now we turn to an example from a study of mammals that also used long-term data.

Figure 12.25. Junco. A small sparrow common in North America.

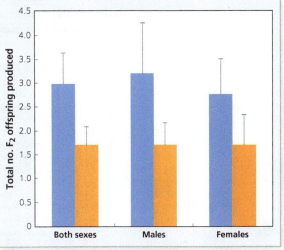

Figure 12.26. Number of junco grand-offspring produced. Mean (+ SE) reproductive success (total F_2 offspring produced) of extra-pair (blue) and within-pair (orange) offspring (*Source:* Gerlach et al. 2012).

Marmot extra-pair mating

Figure 12.27. Alpine marmot. Marmots are large rodents that live in alpine habitats throughout Europe.

Aurélie Cohas and colleagues studied alpine marmots (*Marmota marmota*) for over ten years in La Grande Sassière Nature Reserve in the French Alps (Cohas et al. 2007) (Figure 12.27). These animals are highly social and monogamous but display fairly high levels of extra-pair matings. The social group consists of a dominant male and female (the breeding pair) and their subordinate offspring. Cohas and colleagues tested the same prediction of the genetic quality hypothesis: extra-pair young should have higher fitness than within-pair young.

Each year, the individuals in this population were trapped, sexed, and marked with ear tags. The researchers collected hair or tissue samples for DNA analysis and fitted the marmots with a transponder to determine their location. The social status of all individuals was assessed by observing dominance interactions.

The research team determined paternity of 220 offspring. Of these, 45 were extra-pair young sired by an individual other than the dominant male. The remaining 175 were within-pair young sired by the dominant male. The research team characterized the fitness of young by two factors: their survivorship each year and the probability that an individual would attain dominant status during its lifetime. In alpine marmots, most offspring are born to dominant individuals, and so the attainment of dominant status strongly influences fitness.

The survival of extra-pair juveniles, yearlings, and two-year-olds was 15%, 10%, and 30% higher, respectively, than the survival of within-pair young. In addition, extra-pair young were four times more likely to attain dominant status than within-pair young (Figure 12.28). The researchers determined that most extra-pair mates were dispersing males from other populations that were less genetically similar to the female than her pair-bonded mate (DaSilva et al. 2006). In this species, survivorship is positively related to the level of heterozygosity, and so offspring derived from extra-pair partners had higher fitness than those derived from the pair-bonded mate. Females in this population can improve their fitness by seeking extra-pair partners that will yield more heterozygous offspring than their social mate can.

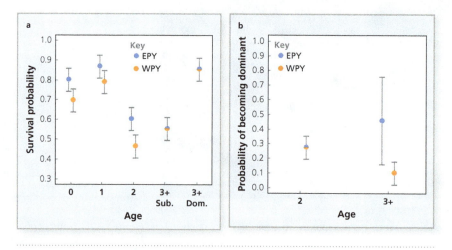

Figure 12.28. Marmot survival. Mean (± SE) probabilities. (a) Young (less than three years) extra-pair young (EPY, blue) have higher survival probability than within-pair young (WPY, orange). (b) Older extra-pair young are more likely to attain dominant status than within-pair young (*Source:* Cohas et al. 2007).

Molecular techniques have revolutionized our understanding of mating systems. It is becoming clear that, in many species, both males and females mate with multiple partners, despite exhibiting very different types of social associations. We expect to see continued research and the development of new hypotheses to explain why social and genetic mating systems differ so often.

Chapter Summary and Beyond

Males and females mate with different numbers of partners, and their mating systems range from a single mate (monogamy) to multiple mates for one or both sexes (polygamy). Two factors best explain the evolution of mating systems. First, males and females experience different fitness benefits and costs of parental care. Second, environmental conditions affect the ability of individuals to monopolize, through successful defense, either resources or mates.

Recent work on the evolution of mating systems has focused on distinguishing resource defense from female defense polygyny (e.g., Seki et al. 2009) and has examined the role of kin selection in the evolution of leks (Reynolds et al. 2009). In many species, the social mating system differs from the genetic mating system, because individuals mate outside of existing pair bonds. Males can increase fitness by mating with multiple partners, because the number of sexual of partners often limits their reproductive fitness. Female reproductive success is rarely limited by the number of sexual partners, making it more difficult to explain why females often mate with multiple partners. Leading hypotheses suggest that females can acquire genetic benefits for their offspring through multiple matings, but debate on this question continues (e.g., Westneat & Stewart 2003; Akçay and Roughgarden 2007; Griffith 2007).

CHAPTER QUESTIONS

1. Consider two species of birds that live in the same environment. One catches small flying insects and the other feeds on seeds on the ground. Imagine that flying insects and seeds on the ground are equally abundant and provide the same energy when eaten. Which species (the insectivore or the granivore) do you predict is more likely to be monogamous, and why?

2. How does an extra-pair mating system differ from promiscuity?

3. Design a study to test the hypothesis that *R. imitator* hatchlings are poor competitors for large phytotelmata.

4. Common crossbills (*Loxia curvirostra*) are socially and genetically monogamous, whereas Gunnison's sage grouse (*Centrocercus minimus*) form leks and are polygynous. What prediction can you make about the degree of variation in male mating success in these two species?

5. The dunnock, or hedge accentor (*Prunella modularis*), displays great variation in mating system. Females defend exclusive territories from each other, but territory size is negatively correlated with resources: they establish large territories when food is scarce but very small territories when food is abundant. Male territory size is less affected by resource level. When food abundance is moderate, monogamy is typically observed, as female and male territory sizes are similar. What mating system would you expect to observe when food is superabundant? What mating system would you expect to observe when food is scarce?

6. Several species of tropical tent-making bats modify a large leaf into a "tent," which is used as a roost that provides protection during the day when the bats are not active. A tent typically houses several adult females and a single adult male that has mated with them. How would you determine the mating system of such species?

Chapter 13

Parental Care

Figure 13.1. Parental care. Male clownfish protect the eggs in their nests from predators.

Each spring we enjoy watching juvenile animals in our backyard. We have observed both male and female Eastern phoebes (*Sayornis phoebe*) feeding their chicks in a nest under our deck and a Virginia opossum (*Didelphis virginiana*) carrying young on her back. Although phoebe offspring receive care from both parents, the opossum young receive care only from one parent, their mother. In contrast, in some fish like clownfish, males can provide all of the care for the developing eggs by fanning the nest to circulate water and engaging in predator protection behavior (Figure 13.1). And, while many birds, mammals, and fish provide care for their young, in many other taxa, care is nonexistent. For example, female spotted salamanders (*Ambystoma maculatum*) lay egg masses in a stream near our house but then depart and provide no care for the growing young.

In this chapter, we first examine variation in parental care across taxa. Next, we discuss why one sex typically provides more care than the other and how parental care behaviors are regulated. Finally, we see how researchers explain variation in parental care among individuals within a species and over their individual lifetimes.

13.1 Parental care varies among species and reflects life history trade-offs

parental care Behaviors by a parent to enhance the fitness of offspring, including incubation, feeding, and defense.

life history traits Traits involved with growth, reproduction, and survivorship.

Parental care, the activities of an adult that enhance the survivorship of offspring, includes nourishing and incubating eggs and young, defending them from predators, and sometimes transporting them. If parental care increases offspring survival, why don't all species provide high levels of care to their young? The answer is found in life history theory.

Animals vary greatly in their **life history traits**—traits involved with growth, reproduction, and survivorship that are the result of natural selection. These traits include the age of first reproduction (sexual maturity), the number and size of offspring, level of parental care, and survival rate throughout life. For example, elephants do not become sexually mature until 13 years of age, reproduce only about every five years, have a gestation period of 22 months, give birth to a single, very large calf (over 100 kg), and provide parental care for several years. Each calf has a fairly high likelihood of surviving because of the enormous amount of parental care it receives and its large size, which decreases the risk of predation. By contrast, fruit flies mature in a few weeks, and females can produce hundreds of eggs over their brief life span, but parents provide no care to their offspring, each of which has a low survivorship. Even more extreme are some marine fish, such as cod, that become sexually mature within a few years and can produce 4 to 6 million small eggs at a single spawning. Again, these eggs receive no parental care, and only a few may survive to reproduce.

Life history theory proposes that natural selection will favor the evolution of behaviors that maximize an individual's lifetime reproductive success. For all animals, energy (or effort) is allocated not only to one's self for growth and maintenance but also to offspring in the form of reproduction and parental care. The limited amount of available energy creates an important life history trade-off: effort allocated toward reproduction reduces effort that can be allocated toward an individual's own growth and survival. Similarly, effort allocated toward current offspring reduces the effort that can be allocated toward future offspring. These trade-offs allow us to understand the evolution of life history traits across species, including parental care behavior (Clutton-Brock 1991). Across species, we often see the evolution of particular sets of life history traits along a continuum. At one end, natural selection can favor individuals that produce many small offspring and provide no parental care, while at the other end, selection can favor the evolution of individuals that produce fewer but larger offspring that receive much parental care (Figure 13.2). In general, taxa that have a longer life span produce a relatively small number of large offspring and exhibit higher levels of parental care than those with a short life span, which tend to produce many small offspring (e.g., Strathman & Strathman 1982; Kolm & Ahnesjö 2005; Summers, Sea, & Heying 2006).

Life History Trait	Species A	Species B	Species C
Fecundity	High	Intermediate	Low
Survivorship	Low	Intermediate	High
Parental care	Low	Intermediate	High

Figure 13.2. Life history continuum. (a) Spotted salamander eggs. (b) Phoebe feeding young. (c) High levels of parental care are typically seen in species that produce few large offspring—those with low fecundity and high survivorship, such as Species C. Species A exhibits low levels of care and has high fecundity and low survivorship. Species B exhibits intermediate life history trait values.

Life history variation in fish

Fish display tremendous variation in life history and illustrate how parental care varies across species. Anna Vila-Gispert and colleagues used this variation to examine life history trait patterns among more than 300 fish species from Europe, North America, and South America (Vila-Gispert, Moreno-Amich, & García-Berthou 2002). For each species, the researchers used published work as well as their own data to estimate life history traits such as age at maturity, maximum body size (standard length), length of breeding season, number of reproductive events per year, average egg size, fecundity (average number of oocytes in a single mature ovary), and level of parental care (amount of protection and nourishment of offspring).

Some traits tend to be positively or negatively correlated, such as larger parental body size and larger offspring. This correlation presents a challenge, because an examination of variation in offspring size among species may be confounded by the correlation with body size. To get around this problem, Vila-Gispert and colleagues used a multivariate statistical technique called principal component analysis in their study. This technique transforms a number of possibly correlated variables (in this case, the life history traits) into a series of uncorrelated principal components that are linear combinations of the original variables.

From the analysis, the researchers found that although there was much variation in life history traits among species, species with larger body size tended to have higher fecundity, later maturation, and fewer reproductive events per year (Figure 13.3). Species with smaller body size showed the opposite trends. Additionally, species with more parental care also tended to have larger eggs. These patterns reflect the basic life history trade-off: species that invest more in individual offspring (large egg size) tend to also provide greater levels of parental care. The researchers

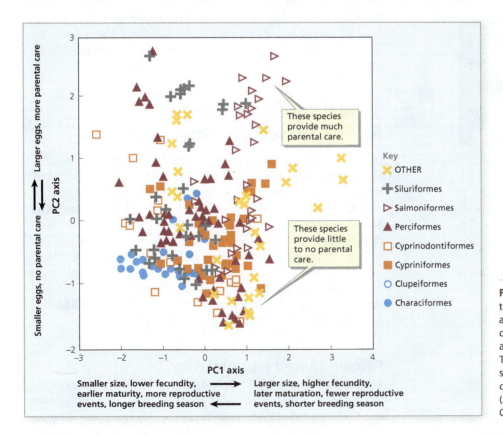

Figure 13.3. Fish life histories. Life history trait patterns in 300 fishes from around the world. Species in the top half of the graph tend to produce larger eggs and provide higher levels of parental care. Those in the bottom half tend to produce smaller eggs and provide little parental care. Each point represents one species (*Source:* Vila-Gispert, Moreno-Amich, & García-Berthou 2002).

TABLE 13.1 Parental care in select vertebrate groups. Numbers are the percent of species (*Source:* Reynolds, Goodwin, & Freckleton 2002 and Cockburn 2006).

TAXON	TYPE OF PARENTAL CARE (PERCENT OF TAXON)			
	NO CARE	FEMALE ONLY	MALE ONLY	BIPARENTAL
Fishes (422 families)	80	6	10	4
Cichlids (Cichlidae) (182 genera)	0	60	<1	40
Anurans (315 genera)	91	4	4	1
Lizards and snakes (938 genera)	97	3	0	0
Crocodilia (21 species)	0	62	0	38
Birds (9450 species)	0	8	1	91
Mammals (1117 genera)	0	91	0	9
Primates (203 species)	0	68	0	32

maternal care Parental care provided by a female to her offspring.

paternal care Parental care that a male provides to offspring.

biparental care Both parents provide care for offspring.

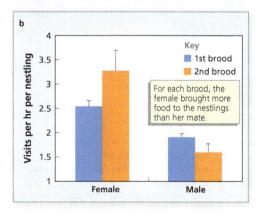

Figure 13.4. Female parental care. (a) Female black-throated blue warbler feeding young. (b) Mean (+ SE) visits to the nest. Female black-throated blue warblers provide more parental care than males in terms of food delivery for their first and second brood (*Source:* Stodola et al. 2009).

also found that these patterns held up over different geographic areas and habitats, as well as across different fish taxa.

The great variation in life history traits across taxa helps explain much variation in care provided across species. This care ranges from no parental care at all to care provided only by females (**maternal care**), care provided only by males (**paternal care**), and to care provided by both sexes (**biparental care**) (Reynolds, Goodwin, & Freckleton 2002). In vertebrates, certain major trends of parental care are evident across taxa: most amphibians and reptiles provide little parental care, birds provide predominantly biparental care, and mammals often provide extensive and mainly maternal care (Table 13.1). Although parental care also exists in invertebrates, it is less common. Males and females may both care for offspring in many species, but rarely do both sexes provide the same level of care. We examine that pattern next.

13.2 Sexual conflict is the basis for sex-biased parental care

In species that provide parental care, we see an interesting pattern: one sex typically provides most or all of the parental care. Often, females provide more care. Even in birds that exhibit biparental care, the female typically provides greater care than her mate. Kirk Stodola and colleagues examined parental care behavior in black-throated blue warblers (*Dendroica caerulescens*). These small songbirds often produce two clutches each summer and exhibit biparental care. Stodola's research team quantified the rate at which males and females brought food to their nestlings. Females consistently brought more food than their mates for both broods raised (Stodola et al. 2009) (Figure 13.4). Why is parental care so often female biased?

Female-biased parental care

To explain the unequal division of parental care, we must examine the fitness benefits and costs of providing parental care for each sex. Both sexes obtain the same fitness benefit from providing parental care: increased offspring survival. However, as we've seen, such care takes time and energy, and it may increase predation risk for the caring adult. These costs create a conflict of interest between the sexes: each sex will benefit if the other pays the cost of care. The difference between the sexes in their fitness interests is known as **sexual conflict** (Trivers 1972; Chapman et al. 2003).

How is this conflict resolved? The sex with the higher cost of care should provide less care, because its net benefit of care will be smaller (Queller 1997). In most species, females have a greater certainty about the identity of their offspring than males, because they lay eggs or give birth. Therefore, the benefit of parental care to females is rather straightforward. What about the benefit for males? Because females can mate with multiple males while fertile, males will always be less certain of whether they sired the offspring of any female they have inseminated. This reduced certainty is an important cost to males, because it results in the possibility that a male will end up providing care to another male's offspring. Thus, natural selection will favor males that reduce their parental care in order to allocate more effort to future offspring. In fact, we can predict that the less certain the paternity, the less care is provided by males (Trivers 1972; Queller 1997). Let's look at one recent test of this prediction in birds.

> **sexual conflict** The differential selection on males and females to maximize their fitness.

Paternity uncertainty and parental care in boobies

Featured Research

The blue-footed booby (*Sula nebouxii*) is a large, colonially nesting seabird (Figure 13.5). In this species, a single male and female form a pair bond and raise offspring together. Parental care is extensive and shared, as they take turns incubating the egg for 40 days and then feed the developing chicks for another 140 days. If a male is uncertain of his paternity, will he provide as much care? Marcela Osorio-Beristain and Hugh Drummond conducted a removal experiment to answer this question (Osorio-Beristain & Drummond 2001).

While fertile, a female copulates frequently with her mate, beginning just before and continuing during egg laying. She may also copulate with other males, however, when her mate is absent. The researchers manipulated certainty of parentage by removing pair-bonded males for 11 hours during the fertile period of the females, which occurs a few days before egg laying. Control males were removed for the same time but only when female fertility was low, about 15 days before egg laying. Female fertility was assessed by visual inspection of the cloaca (the urogenital opening). When the cloaca is open and reddish, females are fertile; when it is closed and not reddish, females are not fertile. After the removal period, all males quickly returned to their mate.

Figure 13.5. Blue-footed booby. A female sits on her clutch of eggs.

Experimental and control males differed in their behavior once eggs were laid. Seven of the 16 experimental males removed the first egg laid by their mate from the nest. (The egg was later destroyed by predators.) No other eggs were removed. None of the 17 control males removed any eggs laid by their mate. The researchers concluded that for experimental males, certainty of paternity was low for the first egg laid, as the male was away during his mate's fertile period and the female may have mated with another male. Because the pair copulates often during egg laying, the male's certainty of paternity for subsequent eggs should be high. In sum, male boobies reduce parental care, even to the point of egg destruction, when their paternity is uncertain.

The evolution of male-only care

In mammals and birds, females typically provide greater levels of care than males because of a reduced certainty of paternity. Although parental care is fairly rare among fishes, amphibians, and reptiles, male-only care does occur in some taxa. Can

determinate growth A pattern in which individuals stop growing at some point in their life.

indeterminate growth Growth that continues throughout the lifetime.

iteroparity A life history strategy in which there are multiple reproductive events throughout the lifetime.

semelparity A life history strategy in which reproduction occurs once in a lifetime.

sexual conflict and uncertainty in paternity explain this difference? Consider the modes of fertilization and growth in these taxa. Mammals and birds have internal fertilization and **determinate growth**: they stop growing when they reach adult size. In fishes, amphibians, and reptiles, many species exhibit external fertilization and **indeterminate growth**: they never stop growing. External fertilization should increase the level of certainty in paternity for males, and so it should reduce the cost of potentially providing care for unrelated offspring. Indeterminate growth matters because the number of eggs a female can produce increases rapidly with body size; larger females can produce many more eggs than small ones (Gross & Sargent 1985; van den Berghe & Gross 1989). The large increase in fecundity with an increase in body size establishes a new cost of parental care for females, because providing care to current offspring requires a sacrifice in feeding and growth rate, which lessens her future reproductive output. For males, the ability to obtain matings also increases with body size, but at a much lower rate than it does for females (Gross 2005). Thus, the net cost of providing care is higher for females than for males, and so selection can favor male-only parental care.

The mode of fertilization affects paternity certainty, and indeterminate growth affects fecundity. Putting these associations together along with sexual conflict, we can predict that male-only care is most likely to evolve in fishes, amphibians, and reptiles with external fertilization and multiple reproduction events (known as **iteroparity**). Judith Mank and colleagues found exactly this pattern in their survey of parental care in teleost fishes (Mank, Promislow, & Avise 2005). For example, smallmouth bass are iteroparous, have external fertilization, indeterminate growth, and male-only parental care. However, in species such as salmon that reproduce only once (known as **semelparity**) and then die, males do not provide care.

Featured Research

Paternity uncertainty and male-only care in sunfish

Does certainty of paternity also affect the level of care in fish with male-only care? Our hypothesis predicts that it should. Bryan Neff examined this question in male bluegill sunfish (*Lepomis macrochirus*) (Neff 2003). Sunfish are common in the lakes and rivers of North America. Large breeding males construct nests along the shoreline. Females visit these nests and deposit eggs, which the male then fertilizes and defends. Smaller "sneaker" males adopt a different breeding strategy by hiding near a breeder male's nest (see Chapter 11). When a female begins depositing eggs, a sneaker may quickly dart into the nest to release sperm and frequently succeeds in fertilizing eggs (Fu, Neff, & Gross 2001). Thus, sneaker males represent a threat to a territorial male's paternity (Scientific Process 13.1).

Neff investigated how males' nest defense behavior was affected by their certainty of paternity. On the morning of spawning, territorial males were divided into two groups. For treatment males, he placed four sneaker males at the edge of their nest in clear plastic containers. Empty containers were placed near the nest of control males. The day after spawning, Neff recorded male nest defense behavior in response to a potential egg predator, a pumpkinseed sunfish (*L. gibbosus*), which was placed in a clear plastic container near the nest. Defense behaviors included lateral displays, opercular flares, and bites directed toward the predator.

Neff observed significantly less egg defense by treatment males than by control males. Apparently, the presence of sneakers at the edge of a nest during spawning reduced a male's certainty of paternity and level of parental care for the eggs in his nest, as predicted.

Sunfish display male-only parental care and external fertilization. Not all species with male-only care fertilize eggs externally. Next, we examine male-only care in an insect with internal fertilization.

SCIENTIFIC PROCESS 13.1

Paternity certainty and parental care in bluegill sunfish
RESEARCH QUESTION: *What affects the level of male parental care in bluegill sunfish?*

Hypothesis: Males should reduce care as their certainty of paternity declines to avoid caring for unrelated offspring.

Prediction: Males who observe sneakers near their nest during spawning should provide less care to eggs than males who do not see sneakers near their nest.

Methods: The researchers:

- divided territorial bluegill sunfish males into two groups on the day of spawning. For treatment males, four sneaker males were placed in clear plastic containers near the nest. For control males, empty plastic containers were placed near the nest.
- placed an egg predator (a pumpkinseed sunfish) in a clear plastic container near each nest the next day for two 30 second periods to simulate a threat to the eggs.
- recorded the amount of defense behavior exhibited by each male to the predator (lateral displays, opercular flares, and bites).

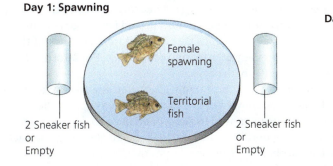
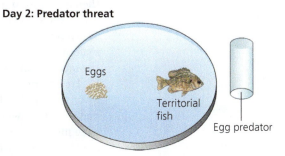

Figure 1. Experimental setup. On Day 1 a sneaker fish or an empty control was placed next to spawning pair; on Day 2 a predator was placed near the male guarding eggs (*Source:* Neff 2003).

Results: Treatment males exhibited significantly lower levels of defense behavior than did controls.

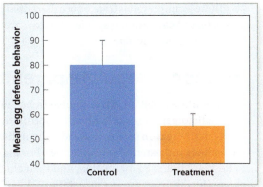

Figure 2. Defense behavior. Treatment males displayed significantly less defense behavior in response to a predator than control males (*Source:* Neff 2003).

Conclusion: Parental care by male bluegill sunfish is affected by their certainty of paternity.

Paternity assurance and male care in water bugs

Figure 13.6. Male water bug with eggs. Males care for eggs that females deposit on their back.

Of the insects that provide care, some species exhibit male-only care and have internal rather than external fertilization. Can sexual conflict and uncertainty in paternity still explain such male-biased care? Yes, according to a classic study by Robert Smith on the giant water bug (*Abedus herberti*) (Smith 1979). Females oviposit eggs on the back of their mate, and males protect and aerate the eggs (Figure 13.6). These behaviors enhance offspring survival but are costly, because such care precludes the male from feeding and mating with other females. In addition, females can mate with several males and store sperm for months. How can a male ensure that he is caring only for those eggs he has fertilized? One way is to mate repeatedly with a female. Smith observed a cyclical round of copulation and oviposition in which the pair mated, the female deposited up to three eggs on the male, and then the pair copulated again. This was repeated over and over: in one case he observed, a pair copulated over 100 times in 36 hours, and a total of 144 eggs were deposited on the male!

To determine the effectiveness of this behavior for ensuring male paternity, Smith conducted several experiments in which he allowed females to mate with two males that differed in a heritable phenotype—males were either unstriped or striped. In one experiment, females were allowed to mate once with one male phenotype and then removed to prevent ovipositing any eggs. After several weeks, the same female was allowed to mate with the other male phenotype and lay all her eggs on his back. Smith recorded the phenotype of nymphs that hatched from the back of the second male. Male paternity was extremely high for the second male. For 20 of the 25 females, the second male fertilized 100% of the eggs.

In a second experiment, Smith repeatedly alternated the two male mates for a given female. The female was allowed to mate and lay three eggs on the back of one male and then was immediately moved to the second male to do the same. Females were moved back and forth until they had oviposited half their eggs on each male. Smith again recorded the phenotype of the nymphs that hatched on each male's back. In this experiment, 245 of 250 nymphs were sired by the male who carried the eggs, demonstrating the very high level of paternity assurance in this species. By and large, males care for their own offspring. In this system, males have high levels of paternity assurance because they frequently copulate with the female whose eggs they accept and then care for. This assurance helps explain the evolution of male-only care.

These examples illustrate how differential costs of care among the sexes are linked to sex-biased care among species. So how are parental care behaviors regulated? We examine this question next.

13.3   Hormones regulate parental care

Many adults exhibit a radical change in behavior over the course of the reproductive season. Before mating, individuals concentrate on finding and attracting a mate. As the reproductive season progresses and more individuals mate successfully, mating behavior declines and parental care increases. For example, birds switch from mating displays to nest building, egg incubation, and then chick feeding. Female mammals undergo a change in physiology to begin lactation. What initiates and regulates these changes?

In mammals, steroid hormones such as estradiol and progesterone vary with pregnancy and can strongly affect courtship and reproductive behavior. The peptide hormone **prolactin**, which is involved in milk synthesis, rises during pregnancy and then falls when offspring are weaned (Zarrow, Gandelman, & Denenberg 1971). It is synthesized in and secreted from specialized cells in the anterior pituitary gland. Does prolactin affect parental care?

prolactin A hormone that influences parental care in vertebrates.

Prolactin and maternal care in rats

Robert Bridges and colleagues conducted a series of classic experiments to examine the effect of prolactin, estradiol, and progesterone on maternal behavior in rats (Bridges et al. 1990). They manipulated prolactin levels in female rats that had never given birth. To set up the experiment, they first treated all individuals with the drug bromocriptine to suppress endogenous prolactin secretion so that the rats could not produce their own prolactin. The research team also removed the ovaries to control the levels of other hormones and inserted a cannula—a small tube—into the brain to administer hormones directly into the central nervous system. All females were "primed" with progesterone and estradiol to mimic the changes in these steroid concentrations in pregnant rats. The researchers then examined the behavior of females treated with different levels of either prolactin or a sodium chloride–sodium bicarbonate control. They placed three unrelated pups into the home cage of each female. Over a period of six days, Bridges and colleagues recorded three aspects of parental care: latency to contact the pups, whether pups were retrieved and carried to the nest, and instances of crouching behavior (huddling over pups to keep them warm) (Figure 13.7).

There was a rapid increase in maternal behavior in treatment females by Day 3. Females treated with prolactin displayed significantly higher levels of parental care behavior than controls (Figure 13.8). In follow-up experiments, the research team examined whether prolactin could stimulate maternal behavior without the steroid regimen. The experimental procedures followed were identical to those of the prior study, except that females were not primed with progesterone and estradiol. Without the steroid hormones, females did not display maternal behavior (Figure 13.8). To determine the action site of prolactin, the researchers injected prolactin subcutaneously but found no difference in maternal behavior compared with that of controls. Subsequent work identified prolactin receptors in the brain and revealed that females with nonfunctional prolactin receptors will exhibit very low levels of maternal care behavior (Lucas et al. 1998). Together, these studies demonstrate that prolactin strongly affects parental care in female rats.

Today, we understand that prolactin affects parental care in mammals and other vertebrates and is sometimes called the "parental hormone" (Schradin & Anzenberger

Figure 13.7. Female rat with young. Females huddle over pups to keep them warm.

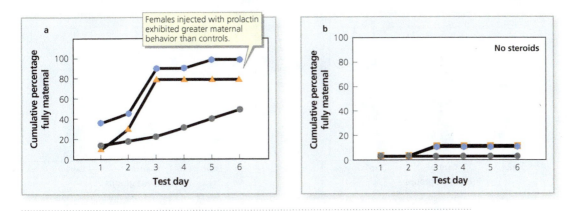

Figure 13.8. Rat maternal behavior. (a) Cumulative percentage of fully maternal behavior is higher in female rats experimentally injected with prolactin (50 µg orange or 10 µg blue) compared to controls (gray). (b) Without the steroid regimen, maternal behavior in female rats experimentally injected with prolactin (blue and orange) was similar to that of controls (gray) (*Source*: Bridges et al. 1990).

1999; Carlson et al. 2006; Bender, Taborsky, & Power 2008; Angelier & Chastel 2009). For example, prolactin plays an important role in regulating maternal care in birds, as we see next (Angelier & Chastel 2009).

Featured Research

Prolactin and incubation in penguins

Incubation of eggs is an important component of parental care in birds, and several studies have shown the importance of prolactin in affecting incubation behavior (e.g., Lormée et al. 1999). Blood prolactin levels tend to increase in birds during egg laying (Vleck 2001), but what is the proximate mechanism that causes this increase? One possibility is that the simple sight of an egg spurs increased prolactin production. Alternatively, birds might require tactile stimulation that occurs when eggs are incubated by contact with the **brood patch**, a featherless area on the belly that is well vascularized for heat transfer to eggs.

brood patch In birds, a featherless area on the belly that is well vascularized for heat transfer to eggs during incubation.

Figure 13.9. Penguin. Female yellow-eyed penguin.

Melanie Massaro, Alvin Setiawan, and Lloyd Davis examined prolactin levels in female yellow-eyed penguins (*Megadyptes antipodes*) at Boulder Beach in New Zealand (Massaro, Setiawan, & Davis 2007) (Figure 13.9). Just before egg laying, the research team placed an artificial egg either in a female's nest or next to the nest in a wire cage. In the first case, the egg provided both tactile and visual stimulation. In the second case, the egg could not be touched, so it provided visual stimulation only. The team used plaster eggs that matched the size, weight, shape, and color of natural yellow-eyed penguin eggs. They divided 33 females into three groups: (1) those sampled for prolactin prior to treatment, on the day the egg was added; (2) those sampled three to four days after the egg was added (experimental birds); and (3) controls, which were sampled at the same time as the treatment group but had no egg placed in their nest. For the first two groups, the researchers recorded the amount of incubating time after the artificial egg was placed in the nest.

When only visual stimulation was provided, there were no differences in prolactin levels, incubation behavior, or brood patch development among the three groups. However, when the artificial egg was placed in the nest, females had higher levels of prolactin, developed a wider brood patch, and spent more time incubating than did controls (Figure 13.10). In other words, for this species, the visual and tactile stimulation resulted in increased prolactin levels, increased maternal care, and an important physiological change: development of a larger brood patch. Additional studies have found that prolactin production is

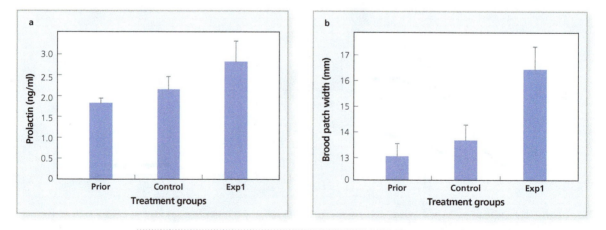

Figure 13.10. Prolactin and brood patch development. Mean (+ SE) (a) prolactin level, and (b) brood patch width in treatment groups. Females with artificial eggs added to their nest (Exp 1) had higher prolactin levels and wider brood patches than females prior to the treatment or controls (*Source:* Massaro, Setiawan, & Davis 2007).

involved in stimulating parental care behavior in many bird species, but our understanding of the mechanism of action in the brain is still limited (Angelier & Chastel 2009).

Juvenile hormones and parental care in earwigs

Although the effects of many hormones on development, the brain, and behavior are rather well studied in vertebrates, much less is known about their relationships in invertebrates. In insects, **juvenile hormones (JH)** are a class of hormones produced by specialized endocrine glands, that affect development and molting and vary in a consistent manner with levels of parental care by females (Trumbo 2002; Mas & Kölliker 2008). For example, in earwigs, females care for offspring by constructing a burrow and then grooming and guarding the developing eggs. In these species, JH increases as a female's oocytes develop and peaks a few days prior to oviposition. JH levels then decline and remain low while she cares for her offspring (Figure 13.11). As offspring reach maturity, JH again rises, parental care ceases, and females reproduce again (Rankin et al. 1995; Trumbo 2002). So in this case, parental care is high when JH levels are low and vice versa.

Susan Rankin, Kelly Fox, and Christopher Stotsky experimentally tested the link between JH level and maternal care in the ring-legged earwig (*Euborellia annulipes*). They topically added JH in acetone to females on the day of oviposition to keep levels artificially high and either did not manipulate or just topically added acetone to controls. Females with enhanced JH exhibited reduced parental care, providing only three days of egg care, compared with the eight days provided by untreated controls and acetone-only individuals (Figure 13.12) (Rankin, Fox, & Stotsky 1995). These results demonstrate an inverse relationship between JH level and female care behavior in this species: parental care is low when JH levels are high.

juvenile hormones In insects, a class of hormones produced by specialized endocrine glands that influence molting, are associated with reproductive maturation, and are known to vary with levels of parental care provided to offspring.

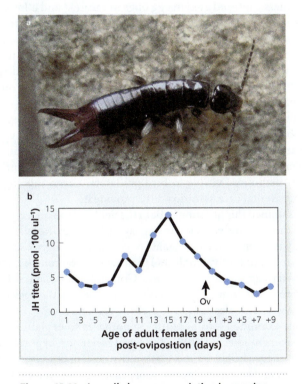

Figure 13.11. Juvenile hormone variation in earwigs. (a) Ring-legged earwig. (b) Juvenile hormone level over time. In earwigs, juvenile hormone increases and peaks a few days prior to oviposition. Juvenile hormone levels remain low while females care for their young (Days + 1 through + 9). OV represents the day of oviposition (*Source:* Rankin et al. 1995).

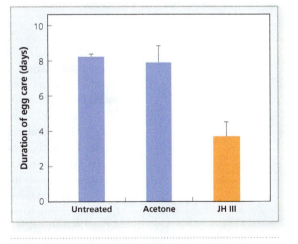

Figure 13.12. Duration of earwig egg care. The mean (+ SE) number of days of egg care by earwigs was five days less for females treated with juvenile hormone and acetone (JH III) than for untreated controls or those treated only with acetone (*Source:* Rankin et al. 1995).

The previous examples illustrate how hormonal regulation can explain variation in parental care behavior over time for an individual. Next, we examine variation in parental care among individuals in a species.

13.4 Parental care involves fitness trade-offs between current and future reproduction

We expect parental care behavior to evolve when the benefits of care exceed the costs. The benefits are straightforward: enhanced survivorship of young. The costs are more complex, because they involve both current and future fitness.

Predation risk and parental care in golden egg bugs

Current fitness costs include factors that reduce current reproduction. One common cost of care is predation risk. Adults that provide care are often more obvious to predators and so suffer higher mortality. Parental care to protect offspring from predators can take many forms. In the golden egg bug (*Phyllomorpha laciniata*), females can lay eggs on the backs of adults of both sexes or on vegetation (Figure 13.13). Eggs carried by an adult receive protection, whereas eggs on plants receive no parental care. How does egg carrying affect the adult's predation risk? What is the cost?

Piedad Reguera and Montserrat Gomendio examined this cost of parental care in a simple experiment (Reguera & Gomendio 1999). They tested the effect of egg carrying on predation risk using a large, outdoor aviary with two great tits (*Parus major*), a common avian predator for these insects. The researchers captured golden egg bugs in the field and allowed females to lay eggs on some adults. They then placed four same-sex golden egg bugs in the aviary. Two bugs carried eggs on their back, and the other two did not. The aviary was checked every 20 minutes for the presence of bugs and to record all predation events. A trial continued until only one prey was left or for a maximum of two hours.

In ten of 11 trials, the last bug alive was one without eggs on its back, suggesting a high cost of egg carrying. Golden egg bugs are a yellow-brown color, and their morphology makes them cryptic. However, their eggs are shiny, which apparently makes them more visible to bird predators. Given this high cost of parental care, why do adults provide it?

Figure 13.13. Golden egg bug. Adults carry eggs on their backs as a form of parental care, but this behavior increases their risk of predation.

Reguera and Gomendio examined this question in both the field and the laboratory. They found a large amount of intraspecific variation in egg-laying behavior. Recall that female golden egg bugs can lay their eggs on either plants or conspecifics. Eggs laid on plants have a much lower probability of survival (only 3%) than eggs laid on conspecifics (25%), and males are the primary egg carriers (Reguera & Gomendio 2002). Females mate with several males, and so a male cannot be certain of paternity for the eggs he carries. Given this uncertainty, why do males provide this costly parental care?

Gomendio and colleagues examined two potential hypotheses to explain egg-carrying behavior in males (Gomendio et al. 2008). The sexual selection hypothesis predicts that males that carry eggs should be more attractive to females (because they are willing to provide care) and so should receive more matings than males that do not carry eggs. However, experimental work revealed that females did not prefer to mate with males that carried eggs compared with those that did not, and so this hypothesis was rejected (Gomendio et al. 2008).

Alternatively, egg carrying can be favored by natural selection when the behavior's benefits exceed the costs for males. This natural selection hypothesis predicts that egg carrying should be more common when it provides a greater net benefit to

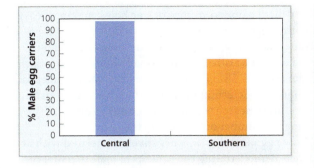

Figure 13.14. Egg carrying in golden egg bugs. More males carry eggs in a population in central Spain compared with a population in southern Spain (*Source:* Reguera & Gomendio 1999).

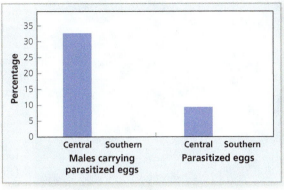

Figure 13.15. Egg parasitism rates. The population in central Spain has more males carrying parasitized eggs and a higher percentage of parasitized eggs compared with a population in southern Spain (*Source:* Reguera & Gomendio 1999).

males. An important cost is predation risk, as we saw previously. What are the benefits? Males sire 10 to 40% of the offspring they carry (García-González et al. 2003; Tay, Miettinen, & Kaitala 2003), and so they carry many other males' eggs as well as some of their own. However, the researchers found a large difference in the proportion of males with eggs and levels of egg mortality as a result of egg parasitism across two study sites. In a population in central Spain, almost all males carried eggs (Figure 13.14), and predation on eggs as a result of egg parasitism was high (Figure 13.15). In a population in southern Spain, fewer males carried eggs, and there was a much lower incidence of egg parasitism. These results are consistent with the natural selection hypothesis: egg carrying by males is favored when lack of care leads to very low offspring survival. So males provide care to eggs, which increases their predation risk, to enhance the survivorship of their offspring. The low level of paternity assurance may be the reason that egg-carrying behavior varies widely across populations. Males provide parental care only when the egg carrying is required for offspring survival.

This work illustrates one important fitness cost of providing care: increased predation risk. A second important factor that reduces current fitness is **opportunity costs**, or the sacrifice that an individual makes by not engaging in a different behavior that may enhance current fitness. For example, by guarding eggs, a male cannot search for new mates. Let's see one example of opportunity costs in frogs that helps to explain variation in parental care among individuals.

opportunity costs The fitness costs of not engaging in other activities.

Egg guarding and opportunity costs of parental care in frogs

Featured Research

For many insects, fish, and amphibians, egg attendance is an important aspect of parental care. One parent remains near the eggs to protect them from predators and remove infected or dead eggs. Experimental removal of the guarding parent typically leads to reduced embryo hatch success, showing the benefits of offspring defense as a form of parental care. Just as we saw in egg carrying, the amount or intensity of egg guarding varies among individuals within a population. For example, in big-thumbed frogs (*Kurixalus eiffingeri*) some males attend eggs nearly 100% of the time until they hatch, whereas others rarely stay near eggs to guard them (Figure 13.16). Wei-Chun Cheng and Yeong-Choy Kam investigated this variation (Cheng & Kam 2010). If egg attendance increases hatch success, why don't all males attend eggs at a high level?

This species, native to Taiwan, has an extended breeding period that lasts from February through August. During mating, the pair engages in **amplexus**. During amplexus, a male grasps a female, which allows him to fertilize the eggs she deposits

amplexus When a male grasps and holds a female prior to copulation.

Figure 13.16. Big-thumbed frog. Males display egg-guarding behavior.

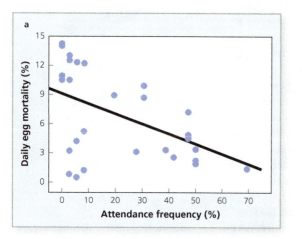

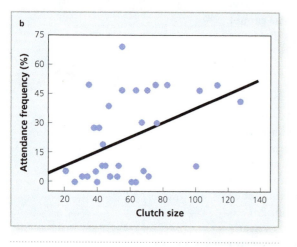

Figure 13.17. Egg attendance in frogs. (a) Higher male attendance led to lower daily egg mortality. (b) Males with large clutches had higher attendance frequency than males with small clutches (*Source:* Cheng & Kam 2010).

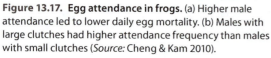

in water-filled bamboo stumps or tree holes, making it easy to identify eggs that a male has fertilized.

Males attending eggs were captured and uniquely marked using a waistband with a small numbered tag. Each night for six days, 38 egg masses were observed six times for ten minutes to determine whether the male was attending the eggs. The research team also noted any egg mortality by counting the number of eggs still in the nest.

Daily egg mortality was negatively correlated with attendance, demonstrating the benefits of male parental care (Figure 13.17). Male attendance was also positively correlated with egg clutch size: males were more likely to attend larger clutches. Why do males with small egg clutches spend less time attending them? The researchers proposed that such males are instead searching for additional mating opportunities to increase their fitness. Time spent attending eggs reduces the ability of a male to search out and mate with additional females. A male with a small clutch of 20 to 30 eggs has less to lose by allocating little time to parental care and more to gain by seeking out other mates. In contrast, males that have already fertilized a large clutch of over 100 eggs can best enhance their fitness by protecting them until they hatch. So here, variation in clutch size and our understanding of life history trade-offs and opportunity costs can explain variation in parental care among males.

In these examples, parental care increased the parent's mortality risk and reduced the time they could spend enhancing current fitness by searching for mates. Parental care can also involve a second fitness trade-off. Energy and effort used to care for current offspring can reduce a parent's ability to invest in future offspring, as we see next.

Parent-offspring conflict theory

The trade-off that adults face between current and future reproduction was first articulated by Robert Trivers and is now known as **parent-offspring conflict theory** (Trivers 1974). This theory states that parents and their dependent offspring are under different selection pressures: parents should maximize their lifetime reproductive fitness, while offspring should maximize the energy and protection they currently receive from their parents to survive to reproductive age. Trivers defined **parental investment** as "any investment by the parents in an individual offspring that increases the offspring's chance of survival at the cost of the parents' ability to invest in other offspring" (Trivers 1974). The benefit of parental investment is increased fitness of the offspring. However, the parent and offspring obtain different benefits from parental investment, because the parent is related to the offspring by 0.5, whereas the offspring is related to itself by 1.0. The offspring therefore receives twice the fitness benefit of the care provided (Lazarus & Inglis 1986). In terms of reduced future offspring, however, the cost of parental investment is the same for both parent and offspring, since both are related to all future offspring by 0.5 (assuming full sibship) (Figure 13.18). Because the optimal level of parental investment is always higher for offspring than for a parent, there is a parent-offspring conflict: the fitness of a parent is maximized at a lower level of investment than the fitness of an offspring. Let's examine two tests of parent-offspring conflict theory.

Figure 13.18. Parent-offspring conflict theory. (a) Eastern bluebird (*Sialia sialis*) chick begging for food. (b) The fitness benefit for offspring (2B) of receiving parental care is twice the fitness benefit for an adult (B) of providing care, because the coefficient of relatedness between parents and offspring is 0.5. C is the cost of parental investment in terms of reduced future offspring. Note that the optimal level of parental investment is higher for the offspring (O) than it is for the parent (P) (*Source:* Trivers 1974).

Parental care trade-off in treehoppers

Featured Research

One way to test parent-offspring conflict theory is to examine how natural variation in care affects both current and future offspring. This theory predicts that parents who invest more in their current offspring should have higher current reproductive success but also reduced future reproductive success.

Andrew Zink tested this prediction in treehoppers, small plant-feeding insects (Zink 2003). Zink studied a common species, *Publilia concava*, at a site in New York (Figure 13.19). Females select a host plant and lay eggs on the underside of mature leaves. Parental care takes the form of egg guarding as females protect their offspring from predators such as mites. In this species, there is tremendous variation among females in guarding behavior: some guard their eggs for three or more weeks until nymphs hatch, whereas others spend no time at all caring for their eggs. There is also variation in the number and size of egg clutches a female can produce in one reproductive season. Are these patterns related?

Figure 13.19. Treehopper. Treehoppers feed on host plants and lay eggs on the underside of leaves.

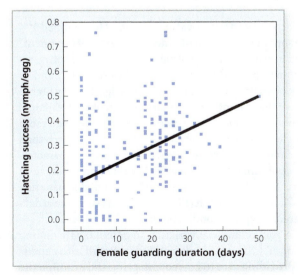

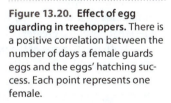

Figure 13.20. Effect of egg guarding in treehoppers. There is a positive correlation between the number of days a female guards eggs and the eggs' hatching success. Each point represents one female.

parent-offspring conflict theory Parents should maximize their lifetime reproductive fitness, while offspring should maximize the energy and protection they currently receive from their parents to survive to reproductive age.

parental investment Any investment by a parent in offspring that enhances the offspring's fitness at the cost of the parent's ability to invest in other offspring.

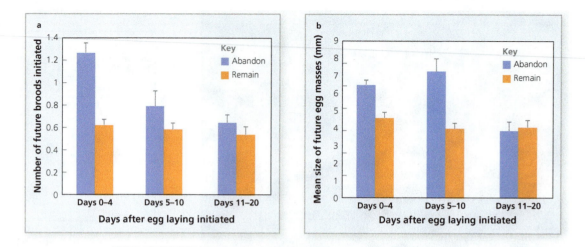

Figure 13.21. Egg guarding and future reproduction in treehoppers. (a) Mean (+ SE) number of future broods initiated and (b) size of future egg masses. Females that abandoned a clutch (blue bars) as opposed to remaining with their eggs (orange bars) in the first ten days after laying initiated more future broods of larger size. Females that abandoned their eggs instead of remaining with them during Days 11 to 20 after egg laying did not differ in future egg mass (*Source:* Zink 2003).

Zink examined how parental care affected both current and future reproduction of individual females. Based on the existence of a trade-off between current and future reproduction, he made two predictions: (1) egg guarding will increase egg survival and thus enhance current reproduction, and (2) increased parental care (guarding) will decrease future reproductive success.

Zink uniquely marked females with four small dots of colored paint. Because females are generally faithful to individual plants, Zink could follow their behavior and the fate of egg clutches through repeated visits. For each egg mass deposited by a female, Zink measured its size to estimate the number of eggs and checked them every two days until the nymphs hatched. At each check, he also recorded whether the female was present (guarding the eggs) or had abandoned them. Hatching success (an estimate of fitness) was calculated as the number of nymphs observed on the leaf divided by the number of eggs estimated in a clutch.

Zink found that about half the females abandoned their eggs within a few days of their deposition while the rest remained to guard them for more than three weeks. He also found a positive correlation between the duration of egg guarding and hatching success (Figure 13.20), demonstrating a benefit of care for current reproduction that supports the first prediction. Although the correlation might be due to successful female defense, it might also result from females abandoning clutches that are damaged or otherwise inviable.

To test these alternatives, Zink conducted a removal experiment. He identified 194 plants with egg clutches that were similar in size. In half—deemed "removal treatment" clutches—the female was removed to reduce the level of parental care that offspring received. The other clutches were unmanipulated controls (female not removed) and so received a higher level of care. Zink observed that hatch success was significantly higher in controls, providing more evidence that parental care enhances current reproduction. Is there a cost of care in terms of future reproduction? Yes. Females that provided more than ten days of care for their first clutch were less likely to initiate a second clutch than females that abandoned their first clutch in the first ten days. Another way to characterize the trade-off involved in the cost of care is to examine the benefit of providing little care: females that abandoned their first clutch early laid larger second clutches than females that provided greater care for their first clutch (Figure 13.21). As such, these data support both predictions of the parent-offspring conflict theory.

In this example, Zink relied in part on natural variation in parental care among females to test parent-offspring conflict theory. An alternative approach is to experimentally manipulate the costs of providing care to current offspring and examine the effect on future fitness, as we see in the next example.

SCIENTIFIC PROCESS 13.2

Parental care costs in eiders

RESEARCH QUESTION: *Does egg incubation affect future reproduction?*

Hypothesis: Incubation costs increase with clutch size and represent a substantial cost of parental care.

Prediction: Individuals with large clutches pay high incubation costs and will experience lower reproductive fitness in the future.

Methods: The researchers:

- manipulated the clutch size of females (either decreased to three eggs or increased to six eggs).
- recorded the body mass of each female at the beginning and end of the incubation period.
- recorded egg-laying date and clutch size for each female in the following year.

Results: Females with high incubation costs lost more mass and had later egg-laying dates and smaller clutches in the next breeding season than those with low costs.

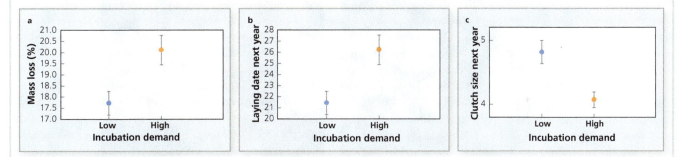

Figure 1. Incubation costs. Mean (± SE) effects of high and low incubation cost on (a) mass loss, (b) laying date, and (c) clutch size the next year in female common eiders (*Source:* Hanssen et al. 2005).

Conclusion: High incubation costs in one year reduce reproductive fitness in the next breeding season.

Incubation of eider eggs as a trade-off

Featured Research

In birds, egg incubation is an important component of chick development. Adults sit on eggs and transfer heat to them through their brood patch. This behavior can be critical when the ambient temperature is low.

Sveinn Hanssen and colleagues were interested in how current incubation costs affect future reproduction in the common eider (*Somateria mollissima*) (Hanssen et al. 2005). Eiders are large sea ducks that live primarily along the cold northern coasts of Europe and North America. Females normally lay a clutch of three to six eggs and incubate the clutch alone without the help of their mate. During the two-week incubation period, females do not leave the nest to eat and so typically lose up to 40% of their body mass—a large parental investment indeed! Assuming that six eggs take more energy to warm than three, Hanssen and colleagues predicted that incubating larger clutches should lead to a greater loss in body mass. Because the eider is a long-lived bird (living up to 12 years), the research team could also examine how mass loss affects future reproduction (Scientific Process 13.2).

APPLYING THE CONCEPTS 13.1

Smallmouth bass defend their nest from exotic predators

Exotic species, particularly predators, can reduce the population size of native species. Usually they do so through predation, but recent work suggests that they can also increase the cost of parental care. Round gobies (*Neogobius melanostomus*) are native to central Eurasia but were unintentionally introduced to Lake Erie in 1993 in ship ballast water and have attained very high abundance there. They can quickly consume an entire unguarded nest of eggs of the native smallmouth bass (*Micropterus dolomieu*) in a few minutes. Think of the moment in catch-and-release fishing when an angler has caught and held a male bass before releasing it. In this time, a male could lose his entire nest to round gobies (Steinhart, Marschall, & Stein 2004).

Smallmouth bass are effective defenders of their nests, but this defense requires a great deal of effort. Geoffrey Steinhart and colleagues looked at two sites: one in Lake Erie, which has an abundance of gobies, and a control site, Lake Opeongo, where gobies do not occur (Steinhart et al. 2005). Bass in Lake Erie engaged in about nine times as many defense chases as bass in Lake Opeongo (Figure 1). Lake Erie males thus spent almost ten times the energy in offspring defense. Because of the trade-off between current and future reproduction, researchers speculate that the high energy demand will reduce the lifetime fitness of bass and could lead to strong population declines in areas where nest predators are common. This decline will be due not to actual predation, but to changes in parental care behavior.

Figure 1. Smallmouth bass nest defense. (a) A male bass defending its nest. (b) Eggs in a nest. (c) Offspring defense (median number of chases per minute) of unhatched embryos and embryos in different lakes. Each box represents the middle 50% of data. Chases were more frequent in Lake Erie (*Source:* Steinhart et al. 2005).

The researchers manipulated clutch size to alter incubation costs. For one group of 24 birds, clutch size was reduced to three eggs (a low incubation demand), whereas in the other group of 30 birds, clutch size was increased to six (a high incubation demand). All birds were marked with colored leg rings and weighed at the beginning and end of the 15-day incubation period.

Females incubating six eggs lost relatively more mass than females incubating only three, confirming the assumption that incubation cost would increase with clutch size. In addition, in the next breeding season, females that experienced high incubation costs laid eggs later and had smaller clutch sizes. The researchers concluded that eiders experience a trade-off between current and future reproductive effort.

Females that lay a large clutch increase current reproduction but pay high incubation costs, which leads to a poorer body condition and reduced future reproduction.

In both treehoppers and eiders, we see that parental care involves a fitness trade-off between current and future reproduction, as predicted by parent-offspring conflict theory. Females that provide high levels of care enhance their current reproduction, but at a cost of reduced future reproduction. The parental care trade-off between current and future reproduction can explain much of the variation observed in parental care behavior among individuals. This theory can also be used to predict how the fitness of individuals will be affected by the introduction of novel predation threats that require greater levels of egg guarding (Applying the Concepts 13.1). We can go even further and use parent-offspring conflict to understand variation in care provided to siblings, as we see next.

Brood reduction and parent-offspring conflict

Parent-offspring conflict theory can also help us understand the evolution of parental care behavior that perhaps seems outright maladaptive. In many birds and some mammals, parents provide different amounts of care to siblings: some receive much more care than others. This difference leads to substantial variation in offspring size and, ultimately, survival. The largest, most well-fed siblings tend to survive, while those that receive little food suffer high mortality, which can result in **brood reduction**, or the death of some siblings.

Why don't parents care equally for their offspring or even bias care toward their smallest young who are at high risk of starvation? At first glance, it might appear that such behavior should increase survivorship of all the offspring. However, parent-offspring conflict theory predicts that brood reduction can be adaptive for a parent. If the costs of parental care for a current offspring exceed the benefits, then parents should cease providing care to that offspring. For example, if resources are very scarce, it may require too much effort on the part of the parents to obtain sufficient food to feed all siblings. A parent can enhance its total lifetime reproduction if it ends care for some offspring now to better provide care to future offspring. While this behavior may seem inconceivable to humans, it does make sense from an evolutionary perspective.

brood reduction The death of some siblings as a result of reduced parental care for the purpose of enhancing the fitness of surviving siblings.

Hatch asynchrony and brood reduction in blackbirds

Featured Research

Many birds lay multiple eggs in a clutch over the course of several days. Often, incubation begins when the first egg is laid, and so offspring begin to develop on different days, resulting in siblings of different ages in the same nest. David Lack suggested that the difference in egg hatch timing, or **hatching asynchrony**, is adaptive because it allows parents to modify their parental care behavior based on current conditions (Lack 1954; Jeon 2008). In years with high levels of resources, all offspring have a high probability of survival, whereas in years of low levels of resources, only the early-hatching, best-developed offspring have a high survival probability. Brood reduction benefits parents who cannot provide sufficient food for all offspring when resources are scarce. In fact, both parents *and* surviving offspring benefit if some siblings do not survive, because this increases the likelihood that some offspring will survive. Parent-offspring conflict theory predicts that (1) later-hatching offspring should suffer higher mortality as a result of reduced care, and (2) resource availability should affect survivorship.

Scott Forbes, Richard Grosshans, and Barb Glassey tested the first prediction in yellow-headed blackbirds (*Xanthocephalus xanthocephalus*) (Forbes, Grosshans, & Glassey 2002). Yellow-headed blackbirds typically lay clutches of three to five eggs over a period of two days. Forbes and colleagues classified the first two eggs laid as "core" and any subsequent eggs as "marginal." They examined the growth rate and survivorship of these egg types from sites near Winnipeg, Canada. They censused nests daily and marked the eggs when laid; each offspring was marked with nontoxic color markers. Each day, nestlings were weighed and mortality noted until chicks were 12 days old, the age at which they normally leave the nest (Scientific Process 13.3).

hatching asynchrony The hatching of offspring in a clutch of eggs on different days.

SCIENTIFIC PROCESS 13.3

Brood reduction in blackbirds

RESEARCH QUESTION: *Does hatch asynchrony affect offspring mortality?*

Hypothesis: Early-hatching offspring are larger than and competitively dominant to their siblings.

Prediction: Mortality will be highest for late-hatching individuals.

Methods: The researchers:

- checked nests (*n* = 245) daily. They uniquely marked eggs and resulting chicks with nontoxic colored markers. The first two chicks hatched in a nest were marked as "core" and the others as "marginal."
- weighed nestlings each day and noted mortality until chicks were 12 days old.

Results:

- Core chicks were heavier than marginal ones in each year of the study.
- Nestling survival was highest for core nestlings and significantly lower for marginal offspring.

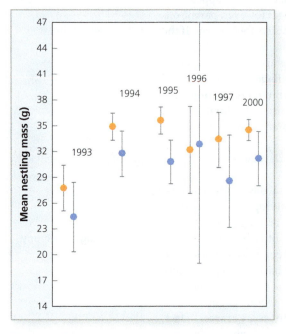

Figure 1. Body mass. Mean (± SE) body mass of core nestlings (orange) was higher than marginal (blue) in all years (*Source:* Forbes et al. 2002).

Figure 2. Survival. Core nestlings had higher mean (+ SE) survival than marginal nestlings (*Source:* Forbes et al. 2002).

Conclusion: Hatching asynchrony facilitates brood reduction by creating offspring of different size and competitive ability.

Very few of the nests (only 16%) fledged their entire clutch of eggs, and asynchronous hatching did affect offspring growth and survival. The mean mass of core nestlings was higher than that of marginal nestlings in most years, which translated into higher survival. This finding supports the first prediction, that later-hatching siblings are at a competitive disadvantage and suffer higher mortality (brood reduction).

What about our second prediction about how resource availability affects survivorship? Evaluating this question requires a long-term study in a variable environment, as we see next.

Brood reduction in fur seals

Featured Research

In some marine mammals as well as some social carnivores, a second offspring is born before the previous one is weaned. This timing creates families with offspring that vary greatly in body size and development, just as in bird clutches with asynchronous hatching. Female Galápagos fur seals (*Arctocephalus galapagoensis*) (Figure 13.22) follow a yearly reproductive cycle. Shortly after giving birth, females copulate and can produce another offspring the following year. However, pups are not weaned until they are two years old. About 35% of females will give birth in two successive years and so have two related offspring of different ages for whom they must care. Lactation for two pups is very costly for the female and leads to high offspring mortality. Why, then, do females give birth in successive years?

These seals live in the eastern Pacific Ocean, where there can be severe annual changes in the environment. Fur seals feed on fish and cephalopods whose population sizes are affected by cold-water upwelling, which has a very high nutrient content. In years of high upwelling and cold water, fish are common and food is plentiful for seals. When water temperatures are warmer (no upwelling), food for seals is scarce. Upwelling is affected by the Pacific tropical climate system El Niño and is unpredictable from one year to the next. Fritz Trillmich and Jochen Wolf hypothesized that survivorship of pups with an older sibling is related to resource availability (Trillmich & Wolf 2008). They predicted that second offspring should suffer high mortality in years of scarce resources and have high survivorship in years of abundant resources.

Trillmich and Wolf studied 188 tagged females on Fernandina Island in the western Galápagos archipelago for ten years. Unweaned older pups were marked with ear tags, and their younger siblings were marked with dye. The survivorship of each pup was noted for its first 30 days of life. To estimate food availability, the researchers observed sea surface temperature (SST) in the eastern Pacific. Years of low SST correspond to high marine productivity and abundant food for fur seals.

The early survivorship of pups with older siblings was related to SST. In cold-water years (with low SST), pup survival was relatively high, approaching the level of survivorship for pups without an older sibling in one year (Figure 13.23). In years of higher SST and lower food availability, the survivorship of pups with an older sibling was dramatically lower. These results support the prediction that brood reduction is a function of food availability. In years of high resources, brood reduction is minimal, and parents can take advantage of the rich food supply. The production of "extra" offspring allows parents to increase their reproductive output. In contrast, in years of low food availability, offspring mortality of the youngest pup is high. Conservation biologists are taking advantage of this relationship to help boost the population of endangered species (Applying the Concepts 13.2).

Figure 13.22. Fur seal with pups. Galápagos fur seal female caring for her pups of different ages.

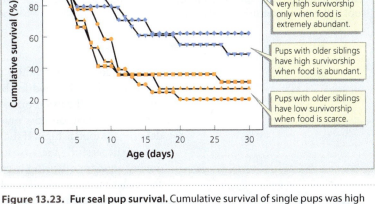

Single pups have high survivorship.

Pups with siblings have very high survivorship only when food is extremely abundant.

Pups with older siblings have high survivorship when food is abundant.

Pups with older siblings have low survivorship when food is scarce.

Figure 13.23. Fur seal pup survival. Cumulative survival of single pups was high in all years (horizontal dashed line). The survival of pups with older siblings was related to water temperature, which was an index of food availability. In cold-water years with high food availability (blue), survival was high. In warm-water years with low food availability (orange), survival was low (*Source*: Trillmich & Wolf 2008).

APPLYING THE CONCEPTS 13.2

Food supplementation reduces brood reduction in endangered eagles

Brood reduction is common in many birds of prey, such as hawks and eagles. One species that exhibits brood reduction is the endangered Spanish imperial eagle (*Aquila adalberti*). Females lay clutches of one to four eggs over a one-week period, but mortality of chicks is high after the first offspring, so typically only about one chick fledges. The Spanish imperial eagle is one of the rarest birds of prey in the world, with a population size of around 200 breeding pairs. Given brood reduction in the species, conservation biologists have considered two strategies to increase population size. First, biologists can remove second-hatched chicks and place them in other nests or hand-raise them and release them into the wild as juveniles (Meyburg 1978). However, these procedures disturb the nest, are costly, and often are ineffective. An alternative is to provide supplemental food to adults with two chicks.

Luis González and colleagues tested the prediction that supplemental feeding would reduce brood reduction in Spanish imperial eagles in western Spain (González et al. 2006). For 13 years, they provided European rabbit carcasses to 22 nests containing more than one chick. Another 37 nests with multiple chicks served as controls.

Supplemental feeding increased chick survival almost every year (Figure 1). In most years, almost twice as many chicks fledged in nests that received supplemental food than did in control nests. The research team concluded that Spanish eagle reproduction is food-limited and that population growth can be enhanced by supplemental feeding. This management tool is cheap, nonintrusive, and effective.

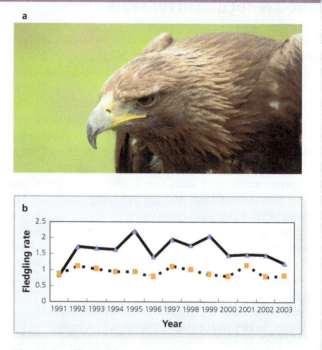

Figure 1. Effect of food supplementation on fledgling rate. (a) Spanish imperial eagle. (b) Fledgling rate (number of fledglings per pair) for food-supplemented (blue) and control (orange) nests in different years (*Source:* González et al. 2006).

Why do parents produce more offspring than they can care for? If they didn't, there would be no need for brood reduction. First, "extra" offspring provide some "insurance": if the first-born offspring dies, a second can live. Second, parents do not have information on how much food will be available in the future when offspring demand for food is high. If the environment contains abundant food, a parent may be able to successfully increase its reproductive fitness for the year. If the benefit of producing an extra offspring is greater than the cost, we can predict that selection will act such that parents will produce "extra" siblings and brood reduction will occur in years of low resources. In this light, brood reduction allows parents to adjust their reproductive effort to environmental conditions.

Brood parasitism

Throughout this section, we have seen that parental care is costly and that the costs are often born disproportionately by females. Such costs can favor the evolution of behaviors that reduce it. One way to reduce the cost of parental care is for a female to lay her eggs in the nest of another female, or a "host," a behavior known as **brood parasitism**. The benefit of this behavior is reduced parental care, which can then result in enhanced reproductive success for the brood parasite. Although the benefit of parasitism is clear, it does entail some risk: a parasitized female may reject the added eggs or chicks, a result that will not increase fitness for the female parasite. How can a brood parasite reduce the likelihood of egg rejection?

brood parasitism A behavior in which a female (brood parasite) lays an egg in the nest of another female.

Some brood parasites lay their eggs in the nest of other species and are known as interspecific brood parasites. Several dozen of these, like the common cuckoo (*Cuculus canorus*), are obligate brood parasites: they do not nest themselves and *only* lay eggs in the nests of other species, thus avoiding all costs of parental care. Because such species never raise their own offspring, many have traits that reduce the risks of egg rejection by a host, including the production of eggs that match the size, color, and shape of host eggs so that the host cannot distinguish its eggs from that of the parasite (Stokke, Moksnes, & Røskaft 2005) (Figure 13.24). The co-evolutionary arms race between interspecific brood parasites and their hosts has been the subject of intense research (Feeney, Welbergen, & Langmore 2012).

The more common way to avoid egg rejection is to parasitize conspecifics; such conspecific brood parasites lay their eggs in the nests of other females of the same species. Rejection by the host will be difficult because eggs of all conspecific females look similar. Conspecific brood parasitism is common in birds, with examples from over 200 species (Yom-Tov 2001), and also occurs in fish and insects (Wisenden 1989; Tallamy 2005).

The cuckoo egg is larger.

Figure 13.24. Parasitic egg in nest. A cuckoo egg (the largest) in a redstart (*Phoenicurus phoenicurus*) nest. Note the similarity in the egg coloration.

To examine the benefits of conspecific brood parasitism, we might envision three reproductive tactics that a female could use (Sorenson 1991):

1. Lay all eggs in her own nest (nonparasite)
2. Lay all eggs in the nest of other females (pure parasite)
3. Lay some eggs in her own nest and some in the nest of others (a mixed strategy of nesting and parasitizing)

Let's examine one study that attempted to quantify the fitness of these three reproductive strategies.

Conspecific brood parasitism in ducks

Matti Åhlund and Malte Andersson examined the prevalence of conspecific brood parasitism in a population of marked female goldeneye ducks (*Bucephala clangula*) nesting in Sweden (Åhlund & Andersson 2001). They used egg albumin protein fingerprinting to identify the maternity of eggs and hatchlings within a nest. Albumin is deposited by the female into the egg and contains over a dozen proteins. Many of these proteins exhibit genetic polymorphisms, so there is much genetic variation among females. Therefore, each female exhibited a unique protein-banding pattern, allowing researchers to identify the eggs of each female (Andersson & Åhlund 2000). Nests were monitored regularly to determine egg laying, and there was no evidence of egg removal by any female. All chicks were marked with numbered tags and their survival recorded over a period of four weeks.

The researchers found that all three strategies existed in the population studied, but females that used the nesting and parasitizing tactic had the highest reproductive success. These females laid about 50% more eggs than nonparasites or pure parasites (Figure 13.25), and the eggs laid in host nests survived as well as nonparasite eggs, indicating that they were not rejected.

Featured Research

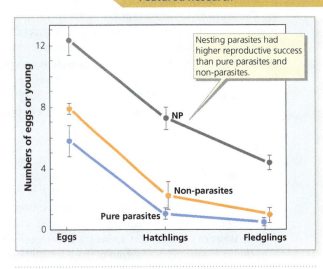

Nesting parasites had higher reproductive success than pure parasites and non-parasites.

Figure 13.25. Brood parasitism in ducks. Mean (± SE) number of eggs or young for females classified as pure parasites (blue), non-parasites (orange), or nesting parasites (NP, gray). The latter laid the most eggs and had the highest number of hatchlings and fledglings survive (*Source:* Åhlund & Andersson 2001).

The researchers observed variation in the reproductive tactics used by a female across years and found no evidence that nesting parasites suffered lower survival or reproduction in subsequent years, despite laying more eggs. Why this occurred is unclear, and so further work is required to understand the reproductive tactics of conspecific brood parasites in this population. Nonetheless, one factor that likely favors the evolution of brood parasitism in this and other species is its minimization of the costs associated with parental care.

Chapter Summary and Beyond

Species differ in life history traits, which can explain the variation in parental care across species: those with a long life span or well-developed offspring at birth tend to provide high levels of parental care. Research has focused on understanding the evolution of parental care in different groups. For instance, parental care in birds appears to have evolved first in their extinct ancestors, the dinosaurs (Varricchio et al. 2009). Other work has examined the evolution of trade-offs involving parental care and other life history traits in insects (Gilbert & Manica 2010) and fish (Kolm & Ahnesjö 2005). In addition, males and females differ in the fitness benefits they obtain from providing parental care because of the differing costs of care, particularly in terms of the certainty of relatedness to offspring. For most species, and especially for species with internal fertilization, females are more certain of their offspring than males. This difference favors greater levels of parental care by females. Reduced parental care allows individuals to seek additional sexual partners or allocate more energy to future reproduction.

Prolactin is an important hormone that regulates parental care in many vertebrates. It initiates lactation and parental care behavior in mammals and also is involved in stimulating parental care in birds, as well as the production of nutrient-rich crop "milk" in mourning doves (*Zenaida macroura*) (Miller, Vleck, & Otis 2009). Much research today investigates how stress affects prolactin production. Increases in stress hormones typically suppress prolactin production. This suppression may be an adaptive response: decreases in parental care behavior allow individuals to concentrate on their own survival, but more work is needed to fully understand this pattern (Angelier & Chastel 2009). Prolactin may also be an important factor in the development of paternal care in mammals and fish, but the evidence is conflicting (Wynne-Edwards & Timonin 2007; Bender, Taborsky, & Power 2008). In insects, juvenile hormone production is often negatively associated with parental care behaviors. Mas and Kölliker (2008) have reviewed work on the physiological aspects of parental care in insects and highlight recent work on chemical communication between mother and offspring that affects the level of care provided.

Adults provide care to increase survivorship of their offspring, but doing so reduces their future reproduction, resulting in a trade-off for parents between allocating parental care effort to current and future offspring. This trade-off explains variation in care among individuals. Recent work has begun to examine how hormones mediate this trade-off (McGlothlin, Jawor, & Ketterson 2007). This trade-off also explains why parents often provide unequal care to siblings and how unequal care can result in brood reduction.

CHAPTER QUESTIONS

1. Consider three female chickadees of the same age and condition. Each successfully raises a clutch of eggs with no help from her mate, but the clutches differ in size. Female A raises four offspring, B raises eight, and C raises 12. What can you predict about the overwinter survivorship and reproductive output of these females next year?

2. Imagine a bird species with biparental feeding of young and equal biparental care. Suppose you manipulate one individual to increase its cost of flight by removing a few of its flight feathers or adding small weights to the bird. How will this manipulation affect its level of parental care in terms of providing food to offspring?

3. In their study of the effects of prolactin on parental care behavior, why did Bridges et al. (1990) first treat all individuals with bromocriptine to suppress endogenous prolactin secretion?

4. Hatching asynchrony is hypothesized to be an adaptation that allows parents to adjust their parental care and number of offspring to varying environmental conditions. Why would asynchronous clutches lead to higher adult fitness than synchronous clutches of the same size when environmental variation is large?

5. Imagine that you discovered two new species of fish. One has internal fertilization and the other fertilizes eggs externally. In both species, only one adult provides parental care for eggs in a nest. In which species is paternal care more likely to have evolved and why?

6. Develop an alternative hypothesis to explain why some big-thumbed male frogs provide little parental care when they fertilize small clutches.

Chapter 14

Social Behavior

Figure 14.1. Sociality. Paper wasps caring for eggs and larvae in a colony.

Around your home or campus, you have probably noted that some animals seem to live in large groups while others appear to be more solitary. On our campus in the winter, large flocks of crows come to roost in the trees each evening. They are very noisy and somewhat messy. In the spring, we often see ants following long foraging trails as they carry food such as seeds or dropped breadcrumbs back to their nest. Wasps and bees, too, are highly social like ants, living in colonies of hundreds or even thousands of individuals—and remarkably, most of them sterile (Figure 14.1). They spend their lives working to ensure the reproductive success of just one or a few queens.

In this chapter, we examine many aspects of group living. First, we examine the benefits and costs of life in groups to better understand variation in the size of groups. We then focus on a behavior unique to social groups: helping. We explore the different ways that individuals help one another and how biologists explain the evolution of helping behavior. We also examine in detail the evolution of sterile workers in highly social species.

14.1 Sociality evolves when the net benefits of close associations exceed the costs

Flocks of birds, schools or shoals of fish, swarms and colonies of insects, and herds or packs of mammals are examples of social groups—sets of individuals that remain in close proximity to one another (Figure 14.2). Across species, there is tremendous variation in the size, composition, and persistence of social groups, or their degree of **sociality**. Some groups, such as in coyotes (*Canis latrans*), contain just a few individuals. Others, such as a flock of European starlings (*Sturnus vulgaris*), a school of Atlantic herring (*Clupea harengus*), or a swarm of Australian plague locust (*Chortoicetes terminifera*), might contain many thousands or even millions of individuals. Group composition can also differ with respect to sex and age. During the nonreproductive season, for instance, many ungulates (hoofed animals) such as caribou (*Rangifer tarandus*) form single-sex groups, and flocks of common ravens (*Corvus corax*) are often composed of a single age class, juveniles. The time that individuals spend in groups can vary seasonally or daily. For example, starlings tend to join large flocks only in the fall and winter and then live as pairs defending territories during the breeding season. In contrast, many primates, parrots, ants, wasps, and coral reef fishes spend their entire life in a social group. Finally, groups can differ in the relatedness of members. Many groups, especially large ones, are composed of unrelated individuals; in contrast, smaller family groups are composed of closely related individuals, often containing siblings of different ages or individuals spanning more than two generations. Can we explain all this variation?

sociality The tendency to live and associate with others.

Sociality can evolve only if it results in an individual in a group obtaining higher fitness than it would living alone. In other words, the net benefits for an individual must be greater than the net costs. Numerous benefits and costs of sociality have been identified (Krause & Ruxton 2002). Benefits include reduced search time to find resources (food or mates), increased feeding success as a result of cooperative hunting, enhanced resource defense, reduced predation risk, lower physiological costs of thermoregulation and movement (Figure 14.3), the division of labor, and communal care of young. Costs associated with sociality involve increased aggression and competition for resources, as well as enhanced disease transmission (Table 14.1). We have already discussed social behavior in regards to the benefit of reduced predation risk (see Chapter 8) and the cost of increased aggression (see Chapter 10). Here, we examine other benefits and costs associated with life in groups.

Figure 14.2. Social groups. (a) A large school of anchovy fish. (b) A single-sex group of male caribou.

Figure 14.3. Physiological benefits. (a) Penguins and (b) bats huddle to minimize heat loss.

TABLE 14.1 Benefits and costs of sociality.

Benefits of sociality
1. Increased diet breadth for predators
2. Increased ability to find resources
3. Decreased search time to find food or mates
4. Decreased risk of predation
5. Decreased physiological costs of movement in air or water
6. Decreased physiological costs of thermoregulation via huddling
7. Division of labor: individuals can specialize on different tasks (e.g., food acquisition, nest defense)
8. Communal care of young: enhanced feeding and defense of offspring

Costs of sociality
1. Increased competition for resources
2. Increased opportunity for aggressive interactions
3. Increased likelihood of pathogen transmission

Foraging benefits: reduced search times for food in minnows

Featured Research

Food is often not conspicuous in the environment, and so animals spend considerable time searching for it. How might sociality increase foraging success? Some birds, for example, have higher encounter rates with insects when the movements of group members flush prey out of hiding (Harsha et al. 2007). In other cases, individuals in groups benefit from the food discoveries of others. In Chapter 5, we saw how individuals can use the presence of others to learn about the location of food. This same mechanism produces a benefit of sociality: individuals in groups can find food patches faster because more individuals are searching. Once hidden food is found, all group members can move to the patch to feed.

Tony Pitcher, Ann Magurran, and Ian Winfield investigated this benefit of sociality in the common minnow (*Phoxinus phoxinus*), a small fish found throughout Europe in freshwater streams and rivers that feeds on a variety of food items buried in the gravel or sand, including mollusks, insects, and crustaceans.

Pitcher and colleagues tested the prediction that as group size increases, the search time for an individual to find hidden food will decrease (Pitcher, Magurran, & Winfield 1982). They conducted experiments in a large aquarium (2 m by 70 cm by 70 cm) that held schools of two, four, six, 12, and 20 fish. In order to examine the effect of group size on an individual's time to find food, the researchers uniquely marked one fish, the test fish, and recorded its behavior. Before the start of a trial, the researchers placed a food patch, consisting of 84 ice cube tray wells filled with gravel, in the aquarium. The researchers randomly added dried fish protein flakes into one of the cubes. They then placed the group in the tank and recorded the amount of time until the test fish found the food. They conducted four replicates, using a different test fish, for each school size.

As predicted, the time before the test fish found the food declined as group size increased (Figure 14.4). This finding

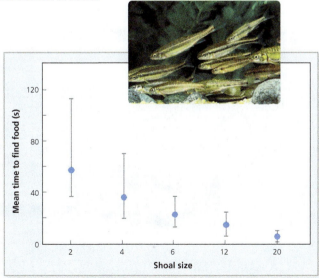

Figure 14.4. Time to find food. Mean (± SE) time to find food. Individual minnows find food faster when they feed in shoals of larger size (*Source:* Pitcher, Magurran, & Winfield 1982). *Inset:* minnow.

demonstrates one advantage of group living: the larger the group, the less time it takes an individual to find food, because others are searching as well.

Featured Research

Foraging benefits: increased diet breadth in coyotes

diet breadth The number of different food types in the diet.

One additional feeding benefit of group living applies to predators. Many predators increase the range of food items they capture, or their **diet breadth**, by feeding in groups, which in turn can increase food intake and fitness. Some large prey items can only be captured successfully by a group. Eric Gese, Orrin Rongstad, and William Mytton examined how group size affected the diet breadth of coyotes (*Canis latrans*) (Gese, Rongstad, & Mytton 1988) (Figure 14.5). Coyotes are medium-sized carnivores that live in small groups of one to four individuals. They feed on a variety of foods, including vegetation, lizards, rodents, rabbits, and both small juvenile and larger adult deer.

In a large study area in Colorado, Gese and colleagues conducted a biweekly census of coyote group size over a period of 33 months. They noted the size of each group and collected fresh scats (or feces) to determine group members' diet. Coyote scats contain large quantities of fur, bone, claws, and teeth that make it fairly easy to identify the prey eaten. The researchers divided the food types into three categories: large (greater than 10 kg), including adult ungulates such as deer; medium (0.5 to 10 kg), including rabbits and ungulate fawns; and small (less than 0.5 kg), including vegetation, rodents, reptiles, and birds.

The research team found that as mean group size increased, coyotes included a greater volume of large prey in their diet—up to 15% (Figure 14.6). In contrast, solitary coyotes and pairs of coyotes rarely fed on large prey. The researchers concluded that increases in coyote group size allowed more successful foraging on large ungulates, which often require several coyotes to capture. The change in diet breadth is likely to have increased a coyote's overall food intake, but this prediction remains to be tested.

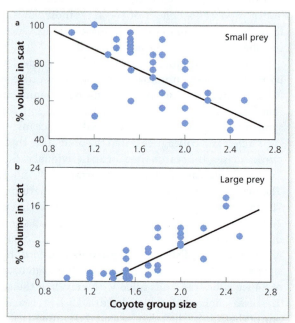

Figure 14.6. Coyote diets. Coyotes in larger groups consume less (a) small prey and more (b) large prey. Note that the y-axis scale differs in the two figures. The x-axis represents mean monthly group size (*Source:* Gese, Rongstad, & Mytton 1988).

Figure 14.5. Coyote group. A group of four coyotes feed on a large ungulate.

Antipredator benefits in honeycreepers

A second benefit of group living is reduced predation risk. Individuals in groups can spot approaching predators sooner than can solitary individuals and so could have more time to take evasive action. In addition, being surrounded by others also allows individuals to spend less time being vigilant, which frees up more time for other activities such as feeding. In Chapter 8, we examined these benefits for individuals living with conspecifics, but these same benefits should also extend to individuals living with other species in **mixed-species groups**.

Patrick Hart and Leonard Freed examined the benefits of sociality in mixed-species flocks of Hawaiian birds. They focused on the Hawaii akepa (*Loxops coccineus*), an endangered Hawaiian honeycreeper that commonly flocks with other species (Figure 14.7). Akepas feed on arthropods in the leaf buds of terminal branches in the canopy, a behavior that exposes feeding birds to hawk predators. Akepas also exhibit variation in sociality: they feed both alone and in mixed-species flocks of up to dozens of birds.

Hart and Freed examined how antipredator vigilance behavior varied as a function of flocking (Hart & Freed 2005). They predicted that the vigilance rate of individuals in mixed-species flocks should decline with increased flock size. The researchers observed birds feeding in the Hakalau Forest National Wildlife Refuge. Hart and Freed conducted focal animal sampling to record a bird's vigilance behavior (the number of scans in a 20-second period), its social status (in a flock or not in a flock), and flock size (the number of individuals within a 10 m radius). They found that as overall group size increased, the level of individual vigilance decreased (Figure 14.8), just as predicted, illustrating that individuals in mixed-species flocks can obtain similar benefits of sociality as those in single-species flocks.

However, Hart and Freed found that the effect of flocking on vigilance rate differed for male and female akepas: males in flocks had lower levels of vigilance than males not in flocks, but females did not. Instead, females exhibited the same level of vigilance whether or not they were in a flock (Figure 14.9). Why? This species exhibits a striking sexual dimorphism: males are bright orange, whereas females are dull brown (Figure 14.7). The researchers hypothesized that males might suffer higher predation risk because predators can spot them more readily. Thus, males not in a flock must spend more time engaging in vigilance behavior than females. This study emphasizes the importance of examining the behaviors of males and females separately, because different selection pressures may exist, as we saw in Chapters 11 through 13.

mixed-species group A social group containing two or more species.

Figure 14.7. Akepa. (a) Female and (b) male akepa.

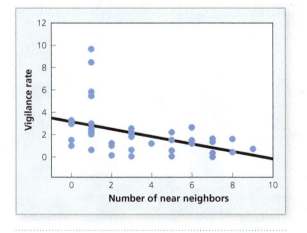

Figure 14.8. Vigilance and group size. Vigilance rate (scans per 20 seconds) declines as group size increases (*Source:* Hart & Freed 2005).

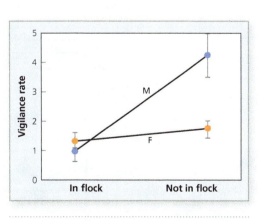

Figure 14.9. Vigilance levels in and out of flocks. Mean (± SE) vigilance rate (scans per 20 sec) for males (M) is significantly lower when in a flock. In contrast, the vigilance level of females (F) is unaffected by their level of sociality (*Source:* Hart & Freed 2005).

Aerodynamic benefit: reduced cost of movement in pelicans

One additional benefit of sociality is the aerodynamic advantage of moving with a group. Many large birds such as geese and cranes fly in a V-formation, with fairly uniform spacing between individuals. Why this specific formation? One possibility is that it provides an aerodynamic advantage that reduces the cost of flying. Henri Weimerskirch and colleagues examined this hypothesis using trained white pelicans (*Pelecanus onocrotalus*) (Weimerskirch et al. 2001). They placed heart rate monitors on the birds and videotaped their flight formations as they followed a motorboat or ultralight airplane. Birds flying in a V-formation had a lower wing-beat frequency and heart rate than birds flying alone, probably because they were flying in the wake of another bird (Figure 14.10). These data demonstrated a clear decrease in energy expenditure for birds flying in formation. Interestingly, smaller bird species like sparrows do not fly in formation, perhaps because their body size does not allow the same aerodynamic advantage, although this idea has not been tested (Hummel 1983).

The costs of sociality

Given the benefits of sociality, why don't all individuals live in social groups? The answer is that sociality also entails costs: individuals in groups are surrounded by

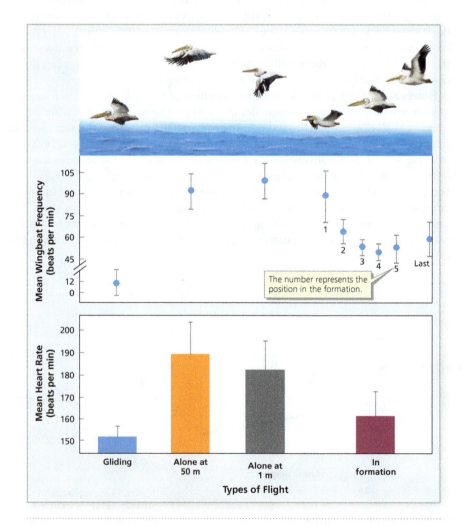

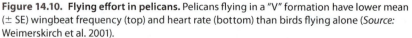

Figure 14.10. Flying effort in pelicans. Pelicans flying in a "V" formation have lower mean (± SE) wingbeat frequency (top) and heart rate (bottom) than birds flying alone (*Source:* Weimerskirch et al. 2001).

close competitors. Competition, when individuals negatively affect one another, can be intense when the individuals in a group require the same resources for survival and reproduction. For example, when a group of minnows finds a food patch, all of them exploit the food. If the food is limited, each individual in the group obtains less food than it would if it fed alone, and its fitness, in terms of energy consumed, is reduced. As group size increases, so does the level of competition and intensity of aggressive interactions. Therefore, competition for resources will often be keen and is a primary factor that limits the size of groups. In addition, diseases are more readily transmitted among individuals that live in groups, and so individual fitness can decline as group size increases. Let's see how researchers study these costs of group living.

Competition in schooling fish

Featured Research

One way to examine the cost of competitive interactions is to compare the fitness of solitary versus social individuals. If competitive interactions are an important cost of sociality, then the fitness of individuals in groups can be lower than the fitness of individuals who live alone. Gábor Herczeg, Abigél Gonda, and Juha Merilä examined this prediction by studying growth rates in young nine-spined sticklebacks (*Pungitius pungitius*) (Herczeg, Gonda, & Merilä 2009). Sticklebacks are small fish that live in lakes, rivers, and coastal areas throughout the Northern Hemisphere. The research team examined the growth rates of fish reared either alone or in a school. Higher growth rates are associated with higher fitness in fish. Adults were obtained from two ponds, one in Sweden and one in Finland. They were brought into the laboratory and allowed to reproduce. Four-day-old fry from each population were divided into two treatment groups: in one set, fish were reared alone in individual 1.4 L aquaria, and in the other set, fish were reared in schools of 100 fish in 140 L aquaria. This design ensured that the average water volume for each individual was identical across treatments. All fish were fed to satiation twice a day; uneaten food had to be removed frequently, indicating that there was little to no competition for food. Fish were measured at 12 and 20 weeks to determine their standard length.

In both populations, individuals raised in a group had lower growth rates than individuals grown alone (Figure 14.11). Because per capita water volume was the same, individuals had unlimited access to food, and other variables such as predation risk and parasite load were controlled, the researchers concluded that differences in growth rate must be related to sociality itself. What aspect of sociality caused the decreased growth rate? One possibility is aggressive interactions among fish, but further research is needed to test this hypothesis.

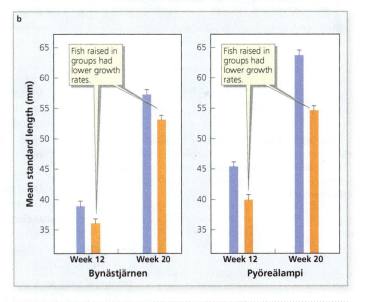

Figure 14.11. Fish growth rate. (a) Nine-spine stickleback. (b) Mean (+ SE) standard length of same-aged individuals at 12 and 20 weeks of age from two study populations. Fish raised alone (blue) had higher growth rates than fish raised in social groups (orange) (*Source:* Herczeg, Gonda, & Merilä 2009).

Group size and competition in primates

Featured Research

If competition for resources often limits group size, we can predict that low-resource environments should have smaller group sizes than high-resource environments. In other words, we can predict that there should be a positive correlation between the resources in a habitat and observed group size. Colin and Lauren Chapman tested this prediction in two species of primates in Uganda (Chapman & Chapman 2000).

Figure 14.12. Red colobus and red-tailed guenon. (a) Red colobus and (b) red-tailed guenons feed in groups.

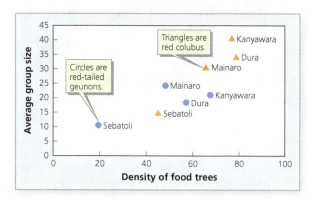

Figure 14.13. Primate group sizes. The average group size in two species of primates correlates positively with the density of food-producing trees. Circles are red-tailed guenons; triangles are red colobus. Names denote populations (*Source:* Chapman & Chapman 2000).

The Chapmans studied red colobus (*Procolobus pennantii*) and red-tailed guenons (*Cercopithecus ascanius*) at four sites in Kibale National Park over a period of two years (Figure 14.12). These medium-sized (3 to 10 kg) primates live in tropical forests in equatorial Africa. Group size ranges from fewer than ten individuals to several dozen. Both species feed on a variety of fruits and vegetation.

In order to examine the influence of competition for resources on sociality in this system, the research team first characterized the specific diets of the species at each site. Both species fed primarily on young leaves, leaf buds, fruits, and flowers from dozens of tree species. Because all food items came from trees, the researchers could estimate food abundance at each site from the density of food trees, the presence of fruits or flowers, and the stage of leaf development. The Chapmans established four to six 200 m paths, or transects, across each site. Each month, the researchers walked along the transects, examining each potential food tree to determine whether it offered food for the monkeys. They also observed the size of all groups at each site for each species.

The Chapmans found a positive relationship between the density of food trees and average group size across the four sites, as predicted (Figure 14.13). This finding supports the hypothesis that group size is limited by competition for food. Because the monkeys' sites were relatively close to each other, the researchers assumed that other ecological variables (such as predators) that might affect group size were similar across sites and so had little effect on their results.

Featured Research

Sociality and disease transmission in guppies

Another important cost of group living involves the transmission of infectious diseases. Many parasites and pathogens reduce fitness (Daszak, Cunningham, & Hyatt 2000) and are transmitted by close contact. Contact between individuals increases with degree of sociality, and so it is not surprising that there is often a positive correlation between group size and the proportion of infected individuals in a group. After all, individuals will frequently be in close proximity when group size is large (Brown & Brown 1986; Côte & Poulin 1995).

E. Loys Richards, Cock van Oosterhout, and Joanne Cable examined how the degree of sociality affects disease transmission in guppies (*Poecilia reticulata*)

(Richards, van Oosterhout, & Cable 2010). Guppies are small tropical fish that live in small shoals (or schools). They are frequently infected with external parasitic worms such as *Gyrodactylus turnbulli* that attach to their fins, gills, and scales. Female guppies tend to spend more time in schools than males, and the researchers took advantage of this variation to examine differences in disease spread in schools of either females or males.

They first removed all external parasites from a population of guppies. Groups of six males or six females were housed in separate tanks. The researchers focused on one randomly selected focal guppy and identified it by its unique color markings. To quantify the degree of sociality, each group was observed daily for 15 minutes for three days. Every 30 seconds, the research team noted the distance of the nearest neighbor from the focal fish and the maximum school size, defined as the number of guppies within four body lengths of one another. At the end of the third day, the focal fish was infected with approximately 100 *G. turnbulli* worms and then placed back with its school to potentially infect others. Three days later, all fish were inspected for parasite infection.

As expected, female groups spent more time schooling and had tighter groups than males—that is, females exhibited a higher degree of sociality (Figure 14.14). After the parasite was introduced, the researchers found that almost 80% of the females in each tank became infected after three days, while only 40% of the males became infected (Figure 14.15). These results suggest that disease transmission in guppies appears to be affected by social behavior: greater time spent near others, as in the female schools, led to higher rates of disease transmission.

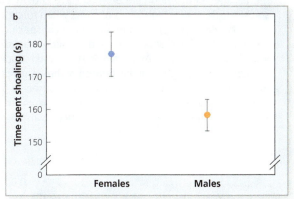

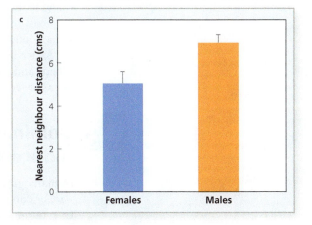

Figure 14.14. Guppy shoaling. (a) School of guppies. (b) Mean (±SE) nearest neighbor distance and (c) mean (+ SE) time spent shoaling for males and females. Female guppies were more tightly spaced and spent more time shoaling than males (*Source:* Richards, van Oosterhout, & Cable 2010).

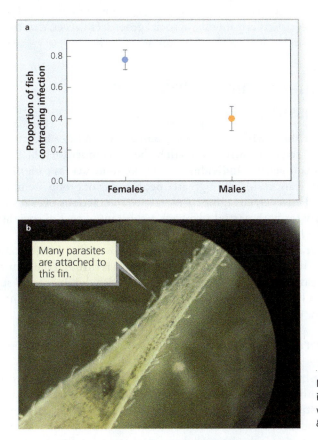

Figure 14.15. Infection rates. (a) Females had higher mean (±SE) infection rates than males in the experiment. (b) *G. turnbulli* parasitic worms attached to the fin of a guppy. (*Source:* Richards, van Oosterhout, & Cable 2010).

APPLYING THE CONCEPTS 14.1

Group size of social species in captivity

Zoos house many social species, including primates and ungulates. What is the best group size for species in captivity? Group size in the wild is often strongly affected by antipredator benefits and competition for food. For captive animals, these factors are irrelevant, meaning that the group size typical in the wild will often not be the best group size for an animal in captivity.

Elizabeth Price and Tara Stoinski have shown that many social species suffer high levels of stress when forced to live in captivity with an inappropriate number of conspecifics (Price & Stoinski 2007). Social species that are housed with too few individuals often exhibit signs of chronic stress and reduced reproduction. For example, gorillas housed as breeding pairs have lower reproductive success than gorillas housed in groups of three or more (Maple & Hoff 1982). Stress, increased levels of aggression, and reduced reproductive success also occur when

densities in captivity are too high. Duikers are an instructive example. These African ungulates live in very small social groups or are solitary. The Los Angeles Zoo initially housed duikers in groups of five, but individuals often developed stress-related jaw abscesses and lived less than five years. When group size was reduced to no more than three, individual health and life span both increased dramatically (Barnes et al. 2002).

Price and Stoinski suggest that naturally observed group size in the wild is rarely the best group size in captivity. To minimize stress and aggression from overcrowding, institutions should create captive environments with greater structural complexity. For example, zoos can provide animals with both visual barriers and escape paths so that individuals can modify their proximity to others. Zoos will also need to regularly assess how the selected group size affects individual well-being.

We have examined several of the many benefits and costs associated with life in groups. When the benefits of sociality are large relative to the costs, we expect to see individuals in groups. When the benefits are low and the costs are high, we expect to see small group sizes or a lack of sociality. This understanding is being used to benefit social groups housed in captivity (Applying the Concepts 14.1).

As we have seen, competitive interactions are an important cost of living with others, and they can be particularly intense in social groups. However, that cost may not apply equally to all individuals within a group, as we see next.

14.2   Dominance hierarchies within groups reduce aggression

In many groups, individuals will vary in their competitive abilities based on age, size, or condition, and so some individuals will be better competitors than others. In general, the most competitive individuals will have more access to limited resources than weaker rivals, resulting in greater net benefits of group membership (Vehrencamp 1983). However, engaging in aggressive interactions over resources runs the risk of injury for both parties involved, a substantial cost of sociality. In most groups, repeated encounters between two individuals usually ends in the same result: the better competitor wins the interaction. Given this predictability of outcomes, one way to minimize aggression within groups is through the formation of a **dominance hierarchy**, an organized social system with dominant and subordinate members. In a **linear dominance hierarchy**, an individual is dominant to each individual below it and subordinate to all above it in rank.

dominance hierarchy An organized social system with dominant and subordinate members.

linear dominance hierarchy An organized social system in which an individual is dominant to each individual below and subordinate to all above it in rank.

Featured Research

Dominance hierarchies and crayfish

Crayfish are small freshwater crustaceans that feed on a wide variety of foods, including living and dead animals and plants (Figure 14.16). They are relatively sedentary and live and feed on stream bottoms. When aggregated in high densities, they often fight over resources and form dominance hierarchies. In aggressive contests, crayfish exhibit both aggressive and submissive behaviors. Aggressive behaviors include approach, in which crayfish move toward an opponent; threat display, in which they spread their claws; and attack, in which they move quickly toward an opponent with claws open. Submissive behaviors include retreat and escape. A retreating individual moves slowly away, whereas an escaping individual makes a rapid movement away from an opponent.

Jens Herberholz, Catherine McCurdy, and Donald Edwards studied the benefits of a social dominance hierarchy in juvenile crayfish (*Procambarus clarkii*) (Herberholz, McCurdy, & Edwards 2007). Because crayfish do not live in permanent social groups, the research team first investigated whether crayfish form dominance hierarchies in the absence of a resource. Second, they asked whether the formation of a hierarchy reduces aggression and determines access to food.

The research team conducted a simple but elegant experiment. They placed three individuals, uniquely marked with dots of color on their carapace, into an aquarium (15 by 30 by 20 cm) that was divided into two compartments by an opaque partition. Individuals were placed in the larger compartment, and all agonistic interactions were recorded for 30 minutes in the absence of any resource. Next, a large food item (a piece of chicken liver) was placed in the other compartment, and the opaque partition was lifted. They recorded the number of aggressive and submissive behaviors exhibited by each individual, both in the absence and presence of the food, to calculate a dominance index (defined as the number of aggressive acts divided by the total number of aggressive plus submissive acts). They also recorded the proximity of each individual to the food (to determine the individual's access to it). This experiment was repeated for ten sets of three crayfish.

In each group, crayfish displayed high levels of aggression. In the first part of the experiment and in the absence of a resource, aggression levels were high and a linear dominance hierarchy was established. The average number of aggressive and submissive behaviors was 5.3 acts per minute, and in each group, the researchers identified a top-ranked alpha individual that possessed the highest dominance index, a second-ranked beta individual, and a third-ranked gamma individual that possessed the lowest dominance index. However, aggression levels were significantly reduced in the second part of the experiment, even though food was present, with the researchers recording a mean of only 1.3 aggressive and submissive acts per minute. As expected, dominance rank strongly affected access to the food. Individuals with higher ranks spent more time in contact with the food than did lower ranked individuals (Figure 14.17). The researchers concluded that dominant crayfish have greater access to resources and that the formation of a dominance hierarchy significantly reduces aggressive interactions among individuals. Once relative dominance ranks are formed, there is little to gain from continued aggression.

Stable dominance hierarchies in baboons

Aggressive interactions are a source of stress, and **glucocorticoids** are hormones that are secreted by the adrenal gland in vertebrates in response to stressful situations. Chronic stress is associated with decreased immune system function and reproduction and thus negatively impacts fitness (Sapolsky 2005). If dominance hierarchies function to reduce aggression within groups, they should also lessen stress. Conversely, any disruption in the dominance hierarchy could result in increased stress levels.

Thore Bergman and colleagues investigated whether an unstable dominance hierarchy could result in increased stress in chacma baboons (*Papio hamadryas*

Figure 14.16. Crayfish. Crayfish live and feed on stream bottoms.

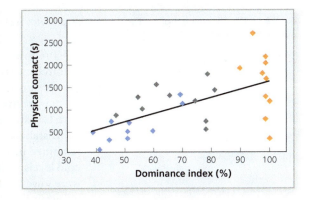

Figure 14.17. Dominance status. The most dominant alpha males (orange) tended to spend the most time in contact with food. Beta individuals (gray) spent more time in contact with food than the lowest ranking gamma individuals (blue) (*Source:* Herberholz, McCurdy, & Edwards 2007).

Featured Research

■ **glucocorticoids** Hormones secreted by the adrenal gland in vertebrates in response to stressful situations.

Figure 14.18. Chacma baboons. These Old World primates are found throughout southern Africa.

ursinus) (Bergman et al. 2005) (Figure 14.18). For over 14 months, the team studied a population in the Moremi Game Reserve in Botswana. This population has been intensively studied for decades (e.g., Bulger & Hamilton 1987), and so the researchers knew that its males form a strict linear dominance hierarchy in which alpha males have sole access to fertile females. However, this population also lives near other groups (which allows males to easily move between groups), and so changes in rank are common, with the identity of the alpha male changing, on average, about every six months. The research team conducted over 1000 ten-minute scans to record male aggression and collected 482 fecal samples to measure glucocorticoid levels (using a radioimmunoassay procedure) (Beehner & Whitten 2004). Each sample was time-coded, and at least seven samples were collected for each male.

The researchers divided the study into time periods in which the dominance hierarchy was either stable (times with no change in alpha or beta males) or unstable (times during which the alpha or beta male changed). Over the 14-month study, there were five periods of instability. These periods of instability were interspersed with two stable hierarchy periods of two to three months each.

Instability strongly increased aggression among males, whereas aggression remained below average during stable periods (Figure 14.19). Male fecal glucocorticoid levels varied with dominance stability, just as predicted. Stress hormone levels were lowest during stable periods but were significantly higher during each period of instability (Figure 14.20). The increase in glucocorticoid levels during periods of instability occurred in both high- and low-ranking males. Therefore, the stable dominance hierarchy has important fitness consequences for all males in the group.

These examples illustrate how the formation of dominance hierarchies can reduce aggression in groups and so reduce an important cost of living with others.

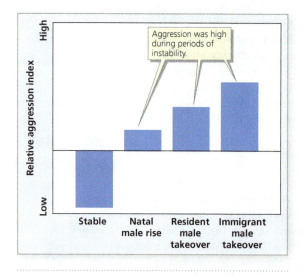

Figure 14.19. Male aggression levels. Male aggression was low when the dominance hierarchy was stable. Aggression levels were higher during periods of instability in the group dominance hierarchy (*Source:* Bergman et al. 2005).

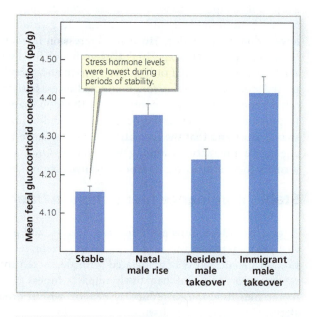

Figure 14.20. Stress hormone levels. Mean (+ SE) male fecal glucocorticoroid levels were lowest (indicating low stress) in stable periods (*Source:* Bergman et al. 2005).

However, in all dominance hierarchies, many individuals are subordinates—so why do they remain social? We examine that question next.

Social queuing in dominance hierarchies in clownfish

Many social animals exhibit dominance hierarchies and, like the chacma baboons, live in permanent groups. Given that subordinates obtain fewer resources than dominants, why do they stay in the group? One reason is that subordinates benefit from associating with others in different ways, such as reduced predation risk. Second, in many groups, subordinates can eventually move up the social ladder as they age, a process called **social queuing**. Social queuing is particularly important with regard to territory inheritance and breeding opportunities (McDonald 1993; Field, Shreeves, & Sumner 1999; Kokko & Johnstone 1999; East & Hofer 2001), and it occurs in a wide range of animals, including insects, fish, birds, and mammals (Kokko & Johnstone 1999).

social queuing A process in which subordinates eventually move up the social ladder as they age.

Peter Buston studied social queuing in clownfish (*Amphiprion percula*). These colorful reef fish live in and around sea anemones, which keep predators at bay (Figure 14.21). Sea anemones are cnidarians: they lack a hard skeleton and possess stinging tentacles that inject a neurotoxin into fish that come into contact with them, an adaptation that is effective in capturing prey. Clownfish have an outer mucus layer that makes them immune to the anemones' sting. These fish live in small groups that consist of a breeding pair and up to four nonbreeders. Adults are protandrous: on maturation, they develop into males but later can develop into females. Within each group, there is a size-based dominance hierarchy in which the breeding female is the largest and most dominant fish, followed by the breeding male and then any nonbreeders. Nonbreeders do not have functioning gonads and so obtain no immediate reproductive benefits from the group, but they often help defend the territory. Why, then, do they stay in a group? Buston predicted that subordinates may obtain future reproductive benefits through social queuing.

To test this idea, he recorded patterns of territory inheritance on three reefs in Papua New Guinea (Buston 2004). Over the course of one year, he observed 97 groups and recorded dominance interactions, the recruitment of new individuals, and changes following the disappearance of one of the breeders. No reversals in rank were observed. Loss of the breeding female always led to a chain reaction: the

Figure 14.21. Clownfish and sea anemones. Clownfish groups live in association with anemones. Note the size difference among group members.

breeding male changed sex and became the dominant breeding female, and the next highest ranking nonbreeder in the group became a sexually mature male. Buston had observed social queuing in action.

Next, he experimentally manipulated 16 groups by removing the breeding male. In each case, nonbreeders moved up the social ladder. The highest ranking nonbreeder became a sexually mature breeding male, and any new recruit always entered the group as the lowest ranking member. Buston never observed an outsider usurp a breeding vacancy.

Buston concluded that nonbreeding group members benefit from group membership by eventually inheriting the territory and thus obtaining reproductive success. Clownfish adults are not strong swimmers and are often preyed upon when they leave their anemones. It is risky to venture out in search of breeding vacancies, away from the protection of anemones. This risk may explain why Buston did not observe outsiders taking over a breeding territory. Biding time within a social queue yields higher fitness for an individual, which helps explain why subordinates remain in groups where they often receive less resources and have fewer mating opportunities than dominant individuals.

So far, we have examined various benefits and costs of social living and have seen how the formation of dominance hierarchies within groups can reduce one important cost: aggression. We have also seen how subordinates can eventually attain breeding status through social queuing. Another way that all group members can benefit by living with others is by receiving their help, a common behavior in some species. How can natural selection favor this behavior when helpers appear to gain nothing and the act of helping may incur a cost in fitness? Next, we examine how evolutionary biologists explain this behavior.

14.3 Helping behavior, or altruism, is often directed toward close kin

altruism A behavior that results in the increased fitness of another individual and involves a cost to the individual performing the behavior.

Helping behavior, or **altruism**, is any behavior that increases the fitness of another and also involves a cost to the individual performing the behavior. The first explanation for altruistic behavior traces its roots back to Darwin (1859) but was formalized by William Hamilton (1964). Imagine that an individual possesses an allele that enhances its likelihood of being altruistic. If that individual helps another individual to reproduce, and if the two individuals are genetically related, there is some probability that both individuals carry the same allele, because they share genetic relatives. Thus, by helping a relative, the altruist is helping to pass its own alleles on to the next generation indirectly. It has increased its **inclusive fitness**, the sum of an individual's own reproductive success plus the success obtained by relatives as a result of the focal individual's behavior. Selection for behaviors that increase the fitness of relatives is called **kin selection**.

inclusive fitness A combination of individual fitness and the fitness obtained by helping close relatives.

kin selection A form of natural selection in which individuals can increase their fitness by helping close relatives, because close relatives share the helper's genes.

Hamilton's rule An equation predicting when altruism should evolve based on the fitness benefits and costs of the altruistic behavior, and the degree of relatedness among an altruist and recipient.

Hamilton's rule

Hamilton proposed that for altruism to evolve, the additional fitness benefits (B) obtained by the recipient must be greater than the fitness cost (C) to the altruist, adjusted for the degree of relatedness (r) between the two individuals (Table 14.2). Hamilton formalized this relationship in an inequality (Hamilton 1964), now known as **Hamilton's rule**:

$$\text{Altruism can evolve when } B \cdot r > C$$

This equation makes predictions about when altruism will evolve, but the fitness benefits, fitness costs, and the degree of relatedness can be difficult to estimate. Many tests of Hamilton's rule therefore examine a more general prediction of kin selection: helping behavior should be most common between close relatives (when r is large), less common between distant relatives (when r is low), and rare between unrelated individuals (when r is 0).

TABLE 14.2 Degree of relatedness. Degree of relatedness (r) for different pairs of diploid animals, like humans.

RELATIONSHIP	COEFFICIENT OF RELATEDNESS (r)
Parent–offspring	0.50
Sibling–sibling	0.50
Grandparent–grandchild	0.25
Aunt/uncle–nephew/niece	0.25
First cousins	0.125
Friends (unrelated)	0

Figure 14.22. Belding ground squirrels. A social mammal that lives in large colonies.

Belding ground squirrel alarm calls

Featured Research

Paul Sherman conducted one of the first studies of kin selection by studying alarm-calling behavior in Belding's ground squirrels (*Spermophilus beldingi*) (Sherman 1977). These rodents live in high-density populations and interact frequently in subalpine meadows in western North America (Figure 14.22). Adult males and females live in very different social environments. Females tend to be surrounded by female relatives, because juvenile females only disperse very short distances from their natal burrow. In contrast, males tend to be surrounded by unrelated squirrels, because juvenile males disperse more than 250 m from their natal burrow, and adult males move an additional 150 m away from the burrow of their mate when the offspring are born.

Ground squirrels are attacked by a variety of predators and produce specific vocalizations, or alarm calls, in response to predators such as long-tailed weasels (*Mustela frenata*) and badgers (*Taxidea taxus*). Alarm calls benefit others, who will typically run to the nearest burrow or scan the environment, behaviors that reduce their risk of predation. Alarm calls also entail a risk: producing a vocalization can attract a predator to the caller. Can kin selection explain the evolution of alarm-calling behavior in ground squirrels?

Sherman studied one population over three years. The population contained hundreds of squirrels that had been marked permanently over the previous five years as well as during the study, so the age and relatedness of all squirrels up to eight years old were known. Sherman recorded 102 occasions when a predator was present while squirrels were active: 22 marked squirrels were subsequently stalked or chased. He also noted the first squirrel to produce an alarm call.

He found that adult females were significantly more likely to give the first alarm call than were adult males or juveniles (Figure 14.23). In fact, adult males rarely gave the first alarm call when a predator was present. Sherman also noted that individuals paid a fitness cost of calling: significantly more calling squirrels were hunted by predators than were noncallers, and more callers were killed. Three that were killed gave an alarm call just before being attacked.

Sherman concluded that kin selection can explain the observed variation in alarm-calling behavior because adult females, and not males, are surrounded by close relatives. If alarm calls help the survival of close relatives, this behavior could provide genetic benefits to the female callers. Alarm calls are costly to produce, because individuals who give them suffer higher predation risk than noncallers, and so adult males rarely produced these calls.

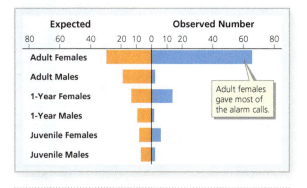

Figure 14.23. Ground squirrel alarm calls. Observed (blue) and expected (orange) number of first alarm calls given based on the numbers of different individuals in each age class in the population. Adult females gave significantly more calls than expected, whereas adult males gave significantly fewer than expected (*Source:* Sherman 1977).

These results demonstrate that kin selection explains variation in helping behavior when animals face predation risk. Next we examine helping behavior in the context of reproduction.

Altruism and helping at the nest in birds

In some birds and mammals, adults physiologically capable of reproducing forgo breeding and instead remain in small groups to help others raise offspring, a behavior known as **cooperative breeding** (Clutton-Brock 2002). Nonbreeders often provide parental care (egg incubation, food provisioning, and antipredator behavior) for the offspring of the breeders. These "helpers at the nest" clearly exhibit altruistic behavior. They pay the energetic costs associated with parental care, sacrifice the chance to breed themselves, and their behavior benefits the fitness of others. Can this behavior be understood using kin selection?

> **cooperative breeding** A situation in social groups in which adults physiologically capable of reproducing forgo breeding and instead help others raise offspring.

Andrew Russell, Ben Hatchwell, and colleagues studied helping behavior in long-tailed tits (*Aegithalos caudatus*) in Great Britian (Russell & Hatchwell 2001; Hatchwell et al. 2004) (Figure 14.24). These small insectivorous birds live in social groups of up to 16 members. In the spring, monogamous pairs form and begin reproductive behavior (nest building, egg laying, and incubation). However, many nests are quickly destroyed by predators, and when that occurs, failed breeders often become helpers by feeding the offspring of another pair. The researchers investigated whether kin selection can explain this behavior, posing the research question: Do helpers feed the young of a close relative?

The team captured and uniquely marked individuals in three populations over the course of several years. The researchers used a combination of known pedigrees and genetic data obtained from blood samples to determine the degree of relatedness among individuals in the study populations. They visited all nests regularly and recorded the success or

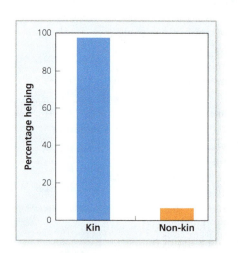

Figure 14.24. Long-tailed tits. A social bird common in Europe.

Figure 14.25. Helping and relatedness. Long-tailed tits helped pairs that contained at least one close relative (kin) more than they helped unrelated pairs (non-kin) (*Source:* Russell & Hatchwell 2001).

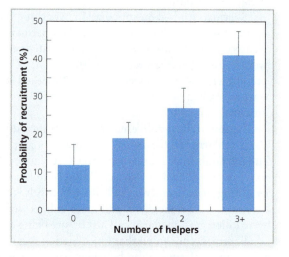

Figure 14.26. Fledgling success. Mean (+ SE) probability of recruitment into the adult population as a function of the number of helpers at the nest. The probability of fledgling recruitment into a social group the following year increased with the number of helpers at their nest (*Source:* Hatchwell et al. 2004).

failure of each nest, all instances of helping behavior, and the long-term survivorship of nestlings, as determined by their recruitment into a social group the following year.

The researchers observed 37 birds that they classified as helpers at the nest of another pair. In all cases, these were birds whose own nesting attempt had failed. Nearly all helpers assisted a pair that contained at least one relative: in over 90% of cases, this individual was a close relative, such as a sibling (Figure 14.25). In addition, the number of helpers at a nest positively affected the probability that a nestling would survive and recruit into a social group the following year (Figure 14.26). These data provide strong evidence that kin selection has played a role in the evolution of helping behavior in this species: helpers do in fact feed the young of a close relative.

Altruism in turkeys

Featured Research

In the examples thus far, helping behavior has been most common among relatives and rare between unrelated individuals, as we should expect based on kin selection. Can we measure all the parameters of Hamilton's rule to more precisely test its prediction? Alan Krakauer attempted just that in turkeys (*Meleagris gallopavo*).

These large birds (over 1 m long in length) live in open fields and woodlands throughout much of North America. Turkeys often form male display partnerships in which a pair of males, called a **coalition**, displays together to court and defend females against other males (Figure 14.27). However, only one of the two males in a coalition mates, the dominant one. Why does the subordinate male display with another male when doing so leads to no direct reproductive fitness benefit? One possibility is that males are closely related and the subordinate individual obtains indirect fitness benefits via kin selection (Watts & Stokes 1971).

Krakauer tested this idea by studying a population at Hastings Natural History Reservation near Carmel, California (Krakauer 2005). He captured and marked 126 turkeys with colored wing tags and collected a blood sample to conduct DNA analysis. DNA was genotyped to determine both relatedness among individuals and

coalition An altruistic partnership between a pair of individuals.

Figure 14.27. Male turkeys. Males form coalitions to attract and defend mates.

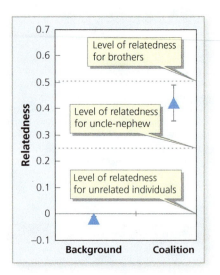

Figure 14.28. Relatedness in turkey coalitions. Males in coalitions are closely related ($r = 0.42$), while random pairs of males in the population are unrelated. Data are mean ± SE (*Source:* Krakauer 2005).

paternity of offspring to quantify the reproductive success of males. In addition, dozens of birds were fitted with radiotransmitters to facilitate rapid location of birds and their nests.

Turkeys were located regularly to observe courtship behavior and associations among males. Coalitions were defined as pairs of males within 2 m of each other that displayed at least twice to a female. The dominant male was identified as the turkey that performed the most complete mating display, strutting. Solitary males were those adults that did not meet these criteria and were observed displaying alone to a female at least twice.

The combination of behavioral observations and genetic data allowed Krakauer to calculate all three parameters in Hamilton's rule: r, B, and C. He determined that coalitions were composed of very close relatives. The mean degree of relatedness, r, between individuals was 0.42, slightly below the value for full brothers (0.50) but significantly higher than that of any two random males in the population (Figure 14.28).

Krakauer found that in coalitions, subordinate males fathered no offspring, but a dominant male fathered, on average, 7.0 offspring. In contrast, solo males sired, on average, only 0.9 offspring (Figure 14.29). We can now plug these results into Hamilton's rule. First, the additional fitness benefit, B, that the dominant obtained from the cooperation it received from a subordinate was $7.0 - 0.9$, or 6.1 offspring per male. The fitness cost of helping, C, was 0.9, the number of offspring that the subordinate could achieve as a solo male. Finally, the average relatedness, r, between coalition members was 0.42. In sum,

$$B \cdot r = 6.1 \cdot 0.42 = 2.56 > 0.9 = C$$

Hamilton's rule is satisfied for subordinate males: $B \cdot r$ is in fact greater than C. The fitness of altruistic subordinates is higher (thanks to kin selection) than the fitness of nonaltruistic solo males. In this case, subordinate males indeed obtain higher fitness by forming a coalition with and helping a close relative reproduce instead of trying to reproduce on their own.

As these examples illustrate, kin selection is an important concept for understanding the evolution of helping behavior among close relatives in a variety of species. This concept has also been used to understand the evolution of an extreme form of helping behavior in many species, as we see next.

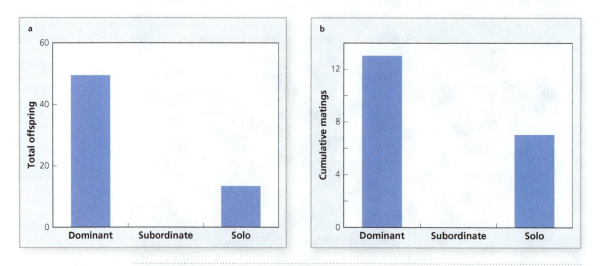

Figure 14.29. Reproductive success in male turkeys. (a) Dominant males sired more offspring and (b) obtained more matings than subordinates and solo males. Note that subordinate males had no direct reproductive success (*Source:* Krakauer 2005).

14.4 Some species exhibit extreme altruism in the form of eusociality

In some social groups, individuals adopt different behavioral roles. We've already seen this tendency in the dominance relationships of clownfish: some individuals reproduce, while others may only help with territory defense. In clownfish, subordinates often become breeders later in life, but in other species, some subordinates never breed and instead only help others to reproduce.

How could such an altruistic lifetime strategy evolve? Remember that Hamilton's rule predicts that individuals should be altruistic when $B \cdot r > C$. Can this inequality explain the evolution of an altruistic nonreproductive strategy? Yes. Hamilton's rule has been used to understand the evolution of helping behavior in species that contain sterile individuals that cannot reproduce. These species include bees, wasps, ants, termites, and even a few mammals that have evolved an extreme form of sociality called eusociality. In **eusocial** species, each social group or colony contains (1) overlapping generations, (2) cooperative brood care by nonparents, and (3) the reproductive division of labor (Batra 1966; Burda et al. 2000). Adults are either reproductive or nonreproductive individuals that care for young and defend the colony. These behaviorally distinct groups, or **castes**, can be morphologically distinct and allow for division of labor (Applying the Concepts 14.2).

One taxon in which thousands of species contain sterile individuals is the Hymenoptera, which includes ants, bees, and wasps. Hymenopteran castes include large reproducing queens and smaller, sterile workers who forage and defend the nest. Ants are a classic example of a eusocial insect. Many ants produce a new generation each year. During the breeding season, winged males and reproductive females (the future queens) fly away from their current nest in a nuptial flight. These females mate with a male or multiple males and then store their sperm. Once mated, a queen loses her wings, establishes a new nest colony, and begins to produce eggs that will develop into workers. Workers are sterile females that may be morphologically specialized for their different roles (Figure 14.30). They help feed and maintain the queen and her brood, forage for food, and defend the colony.

eusocial A species that lives in social groups that contain overlapping generations, have cooperative offspring care by nonparents, and have reproductive division of labor.

castes Morphologically and behaviorally distinct individuals within a social group.

haplodiploid genetic system A genetic system with haploid males and diploid females.

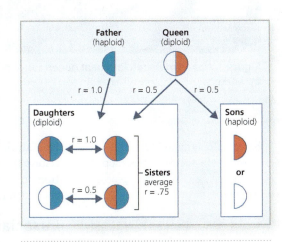

Figure 14.30. **Ant castes.** Major and minor workers of *Pheidole barbata*.

Eusociality and haplodiploidy

William Hamilton proposed an explanation for the prevalence of sterile worker castes in Hymenoptera (Hamilton 1963; Hamilton 1964). In addition to their unusual social system, many ants and other Hymenopterans also have a unique genetic system. This **haplodiploid genetic system** is characterized by haploid males that develop from unfertilized eggs and diploid females that develop from fertilized eggs, and it results in unusual genetic relationships among individuals. Diploid female offspring display the usual degree of relatedness to their queen mother of 0.5. However, daughters inherit their haploid father's entire genome, and so their degree of relatedness to their father is 1.0 (Figure 14.31). How can we explain the evolution of diploid daughter workers that forgo reproduction? All daughters share the same genetic makeup from their haploid father—half of their genome. The other half of their genome comes from their mother, who, because she is diploid, provides two possible genotypes for her daughters. As a result, sisters share the half of their genome that is inherited from their father ($1 \cdot 0.5$) but only a quarter of their genome (on average) that is inherited from their mother ($0.5 \cdot 0.5$). Therefore, although females are related to their offspring by 0.50, sisters are, on average, related to each other by 0.75:

$$(1 \cdot 0.5) + (0.5 \cdot 0.5) = 0.75$$

Figure 14.31. **Haplodiploidy.** Daughters share half of their genome with their mothers but inherit all of their father's genome. Sons share half of their genome with their mother and none with their father. Sisters share (on average) three-quarters of their genome.

APPLYING THE CONCEPTS 14.2

Are naked mole rats eusocial?

Naked mole rats (*Heterocephalus glaber*) are odd-looking burrowing rodents found only in East Africa and perhaps your local zoo (Figure 1). They live in large, complex underground burrows where they feed on tubers (underground plant structures that store nutrients). Naked mole rats are well adapted for life underground, with large front teeth that function as shovels to dig tunnels and bodies that can run fast both forwards and backwards. They live in social groups of up to several hundred, but only the queen reproduces. Other members of the group help care for her litters.

Jarvis (1981) was the first to suggest that naked mole rats might be a eusocial mammal. Recall that we defined eusocial species as having (1) overlapping generations, (2) cooperative brood care by nonparents, and (3) a reproductive division of labor, with either reproductive or nonreproductive castes that care for young and defend the colony. Naked mole rats satisfy most of these criteria, but unlike eusocial insects, their variation in body size is gradual, not discrete, and some workers are similar in size to the queen. This lack of distinct castes has led some to propose that naked mole rats are not eusocial after all but are instead more similar to cooperative breeding species. So which are they—eusocial or not?

Justin O'Riain and colleagues examined the internal morphology of naked mole rat queens and workers to determine whether they exhibit any distinct anatomical differences like those found in Hymenoptera castes (O'Riain et al. 2000). The researchers measured six skeletal traits, including the skull, vertebrae, and femur. Most of these traits simply increased continuously with body size, except for one: the lumbar vertebrae L5 (Figure 2), which was over 50% longer in queens than in other colony members. The researchers speculate that increased vertebral length allows for an increase in the size of the reproductive tract and thus litter size (naked mole rat queens can give birth to 28 pups in one litter). These data support arguments that naked mole rats do indeed have discrete morphological castes—enough to satisfy the definition of eusociality. These animals also happen to live very long lives (up to almost 30 years) and seem to be resistant to cancer. Researchers are currently using them as a model organism to better understand senescence (aging) and cancer.

Figure 1. Naked mole rat. Pregnant queen surrounded by workers.

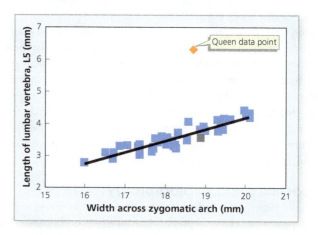

Figure 2. Mole rat vertebral lengths. The length of lumbar vertebrae L5 is 50% larger in queens (orange) than in workers (blue) and the breeding male (gray) (*Source:* O'Riain et al. 2000).

haplodiploidy hypothesis The evolution of eusociality in Hymenoptera is based on their haplodiploid genetic system.

In Hamilton's **haplodiploidy hypothesis**, a high degree of relatedness between sisters ($r = 0.75$) explains why females benefit from forgoing their own reproduction and instead helping the queen produce more offspring (which are genetically their sisters).

Featured Research

Eusociality in ants

Can we evaluate Hamilton's hypothesis? Katja Bargum and colleagues examined the relatedness of workers in the ant *Formica fusca* (Bargum, Helanterä, & Sundström 2007). They collected queens and workers from 56 colonies in Finland and genotyped them using six polymorphic DNA microsatellite loci. For about one-third of the colonies, the average degree of relatedness among workers was 0.70, not significantly different from the 0.75 predicted by the haplodiploidy hypothesis.

However, in the rest of the colonies, the degree of relatedness among workers was approximately 0.3. Why this discrepancy? Some colonies had a single queen, and these conformed to the haplodiploidy hypothesis prediction. The other colonies had multiple queens, which resulted in a lower average relatedness among workers. In some eusocial species, queens mate with multiple males, which also reduces the degree of relatedness among workers. If relatedness among workers is significantly lower than that predicted by the haplodiploidy hypothesis, kin selection may not fully explain the evolution of eusociality in Hymenopterans. Kin selection and the haplodiplody hypothesis can explain the evolution of eusociality only when a colony has one queen who mates with a single male.

Evolution of eusociality and kin selection in Hymenoptera

Featured Research

How might we resolve the discrepancy between the predictions of kin selection and eusocial species with low degrees of relatedness among individuals? One way is to use the comparative method to examine historical patterns of reproduction in this group.

William Hughes and colleagues examined the evolution of breeding systems in 267 species of eusocial bees, wasps, and ants, focusing on female mating behavior (Hughes et al. 2008). They mapped female reproductive behavior of each species onto a phylogeny of the Hymenoptera. For each species, they determined whether a female had just one mate (monandry) or two or more mates (polyandry). In the latter case, they determined whether females exhibited low polyandry (fewer than two mates, on average) or high polyandry (more than two mates, on average).

The phylogeny indicated that eusociality evolved independently eight times, and that monandry was always the ancestral state (Figure 14.32). The research team concluded that polyandry was a derived trait, especially in species that today exhibit high levels of polyandry. This finding means that the original state was monandry and that polyandry evolved later. Therefore, the researchers concluded that kin selection likely played an important role in the initial evolution of eusociality in the Hymenoptera. Polyandry, they argue, evolved in several species only after the reversion from eusociality was impossible (Hughes et al. 2008).

Not all researchers agree. Edward O. Wilson has proposed that ecological factors play a more important role than kinship in the evolution of eusociality (Wilson 2008). In insects, the evolution of eusociality is typically observed in species that must defend a nest site located within foraging distance of a persistent food source. Wilson argues that these conditions favor groups that remain together and so can benefit from communal defense. Over time, multilevel selection (see Chapter 2) will favor some groups over others—particularly those that more efficiently harvest resources and defend the nest—thanks to a division of labor (Wilson 2008; Nowak, Tarnita, & Wilson 2010). In Wilson's scenario, relatedness within a colony is a consequence, rather than a cause, of sociality. The debate continues today.

As we have seen, kin selection explains altruistic behavior between related individuals and even eusociality. However, nonrelated individuals also help one another, and so we need another explanation. We examine this behavior next.

14.5 Helping between unrelated individuals requires reciprocity

In many groups, unrelated individuals help one another, a behavior that cannot be explained by kin selection. For example, many birds, ungulates, and primates groom unrelated individuals to remove ectoparasites, a behavior called **allogrooming**. This behavior is altruistic because it benefits the recipient and requires time and

allogrooming One individual removing ectoparasites from another individual.

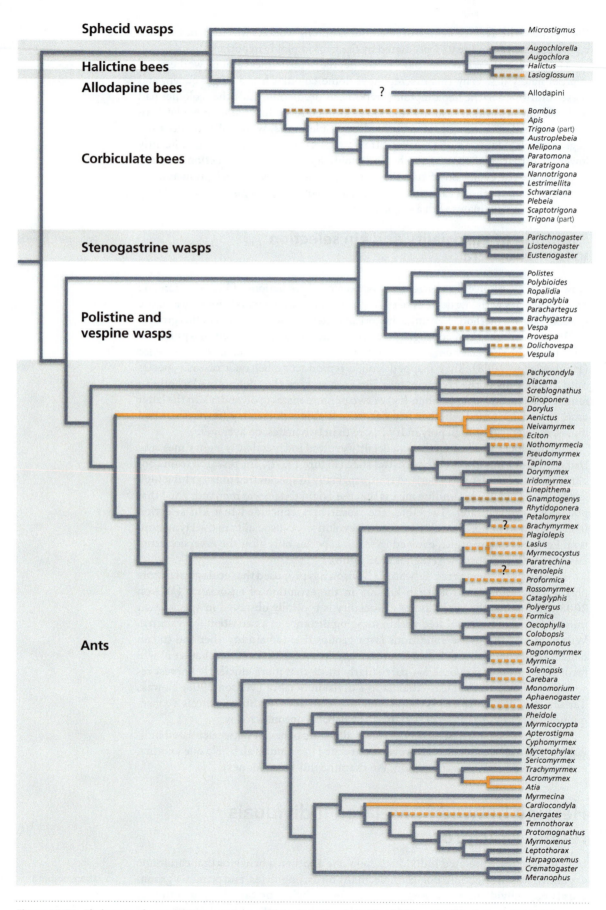

Figure 14.32. Phylogeny of eusocial Hymenoptera and female mating behavior. Solid orange lines indicate species that are polyandrous; dashed orange lines indicate low-polyandry clades. Blue lines indicate monandrous clades. Shading or non-shading indicates different major groups (*Source:* Hughes et al. 2008).

effort on the part of the groomer. How can such behaviors evolve? Robert Trivers offered an answer: altruism between nonrelated individuals can evolve through **direct reciprocal altruism**, or direct reciprocity (Trivers 1971), a behavior from which an altruist can benefit if the recipient of an altruistic act reciprocates and helps the altruist in the future.

Trivers argued that helping behavior can evolve between unrelated individuals only when the benefit of helping exceeds the cost. For Trivers, the evolution of reciprocal altruism requires two individuals to interact repeatedly so that over many interactions, each individual helps and subsequently is helped. If the benefit received exceeds the cost of helping, individuals that help will obtain higher fitness through the repeated interaction than will nonhelpers, and so helping behavior can spread in a population. Trivers also recognized that reciprocal altruism is subject to cheating: an individual that receives help but then does not reciprocate will obtain a fitness benefit but pay no cost. The temptation to cheat can in fact prevent the evolution of helping, as we see next.

The prisoner's dilemma

An illustration of the temptation to cheat is the **prisoner's dilemma**, an example of a situation analyzed using a game theory approach popularized in economics and political science that illustrates why the evolution of cooperation is difficult (Trivers 1971). Imagine that two criminals have broken into a museum and stolen a valuable piece of art. The criminals agree not to rat on each other if they are captured; both will remain silent about the crime when interrogated. They are captured, but the police have only enough evidence to convict them of a lesser crime (breaking into the museum). The police need the criminals to testify against each other in order to charge them with a major crime (stealing the art).

Each prisoner is interrogated in a separate room, and the police offer him freedom if he agrees to testify against the other. Each can remain loyal to their agreement and refuse to confess, or he can testify against his partner. If both stay silent, they will each spend one year in prison. If both betray one another, the police will renege on their offer of freedom and convict each suspect to a three-year prison term. If one suspect betrays his partner and the other stays silent, the betrayer will go free while his partner will go to jail for ten years.

Will the prisoners stick to their agreement and remain silent? To see why an agreement is likely to break down, we can imagine the prisoner's dilemma as a kind of game with winners, losers, and strategies for winning. In this game, each player can play one of two strategies: cooperate with his partner (stay silent) or betray his partner (testify against him). The payoffs for Partner A look like this and are the same as the payoffs for Partner B:

> **direct reciprocal altruism** An altruist can benefit if the recipient of an altruistic act reciprocates and helps the altruist in the future.

> **prisoner's dilemma** A game theory model that illustrates why the evolution of cooperation (altruism) is difficult.

Prisoner's dilemma payoff matrix			
		Partner B	
		Cooperate	Betray
Partner A	Cooperate Payoff	Cooperation (C): −1	Sucker (S): −10
	Betray Payoff	Temptation to cheat (T): 0	Betrayal (B): −3

A negative number represents the number of years in jail, so 0 is the best payoff (no jail time) and larger negative numbers are lower payoffs (more jail time). Obviously, if both players cooperate and remain silent, their payoff is relatively good, because they will only spend one year in prison (payoff = −1). However, there is a clear temptation to cheat: an individual can do better by ratting on his partner when his partner stays silent. Such an individual receives the highest payoff (0) and avoids jail time.

How should an individual behave? What is the evolutionary stable strategy? For a single play of the game, the best strategy is for an individual to betray the other. Such a strategy always avoids the −10 payoff (Sucker) that a player will receive if he remains silent and then is betrayed. So we see that for the one-time play of the game, cooperation is not favored, even though mutual cooperation would lead to less jail time. More formally, T is the temptation to cheat (the highest payoff), C is the payoff for cooperation, B is the payoff if both betray, and S is the sucker (and lowest) payoff. Although in this game $T > C > B > S$, the most likely outcome is still neither T nor C, but B.

The same reasoning can apply to direct reciprocity in animals. Here, T (temptation to cheat), C (both cooperate), B (betrayal, or neither cooperate), and S (sucker) are the outcomes in fitness of each combination of behaviors. When animals interact, fitness payoffs can be positive or negative, so we use different payoff values. Assume a cost of helping is −1 and a fitness payoff for receiving help is 3. The payoff when both help is 2, because each individual also pays the cost of helping. If neither helps, the fitness payoff is 0, as neither gains a benefit nor pays a cost. Again, the ESS is for neither individual to cooperate or help, because of the temptation to cheat.

Direct reciprocity payoff matrix

		Individual B	
		Cooperate	Not Cooperate
Individual A	Cooperate Payoff	Cooperation (C): 2	Sucker (S): −1
	Not Cooperate Payoff	Temptation to cheat (T): 3	Betrayal (B): 0

So how can cooperation evolve? Cooperation can be favored if individuals interact not just once, but repeatedly, providing the opportunity for reciprocity (Trivers 1971). Robert Axelrod and William Hamilton proposed one solution to the prisoner's dilemma game called the **tit-for-tat** strategy, a form of reciprocal altruism (Axelrod & Hamilton 1981). Here, a player always cooperates the first time it interacts with an opponent and then simply matches the behavior of the other player from the previous interaction. When two players use this strategy, they always cooperate on the first encounter (and all subsequent encounters) and so achieve the highest payoff. If a player using tit-for-tat interacts with a betrayer, the player receives the lowest payoff on the first interaction but then does not cooperate with that individual on all subsequent interactions to avoid receiving the sucker payoff again. If tit-for-tat is common, another strategy cannot take hold in the population (Axelrod & Hamilton 1981).

tit-for-tat In this strategy, a player always cooperates the first time it interacts with an opponent and then simply matches the behavior of the other player from the previous interaction.

Featured Research

Reciprocal altruism in vampire bats

Given this model, we can outline the conditions required for the evolution of reciprocal altruism: (1) individuals must have the opportunity to interact repeatedly, (2) the fitness benefit of receiving help must exceed the cost of providing help, and (3) individuals must be able to recognize one another so that they can reciprocate the benefit from a past helper and not help a past cheater.

Gerald Wilkinson thought that common vampire bats (*Desmodus rotundus*) might be one species in which these conditions are satisfied (Wilkinson 1984; Wilkinson 1990) (Figure 14.33). These bats live in relatively stable social groups with a life span of up to 18 years, providing the opportunity for repeated interactions. A social group, or cluster, living in a tree hollow typically contains eight to 12 breeding females and a breeding male. Females tend to associate with one another in distinct clusters, and female offspring remain in their natal cluster. Many females within a cluster are thus commonly related, but unrelated females do occasionally join clusters. A cluster will sometimes roost in different trees, so females within a cluster often mate with different males. Wilkinson determined that about 50% of the offspring from a cluster share the same father. Thus, relatedness between females is high but variable within a cluster.

Vampire bats feed on blood from vertebrates and are known to share blood meals. Females will often regurgitate blood to their offspring as a form of parental care. They also share food with other female adults. Food sharing among adults could be explained by kin selection or direct reciprocity. Why do they share food? Vampire bats starve if they do not feed on blood for 60 hours, and 7 to 30% of bats within a cluster fail to find a blood meal each night. Females that fail to find a blood meal will often solicit a meal from a successful roostmate by "begging."

Do unrelated roostmates help each other reciprocally? To examine this question, Wilkinson captured four individuals from each of two natural clusters that lived 50 km apart and put them together in captivity. With one exception of a grandmother-granddaughter pair, individuals within each cluster were not close relatives. Each night, one of the eight individuals was removed and deprived of food while all the other bats were fed a blood meal. The

Figure 14.33. Vampire bat. These social species feed on vertebrate blood and sometimes share food with others in their group.

next morning, the hungry bat was returned to the group and subsequent interactions were observed. This process was repeated in turn for each bat.

Wilkinson observed that food sharing within the experimental groups was common. With just one exception, it occurred only between individuals of the same natural cluster. Females preferentially shared blood with individuals with whom they lived, even though they were unrelated. Wilkinson also observed examples of reciprocity. Starved bats that received a blood meal would reciprocate and provide a blood meal to a previous donor when it was starving, more often than expected by chance. He estimated that a starved bat's benefit from receiving blood—reducing its risk of starvation—exceeded the cost of the well-fed bat's donation of blood. This situation therefore satisfied an important condition required for reciprocity to evolve.

Wilkinson's experimental data suggest that food sharing in vampire bats may have evolved as a form of direct reciprocity because unrelated adults helped one another. However, Wilkinson's observations of bats in natural clusters indicated that food sharing most often occurred between close relatives. Of 110 cases of food sharing recorded in the wild, only one case involved bats that were unrelated (Wilkinson 1984). Thus, Wilkinson concluded that both kin selection and reciprocity likely played a role in the evolution of food-sharing behavior in these bats.

Allogrooming in Japanese macaques

More recently, Gabriele Schino, Eugenia Polizzi de Sorrentino, and Barbara Tiddi studied direct reciprocal altruism in Japanese macaques (*Macaca fuscata*) (Figure 14.34). These macaques, native to Japan, live in large social groups. Two altruistic behaviors occur between adult females. First, macaques frequently exhibit allogrooming. Second, when agonistic interactions occur between group members, adult females will occasionally intervene and provide support for one of the animals. Their support benefits the recipient, which tends not to become injured or lose the encounter. However, such support also takes time, energy, and a risk of injury on the part of the altruistic supporter.

The research team studied a captive population of 23 adult females, all of whom were uniquely identifiable, in the Rome Zoo (Schino, Polizzi de Sorrentino, & Tiddi 2007). They used demographic records to determine the macaques' maternal relatedness—sisters, mothers, grandmothers, or aunts. Over the course of a year, the research team conducted over 1000 30-minute observations, recording all agonistic interactions, the aggressor and recipient, and any support provided by a third group member. They also conducted one group scan per observation to record allogrooming, noting both the donor and the recipient.

They found that an individual's helping behavior was not correlated with the level of relatedness between individuals, and therefore, kin selection could not provide an explanation. Instead, females chose to groom others based on the level of grooming they had received from that individual. In addition, females that had received support during an aggressive encounter were more likely to groom the supporting individual (Figure 14.35). Similarly, individual females chose to support others who had supported them in the past and from whom they had received grooming (Figure 14.36). Reciprocal altruism seems to be the best explanation for allogrooming and agonistic support in this case.

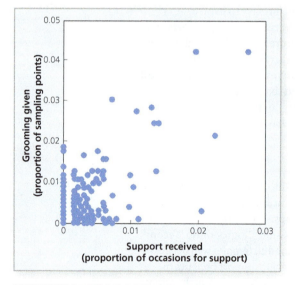
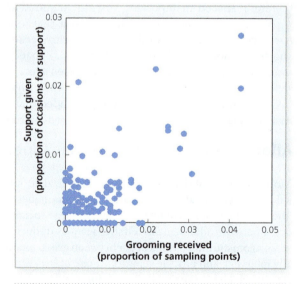

Figure 14.34. Japanese macaques. Japanese macaques exhibit many social behaviors, including (a) grooming and (b) aggressive interactions.

Figure 14.35. Macaque interactions. There was a positive relationship between "grooming given" and "support received" with the individual who supported them in the aggressive encounter. Each point represents a different dyad (*Source:* Schino, Polizzi de Sorrentino, & Tiddi 2007).

Figure 14.36. Third-party support. Macaques exhibited a positive correlation between "support given" and "grooming received" from the supporting individual. Each point represents a different dyad (*Source:* Schino, Polizzi de Sorrentino, & Tiddi 2007).

Although these examples suggest that direct reciprocity has played a role in the evolution of helping behavior, they do not involve evidence of the tit-for-tat strategy. Do animals cooperate as long as other animals cooperate as well but stop helping a cheater? A recent study on blackbirds suggests the answer is yes.

Tit-for-tat in red-winged blackbirds

Featured Research

Red-winged blackbirds (*Agelaius phoeniceus*) nest in dense aggregations in vegetation that grows at the edge of ponds. Males defend territories and often breed in the same territory for successive years. This behavior allows neighboring males to live side by side for multiple years, which provides the opportunity for repeated interactions. Crows (*Corvus brachyrhynchos*) are common nest predators of blackbirds. When they approach a nesting site, neighboring red-winged blackbirds often engage in cooperative nest defense in the form of mobbing (see Chapter 8) to drive them away (Figure 14.37).

Do neighboring red-winged blackbirds use a tit-for-tat strategy in cooperative nest defense? Robert Olendorf, Thomas Getty, and Kim Scribner manipulated the cooperative behavior of a neighbor in response to a threatening crow to address this question (Olendorf, Getty, & Scribner 2004). They first identified the territorial boundaries of neighboring males. One male was deemed the focal male and the other the neighbor. Next, they presented a stuffed crow predator at the territorial boundary of pairs of males in three phases of the experiment. In the first phase, the researchers recorded the territory defense behavior of both males in response to the crow. A predator at the territorial boundary represents an equal threat to both territory owners and should encourage cooperative mobbing.

In the second phase, half the focal males were categorized as controls. The others were assigned to the "simulated neighbor defector" treatment. These treatment males were captured and held in a building during the simulated defection, when the crow was again presented at the territory boundary. During the simulated defection, vocalizations of the defecting neighbor were played from his territory, which provided auditory information that the territory owner was present but not assisting in territorial defense. The researchers again recorded the behavior of focal males during this phase (and for controls, both focal and neighboring males). After data collection, the defecting neighbor was released, and he quickly returned to his territory. The next day, during the third phase, there was another presentation of the crow at the territorial boundary of the males, and their defense behavior recorded. Territorial defense behaviors were identified as approaches toward the crow, as well as strikes and dives at it. These behaviors were combined into a single value representing overall defense behavior.

During the first phase, focal and neighboring males exhibited similar high levels of defense behavior. In the simulated defection phase, however, focal males whose neighbor defected exhibited significantly less defense behavior than control males (Figure 14.38). This decrease possibly occurred because the focal males had a higher risk of injury when defending alone against a large predator such as a crow. More important, in the third phase, the reduced level of defense behavior persisted for treatment males whose neighbor had defected, compared with the level exhibited by control males. Males

Figure 14.37. Red-winged blackbird and crow. Crows are common nest predators. Red-winged blackbirds aggressively chase them from their territories during the breeding season.

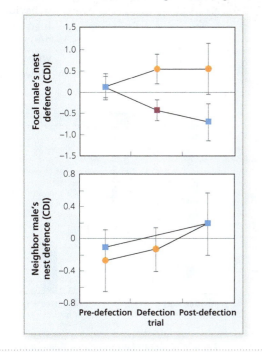

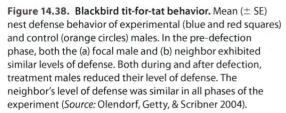

Figure 14.38. Blackbird tit-for-tat behavior. Mean (± SE) nest defense behavior of experimental (blue and red squares) and control (orange circles) males. In the pre-defection phase, both the (a) focal male and (b) neighbor exhibited similar levels of defense. Both during and after defection, treatment males reduced their level of defense. The neighbor's level of defense was similar in all phases of the experiment (*Source:* Olendorf, Getty, & Scribner 2004).

whose partner defected during the simulated defection phase allocated less time and effort toward cooperative mobbing the following day.

These data suggest that male red-winged blackbirds assess the level of cooperation of their neighbor and adjust their level of nest defense accordingly, as described by the tit-for-tat strategy. When both birds engage a predator, cooperative nest defense behavior is high. However, when one partner betrays and fails to defend, its neighbor immediately reduces its level of defense, and this reduction lasts through a second interaction the following day.

Indirect reciprocity

The prisoner's dilemma and the temptation to cheat may explain why direct reciprocal altruism has rarely been observed in animals. However, Martin Nowak and Karl Sigmund suggested that altruism between unrelated individuals can also evolve through **indirect reciprocity**, in which Individual A helps Individual B because A observed B helping a third individual in the past (Nowak & Sigmund 1998). For indirect reciprocity to evolve, it is crucial that individuals know something about the past helping behavior of other individuals. They must also keep track of others' reputations for helping, their **image score**. If A knows that B has helped others often, then A will have a high image score for B (Nowak & Sigmund 1998; Nowak & Sigmund 2005). Reciprocity spreads through the population indirectly, because helping individuals are more likely to receive help in the future (Nowak & Sigmund 1998; Nowak & Sigmund 2005; Leimar & Hammerstein 2001).

■ **indirect reciprocity** The idea that individuals are more likely to receive help in the future if they help others.

■ **image score** An assessment of an individual's propensity for helping others based on its observed prior behavior.

Featured Research

Reputations and cleaner fish

Is there any evidence that animals use their knowledge of reputations to modify their behavior? To address this question, Redouan Bshary and Alexandra Grutter examined behavioral interactions between cleaner fish and their clients (Bshary & Grutter 2002; Bshary & Grutter 2006; Grutter & Bshary 2003). Cleaner fish, like the cleaner wrasse *Labroides dimidiatus*, feed on ectoparasites on client fish bodies (see Chapter 6). However, they actually prefer to feed on client tissue, such as their mucus, because it is high in calories and nutrients (Grutter & Bshary 2003). Client fish approach cleaners to solicit an interaction, and so when cleaners interact with a client, they can "cooperate" by feeding on ectoparasites or "cheat" by feeding on mucus. When a cleaner cheats, clients shake their body as if jolted and then typically chase away the cleaner (Bshary 2002; Bshary & Grutter 2002). Other potential clients can observe these interactions. Bshary and Grutter investigated whether clients form image scores of cleaners based on such observations.

The research team observed interactions between individual cleaners and their clients in the field. For each interaction, they determined whether cleaners cooperated or cheated, based on the client's behavioral response. Did it jolt and chase, or did it swim slowly away? Bshary and Grutter found that when the next client arrived within six seconds, its behavior was affected by the preceding interaction. If the cleaner had previously "cooperated," the next client invited inspection by a cleaner almost all of the time. However, if the cleaner had just "cheated," the next client invited inspection less than 30% of the time (Figure 14.39). These data suggest that client fish observe cleaners and form image scores based on their recent interactions.

Bshary and Grutter further tested this hypothesis in a laboratory experiment that was repeated in two different years (Bshary & Grutter 2006). A client fish, the two-line monocle beam (*Scolopsis bilineatus*),

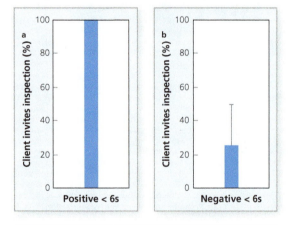

Figure 14.39. Client–cleaner interactions. Mean (+ SE) proportion of interactions in which a client fish invited inspection by a cleaner. Client fish tend to invite interactions with cleaners who have ended a (a) positive interaction within six seconds. They rarely invite interactions with cleaners who have ended a (b) negative interaction within the last six seconds (*Source:* Bshary 2002).

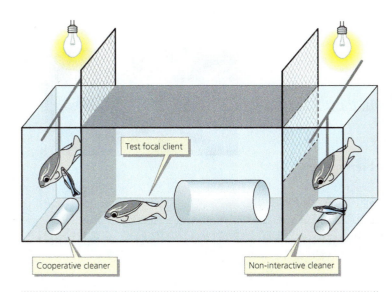

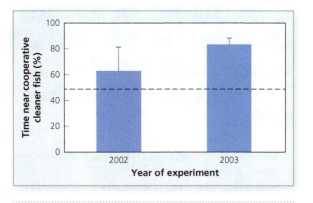

Figure 14.41. Client fish prefer cooperative cleaners. Mean (+ SE) time near the cooperative cleaner. In both years of the experiment, the client fish spent more time near the cooperative cleaner. The dash represents 50%, the null expectation (*Source:* Bshary & Grutter 2002).

Figure 14.40. Cleaner–client experimental design. A central focal client observed two cleaners, one at each end of its tank. One cleaner interacted cooperatively with a model (left), whereas the other did not interact with its model (right) (*Source:* Bshary & Grutter 2006).

was allowed to view two cleaner fish, placed at opposite ends of an aquarium, through one-way mirrors (Figure 14.40). One cleaner, the altruist, participated in a cooperative interaction with a model client fish, while the other cleaner did not interact with its model. The researchers found that the client fish spent significantly more time near the altruistic cleaner than they did with the unresponsive cleaner (Figure 14.41). Together, these results indicate that client fish can learn about altruistic tendencies by observing cleaners' behavior to form image scores, which then affect the client fish's subsequent behavior.

Chimpanzee image scores

Featured Research

Do other animals form reputations and keep track of image scores? Recent work suggests that some do. Yvan Russell, Josep Call, and Robin Dunbar examined whether captive chimpanzees (*Pan troglodytes*) formed image scores of human subjects (Russell, Call, & Dunbar 2008). They allowed chimpanzees to witness interactions among three human subjects with whom they had had no previous contact. Two of the human subjects had food, a box of grapes in a transparent container, and one had no grapes and begged for food. The two individuals with food responded differently to begging. The "nice" person always gave the beggar a grape, but the "nasty" person never did. Immediately following these interactions, the chimpanzee subject was moved to a different compartment, where the "nice" and "nasty" people now sat side by side with their box of grapes in front of two Plexiglas windows. The researchers recorded the amount of time the chimpanzee spent at each window (Scientific Process 14.1).

The chimpanzees spent significantly more time in front of the window facing the "nice" person than they did in front of the window facing the "nasty" one. These results suggest that chimpanzees can form image scores for human altruism.

You may not be surprised to learn that humans also care about their reputations for helping others (Applying the Concepts 14.3). This concern may explain the high degree of altruism observed in our species—and not just toward kin.

SCIENTIFIC PROCESS 14.1

Chimpanzee image scoring

RESEARCH QUESTION: *Do chimpanzees learn about the altruistic tendencies of others?*

Hypothesis: Chimpanzees form image scores of others by observing their interactions.

Prediction: Individuals will prefer to associate with individuals that they have observed being altruistic toward others.

Methods: The researchers:

- recorded the behavior of captive chimpanzees.
- allowed a focal individual to observe interactions between three humans in the human-only area. The "nice" human always gave the beggar grapes, whereas the "nasty" human never gave food to the beggar.
- recorded the amount of time the focal chimpanzee spent in front of either the "nasty" or "nice" person.

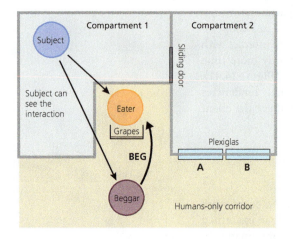

Figure 1. Testing room. Diagram of the experimental testing room (*Source:* Russell, Call, & Dunbar 2008).

Results: Most chimpanzees (14 of 17) spent more time in front of the "nice" person.

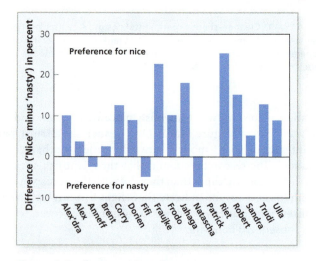

Figure 2. Preference. Most chimpanzees spent more time in front of the nice person's window (Nice > Nasty) (*Source:* Russell, Call, & Dunbar 2008).

Conclusion: Chimpanzees form image scores of humans by observing their interactions.

APPLYING THE CONCEPTS 14.3

Human altruism and reputations

Humans often help unrelated individuals, and indirect reciprocity may explain why. Our reputations depend in part on our altruistic behavior (Nowak & Sigmund 1998; Panchanathan & Boyd 2004). In laboratory experiments, humans are more altruistic when they know they are being watched (Milinski, Semmann, & Krambeck 2002a; Milinski, Semmann, & Krambeck 2002b)—evidence that they care about their own reputations. How might this work in the real world?

Melissa Bateson, Daniel Nettle, and Gilbert Roberts examined human altruism in a real-world setting (Bateson, Nettle, & Roberts 2006). They examined the propensity for people to donate money for tea, coffee, and milk in a workplace coffee room by placing money in an "honesty box." Information about costs and donations were posted at eye level on a cupboard above a counter with the box. The notice contained

a 15 by 3.5 cm banner whose image alternated each week for ten weeks between a pair of eyes and an image of flowers (Figure 1). Each week, the researchers recorded the amount of money donated and the amount of milk consumed, calculating the money collected per liter of milk. This procedure controlled for weekly variation in consumption. Most, if not all, participants added milk to their tea and coffee in this study site in the United Kingdom.

On average, people donated almost three times more money in weeks when the image consisted of eyes than they did in weeks with an image of flowers. The researchers concluded that the image of eyes motivated people to be more generous, because it created a perception of being watched, and people care about their reputation for altruism.

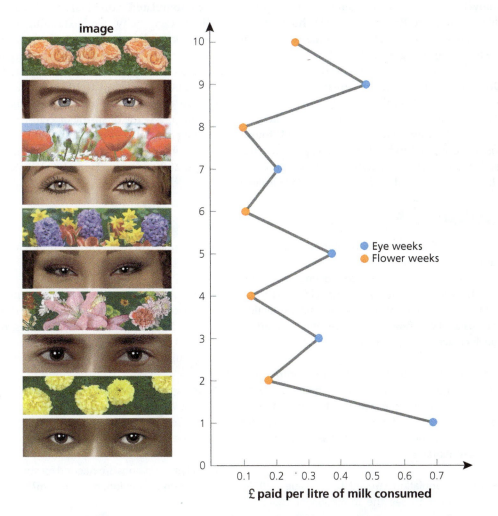

Figure 1. Human altruism. The amount of money donated per week (x-axis) increased when an image of human eyes (blue) was present compared with when images of flowers (orange) were present. The weekly image displayed is presented on the y-axis (*Source:* Bateson, Nettle, & Roberts 2006).

Both direct and indirect reciprocity can explain the evolution of helping behavior between unrelated individuals. Direct reciprocity requires repeated interactions between the same two individuals, whereas indirect reciprocity involves information about the altruistic tendencies of others. There is evidence for the presence of both behaviors among a limited number of species.

Chapter Summary and Beyond

Sociality benefits individuals through reduced search time for food, increased diet breadth, and greater safety from predators. However, it also comes with important costs, such as increased competition for resources, a higher probability of aggressive interactions, and enhanced disease transmission. Variation in costs and benefits helps explain observed variation in sociality. Recent work has examined how individuals assess differences in group size (Buckingham, Wong, & Rosenthal 2007), how groups make decisions (Conradt et al. 2009), and the evolution of cooperative breeding (Leggett et al. 2012).

Many social groups are structured by dominance hierarchies. Dominant individuals gain greater access to resources, while subordinates suffer less aggression once the hierarchy is established. Stable dominance hierarchies also reduce stress. However, Gesquiere et al. (2011) found that the most dominant individual can experience the highest level of stress within a stable hierarchy, so more work on this subject is needed. Eventually, subordinates can become dominants thanks to social queuing. Recent work has focused on understanding how queues are maintained and how queue position affects individual behavior (e.g., Field & Cant 2009).

Individuals in social groups often help one another. Helping behavior is quite common when two individuals are related and is explained by kin selection. Helping between nonrelated individuals can occur through direct or indirect reciprocity. Understanding the relative importance of kin selection and reciprocity to altruistic behavior has been the focus of much recent work, particularly among primatologists (Leinfelder et al. 2001; Payne, Lawes, & Henzi 2003; Manson et al. 2004; Schino & Aureli 2010). In eusocial species, only a few individuals reproduce, and most help raise the offspring of a close relative. Our understanding of the evolution of eusociality is far from complete and is a vigorous area of research (Hunt & Amdan 2005; Pennisi 2009; Nowak, Tarnita, & Wilson 2010).

CHAPTER QUESTIONS

1. While observing flocks of starlings that frequently feed on a large lawn on your campus, you note that average group size is six birds. Imagine conducting an experiment in which you periodically fly a model falcon predator near this area on random occasions. What prediction can you make about how this manipulation will affect average flock size when starlings are feeding?

2. Imagine that you are studying group size and goldfish feeding in a large pond in two experimental treatments. In one treatment, fish can see the location of food patches, and in the other, food patches are hidden under gravel. What prediction can you make about group size in these treatments?

3. In the turkey population studied by Krakauer (2005), would you expect to see the formation of coalitions composed of two male first cousins for whom relatedness $r = 0.125$?

4. Imagine that you are studying a group of wolves and notice that one of the individuals loses all of its fights over food. Propose two hypotheses to explain why this most subordinate individual chooses to stay in its group.

5. In a group of parrots, you frequently observe allogrooming. What data would allow you to determine whether kin selection or reciprocity was a better explanation for these observations?

6. What conditions are needed for the tit-for-tat strategy to explain the evolution of altruism?

7. Researchers have occasionally observed unrelated helpers in a cooperative breeding species. Such behavior cannot be explained by kin selection. Propose a hypothesis to explain these observations.

8. Richards, van Oosterhout, and Cable (2010) concluded that disease transmission in guppies appears to be affected by degree of sociality: greater time spent near others, as in the female shoals, leads to higher rates of disease transmission. Propose an alternative hypothesis for these data.

9. How would you distinguish a cooperative breeding species from one that is eusocial?

Scientific Literature

Dozens of scientific journals publish research on animal behavior. Here we list some of the most common sources of primary (i.e., peer-reviewed) and secondary (interpretations of the research) literature. Scientific journals differ in their generality, discipline, and taxonomic focus, which are described in their statement of mission and scope. Some journals report on all of the sciences for a wide audience, and these tend to focus on very novel research that may have broad implications. Discipline-specific journals report on science that is related only to that discipline, such as behavior, ecology, or evolution. Finally, some journals focus on taxa-specific research, such as research on only fish or only mammals.

The secondary literature typically refers to publications that report on the primary literature. This ranges from journals, magazines, and popular books designed for nonscientists (e.g., *Discover* or *Scientific American*) to journals designed for a wide scientific audience (e.g., *American Scientist*) and textbooks.

Primary literature resources

Scientific journals that publish a broad spectrum of research that can include behavior

American Naturalist
American Zoologist
Ecology
Integrative and Comparative Biology
Journal of Comparative Physiology A: Neuroethology, Sensory, Neural and Behavioral Physiology
Journal of Experimental Biology
Nature
PLoS (Public Library of Science) Biology
Proceedings of the National Academy of Sciences USA
Proceedings of the Royal Society B
Science

Scientific journals that focus primarily on behavior

The following journals publish studies that focus on behavior. This list is not exhaustive but does indicate the breadth and interest in research on behavior.

Aggressive Behavior
Animal Behaviour
Animal Cognition
Animal Learning & Behavior
Applied Animal Behaviour Science
Behavioral and Brain Sciences
Behavioral Ecology
Behavioral Ecology and Sociobiology
Behavioral Neuroscience
Behavior Genetics
Behavior Research Methods
Behaviour
Behavioural Brain Research
Behavioural Processes
Brain Behavior and Evolution
Ethology
Ethology Ecology & Evolution
Hormones and Behavior
Journal of Ethology

Journal of the Experimental Analysis of Behavior
Journal of Experimental Psychology: Animal Behavior Processes
Journal of Comparative Psychology
Learning & Memory
Learning & Motivation
Physiology & Behavior
Zoo Biology

Scientific journals that focus on specific taxa and may publish behavioral studies

Many journals publish papers only on a specific group of organisms. Such journals publish papers on all aspects of a particular taxon, including information on evolution, physiology, morphology, ecology, and behavior. This list is not exhaustive but does offer a representation of the diversity of taxa-related journals.

BIRDS

Auk
Bird Behavior
Condor
The Wilson Bulletin

FISHES

Copeia
Environmental Biology of Fishes
Journal of Fisheries Research
Transactions of the American Fisheries Society

REPTILES AND AMPHIBIANS

Copeia
Herpetologica
Journal of Herpetology

MAMMALS

American Journal of Primatology
Journal of Mammalogy

INVERTEBRATES

Journal of Insect Behavior

Secondary literature

Popular science magazines that may report on animal behavior

The following is a short list of some major secondary literature sources.

Discover
Natural History
Scientific American
Smithsonian

Writing a Scientific Paper

You may be asked to read the primary literature at some point in this course. You may even be asked to engage in scientific research and report your findings. To read or write a scientific paper, it is important to understand its format.

Scientific papers are written in a specific format—title, abstract, introduction, methods, results, discussion or conclusions, and references or literature cited—and in a clear and concise writing style. This standardization helps readers find information. For example, the methods section explains the specific techniques used to collect and analyze data. Before attempting to write a scientific paper, it is very useful to read several papers in the primary literature to become familiar with both the format and the style of writing.

Although sections must appear in a particular order in the final paper, it is often useful to write the paper in the order that you conduct the research. Many authors start by writing the methods section. Once they have analyzed their data, they often create all their figures and tables and then write their results. Next they write the introduction and discussion. The abstract is often written last, so that it will best reflect the complete paper.

Title

The title of the paper is a single sentence that clearly describes the research, so that readers can quickly determine whether it is relevant to their interests. The title, which should be short and informative, often includes both the common and scientific names of the study organism, along with a description of the hypothesis tested. The first letter of the genus name should be capitalized, and the entire genus and species name should be italicized (e.g., *Homo sapiens*). In addition to the title, the names and academic addresses of the authors are included so that the reader knows who conducted the research.

Abstract

The abstract is a brief synopsis of the entire paper. Its first sentence should summarize the ideas presented in the introduction. The next sentences will summarize the methods and important results, and the final sentences should contain the most important conclusions. Busy scientists often rely on abstracts to decide which papers to read in full. So many papers are published each week that it is impossible to read them all.

Introduction

The introduction has two purposes:

1. To provide the reader with background on the research question(s) addressed
2. To state and justify the hypotheses tested

The introduction provides the reader with the background needed to understand the context of the research question. Introductions usually begin with a description of the research area, with citations from the primary (and sometimes secondary) literature. The background information should start by introducing the broad research area and then narrow to the research question and hypothesis tested. A good introduction will provide information on what is already known about a research area and what is not, based on the literature cited. This information allows the reader to understand why it was important to conduct the research.

Any information that is not common knowledge within the discipline requires a reference, usually from the primary literature. Although footnotes are commonly used in the humanities, they are almost never used in science papers. Instead, most biology journals cite sources using the **name-year system**, such as "(Nordell & Valone 1998)." (The full citation is given in the literature cited or references section at the end of the paper.) This system is also called **CSE style**, after the *CSE Manual for Authors, Editors, and Publishers*, prepared by the Council of Science Editors Style

Manual Committee. This council was formed in 1957 by the National Science Foundation and the American Institute of Biological Sciences to promote excellence in the communication of science. Why do disciplines have such specific guidelines? Information is more easily located if it is presented in a standard format.

Referencing other research papers is important because science is a cumulative enterprise. All scientific knowledge builds on prior research to augment or modify our understanding and requires the ability to synthesize past findings concisely. (Direct quotations are rarely used.) After reading a research paper, it is often useful to summarize the key findings in a few sentences. Many students do this by first reading the entire paper and then turning it over and writing out a few key findings of the paper. This approach allows you to test your understanding and minimizes the potential for plagiarism.

Methods

The methods section describes how the research was conducted, known as research protocols. This section often begins with information on the study organism, its location or environment, and, for a laboratory study, care of the animals. Next, a detailed description of any experimental treatments and controls is provided. Sometimes figures are used to illustrate an experimental setup. Finally, the protocol for collecting data and the statistical analysis of the data are described. Subheadings for each of these topics can be helpful. References may be made to studies that used a similar protocol or that explain the relevant statistical analyses.

Results

The results section has one major goal: the presentation of the data as they relate to the hypothesis. The results typically begin with a summary of the number of animals studied and perhaps a description of them (age, number of each sex, body size, and so on). All statements should be supported with the appropriate statistical analysis. A variety of statistical tests can be used. For example, the overall distribution of the data is presented using what are called measures of central tendency. These describe the center of the data distribution (mean, median, or mode) and its dispersion, or amount of variability (standard deviation, variance, or standard error).

The most important results are often presented graphically. Tables contain words and numbers and might show, say, the mean grams of food eaten by males and females each month, while figures allow a visual representation in a picture or graph of the study site and experimental design or relationships between variables (e.g., between age and the number of offspring). Each figure is accompanied by a legend to describe the data.

Discussion

The discussion section states whether the tested hypothesis is supported or rejected and summarizes important findings. It includes conclusions, alternative interpretations of the data, and comparisons to other research findings. It may also briefly describe future research needed.

When the results support the alternate hypothesis, it is important to discuss the generalizability of the findings. Is there the potential to apply the knowledge gained in this study to other species? If the results do not support the alternative hypothesis, it is important to acknowledge this and discuss the implications. Did the study organism, study site, protocol, or alternative hypothesis make it difficult to test the hypothesis? If so, why? Often, the original rationale for the study is reiterated, followed by a discussion of how the study might need to be revised.

After the main conclusions are stated, alternative explanations (additional alternate hypotheses) are critically evaluated. Could any of these better account for the findings? How could future research help to clarify the situation?

In the discussion, it is important that the research findings be placed in the context of prior research, as presented in the introduction. Do the results of this study support or contradict prior research?

Acknowledgments

Any people or organizations that helped with the design or execution of the project are acknowledged and thanked, as are those that provided funding or resources.

Literature Cited

Here the references cited are listed alphabetically by last name, followed by first name or initials, the publication year for a journal article, article title, publication, and volume and page numbers. The format for books cited is only slightly different. For more information, see the Council of Science Editors website at http://www .councilscienceeditors.org/publications/style.cfm.

Animal Care Information

Each country develops policy regarding ethical animal care and use. Here we list some examples of organizations and governmental institutions or departments that provide such information online.

United States

American Association for Lab Animal Science

Association for the Assessment and Accreditation of Laboratory Animal Care International

National Institutes of Health Office of Animal Care and Use

United States Department of Agriculture—Animal and Plant Health Inspection Service

Canada

Canadian Council on Animal Care

Europe

Council of Europe Conventions on the Protection of Animals

European Biomedical Research Association

Australia

Australian Government National Health and Medical Research Council

Japan

Science Council of Japan

New Zealand

New Zealand Ministry of Agriculture and Forestry National Animal Welfare Committee

Glossary

abiotic   Nonliving aspects of the environment.

adaptation   A trait that enhances fitness (survivorship and reproduction). Also an evolutionary process that results in a population of individuals with traits best suited to the current environment.

additive effects (A)   The average effect of individual alleles on the phenotype.

aggressive mimicry   A situation in which a predator mimics a nonthreatening model.

alarm call   Unique vocalizations produced by social animals when a predator is nearby.

alarm signal hypothesis   The hypothesis that advertisement behavior functions to warn nearby conspecifics of danger.

Allee effect   Occurs when the fitness of individuals increases with increased density.

allogrooming   One individual removing ectoparasites from another individual.

alternate hypothesis   The statistical hypothesis that the proposed explanation for observations does have a significant effect.

alternative mating tactics   Multiple behavioral mating phenotypes in a population.

altruism   A behavior that results in the increased fitness of another individual and involves a cost to the individual performing the behavior. Also known as helping behavior.

amplexus   When a male grasps and holds a female prior to copulation.

ancestral trait   A trait found in the common ancestor of two or more species. Also called a **plesiomorphic trait**.

animal behavior   Any internally coordinated, externally visible pattern of activity that responds to changing external or internal conditions.

animal culture   Differences in multiple behavioral traditions among populations.

animal personalities   Consistent relative differences in behavior among individuals over time or across different environmental contexts.

anisogamy   The existence of different sized gametes (small and large) in the different sexes.

anthropomorphism   Attributing human motivations, characteristics, or emotions to animals.

apomorphic trait   A trait found in an organism that was not present in the last common ancestor of a group of two or more species. Also known as a **derived trait**.

aposematic coloration   Brightly colored morphology in a species that stands out from the environment and is associated with noxious chemicals or poisons that make them unpalatable or dangerous prey.

asymmetric game   In game theory, a situation in which the strategies or payoffs for each player differ.

attenuation   The gradual loss in intensity of a signal as it travels though the environment.

audience effect   Occurs when the presence of bystanders influences the behavior of a signaler.

auditory receptor   Specialized nerve cell that detects auditory signals.

author-year system   An in-text citation system that indicates the author(s) and year of publication of a reference in parentheses.

Bateman's hypothesis   Female reproductive success is most strongly limited by the number and success of eggs that she can produce, while male reproductive success is limited by the number of mates he obtains.

Batesian mimicry   A palatable mimic resembles an unpalatable model that predators have learned to avoid.

Bayesian estimation   A process of combining sample information and prior knowledge using Bayes' theorem.

behavioral ecology   An approach that focuses on the ecology and evolution of behavior and its fitness consequences.

behavioral genetics   A field of study that examines how genes and the environment contribute to the differences in behavior by partitioning these effects.

behavioral trade-off   Sacrificing one activity for another.

behavioral traditions   Differences in behavior among populations, transmitted across generations through social learning.

behaviorism   A school of comparative psychology that studies behavior independent of animal mental states or consciousness.

bicoordinate navigation   The ability to identify a geographic location using two varying environmental gradients.

biotic   Living aspects of the environment.

biparental care   Both parents provide care for offspring.

bourgeois tactic   A reproductive tactic in which competitive males defend a territory to monopolize its resources for females or possess traits attractive to females.

breeding dispersal   Abandoning one breeding site and moving to another.

broad-sense heritability   The proportion of phenotypic variation in a population that is due to genetic variation.

brood parasitism   A behavior in which a female (brood parasite) lays an egg in the nest of another female.

brood patch   In birds, a featherless area on the belly that is well vascularized for heat transfer to eggs during incubation.

brood reduction   The death of some siblings as a result of reduced parental care.

bystander   A third-party individual that detects a signal transmitted between a signaler and a receiver. Also called an **eavesdropper**.

cache   Food stored in a hidden location for later retrieval.

candidate genes   Major genes suspected of contributing to a large amount of the phenotypic variation in a specific trait.

castes   Morphologically and behaviorally distinct individuals within a social group.

challenge hypothesis   The hypothesis that male-male interactions increase plasma testosterone and thus sustain subsequent aggressive behavior.

chemical alarm substances   Chemicals released from damaged epidermal cells in fish that function as an alarm cue for others.

chemoreception   The process by which an animal detects chemical stimuli.

chemoreceptor   A sensory receptor that detects chemical stimuli.

chimera   Individuals that develop from two kinds of stem cells: one that contains a new mutation and one that does not.

clade   A group of organisms that are all descended from a common ancestor.

cladistics   A classification system that groups species based on shared derived traits.

cladogram   A diagram showing evolutionary ancestor–descendent relationships among a set of organisms.

classical conditioning   A type of learning in which a novel neutral stimulus is paired with an existing stimulus that elicits a particular response. Eventually the novel neutral stimulus alone elicits the same response as the existing stimulus.

cleaner fish   Fish that feed on ectoparasites and the dead skin of other fish.

closed-ended learners   Individuals that must hear a tutor sing its conspecific song shortly after hatching in order to learn the song correctly.

coalition   An altruistic partnership between a pair of individuals.

cognition   The ability to generate and store mental representations of the physical and social environment to motivate behavior or solve problems.

cognitive ethology   An approach that seeks to understand how natural selection has acted on mental processes and cognition.

communication   The process in which a specialized signal produced by one individual affects the behavior of another.

communication network   Communication systems involving a signaler, a receiver, and at least one bystander.

comparative method   An approach that examines differences and similarities between species to understand the evolution of behavioral traits.

comparative psychology   A branch of psychology that studies animal behavior, often in a comparative manner, across species.

competition hypothesis   The hypothesis that dispersal functions to reduce competition for resources.

conditional response   A learned response to a previous neutral stimulus.

conditional stimulus   A previously neutral stimulus that eventually evokes a conditional response.

conditional strategy   The use of a particular strategy based on an individual's condition.

cones   A type of photoreceptor for color vision under bright light conditions.

conspecific attraction   A phenomenon in which individuals are attracted to others of their species, particularly during habitat selection.

conspecific cueing   A hypothesized mechanism to explain attraction to members of the same species. A settler uses the presence of another individual as a cue to habitat quality.

control group   In an experiment, the group that does not experience a manipulation and thus provides a comparison to experimentally manipulated groups.

cooperative breeding   A situation in social groups in which adults physiologically capable of reproducing forgo breeding and instead help others raise offspring.

correlation   Two variables that vary together predictably.

cost-benefit approach   A method used to study behavioral adaptations in which the fitness benefits and costs of different traits are examined to determine which has the highest net benefit (benefit – cost).

cryptic coloration   Morphological coloration that matches the color of the environment to reduce detection by predators.

cryptic female choice   A situation in which a female influences the fertilization success of sperm from one male over others.

dear enemy hypothesis   A hypothesis stating that territory owners will show reduced aggressive interactions toward territorial neighbors, compared to strangers.

degradation   The loss of signal quality as it travels through an environment.

dendritic spines   Small protuberances on a dendrite that typically receive synaptic inputs.

dependent variable   The variable measured in an experiment in response to variation in the independent variable.

derived trait   A trait found in an organism that was not present in the last common ancestor of a group of two or more species. Also known as an **apomorphic trait**.

descriptive statistics   Information that provides summaries of the main features of a dataset.

determinate growth   A pattern in which individuals stop growing at some point in their life.

diet breadth   The number of different food types in the diet.

dilution effect   A reduction in the probability of death as a result of associating with others.

diminishing returns   A decline in instantaneous harvest rate as a food patch is depleted.

direct material benefits   Material resources obtained by a female from mating with a particular male.

direct reciprocal altruism   A hypothesis to explain the evolution of altruism in which an altruist can benefit if the recipient of an altruistic act reciprocates and helps the altruist in the future. Also known as **direct reciprocity**.

direct reciprocity   See **direct reciprocal altruism**.

directional selection   A situation in which individuals in a population with an extreme trait value at one end of the spectrum possess the highest fitness.

dispersal   A relatively short-distance, one-way movement away from a site.

disruptive selection   A situation in which individuals in a population with extreme trait values on both ends of the spectrum have the highest fitness.

dominance effects (D)   The interaction between alleles at one locus.

dominance hierarchy   An organized social system with dominant and subordinate members.

dove   A strategy in the hawk-dove game of aggression. Doves use a no-cost display in aggressive contests and so never fight or pay the cost of aggression.

eavesdropper   A third-party individual that detects a signal transmitted between a signaler and a receiver. Also called a **bystander**.

endotherm   An organism that generates its own body heat through metabolism to maintain its body temperature.

energetic costs of foraging   The energy used to exploit a patch and the metabolic costs incurred while feeding.

energy intake rate   The energy acquired while feeding divided by total feeding time.

environment (E)   The conditions an individual has experienced and its effect on its behavioral phenotype.

episodic memory   Memory of a specific object, place, and time.

epistasis (I)   Interactions between genes at different loci.

ethogram   A formal description or inventory of an animal's behaviors.

eusocial   A species that lives in social groups that contain overlapping generations, have cooperative offspring care by nonparents, and have reproductive division of labor.

evolution   Changes in allele frequency in a population over time.

evolutionary arms race   The back-and-forth process of adaptation in one species favoring counteradaptation in another.

evolutionary psychology   An approach to understanding human thinking and behavior that assumes natural selection has shaped brain architecture and thought processes in an adaptive manner.

evolutionary stable strategy (ESS)   A strategy that, if adopted by a population, cannot be trumped by another strategy because it yields the highest fitness.

experimental method   An approach in which scientists manipulate or change a variable to examine how it affects the behavior of an animal.

extra-pair copulations   Copulations of a pair-bonded individual with a third individual outside the pair bond.

extra-pair mating system   A situation in which social and genetic mating systems differ because of extra-pair paternity.

extra-pair young   Offspring of a pair-bonded female produced outside the pair bond by a third-party male.

family groups   A social group composed of closely related individuals, often containing siblings of different ages or individuals spanning more than two generations.

female defense polygyny   A single male monopolizes and mates with two or more females.

filial imprinting   The learning of the phenotype and identity of parents by offspring.

fitness   The relative survivorship and reproductive success (ability to produce viable offspring) of an individual.

fixed action pattern   Behaviors that are invariant and unlearned. Once initiated, they are brought to completion.

flash disappearance hypothesis   A hypothesis to explain tail flagging by deer. It states that the bright white of the underside of the tail gives the predator an easily observed target. The sudden disappearance of this target against the dark background of dense woods makes it difficult for the predator to locate its prey and thus encourages it to give up its attack.

focal animal sampling   A data collection technique in which a single focal individual is randomly selected and observed for a specified time period and all pertinent behaviors performed by the individual are recorded. Typically, many focal individuals are observed, each in a different session.

food patch   A location that contains food.

frequency-dependent selection   An evolutionary process in which the fitness of a trait is related to its frequency in a population.

game theory   A cost-benefit modeling approach used to understand the evolution of behavior when the fitness of an individual is affected by how other individuals behave.

gene environment interaction (GEI)   Interactions between a genotype (G) and an environment (E) that influence the behavioral phenotype.

gene expression   The process by which gene products are produced.

gene expression profiling   Measurement of the activity of many genes by quantifying gene products.

generalist   A forager that consumes a wide variety of items in its diet.

genetic marker   Known short sequences of DNA.

genetic mating system   The actual number of sexual partners of each sex in a social mating system that contribute to a set of offspring.

genetic predisposition   Genetic influence on the phenotype that can be influenced by the environment.

genetic quality hypothesis   The hypothesis that females that engage in multi-male matings can improve the fitness of their offspring via genetic mechanisms.

genotype (G)   All the alleles of an individual.

geomagnetic compass   The ability to orient using the earth's magnetic field.

giving-up density (GUD)   The density of food items left in a food patch after being exploited by a forager.

glucocorticoids　Hormones secreted by the adrenal gland in vertebrates in response to stressful situations.

good genes　The alleles of a high-quality individual.

group selection　Selection that favors particular groups of individuals over other such groups of the same species.

group size effect　A commonly observed pattern in which the vigilance behavior of an individual declines as group size increases.

gustation　The detection of dissolved chemicals, often within the mouth.

habituation　A simple form of learning; the reduction and then lack of response to a stimulus over time.

Hamilton's rule　An equation formulated by William Hamilton predicting when altruism should evolve based on the fitness benefits and costs of the altruistic behavior, and the degree of relatedness among an altruist and recipient.

Hamilton-Zuk hypothesis　A hypothesis that parasites and pathogens play an important role in sexual selection when secondary sexual traits are costly and condition dependent.

handicap principle　Well-developed secondary sexual characteristics are costly to survival but reliable signals of fitness.

handling time　The amount of time to manipulate a food item so that it is ready to eat.

haplodiploid genetic system　A genetic system with haploid males and diploid females.

haplodiploidy hypothesis　A hypothesis that explains the evolution of eusociality in Hymenoptera based on their haplodiploid genetic system.

hatching asynchrony　The hatching of offspring in a clutch of eggs on different days.

hawk　A strategy in the hawk-dove game of aggression. Hawks are always willing to escalate into an intense, prolonged interaction in order to win.

hawk-dove model　A game theory model that examines the behavior of two individuals engaged in a contest over a resource.

heritability (h^2)　(broad-sense) The proportion of phenotypic variation in a population that is due to genetic variation.

heritable　A trait that can be passed from parents to their offspring because of heredity.

home range　An area of repeated use by an individual that is not defended from others.

homology　A trait that is similar among species because they share a common ancestor and not because of convergent evolution.

hotshot hypothesis　One hypothesis to explain the evolution of leks. It postulates that low-ranking males can benefit by aggregating around high-ranking males because females are more likely to visit attractive males. This increases their encounter rate with females and enhances their reproductive success.

hotspot hypothesis　One hypothesis to explain the evolution of leks. It postulates that all males can benefit by aggregating in a location where they are likely to encounter many females.

hypothesis　An explanation based on assumptions that makes a testable prediction.

ideal free distribution (IFD) model　A model that explains how animals distribute themselves among habitats or food patches.

image score　An assessment of an individual's propensity for helping others based on its observed prior behavior.

imprinting　Rapid learning in young animals through observation. Individuals are typically attracted to objects they imprint on.

inbreeding avoidance hypothesis　Natal dispersal behavior minimizes the likelihood of mating with relatives.

inbreeding depression　A reduction in fitness as a result of mating with close relatives.

inclusive fitness　A combination of individual fitness and the fitness obtained by helping close relatives.

independent variable　In an experiment, the variable that is changed by the researcher.

indeterminate growth　Growth that continues throughout the lifetime.

indirect genetic benefits　Genetic benefits females can obtain for their offspring by mating with males that have high genetic quality.

indirect reciprocity　A hypothesis to explain the evolution of altruism based on the idea that helping individuals are more likely to receive help in the future because they obtain high image scores by helping others.

individual selection　Natural selection acting on individuals.

inferential statistics   Used in hypothesis testing to determine if there is a statistical difference between sets of data. Used to make inferences about a population from a sample.

innate   Behaviors that are performed the same way each time, are fully expressed the first time they are exhibited, and are present even in individuals raised in isolation. Also called **instinct**.

innovation   A novel, learned behavior.

insight learning   Spontaneous problem solving without the benefit of trial-and-error learning.

instantaneous sampling   A method of collecting data in which data are collected from all individuals at regular time intervals.

instinct   Behaviors that are performed the same way each time, are fully expressed the first time they are exhibited, and are present even in individuals raised in isolation.

instrumental conditioning   Learning associations between learned behaviors and outcomes.

intergenerational migration   Migration that occurs over more than one generation.

intersexual selection   Selection by one sex for members of the other sex for reproduction. Also known as **mate choice**.

intrasexual selection   Selection in which one sex competes with other members of the same sex for access to the other sex for reproduction. Also known as **mate competition**.

isodar   The number of individuals in one habitat plotted against the number of individuals in another habitat at several points in time.

isogamy   The production of gametes of the same size by all individuals.

iteroparity   A life history strategy in which there are multiple reproductive events throughout the lifetime.

juvenile hormones   In insects, a class of hormones produced by specialized endocrine glands that influence molting, are associated with reproductive maturation, and are known to vary with levels of parental care provided to offspring.

kin selection   A form of natural selection in which individuals can increase their fitness by helping close relatives, because close relatives share the helper's genes. Often used to explain altruism.

knockdown technique   A procedure that reduces the expression of a gene.

knockout technique   A procedure in which a single gene is rendered nonfunctional.

lactation   The secretion of milk by mammary glands in female mammals to nourish offspring.

landmarks   Physical cues such as coastlines, islands, river systems, or distant mountains that can provide directional information for long-distance movements.

lateral line system   In fish, mechanoreceptors that provide hydrodynamic information.

learning   A relatively permanent change in behavior as a result of experience.

learning curve   A graphical representation of a change in learning over time.

lek   A location where males aggregate and display to females.

life history theory   A theory that assumes that natural selection will favor the evolution of behaviors and life history traits that maximize an individual's lifetime reproductive success.

life history traits   Traits involved with growth, reproduction, and survivorship.

linear dominance hierarchy   An organized social system in which an individual is dominant to each individual below and subordinate to all above it in rank.

linkage map   A genetic map of the relative positions of genetic markers on chromosomes.

Linnean classification system   A method of taxonomy that classifies organisms in a nested hierarchical system ranging from the largest groups (domains or kingdoms) to smaller and smaller groups, including phylum, class, order, family, genus, and species.

local enhancement   The direction of an individual's focus to a particular part of the environment by the presence of another.

magnetoreception   The ability to detect magnetic fields and orient by using them.

major gene   Individual gene that is responsible for a large fraction of phenotypic variation.

male dominance polygyny   In a lek system, only a few males mate with most females.

mandibulate   The maneuvering of food in the bill for several seconds to prepare it for consumption.

marginal benefit   In foraging, the benefit obtained by feeding for one more instant. Often measured as the instantaneous harvest rate.

marginal value theorem   An optimal foraging model that predicts how long an individual should exploit a food patch.

mate choice   Selection by one sex for members of the other sex for reproduction.

mate choice copying   A situation in which one individual observes and copies the mating decisions of another individual.

mate competition   One sex competes with other members of the same sex for access to the other sex for reproduction.

mate guarding   A behavior in which a male follows his mate to prevent her from mating with rivals.

mate guarding hypothesis   A hypothesis that explains the evolution of monogamy in the absence of biparental care. It posits that selection favors males that mate with and guard one female over one or more reproductive cycles by remaining in close association with her.

maternal care   Parental care provided by a female to her offspring.

mating system   A description of the social associations and number of sexual partners an individual has during one breeding season.

measures of central tendency   Ways to characterize the centrality of a dataset. The most common measure is the mean, or average.

measures of dispersion   Ways to characterize the spread or variation in a dataset.

mechanoreceptors   Sensory receptors sensitive to changes in pressure.

median   The middle value of a set of ordered data.

memory   The retention of learned experiences.

mental map   A spatial representation of the world. The ability to identify a geographic location using two varying environmental gradients.

microarray analysis   Measurement of the activity of many genes by quantifying gene products.

migration   Relatively long-distance two-way movements.

mimicry   The adaptive resemblance of one species to another so that a third species cannot distinguish them.

minor gene   Individual gene that contributes to small amounts of variation in the phenotype.

missed opportunity costs of foraging   The fitness costs of not engaging in other activities while feeding.

mixed ESS   An evolutionary stable strategy in which a fraction of individuals play different strategies, or else each individual plays a particular strategy some fraction of the time.

mixed-species group   A social group containing two or more species.

mobbing behavior   A behavior in which two or more individuals harass a predator.

mode   The most common value in a dataset.

modes of natural selection   Descriptions of how populations change as a result of the relative fitness values of different trait values.

monogamy   A mating system in which one male mates with one female.

Morgan's canon   The idea that the simplest psychological process possible should be used to interpret an animal's behavior.

move-on hypothesis   Predator harassment functions to cause a predator to abandon an area.

multilevel selection   A form of natural selection that involves both selection on groups and selection on individuals.

narrow-sense heritability   The proportion of phenotypic variance solely as a result of additive genetic values.

natal dispersal   A one-time movement away from an individual's place of birth.

natural selection   Differential reproduction and survivorship among individuals within a population. The mechanism that results in adaptive evolution.

navigation   The process of determining a particular location and moving toward it.

negative density-dependence   A situation in which the fitness of individuals declines as density increases.

negative punishment   The removal of a desired stimulus that causes a behavior to decline.

negative reinforcement   The removal of an aversive stimulus that causes a behavior to increase.

negative results   A situation in which one does not reject the null hypothesis (and thus one rejects the alternate hypothesis).

neural plasticity   Structural changes in the brain, especially in the number of synapses and the strength of chemical synapses between neurons.

neuromasts   A mechanoreceptive organ that provides information about water velocity and acceleration as well as the direction of water movements. The major functional unit of the lateral line system.

null hypothesis   The statistical hypothesis that observations result from chance. The hypothesis of no effect.

numerical competency   The ability to recognize numerical quantities.

nuptial gift   A physical resource such as a food item that a male provides to a female to enhance his mating success.

observational method   An approach in which scientists observe and record the behavior of an organism without manipulating the environment or the animal.

odorant   A gaseous compound that is perceived as odorous.

olfaction   The detection of airborne chemical stimuli.

open-ended learners   Individuals that can acquire new song elements throughout life.

operant chamber   An enclosure used to study behavioral conditioning.

operant conditioning   A learning process in which an animal learns to associate a behavior with a particular consequence.

opportunity costs   The fitness costs of not engaging in other activities. Also known as **missed opportunity costs**.

optimal behavior   The behavior that maximizes fitness in an optimality model.

optimal diet model   An optimal foraging model that predicts the diet that maximizes an individual's energy intake rate.

optimal foraging theory   An approach to studying feeding behavior that assumes that natural selection has favored feeding behaviors that maximize fitness.

optimal patch-use model   The optimal foraging theory model that predicts how long a forager should exploit a food patch.

optimal trait value   The trait value that confers the highest fitness in a population in a particular environment.

orientation   The determination and maintenance of a proper direction.

ornaments   Exaggerated morphological traits used to attract females.

outgroup   A species distantly related to the focal group. Used when creating a phylogeny of a focal group of organisms.

parasitic tactic   An alternative mating tactic used by less competitive males to usurp matings from highly competitive males.

parental care   Behaviors by a parent to enhance the fitness of offspring, including incubation, feeding, and defense.

parental investment   Any investment by a parent in offspring that enhances the offspring's fitness at the cost of the parent's ability to invest in other offspring.

parental investment theory   The sex that has a greater investment in offspring production should be choosier when it comes to mates.

parent-offspring conflict theory   A theory that assumes that parents and their dependent offspring are under different selection pressures: parents should maximize their lifetime reproductive fitness, while offspring should maximize the energy and protection they currently receive from their parents to survive to reproductive age.

parent-offspring regression   A statistical technique used to examine the similarity between parents and their offspring in the traits they possess.

parsimony   A principle used in phylogenetics to construct a phylogenetic hypothesis; the hypothesis that is based on the fewest evolutionary changes.

partial preference   Acceptance of a food item some fraction of the time.

patch variance   The variation in patch types in an environment.

paternal care   Parental care that a male provides to offspring.

Pavlovian conditioning   A type of learning in which a novel neutral stimulus is paired with an existing stimulus and elicits a particular response. Eventually the novel neutral stimulus alone elicits the same response as the existing stimulus.

peer review   A process in which editors of scientific journals use experts to help decide whether to accept or reject a paper for publication.

phenotype (P)   The observed traits of an individual.

pheromones   Volatile compounds that are species specific; these organic compounds affect the behavior of another individual of the same species.

photoreceptor   A specialized neuron that is sensitive to light; a visual receptor.

phylogenetic classification system   A classification system that groups species based on shared derived traits.

phylogenetic tree   A branching diagram showing hypothesized evolutionary relationships among species.

phylogeny   A diagram showing evolutionary ancestor-descendent relationships among a set of organisms.

piscivorous   Fish-eating.

plesiomorphic trait   A trait found in the common ancestor of two or more species. Also called an **ancestral trait**.

plural breeding   A mating system of social species in which multiple males and multiple females within one social group produce young in one breeding attempt.

polyandry   A mating system in which a single female associates and mates with multiple males.

polygamy   Any mating system in which one sex mates with multiple partners.

polygynandry   A mating system of social species in which multiple males and multiple females associate with one another and produce young in one breeding attempt.

polygyny   A mating system in which a single male associates and mates with multiple females.

polygyny threshold model   A model that predicts the occurrence of polygyny based on the amount of resources available to females in male territories.

positive reinforcement   An increase in a behavior as a result of its association with the addition of a desired stimulus.

predation risk   The likelihood of being killed by a predator.

predation risk cost   The fitness cost associated with being killed by a predator. Often equated with the probability of being killed by a predator.

predator harassment   Interactions with a predator to deter attack. These can involve rapid movements, loud vocalizations, and throwing or kicking objects at the predator.

primary literature   The original source of scientific information, typically peer-reviewed scientific journals.

primary sexual characteristics   The genitalia and organs of reproduction.

prior knowledge   In Bayesian estimation, information about a parameter prior to sampling.

In foraging, often the distribution of patch types in the environment.

prisoner's dilemma   A game theory model that illustrates why the evolution of cooperation (altruism) is difficult.

process of science   Observing events, organizing knowledge, and providing explanations through the formulation and testing of hypotheses.

profitability   For a food item, the energy it contains divided by its handling time.

prolactin   A hormone that influences parental care in vertebrates.

promiscuity   A mating system in which both males and females mate with multiple partners in the absence of pair bonds.

proximate explanation   An explanation that focuses on understanding the immediate causes of a behavior. This contrasts with ultimate explanations of behavior.

public information   Information obtained from the activity or performance of others about the quality of an environmental parameter or resource.

pure strategy   In game theory, a situation in which all individuals adopt the same strategy in the game.

pursuit-deterrence hypothesis   The hypothesis that advertisement behavior informs a predator that it has been detected and so pursuit is not likely to be successful.

QTL mapping   A statistical technique that combines genetic information with trait information to determine which regions of the genome contain the genes that influence the trait QTLs.

quantitative trait loci (QTL)   Stretches of DNA that either contain or are linked to genes influencing a trait such as behavior.

quitting harvest rate   The forager's instantaneous harvest rate when it leaves a food patch.

range   One measure of dispersion in a dataset. The difference between the highest and lowest measurements.

reaction norm   The range of behaviors expressed by a single genotype in different environments.

reduction   An ethical guideline that promotes limiting the number of animals subject to disturbance in research or teaching.

refinement   An ethical guideline that involves improving procedures and techniques to minimize pain and stress for animals.

reflex   One kind of innate behavior; an involuntary movement in response to a stimulus.

releaser stimuli   Stimuli that initiate a fixed action pattern.

replacement   An ethical guideline that encourages the use of computer modeling, videotapes, or other approaches in place of actual research animals in the laboratory.

research question   The first methodological step in the process of science. A formal statement of an unknown that one would like to understand.

resource holding power   The ability to win an aggressive encounter.

rich patch   A food patch that contains abundant food.

rod   A type of photoreceptor, sensitive to low light levels.

rover   One type of genetic and behavioral variant in fruit flies. Larval rovers have longer foraging trails than sitters in the presence of food and are more likely to leave a food patch.

runaway process   An evolutionary process in which a male trait co-evolves with a female preference for it and becomes increasingly exaggerated.

runaway selection   A genetic formalization of the runaway process. It requires that genes for the male trait and genes for female preference for the trait be genetically linked through linkage disequilibrium.

sample   A subset of a population or group that is chosen to be representative of the entire population.

sample information   In Bayesian estimation, the information obtained by sampling an unknown parameter such as a food patch.

sample mean   The mean of a sample. It provides an estimate of the population mean.

satellite males   An alternative, parasitic mating tactic in which a male remains near a bourgeois male to intercept females that are attracted to the bourgeois male.

scan sampling   A method of collecting data in which data are collected from all individuals at regular time intervals.

scientific literacy   The ability to evaluate scientific information critically and ascertain its validity.

scientific method   A formalized process that involves the testing of hypotheses. It involves the formulation of a research question, a hypothesis that makes a prediction(s), a test of the prediction(s), and evaluation of the hypothesis based on the outcome of the test.

scientific misconduct   Violation of ethical behavior in science. It includes the falsification or fabrication of data, purposeful inappropriate analysis of data, and plagiarism.

scientific theories   Hypotheses that make many predictions, have been tested many times by many different scientists, and have not been rejected. Scientific theories provide a conceptual framework that explains many phenomena and are well supported by observations and experimental tests.

scramble competition   A competition in which individuals compete indirectly with each other to find and secure copulations with multiple mates.

search image   The visual distinctive features of a single prey type that, once learned, can enhance prey detection.

secondary literature   A report that summarizes and interprets the primary literature. Often reported in newspapers, magazines, and books.

secondary sexual characteristics   Morphological differences between the sexes that are not directly involved in reproduction.

selection experiment   An experiment in which different groups of individuals are subjected to differential selection on a trait.

selfish herd hypothesis   The term used to describe the tight clustering of individuals in a group. A predator is more likely to kill a member on the outside or periphery of a group of prey, because it will encounter outside individuals first. Therefore, individuals can reduce their predation risk by moving to the center of a group.

semelparity   A life history strategy in which reproduction occurs once in a lifetime.

sensory bias hypothesis   The hypothesis that female mating preferences are a byproduct of preexisting biases in a female's sensory system.

sensory receptor   Nerve endings that respond to an internal or external environmental stimulus.

sequential assessment model   An asymmetric game involving a contest over a resource. Individuals assess their opponent's resource holding power, relative to their own, in order to determine how much effort to exert.

sex role-reversed species   A species in which females compete for males, who invest heavily in parental care.

sexual conflict   The differential selection on males and females to maximize their fitness.

sexual deception hypothesis   The hypothesis that states that males will produce deceptive signals to females in order to enhance their own reproduction.

sexual dimorphism   Morphological differences between the sexes.

sexual selection   A form of natural selection that acts on heritable traits that affect reproduction.

signal   A packet of energy or matter generated from a display or action of a signaler that travels to a receiver.

signaler   An individual that produces a signal.

signal receiver   An individual that detects a signal.

sister species   Two species that are more closely related to one another than to any other species; species that share a recent, common ancestor.

site fidelity   Individuals that remain at or return to a previous location to breed.

sitter   One type of genetic and behavioral variant in fruit flies. Larval sitters have shorter foraging trails than rovers in the presence of food and are less likely to leave a food patch.

Skinner box   Operant chamber to study behavioral conditioning. Also known as an **operant chamber**.

sneaker male   An alternative male reproductive tactic in which a male attempts to avoid detection so that he can quickly enter a bourgeois territory to fertilize eggs being deposited in a nest.

social group   A set of individuals that live near and associate with one another.

social information   Any information obtained from others about the environment.

social learning   Learning by observing or interacting with other individuals.

social mating system   The social associations and presumed mating behavior of individuals based on those associations.

social queuing   A process in which subordinates eventually move up the social ladder as they age.

sociality   The tendency to live and associate with others.

song dialects   Characteristic differences in songs that vary geographically among populations.

song system   In birds, the enlarged and interconnected areas of the brain involved in song memory and production.

specialist   A forager that has a narrow diet.

spectrogram (sonogram)   Two-dimensional representations of sound that allow researchers to characterize the acoustic structure of vocalizations.

sperm competition   The competition between sperm of different males to fertilize eggs.

stabilizing selection   A situation in which individuals in a population with intermediate trait values have the highest fitness in a particular environment.

standard deviation   A measure of dispersion in a set of data. The square root of the variance.

standard error   A measure of dispersion of the sample mean. The standard deviation divided by the square root of the sample size.

star compass   The use of stars or constellations to orient.

statistics   A branch of applied mathematics that organizes, analyzes, and interprets data. Often uses probability theory to produce conclusions from data.

stot (stotting)   A high jump into the air with all four feet off the ground in which an individual moves forward.

sun compass   The use of the sun for orientation.

synapomorphies   Shared derived traits among species.

tail flagging   Rapid wagging of the tail, especially when a predator is nearby.

territorial cooperation hypothesis   One hypothesis to explain the evolution of monogamy in the absence of biparental care. It posits that that two individuals (one of each sex) can better defend a critical resource, such as a safe refuge, and this can select for pair formation and shared defense.

territory   An area defended to obtain the exclusive use of the resources it contains.

testosterone   A steroid hormone produced in the gonads and regulated by the hypothalamus and pituitary gland.

thigmotaxis   A tendency to seek out physical contact with objects or enclosed spaces. The avoidance of open areas.

time budget   A summary of the total time and relative frequency of different behaviors of an individual.

tit-for-tat   One strategy in the prisoner's dilemma game. In this strategy, a player always cooperates the first time it interacts with an opponent and then simply matches the behavior of the other player from the previous interaction.

totipotency   The ability of a cell to develop into any type of cell.

transgenic   Animals that have had foreign DNA inserted into their genome.

trial-and-error learning   Learning through repetition that results in rewards or progress toward a goal.

ultimate explanation   An explanation of behavior that requires evolutionary reasoning and analyses. This contrasts with proximate explanations.

unconditional response   An unlearned response that occurs naturally in response to a particular stimulus.

unconditional stimulus   A stimulus that does not require learning to trigger a response.

variance   A measure of dispersion in a set of data.

vigilance   A behavior in which an animal scans the environment for predators.

volatile   A chemical that can evaporate; that is, become gaseous.

waggle dance   A behavior performed by a honeybee scout that recruits workers to a distant food source.

weapons   Exaggerated morphological traits used in male-male competition.

wild type   The typical form of an organism or gene that occurs in nature.

winner effect   A phenomenon in which winning an aggressive interaction enhances the likelihood of winning a subsequent interaction.

winner-challenge effect   Winning an aggressive interaction increases plasma testosterone levels and so enhances aggressive behavior and the likelihood of winning subsequent interactions.

win-stay lose-shift dispersal   A common behavioral pattern in which individuals return to a previous breeding location after reproductive success there, but disperse to a different location after a reproductive failure.

zero-one rule   In the optimal diet model, the prediction that each food type should either be always eaten when found or always rejected.

Bibliography

Abrahams, M. V. 1989. Foraging guppies and the ideal free distribution: The influence of information on patch choice. *Ethology* 82:116–126.

Abrahams, M. V., Robb, T. L., & Hare, J. F. 2005. Effect of hypoxia on opercular displays: Evidence for an honest signal? *Animal Behaviour* 70:427–432.

Adkin-Regan, E., & MacKillop, E. 2003. Japanese quail (*Coturnix japonica*) inseminations are more likely to fertilize eggs in a context predicting mating opportunities. *Proceedings of the Royal Society B* 270:1685–1689.

Ahlering, M. A., & Faaborg, J. 2006. Avian habitat management meets conspecific attraction: If you build it will they come? *Auk* 123:301–312.

Åhlund, M., & Andersson, M. 2001. Female ducks can double their reproduction. *Nature* 414:600–601.

Akçay, E., & Roughgarden, J. 2007. Extra-pair paternity in birds: Review of the genetic benefits. *Evolutionary Ecology Research* 9:855–868.

Åkesson, S., & Hedenström, A. 2007. How migrants get there: Migratory performance and orientation. *Bioscience* 57:123–133.

Akey, J. M., Ruhe, A. L., Akey, D. T., Wong, A. K., Connelly, C. F., Madeoy, J., Nicholas, T. J., & Neff, M. W. 2010. Tracking footprints of artificial selection in the dog genome. *Proceedings of the National Academy of Sciences USA* 107:1160–1165.

Alatalo, R. V., Höglund, J., Lundberg, A., & Sutherland, W. J. 1992. Evolution of black grouse leks: Female preferences benefit males in larger leks. *Behavioral Ecology* 3:53–59.

Alerstram, T., Hedenström, A., & Åkesson, S. 2003. Long-distance migration: Evolution and determinants. *Oikos* 103:247–260.

Allee, W. C. 1931. Co-operation among animals. *American Journal of Sociology* 37:386–398.

Allen, C., & Bekoff, M. 2007. Animal minds, cognitive ethology, and ethics. *Journal of Ethics* 11:299–317.

Allen, T., & Clarke, J. A. 2005. Social learning of food preferences by white-tailed ptarmigan chicks. *Animal Behaviour* 70:305–310.

Alonso, S. H. 2008. Female mate choice copying affects sexual selection in wild populations of the ocellated wrasse. *Animal Behaviour* 75:1715–1723.

Altmann, J. 1974. Observational study of behavior: Sampling methods. *Behaviour* 49:227–267.

American Veterinary Medical Association. 2012. http://www.avma.org/KB/resources/statistics/Pages/default.aspx (accessed January 26, 2013).

American Veterinary Society of Animal Behavior (AVSAB). 2008. American Veterinary Society of Animal Behavior position statement–puppy socialization. http://www.avsabonline.org/avsabonline/images/stories/Position_Statements/puppy%20socialization.pdf (accessed January 2, 2012).

Anderson, W. W., Kim, Y. K., & Gowaty, P. A. 2007. Experimental constraints on mate preferences in *Drosophila pseudoobscura* decrease offspring viability and fitness in mated pairs. *Proceedings of the National Academy of Sciences USA* 104:4484–4488.

Andersson, M. 1995. Evolution of reversed sex roles, sexual size dimorphism, and mating system in coucals (*Centropodidae: Aves*). *Biological Journal of the Linnean Society* 54:173–181.

———. 2005. Evolution of classical polyandry: Three steps to female emancipation. *Ethology* 111:1–23.

Andersson, M., & Åhlund, M. 2000. Host-parasite relatedness shown by protein fingerprinting in a brood parasitic bird. *Proceedings of the National Academy of Sciences USA* 97:13,188–13,193.

Andersson, M., & Simmons, L. W. 2006. Sexual selection and mate choice. *Trends in Ecology and Evolution* 21:296–302.

Angelier, F., & Chastel, O. 2009. Stress, prolactin and parental investment in birds: A review. *General and Comparative Endocrinology* 163:142–148.

Anholt, R. H., & Mackay, T. F. C. 2010. *Principles of behavioral genetics*. London: Elsevier.

Animal Behavior Society. 2006. Guidelines for the treatment of animals in behavioural research and teaching. *Animal Behaviour* 71:245–253.

Aparicio, J. M., Bonal, R., & Muñoz, A. 2007. Experimental tests on public information use in the colonial lesser kestrel. *Evolutionary Ecology* 21:783–800.

Arbelle, S., Benjamin, J., Golin, M., Kremer, I., Belmaker, R. H., & Ebstein, R. P. 2003. Relation of shyness in grade school children to the genotype for the long form of the serotonin transporter promoter region polymorphism. *American Journal of Psychiatry* 160:671–676.

Archer, J. 2006. Testosterone and human aggression: An evaluation of the challenge hypothesis. *Neuroscience and Biobehavioral Reviews* 30:319–345.

Archer, J., Graham-Kevan, N., & Davies, M. 2005. Testosterone and aggression: A re-analysis of Book, Starzyk and Quinsey's (2001) study. *Aggression and Violent Behavior* 10:241–261.

Arens, P., Bugter, R., van't Westende, W., Zollinger, R., Stronks, J., Vos, C. C., & Smulders, M. J. M. 2006. Microsatellite variation and population structure of a recovering tree frog (*Hyla arborea* L.) metapopulation. *Conservation Genetics* 7:825–835.

Arnott, G., & Elwood, R. W. 2008. Information gathering and decision making about resource value in animal contests. *Animal Behaviour* 76:529–542.

Asa, C. S., Traylor-Holzer, K., & Lacy, R. C. 2011. Can conservation-breeding programmes be improved by incorporating mate choice? *International Zoo Yearbook* 45:203–212.

Ascunce, M. S., Yang, C. C., Oakey, J., Calcaterra, L., Wu, W. J., Shih, C. J., Goudet, J., Ross, K. G., & Shoemaker D. 2011. Global invasion history of the fire ant *Solenopsis invicta*. *Science* 331:1066–1068.

Augustsson, H., & Meyerson, B. J. 2004. Exploration and risk assessment: A comparative study of male house mice (*Mus musculus musculus*) and two laboratory strains. *Physiology & Behavior* 81:685–698.

Axelrod, R., & Hamilton, W. D. 1981. The evolution of cooperation. *Science* 211:1390–1396.

Ballantine, B. 2009. The ability to perform physically challenging songs predicts age and size in male swamp sparrows, *Melospiza georgiana*. *Animal Behaviour* 77:973–978.

Balter, M. 2010. Animal communication helps reveal roots of language. *Science* 328:969–971.

Baptista, L. F., & Petrinovich, L. 1984. Social interaction, sensitive phases and the song template hypothesis in the white-crowned sparrow. *Animal Behaviour* 32:1359–1371.

Barber, N. A., Marquis, R. J., & Tori, W. P. 2008. Invasive prey impacts the abundance and distribution of native birds. *Ecology* 89:2678–2683.

Bargum, K., Helanterä, H., & Sundström. L. 2007. Genetic population structure, queen supersedure and social polymorphism in a social Hymenoptera. *Journal of Evolutionary Biology* 20:1351–1360.

Barnes, R., Greene, K., Holland, J., & Lamm, M. 2002. Management and husbandry of duikers at the Los Angeles Zoo. *Zoo Biology* 21:107–121.

Barry, K. L., & Kokko, H. 2010. Male mate choice: Why sequential assessment can make its evolution difficult. *Animal Behaviour* 80:163–169.

Basolo, A. L. 1990. Female preference predates the evolution of the sword in swordtail fish. *Science* 250:808–881.

Bastock, M. 1956. A gene mutation which changes a behavior pattern. *Evolution* 10:421–439.

Bateman, A. J. 1948. Intra-sexual selection in *Drosophila*. *Heredity* 2:349–368.

Bates, H. W. 1862. Contributions to an insect fauna of the Amazon valley. Lepidoptera: Heliconidae. *Transactions of the Linnaean Society of London* 23:495–566.

Bateson, P., & Martin, P. 2007. *Measuring behavior: An introductory guide*. Cambridge: Cambridge University Press.

Bateson, M., Nettle, D., & Roberts, G. 2006. Cues of being watched enhance cooperation in a real-world setting. *Biology Letters* 2:412–414.

Batra, S. W. T. 1966. Nests and social behavior of halictine bees of India. *Indian Journal of Entomology* 28:375–393.

Beard, C. 1913. *An economic interpretation of the Constitution of the United States*. New York: MacMillan Co.

Beck, W. S., Liem, K. F., & Simpson, G. G. 1991. *Life, an introduction to biology*. 3rd ed. New York: HarperCollins.

Bednekoff, P. A. 2007. Foraging in the face of danger. In *Foraging*, ed. D. W. Stephens, J. S. Brown, & R. C. Ydenberg, 305–329, Chicago: University of Chicago Press.

Bednekoff, P. A., & Kotrschal, T. 2002. Observational learning and the raiding of food caches in ravens, *Corvus corax*: Is it "tactical" deception? *Animal Behaviour* 64:185–195.

Beecher, M. D., & Brenowitz, E. A. 2005. Functional aspects of song learning in songbirds. *Trends in Ecology and Evolution* 20:143–149.

Beecher, M. D., Burt, J. M., O'Loghlen, A. L., Templeton, C. N., & Campbell, S. E. 2007. Bird song learning in an eavesdropping context. *Animal Behaviour* 73:929–935.

Beehler, B. M., & Foster, M. S. 1988. Hotshots, hotspots, and female preference in the organization of lek mating systems. *American Naturalist* 131:203–219.

Beehner, J. C., & Whitten, P. L. 2004. Modifications of a field method for fecal steroid analysis in baboons. *Physiology & Behavior* 82:269–277.

Bekoff, M., Allen, C, & Burghardt, G. M. 2002. *The cognitive animal: Empirical and theoretical perspectives on animal cognition*. Cambridge, MA: MIT Press.

Bekoff, M., & Sherman, P. W. 2004. Reflections on animal selves. *Trends in Ecology and Evolution* 4:176–180.

Bell, G. 1978. The evolution of anisogamy. *Journal of Theoretical Biology* 73:247–270.

Bell, G., Lefebvre, L., Giraldeau, L.-A., & Weary, D. 1984. Partial preferences of insects for male flowers of an annual herb. *Oecologia* 64:287–294.

Belthoff, J. R., & Ritchison, G. 1989. Natal dispersal of Eastern screech-owls. *Condor* 91:254–265.

Bender, N., Taborsky, M., & Power, D. M. 2008. The role of prolactin in the regulation of brood care in the cooperatively breeding fish *Neolamprologus pulcher*. *Journal of Experimental Zoology* 309A:515–524.

Bendesky, A., & Bargmann, C. I. 2011. Genetic contributions to behavioural diversity at the gene-environment interface. *Nature Reviews Genetics* 12:809–820.

Bendich, A. 1989. Carotenoids and the immune response. *Journal of Nutrition* 119:112–115.

Bengtsson, B. O. 1978. Inbreeding avoidance—at what cost? *Journal of Theoretical Biology* 73:439–444.

Bengttson, G., Hedlund, K., & Rundgren, S. 1994. Food- and density-dependent dispersal: Evidence from a soil collembolan. *Journal of Animal Ecology* 63:513–520.

Berec, L., & Křivan, V. 2000. A mechanistic model for partial preferences. *Journal of Theoretical Biology* 58:279–289.

Berglund, A., & Rosenqvist, G. 1993. Selective males and ardent females in pipefishes. *Behavioral Ecology and Sociobiology* 32:331–336.

———. 2001. Male pipefish prefer ornamented females. *Animal Behaviour* 61:345–350.

Berglund, A., Rosenqvist, G., & Svensson, I. 1986. Mate choice, fecundity and sexual dimorphism in two pipefish species (*Syngnathidae*). *Behavioral Ecology and Sociobiology* 19:301–307.

———. 1989. Reproductive success of females limited by males in two pipefish species. *American Naturalist* 133:506–516.

Bergman, T. J., Beehner, J. C., Cheney, D. L., Seyfarth, R. M., & Whittens, P. L. 2005. Correlates of stress in free-ranging male chacma baboons *Papio hamadryas ursinus*. *Animal Behaviour* 70:703–713.

Berthold, P. 1993. *Bird migration: A general survey*. Oxford, UK: Oxford University Press.

———. 1999. A comprehensive theory for the evolution, control and adaptability of avian migration. *Ostrich* 70:1–11.

Bielski, I. F., Hu, S-B., Szegda, K. L., Westphal, H., & Young, L. J. 2004. Profound impairment in social recognition and reduction in anxiety-like behavior in vasopressin V1a receptor knockout mice. *Neuropsychopharmacology* 29:483–493.

Biernaskie, J. M., Walker, S. C., & Gegear, R. J. 2009. Bumblebees learn to forage like Bayesians. *American Naturalist* 174:413–423.

Bingman, V. P., & Chen, K. 2005. Mechanisms of animal global navigation: Comparative perspectives and enduring challenges. *Ethology Ecology & Evolution* 17:295–318.

Bird, B. L., Branch, L. C., & Miller, D. L. 2004. Effects of coastal lighting on foraging behavior of beach mice. *Conservation Biology* 18:1435–1439.

Blackledge, T. A., Scharff, N., Coddington, J. A., Szüts, T., Wenzel, J. W., Hayashi, C. Y., & Agnarsson, I. 2009. Reconstructing web evolution and spider diversification in the molecular era. *Proceedings of the National Academy of Sciences USA* 106:5229–5334.

Blas, J., Bortolotti, G. R., Tella, J. L., Baos, R., & Marchant, T. A. 2007. Stress response during development predicts fitness in a wild, long lived vertebrate. *Proceedings of the National Academy of Sciences USA* 104:8880–8884.

Blomqvist, D., Fessl, B., Hoi, H., & Kleindorfer, S. 2005. High frequency of extra-pair fertilizations in the moustached warbler, a songbird with a variable mating system. *Behaviour* 142:1133–1148.

Blumenrath, S. H., & Dabelsteen, T. 2004. Degradation of great tit (*Parus major*) song before and after foliation: Implications for vocal communication in a deciduous forest. *Behaviour* 141:935–958.

Blyth, J. E., & Gilburn, A. S. 2006. Extreme promiscuity in a mating system dominated by sexual conflict. *Journal of Insect Behavior* 19:447–455.

Boal, J. G., Dunham, A. W., Williams, K. T., & Hanlon, R. T. 2000. Experimental evidence for spatial learning in octopuses (*Octopus bimaculoides*). *Journal of Comparative Psychology* 114:246–252.

Boggs, C. L. & Dau, B. 2004. Resource specialization in puddling Lepidoptera. *Ecological Entomology* 33:1020–1024.

Bolhuis, J. J., Okanoya, K., & Scharff, C. 2010. Twitter evolution: Converging mechanisms in birdsong and human speech. *Nature Reviews Neuroscience* 11:747–759.

Bolhuis, J. J., & Wynne, C. D. L. 2009. Can evolution explain how minds work? *Nature* 458:832–833.

Bolhuis, J. J., Zijlstra, G. G. O., den Boer-Visser, A. M., & Van der Zee, E. 2000. Localized neuronal activation in the zebra finch brain is related to the strength of song learning. *Proceedings of the National Academy of Sciences USA* 97:2282–2285.

Bollinger, E. K., Harper, S. J., & Barrett, G. W. 1993. Inbreeding avoidance increases dispersal movements of the meadow vole. *Ecology* 74:1153–1156.

Bond, A. 2007. The evolution of color polymorphism: Crypticity, searching images, and apostatic selection. *Annual Review of Ecology, Evolution, and Systematics* 38:489–514.

Bond, A. B., & Kamil, A. C. 1999. Searching image in blue jays: Facilitation and interference in sequential priming. *Animal Learning & Behavior* 27:461–471.

Bonine, K. E., & Garland, T. Jr. 1999. Sprint performance in phyrnosomatid lizards, measured on a high-speed treadmill, correlates with hindlimb length. *Journal of Zoology* 248:255–265.

Bonte, D., Bossuyt, B., & Lens, L. 2007. Aerial dispersal plasticity under different wind velocities in a salt marsh wolf spider. *Behavioral Ecology* 18:438–443.

Boogert, N. J., Fawcett, T. W., & Lefebvre, L. 2011. Mate choice for cognitive traits: A review of the evidence in nonhuman vertebrates. *Behavioral Ecology* 22:447–459.

Book, A. S., Starzyk, K. B., & Quinsey, V. L. 2001. The relationship between testosterone and aggression: A meta-analysis. *Aggression and Violent Behavior* 6:579–599.

Boone, R. B., Thirgood, S. J., & Hopcraft, J. G. C. 2006. Serengeti wildebeest migratory patterns modeled from rainfall and new vegetation growth. *Ecology* 87:1987–1994.

Boonstra, R., Krebs, C. J., Gaines, M. S., Johnson, M. L., & Craine, I. T. M. 1987. Natal philopatry and breeding systems in voles (*Microtus* spp.). *Journal of Animal Ecology* 56:655–673.

Booth, A., Shelley, G., Mazur, A., Tharp, G., & Kittok, R. 1989. Testosterone, and winning and losing in human competition. *Hormones and Behavior* 23:556–571.

Boulinier, T., & Danchin, E. 1997. The use of conspecific reproductive success for breeding patch selection in territorial migratory species. *Evolutionary Ecology* 11:505–517.

Boulinier, T., McCoy, K. D., Yoccoz, N. G., Gasparini, J., & Tveraa, T. 2008. Public information affects breeding dispersal in a colonial bird: Kittiwakes cue on neighbours. *Biology Letters* 4:538–540.

Boutin, M. E. 2007. *Learning and behavior: A contemporary synthesis.* Sunderland, MA: Sinauer Associates.

Boyle, W. & Conway, C. 2007. Why migrate? A test of the evolutionary precursor hypothesis. *American Naturalist* 169:344–359.

Boysen, S. T., Berntson, G. G., & Mukobi, K. L. 2001. Size matters: Impact of item size and quantity on array choice by chimpanzees (*Pan troglodytes*). *Journal of Comparative Psychology* 115:106–110.

Bracis, C., & Anderson, J. J. 2012. An investigation of the geomagnetic imprinting hypothesis for salmon. *Fisheries Oceanography* 21:170–181.

Bradbury, J. W., Gibson, R. M., & Tsai, I. M. 1986. Hotspots and the evolution of leks. *Animal Behaviour* 34:1694–1709.

Bradbury, J. W., & Vehrencamp, S. L. 1998. *Principles of animal communication.* Sunderland, MA: Sinauer Associates.

Brandstaetter A. S., Endler, A., & Kleineidam, C. J. 2008. Nestmate recognition in ants is possible without tactile interaction. *Naturwissenschaften* 95:601–608.

Bridges, R. S., Numan, M., Ronsheim, P. M., Mann, P. E., & Lupini, C. E. 1990. Central prolactin infusions stimulate maternal behavior in steroid-treated, nulliparous female rats. *Proceedings of the National Academy of Sciences USA* 87:8003–8007.

Bro-Jorgensen, J., & Pangle, W. M. 2010. Male topi antelopes alarm snort deceptively to retain females for mating. *American Naturalist* 176:E33–E39.

Brown, J. L. 1964. The evolution of diversity in avian territorial systems. *Wilson Bulletin* 76:160–169.

———. 1997a. A theory of mate choice based on heterozygosity. *Behavioral Ecology* 8:60–65.

Brown, J. L. 1997b. Long-term memory of an auditory stimulus for food in a natural population of the Mexican jay. *Wilson Bulletin* 109:749–752.

Brown, J. S. 1988. Patch use as an indicator of habitat preference, predation risk, and competition. *Behavioral Ecology and Sociobiology* 22:37–47.

———. 1999. Vigilance, patch use, and habitat selection: Foraging under predation risk. *Evolutionary Ecology Research* 1:49–71.

Brown, C. R., & Brown, M. B. 1986. Ectoparasitism as a cost of coloniality in cliff swallows (*Hirundo pyrrhonota*). *Ecology* 67:1206–1218

Brown, J. S., & Kotler, B. P. 2004. Hazardous pay duty and the foraging costs of predation. *Ecology Letters* 7:999–1014.

Brown, J. L., Morales, V., & Summers, K. 2010. A key ecological trait drove the evolution of biparental care and monogamy in an amphibian. *American Naturalist* 175:436–446.

Brown, J. L., Twomey, E., Morales, V., & Summers, K. 2008. Phytotelm size in relation to parental care and mating strategies in two species of Peruvian poison frogs. *Behaviour* 145:1139–1165.

Bshary, R. 2002. Biting cleaner fish use altruism to deceive image-scoring client reef fish. *Proceedings of the Royal Society of London B* 269:2087–2093.

Bshary, R., & Grutter, A. S. 2002. Asymmetric cheating opportunities and partner control in a cleaner fish mutualism. *Animal Behaviour* 63:547–555.

———. 2006. Image scoring and cooperation in a cleaner fish mutualism. *Nature* 441:975–978.

Bucher, T. A., Ryan, M. J., & Bartholomew, G. A. 1982. Oxygen consumption during resting, calling, and nest building in the frog *Physalaemus pustulosus*. *Physiological Zoology* 55:10–22.

Buckingham, J. N., Wong, B. B. M., & Rosenthal, G. G. 2007. Shoaling decisions in female swordtails: How do fish gauge group size? *Behaviour* 144:1333–1346.

Bulger, J. B., & Hamilton, W. J. 1987. Rank and density correlates of inclusive fitness measures in a natural chacma baboon (*Papio ursinus*) population. *International Journal of Primatology* 8:635–650.

Bulmer, M. G., & Parker, G. A. 2002. The evolution of anisogamy: A game-theoretic approach. *Proceedings of the Royal Society B* 269:2381–2388.

Burda, H., Honeycutt, R. L., Begall, S., Locker-Grütjen, O., & Scharff, A. 2000. Are naked mole-rats eusocial and if so, why? *Behavioral Ecology and Sociobiology* 47:293–303.

Burghardt, G. 1985. Animal awareness: Current perception and historical perspective. *American Psychologist* 405:905–919.

Buston, P. M. 2004. Territory inheritance in clownfish. *Proceedings of the Royal Society B* 271 (suppl): S252–S254.

Cade, T. J., & Jones, C. G. 1993. Progress in restoration of the Mauritius kestrel. *Conservation Biology* 7:169–175.

Caillaud, M. C., & Via, S. 2000. Specialized feeding behavior influences both ecological specialization and assortative mating in sympatric host races of pea aphids. *American Naturalist* 156:606–621.

——. 2012. Quantitative genetics of feeding behavior in two ecological races of the pea aphid, *Acyrthosiphon pisum*. *Heredity* 108:211–218.

Calsbeek, R., Alonzo, S. H., Zamudio, K., & Sinervo, B. 2002. Sexual selection and alternative mating behaviors generate demographic stochasticity in small populations. *Proceedings of the Royal Society B* 269:157–164.

Calsbeek, R, & Sinervo, B. 2002. Uncoupling direct and indirect components of female choice in the wild. *Proceedings of the National Academy of Sciences USA* 99:14,897–14,902.

Cant, M. A. 2000. Social control of reproduction in banded mongooses. *Animal Behaviour* 59:147–148.

Carazo, P., & Font, E. 2010. Putting information back into biological communication. *Journal of Evolutionary Biology* 23:661–669.

Carlson, A. A., Russel, A. F., Young, A., Jordan, N. R., McNeilly, A. S., & Clutton-Brock, T. 2006. Elevated prolactin levels immediately precede decisions to babysit by male meerkat helpers. *Hormones and Behavior* 50:94–100.

Caro, T. 1995. Pursuit-deterrence revisited. *Trends in Ecology and Evolution* 12:500–503.

——. 2005. *Antipredator defenses in birds and mammals*. Chicago: University of Chicago Press.

Caro, T. M., Lombardo, L., Goldizen, A. W., & Kelly, M. 1995. Tail-flagging and other antipredator signals in white-tailed deer: New data and synthesis. *Behavioral Ecology* 6:442–450.

Carré, J. M., & Putnam, S. K. 2010. Watching a previous victory produces an increase in testosterone among elite hockey players. *Psychoneuroendocrinology* 35:475–479.

Carter, R.J., Morton, J., & Dunnett, S.B. 2001. Motor coordination and balance in rodents. *Current Protocols in Neuroscience* 2001:8–12.

Chaiken, M., Böhner, J., & Marler, P. 1993. Song acquisition in European starlings *Sturnus vulgaris*. A comparison of the songs of live-tutored, tape-tutored, untutored, and wild males. *Animal Behaviour* 46:1079–1090.

Champion de Crespigny, F. E., & Hoske, D. J. 2007. Sexual selection: Signals to die for. *Current Biology* 17:R853–R855.

Chapman, T., Arnqvist, G., Bangham, J., & Rowe, L. 2003. Sexual conflict. *Trends in Ecology and Evolution* 18:41–47.

Chapman, B. B., Brönmark, C., Nilsson, J-A., & Hansson, L-A. 2011. The ecology and evolution of partial migration. *Oikos* 120:1764–1775.

Chapman, C. A., & Chapman, L. J. 2000. Constraints on group size in red colobus and red-tailed guenons: Examining the generality of the ecological constraints model. *International Journal of Primatology* 21:565–585.

Charnov, E. L. 1976. Optimal foraging: Attack strategy of a mantid. *American Naturalist* 110:141–151.

Chen, S., Lee, A. Y., Bowens, N. M., Huber, R., & Kravitz, E. A. 2002. Fighting fruit flies: A model system for the study of aggression. *Proceedings of the National Academy of Sciences USA* 99:5564–5568.

Cheney, K. L., & Côté, I. M. 2005. Mutualism or parasitism? The variable outcome of cleaning symbioses. *Biology Letters* 1:162–165.

——. 2007. Aggressive mimics profit from a model-signal receiver mutualism. *Proceedings of the Royal Society B* 274:2087–2091.

Cheney, K. L., Grutter, A. S., Blomberg, S. P., & Marshall, N. J. 2009. Blue and yellow signal cleaning behavior in coral reef fishes. *Current Biology* 19:1283–1287.

Cheney, D. L., & Seyfarth, R. M. 1980. Vocal recognition in free-ranging vervet monkeys. *Animal Behaviour* 28:362–367.

——. 1985. Vervet monkey alarm calls: Manipulation through shared information? *Behaviour* 94:150–166.

Cheney, D., Seyfarth, R., & Smuts, B. 1986. Social relationships and social cognition in nonhuman primates. *Science* 234:1361–1366.

Cheng, W-C., & Kam, Y-C. 2010. Paternal care and egg survivorship in a low nest-attendance Rhacophorid frog. *Zoological Studies* 49:304–310.

Chernetsov, N., Kishkinev, D., & Mouritsen, H. A. 2008. A long-distance avian migrant compensates for longitudinal displacement during spring migration. *Current Biology* 18:188–190.

Chitty, D. 1996. *Do lemmings commit suicide?* New York: Oxford University Press.

Chuang, H. C., Webster, M. S., & Holmes, R. T. 1999. Extra-pair paternity and local synchrony in the black-throated blue warbler. *Auk* 116:726–736.

Chuang-Dobbs, H. C., Webster, M. S., & Holmes, R. T. 2001. The effectiveness of mate guarding by male black-throated blue warblers. *Behavioral Ecology* 12:541–546.

Citta, J. J., & Lindberg, L. S. 2007. Nest-site selection of passerines: Effects of geographic scale and public and personal information. *Ecology* 88:2034–2046.

Clark, D. L., Roberts, J. A., & Uetz, G. W. 2012. Eavesdropping and signal matching in visual courtship displays of spiders. *Biology Letters* 8:375–378.

Clayton, D. F., Balakrishnan, C. N., & London, S. E. 2009. Integrating genomes, brain and behavior in the study of songbirds. *Current Biology* 19:R865–R873.

Clobert, J., Le Galliard, J.-F., Cote, J., Meylan, S., & Massot, M. 2009. Informed dispersal, heterogeneity in animal dispersal syndromes and the dynamics of spatially structured populations. *Ecology Letters* 12:197–209.

Clutton-Brock, T. H. 1989. Mammalian mating systems. *Proceedings of the Royal Society B* 236:339–372.

———. 1991. *The evolution of parental care*. Princeton, NJ: Princeton University Press.

———. 2002. Breeding together: Kin selection and mutualism in cooperative vertebrates. *Science* 296:69–72.

———. 2007. Sexual selection in males and females. *Science* 318:1882–1885.

Clutton-Brock, T. H., & Albon, S. D. 1979. The roaring of red deer and the evolution of honest advertisement. *Behaviour* 69:145–170.

———. 1989. *Red deer in the highlands*. Oxford, UK: Blackwell Scientific.

Cobb, M. 2007. A gene mutation which changed animal behaviour: Margaret Bastock and the yellow fly. *Animal Behaviour* 74:163–169.

Cockburn, A. 2006. Prevalence of different modes of parental care in birds. *Proceedings of the Royal Society B* 273:1375–1383.

Cockrem, J. F., & Silverin, B. 2002. Sight of a predator can stimulate a corticosterone response in the great tit (*Parus major*). *General and Comparative Endocrinology* 125:248–255.

Cohas, A., & Allainé, D. 2009. Social structure influences extra-pair paternity in socially monogamous mammals. *Biology Letters* 5:313–316.

Cohas, A., Bonenfant, C., Gaillard, J.-M., & Allainé, D. 2007. Are extra-pair young better than within-pair young? A comparison of survival and dominance in alpine marmots. *Journal of Animal Ecology* 76:771–781.

Collett, T. S., & Graham, P. 2004. Animal navigation: Path integration, visual landmarks and cognitive maps. *Current Biology* R475–R477.

Conner, J. K., & Hartl, D. L. 2004. *A primer of ecological genetics*. Sunderland, MA: Sinauer Associates, Inc.

Conradt, L., Krause, J., Couzin, I. D., & Roper, T. J. 2009. "Leading according to need" in self organizing groups. *American Naturalist* 173:304–312.

Contreras-Garduño, J., Canales-Lazcano, J. & Córdoba-Aguilar, A. 2006. Wing pigmentation, immune ability, fat reserves and territorial status in males of the rubyspot damselfly *Hetaerina americana*. *Journal of Ethology* 24:165–173.

Coolen, I., Ward, A. J. W., Hart, P. J. B., & Laland, K. N. 2005. Foraging nine-spined sticklebacks prefer to rely on public information over simpler social cues. *Behavioral Ecology* 16:865–870.

Coombs, S., & Montgomery, J. C. 1999. The enigmatic lateral line system. In *Comparative hearing: Fish and amphibians*, Springer Handbook of Auditory Research Series, vol. 11, ed. R. R. Fay & A. N. Popper, 319–362. New York: Springer.

Corcoran, A. J., Barber, J. R., & Conner, W. E. 2009. Tiger moth jams bat sonar. *Science* 325:325–327.

Córdoba-Aguilar, A., & Cordera-Rivera, A. 2005. Evolution and ecology of Calopterygidae (Zygoptera: Odonta): Status of knowledge and research perspectives. *Neotropical Entomology* 34:861–879.

Costall, A. 1993. How Lloyd Morgan's canon backfired. *Journal of the History of the Behavioral Sciences* 29:113–122.

Côte, I. M., & Poulin, R. 1995. Parasitism and group size in social animals: A meta-analysis. *Behavioral Ecology* 6:159–163.

Courter, J. R., & Ritchison, G. 2010. Alarm calls of tufted titmice convey information about predator size and threat. *Behavioral Ecology* 5:936–942.

Cox, G. W. 1985. The evolution of avian migration systems between temperate and tropical regions of the New World. *American Naturalist* 126:451–474.

Cratsley, C. K., & Lewis, S. M. 2003. Female preference for male courtship flashes in *Photinus ignitus* fireflies. *Behavioral Ecology* 14:135–140.

Cristol, D. A., Baker, M. C., & Carbone, C. 1999. Differential migration revisited: Latitudinal segregation by age and sex class. *Current Ornithology* 15:33–88.

Crowe, S. A., Kleven, O., Delmore, K. E., Laskemoen, T., Nocera, J. J., Lifjeld, J. T., & Robertson, K. J. 2009. Paternity assurance through frequent copulations in a wild passerine with intense sperm competition. *Animal Behaviour* 77:183–187.

Curio, E., Ernst, U., & Vieth, W. 1978. Cultural transmission of enemy recognition: One function of mobbing. *Science* 202:899–901.

Cynx, J., Lewis, R., Tavel, B., & Tse, H. 1998. Amplitude regulation of vocalizations by a songbird, *Taeniopygia guttata*. *Animal Behaviour* 56:107–113.

Dada, M., Pilastro, A., & Bisazza, A. 2005. Male sexual harassment and female schooling behavior in the eastern mosquitofish. *Animal Behaviour* 70:463–471.

Daly, J. W., & Myers, C. W. 1967. Toxicity of Panamanian poison frogs (Dendrobates): Some biological and chemical aspects. *Science* 156:970–973.

Daly, J. W., Spande, T. F., & Garraffo, H. M. 2005. Alkaloids from amphibian skin: A tabulation of over eight hundred compounds. *Journal of Natural Products* 68:1556–1575.

Danchin, E., Boulinier, T. & Massot, M. 1998. Conspecific reproductive success and breeding habitat selection: Implications for the study of coloniality. *Ecology* 79:2415–2428

Darwin, C. 1859. *On the origin of species by means of natural selection*. London: John Murray.

———. 1871.*The descent of man, and selection in relation to sex*. London: John Murray.

———. 1872. *Expression of the emotions in man and animals*. London: John Murray.

DaSilva, A., Luikart, G., Yoccoz, N., Cohas, A., & Allainé, D. 2006. Genetic diversity-fitness correlation revealed by microsatellite analyses in European alpine marmots (*Marmota marmota*). *Conservation Genetics* 3:371–382.

Daszak, P., Cunningham, A. A., & Hyatt, A. D. 2000. Emerging infectious diseases to wildlife— threats to biodiversity and human health. *Science* 287:443–449.

David, P., Bjorksten, T., Fowler, K., & Pomiankowski, A. 2000. Condition-dependent signalling of genetic variation in stalk-eyed flies. *Nature* 406:186–188.

Davies, N. B., Krebs, J. R., & West, S. A. 2012. *An introduction to behavioral ecology*. 4th ed. London: Wiley-Blackwell.

Davis, A. K., & Rendón-Salinas, E. 2010. Are female monarch butterflies declining in eastern North America? Evidence of a 30-year change in sex ratios at Mexican overwintering sites. *Biology Letters* 6:45–47.

Dawkins, M. S. 2008. The science of animal suffering. *Ethology* 114:937–945.

Dawkins, M. S., & Guilford, T. 1991. The corruption of honest signaling. *Animal Behaviour* 41:865–873.

Dawkins, R., & Krebs, J. R. 1979. Arms races within and between species. *Proceedings of the Royal Society B* 205:489–511.

DeBose, J. L., & Nevitt, G. A. 2008. The use of odors at different spatial scales: Comparing birds with fish. *Journal of Chemical Ecology* 34:867–881.

De Sousa, V. T. T., Teresa, F. B., & de Cerqueirra Rossa-Feres, D. 2011. Predation risk and jumping behavior in *Pseudopaludicola* aff. *falcipes* tadpoles. *Behavioral Ecology* 22:940–946.

Deutschlander, M. E., Phillips, J. B., & Borland, S. C. 1999. The case for light-dependent magnetic orientation in animals. *Journal of Experimental Biology* 202:891–908.

Dewsbury, D. A. 1985. *Studying animal behavior: Autobiographies of the founders*. Chicago: University of Chicago Press.

Dias, R. I. 2006. Effects of position and flock size on vigilance and foraging behaviour of the scaled dove *Columbina squammata*. *Behavioral Processes* 73:248–252.

Dingemanse, N. J., Both, C., Drent, P. J., Van Oers, K., & Van Noordwijk, A. J. 2002. Repeatability and heritability of exploratory behavior in great tits from the wild. *Animal Behaviour* 64:929–938.

Dittman, A., & Quinn, T. 1996. Homing in Pacific salmon: mechanisms and ecological basis. *The Journal of Experimental Biology* 199:83–91.

Dobzhansky, T. 1973. Nothing in biology makes sense except in the light of evolution. *American Biology Teacher* 35:125–129.

Doligez, B., Cadet, C., Danchin, E., & Boulinier, T. 2003. When to use public information for breeding habitat selection? The role of environmental predictability and density dependence. *Animal Behaviour* 66:973–988.

Doligez, B., Danchin, E., & Clobert, J. 2002. Public information and breeding habitat selection in a wild bird population. *Science* 297:1168–1170.

Donahue, M. J. 2006. Allee effects and conspecific cueing jointly lead to conspecific attraction. *Oecologia* 149:33–43.

Doutrelant, C., McGregor, P. K., & Oliveira, R. F. 2001. The effect of an audience on intrasexual communication in male Siamese fighting fish, *Betta splendens*. *Behavioral Ecology* 12:283–286.

Drent, P. J., & Marchetti, C. 1999. Individuality, exploration and foraging in hand raised juvenile great tits. In *Proceedings of the 22nd International Ornithological Congress*, ed. N. J. Adams & R. H. Slotow, 896–914. Johannesburg: Birdlife South Africa.

Drent, P. J., van Oers, K., & van Noordwijk, A. J. 2003. Realized heritability of personalities in the great tit (*Parus major*). *Proceedings of the Royal Society B* 270:45–51.

Drickamer, L. C., Gowaty, P. A., & Holmes, C. M. 2000. Free female mate choice in house mice affects reproductive success and offspring viability and performance. *Animal Behaviour* 59:371–378.

Dubois, F., Drullion, D., & Witte, K. 2012. Social information use may lead to maladaptive decisions: A game theoretic model. *Behavioral Ecology* 23:225–231.

Dugatkin, L. A. 1992. Sexual selection and imitation: Females copy the mate choice of others. *American Naturalist* 139:1384–1389.

Dugatkin, L. A., & Godin, L.-G. 1992. Reversal of female mate choice in the guppy (*Poecilia reticulata*). *Proceedings of the Royal Society B* 249:179–184.

———. 1993. Female mate copying in the guppy (*Poecilia reticulata*): Age dependent effects. *Behavioral Ecology* 4:289–292.

Dugdale, H. L., MacDonald, D. W., Pope, L. C., & Burke, T. 2007. Polygynandry, extra-group paternity and multiple-paternity litters in European badger (*Meles meles*) social groups. *Molecular Ecology* 16:5294–5306.

Dukas, R. 2008. Life history of learning: Performance curves of honeybees in the wild. *Ethology* 114:1195–1200.

Dunk, J. R., & Cooper, R. J. 1994. Territory-size regulation in black-shouldered kites. *Auk* 111:588–595.

Dunlap, A. S., & Stephens, D. W. 2009. Components of change in the evolution of learning and unlearned preference. *Proceedings of the Royal Society B* 276:3201–3208.

Dziuk, P. J. 1996. Factors that influence the proportion of offspring sired by a male following heterspermatic insemination. *Animal Reproduction Science* 43:65–88.

East, M. L., & Hofer, H. 2001. Male spotted hyenas (*Crocuta crocuta*) queue for status in social groups dominated by females. *Behavioral Ecology* 12:558–568.

Eifler, D. A., Eifler, M. A., & Harris, B. R. 2008. Foraging under risk of predation in desert grassland whiptail lizards (*Aspidoscelis uniparens*). *Journal of Ethology* 26:219–223.

Elias, D. O., Kasumovic, M. M., Punzalon, D., Andrade, M. C. B., & Mason, A. C. 2008. Assessment during aggressive contests between male jumping spiders. *Animal Behaviour* 76:901–910.

Ellis, L. L., & Carney, G. E. 2011. Socially-responsive gene expression in male *Drosophila melanogaster* is influenced by the sex of the interacting partner. *Genetics* 187:157–169.

Ellis, D. H., Sladen, W. J. L., Lishman, W. A., Clegg, K. R., Duff, J. W., Gee, G. F., & Lewis, J. C. 2003. Motorized migrations: The future or mere fantasy. *Bioscience* 53:260–264.

Emlen, S. T. 1967a. Migratory orientation in the indigo bunting, *Passerina cyanea*. Part I: Evidence for use of celestial cues. *Auk* 84:309–342.

————. 1967b. Migratory orientation in the indigo bunting, *Passerina cyanea*. Part II: Mechanism of celestial orientation. *Auk* 84:463–489.

————. 1970. Celestial rotation: Its importance in the development of migratory orientation. *Science* 170:1198–1201.

————. 1994. Environmental control of horn length dimorphism in the beetle *Onthophagus acuminatus* (Coleoptera: Scarabaeidae). *Proceedings of the Royal Society B* 256:131–136.

————. 2008. The evolution of animal weapons. *Annual Review of Ecology, Evolution and Systematics* 39:387–413.

Emlen, S. T., & Emlen, J. T. 1966. A technique for recording migratory orientation of captive birds. *Auk* 83:361–367.

Emlen, S. T., & Oring. L. W. 1977. Ecology, sexual selection, and the evolution of mating systems. *Science* 197:215–223.

Endler, J. A. 1980. Natural selection on color patterns in *Poecilia reticulata*. *Evolution* 34:76–91.

Engh, A. L., Funk, S. M., Van Horn R. C., Scribner, K. T., Bruford, M. W., Libants, S., Szykman, M., Smale, L., & Holekamp, K. E. 2002. Reproductive skew among males in a female-dominated mammalian society. *Behavioral Ecology* 13: 193–200.

Enquist, M., & Leimar, O. 1983. Evolution of fighting behavior: Decision rules and assessment of relative strength. *Journal of Theoretical Biology* 102: 387–410.

————. 1987. Evolution of fighting behavior: The effect of variation in resource value. *Journal of Theoretical Biology* 127:187–205.

Fausto da Silva, J. J. R., & Williams, R. J. P. 2001. *The biological chemistry of the elements: The inorganic chemistry of life*. Oxford, UK: Oxford University Press.

Feeney, W. E., Welbergen, J. A., & Langmore, N. E. 2012. The frontline of avian brood parasite-host coevolution. *Animal Behaviour* 84:3–12.

Fernández-Juricic, E., Beauchamp, G., & Bastain, B. 2007. Group-size and distance-to-neighbor effects on feeding and vigilance in brown-headed cowbirds. *Animal Behaviour* 73:771–778.

Ferrari, M. C. O., Sih, A., & Chivers, D. P. 2009. The paradox of risk allocation: A review and prospectus. *Animal Behaviour* 78:579–585.

Field, J., & Cant, M. A. 2009. Social stability and helping in small animal societies. *Philosophical Transactions of the Royal Society B* 364:3181–3189.

Field, J., Shreeves, G., & Sumner, S. 1999. Group size queuing and helping decisions in facultatively eusocial hover wasps. *Behavioral Ecology and Sociobiology* 45:378–385.

Fiorito, G., & Scotto, P. 1992. Observational learning in *Octopus vulgaris*. *Science* 256:545–547.

Fisher, R. A. 1930. *The genetical theory of natural selection*. Oxford: Clarendon Press.

————. 1966. *The design of experiments*. 8th ed. Edinburgh: Hafner.

Fitzpatrick, M. J., Ben-Shahar, Y., Smid, H. M., Vet, L. E. M., Robinson, G. E., & Sokolowski, M. B. 2005. Candidate genes for behavioural ecology. *Trends in Ecology and Evolution* 20:96–104.

Fitzpatrick, B. M., Shook, K., & Izally, R. 2009. Frequency-dependent selection by wild birds promotes polymorphism in model salamanders. *BMC Ecology* 9:12.

Flint, J. 2003. Analysis of quantitative trait loci that influence animal behavior. *Journal of Neurobiology* 54:46–77.

Foerder, P., Galloway, M., Barthel, T., Moore, D. E. III, & Reiss, D. 2011. Insightful problem solving in an Asian elephant. *PLoS ONE* 6:e23251.

Forbes, S., Grosshans, R., & Glassey, B. 2002. Multiple incentives for parental optimism and brood reduction in blackbirds. *Ecology* 83:2529–2541.

Foster, W. A., & Treherne, J. E. 1981. Evidence for the dilution effect in the selfish herd from fish predation on a marine insect. *Nature* 293:466–467.

Fretwell, S. D., & Lucas, H. L. 1969. On territorial behavior and other factors influencing habitat distributions of birds. *Acta Biotheoretica* 19:16–36.

Friedl, T. W. P., & Klump, G. M. 2005. Sexual selection in the lek-breeding European treefrog: Body size, chorus attendance, random mating and good genes. *Animal Behaviour* 70:1141–1154.

Frisch, K. von. 1956. *Bees: Their vision, chemical senses, and language.* Ithaca, NY: Cornell University Press.

Frommen, J. G., Hiermes, M., & Bakker, T. C. M. 2009. Disentangling the effects of group size and density on shoaling decisions of three-spined sticklebacks (*Gasterosteus aculeatus*). *Behavioral Ecology and Sociobiology* 63:1141–1148.

Fu, P., Neff, B. D., & Gross, M. R. 2001. Tactic-specific success in sperm competition. *Proceedings of the Royal Society B* 268:1105–1112.

Galef, B. G. Jr. 1998. Edward Thorndike: Revolutionary psychologist, ambiguous biologist. *American Psychologist* 53:1128–1134.

García-González, F., Núñez, Y., Ponz, F., Roldán, E. R. S., & Gomendio, M. 2003. Sperm competition mechanisms, confidence of paternity, and the evolution of paternal care in the golden egg bug (*Phyllomorpha laciniata*). *Evolution* 57:1078–1088.

García-Navas, V., Ortega, J., & Sanz, J. J. 2009. Heterozygosity-based assortative mating in blue tits (*Cyanistes caeruleus*): Implications for the evolution of mate choice. *Proceedings of the Royal Society B* 276:2931–2940.

Gardner, R. A., & Gardner, B. T. 1969. Teaching sign language to a chimpanzee. *Science* 165:664–672.

Garland, T. Jr. 1994. Quantitative genetics of locomotor behavior and physiology in a garter snake. In *Quantitative genetic studies of behavioral evolution,* ed. C. R. B. Boake, 251–277. Chicago: University of Chicago Press.

Gazit, I., Goldblatt, A., & Terkel, J. 2005. Formation of an olfactory search image for explosives in sniffer dogs. *Ethology* 111:669–680.

Gegear, R. J., Foley, L. E., Casselman, A., & Reppert, S. M. 2010. Animal cryptochromes mediate magnetoreception by an unconventional photochemical mechanism. *Nature* 463:804–807.

Gerhardt, H. C. 1987. Evolutionary and neurobiological implications of selective phonotaxis in the green treefrog (*Hyla cinerea*). *Animal Behaviour* 35:1479–1489.

Gerlach, N. M., McGlothlin, J. W., Parker, P. G., & Ketterson, E. D. 2012. Promiscuous mating produces offspring with higher lifetime fitness. *Proceedings of the Royal Society B* 279:860–866.

Gese, E. M., Rongstad, O. J., & Mytton, W. R. 1988. Relationship between coyote group size and diet in southeastern Colorado. *Journal of Wildlife Management* 52:647–653.

Gesquiere, L. R., Learn, L. H., Simao, M. C. M., Onyango, P. O., Alberts, S. C., & Altmann, J. 2011. Life at the top: Rank and stress in wild male baboons. *Science* 333:357–360.

Gilbert, J. D. J., & Manica, A. 2010. Parental care trade-offs and life-history relationships in insects. *American Naturalist* 176:212–226.

Gill, R. D. 2011. The Monty Hall problem is not a probability puzzle (it's a challenge in mathematical modeling). *Statistica Neerlandica* 65:58–71.

Gillis, E. A., Green, D. J., Middleton, H. A., & Morrissey, C. A. 2008. Life history correlates of alternative migratory strategies in American dippers. *Ecology* 89:1687–1695.

Gleason, E. D., Fuxjager, M. J., Oyegbile, T. O., & Marler, C. A. 2009. Testosterone release and social context: When it occurs and why. *Frontiers in Neuroendocrinology* 30:460–469.

Goerlitz, H. R., & Siemers, B. M. 2007. Sensory ecology of prey rustling sounds: Acoustical features and their classification by wild grey mouse lemurs. *Functional Ecology* 21:143–153.

Gomendio, M., García-González, F., Reguera, P., & Rivero, A. 2008. Male egg carrying in *Phyllomorpha laciniata* is favoured by natural not sexual selection. *Animal Behaviour* 75:763–770.

González, L. M., Margalida, A., Sánchez, R., & Oria, J. 2006. Supplmentary feeding as an effective tool for improving breeding success in the Spanish imperial eagle (*Aquila adalberti*). *Biological Conservation* 129:477–486.

Gould, J. L. 2008. Animal navigation: The evolution of magnetic orientation. *Current Biology* 18: R482–R485.

————. 2010. Magnetoreception. *Current Biology* 20:R431–R435.

Gould, J. L., & Gould, C. G. 1988. *The honey bee.* New York: W. H. Freeman.

Graf, S. A., & Sokolowski, M. B. 1989. Rover/sitter *Drosophila melanogaster* larval foraging polymorphism as a function of larval development, food-patch quality, and starvation. *Journal of Insect Behavior* 2:310–313.

Grafen, A. 1990. Biological signals as handicaps. *Journal of Theoretical Biology* 144:517–546.

Gray, D. A., & Cade, W. H. 2000. Sexual selection and speciation in field crickets. *Proceedings of the National Academy of Sciences USA* 97: 14,449–14,454.

Grayson, K. L., & Wilbur, H. M. 2009. Sex- and context-dependent migration in a pond-breeding amphibian. *Ecology* 90:306–312.

Greenfield, M. D., & Shelly, T. E. 1985. Alternative mating strategies in a desert grasshopper: Evidence of density-dependence. *Animal Behaviour* 33:1192–1210.

Greenfield, M. D., Shelly, T. E., & Downum, K. R. 1987. Variation in host-plant quality: Implications for territoriality in a desert grasshopper. *Ecology* 68:828–838.

Greenwood, P. J. 1980. Mating systems, philopatry and dispersal in birds and mammals. *Animal Behaviour* 28:1140–1162.

Griffin, D. R. 1976. *The question of animal awareness: Evolutionary continuity of mental experience.* New York: Rockefeller University Press.

_____. 1984. *Animal thinking.* Cambridge, MA: Harvard University Press.

Griffin, D. R., & Speck, G. B. 2004. New evidence of animal consciousness. *Animal Cognition* 7:15–18.

Griffith, S. C. 2007. The evolution of infidelity in socially monogamous passerines: Neglected components of direct and indirect selection. *American Naturalist* 169:274–281.

Griffith, S. C., Owens, I. P. F., & Thuman, K. A. 2002. Extra pair paternity in birds: A review of interspecific variation and adaptive function. *Molecular Ecology* 11:2195–2212.

Griffith, S. C., Pryke, S. R., & Buttemer, W. A. 2011. Constrained mate choice in social monogamy and the stress of having an unattractive partner. *Proceedings of the Royal Society B* 278:2798–2805.

Griffiths, P. E. 2008. Ethology, sociobiology, evolutionary psychology. In *A companion to the philosophy of biology,* ed. S. Sarkar & A. Plutyinski, 393–414. Blackwell Companions to Philosophy. Oxford, UK: Wiley-Blackwell.

Grimm, D. 2011. Are dolphins too smart for captivity? *Science* 332:526–529.

Gross, M. R. 1996. Alternative reproductive strategies and tactics: Diversity within sexes. *Trends in Ecology and Evolution* 11:92–98.

_____. 2005. The evolution of parental care. *Quarterly Review of Biology* 80:37–45.

Gross, M. R., & Sargent, R. C. 1985. The evolution of male and female parental care in fishes. *American Zoologist* 25:807–822.

Grutter, A. S. 1994 Spatial and temporal variations of the ectoparasites of 7 reef fish species from Lizard Island and Heron Island, Australia. *Marine Ecology Progress Series* 115:21–30.

Grutter, A. S., & Bshary, R. 2003. Cleaner wrasse prefer client mucus: Support for partner control mechanisms in cleaning interactions. *Proceedings of the Royal Society B* 270 (suppl): S242–S244.

Grütter, C., Balbuena, M. S., & Farina, W. M. 2008. Informational conflicts created by the waggle dance. *Proceedings of the Royal Society B* 275:1321–1327.

Grütter, C., & Farina, W. M. 2009. The honeybee waggle dance: Can we follow the steps? *Trends in Ecology and Evolution* 24:242–247.

Gubernick, D. J., & Alberts, J. R. 1987. The biparental care system of the California mouse, *Peromyscus californicus. Journal of Comparative Psychology* 101:169–177.

Gubernick, D. J., & Teferi, T. 2000. Adaptive significance of male parental care in a monogamous mammal. *Proceedings of the Royal Society B* 267:147–150.

Gwynne, D. T. 2008. Sexual selection over nuptial gifts in insects. *Annual Review of Entomology* 53: 83–101.

Haesler, S., Rochefort, C., Georgi, B., Licznerski, P., Osten, P., & Scharff, C. 2007. Incomplete and inaccurate vocal imitation after knockdown of FoxP2 in songbird basal ganglia nucleus area X. *PLoS Biology* 5:e321.

Haesler, S., Wada, K., Nshdejan, A., Morrisey, E. E., Lints, T., Jarvis, E. D., & Scharff, C. 2004. FoxP2 expression in avian vocal learners and non-learners. *Journal of Neuroscience* 24:3164–3175.

Haesler, A. D., & Wisby, W. J. 1951. Discrimination of stream odors by fishes and its relation to parent stream behavior. *American Naturalist* 85:223–238.

Hahn, B. A., & Silverman, E. D. 2006. Social cues facilitate habitat selection: American redstarts establish breeding territories in response to song. *Biology Letters* 2:337–340.

Hamilton, W. D. 1963. The evolution of altruistic behavior. *American Naturalist* 97:354–356.

_____. 1964. The genetical evolution of social behavior. *Journal of Theoretical Biology* 7:1–52.

_____. 1971. Geometry for the selfish herd. *Journal of Theoretical Biology* 31:295–311.

Hamilton, W. D., & May, R. M. 1977. Dispersal in stable habitats. *Nature* 269:578–581.

Hamilton, W. D., & Zuk, M. 1982. Heritable true fitness and bright birds: A role for parasites? *Science* 218:384–387.

Hanauske, M., Kunz, J., Bernius, S., & König, W. 2009. Doves and hawks in economics revisited. An evolutionary quantum game theory-based analysis of financial crises. *Physica A: Statistical Mechanics and Its Applications* 389:5084–5102.

Handley, L. J. L., & Perrin, N. 2007. Advances in our understanding of mammalian sex-biased dispersal. *Molecular Ecology* 16:1559–1578.

Hansen, B. T., Erik, L., & Slagsvold, T. 2008. Imprinted species recognition lasts for life in free-living great tits and blue tits. *Animal Behaviour* 75:921–927.

Hanssen, S. A., Hasselquist, D., Folstad, I., & Erikstad, K. E. 2005. Cost of reproduction in a long-lived bird: Incubation effort reduces immune function and future reproduction. *Proceedings of the Royal Society B* 272:1039–1046.

Harsha, S., Satischandra, K., Kudavidanage, E. P., Kotagama, S. W., & Goodale, E. 2007. The benefits of joining mixed-species flocks for greater racket-tailed drongos *Dicrurus paradiseus*. *Forktail* 23:145–148.

Hart, P. J., & Freed, L. A. 2005. Predator avoidance as a function of flocking in the sexually dichromatic Hawaii akepa. *Journal of Ethology* 23:29–33.

Hartl, D. 2000. *A primer of population genetics*. 3rd ed. Sunderland, MA: Sinauer Associates.

Hatchwell, B. J., Russell, A. F., MacColl, A. D. C., Ross, D. J., Fowlie, M. K., & McGowan, A. 2004. Helpers increase long-term but not short-term productivity in cooperatively breeding long-tailed tits. *Behavioral Ecology* 15:1–10.

Haugen, T. O., Winfield, I. J., Asbjörn Vøllestad, L., Fletcher, J. M., James J. B., & Stenseth, N. C. 2006. The ideal free pike: 50 years of fitness-maximizing dispersal in Windermere. *Proceedings of the Royal Society B* 273:2917–2924.

Hauser, M. D., Comsky, N., & Fitch, W. T. 2002. The faculty of language: What is it, who has it, and how did it evolve? *Science* 298:1569–1579.

Hebets, E. A., & Papaj, D. R. 2004. Complex signal function: Developing a framework of testable hypotheses. *Behavioral Ecology and Sociobiology* 57:197–214.

Hedrick, A. V. 1994. The heritability of mate-attractive traits: A case study on field crickets. In *Quantitative genetic studies of behavioral evolution*, ed. C. R. B. Boake, 228–250. Chicago: University of Chicago Press.

Herberholz, J., McCurdy, C., & Edwards, D. H. 2007. Direct benefits of social dominance in juvenile crayfish. *Biological Bulletin* 213:21–27.

Herbranson, W. T., & Schroeder, J. 2010. Are birds smarter than mathematicians? Pigeons (*Columbia livia*) perform optimally on a version of the Monty Hall dilemma. *Journal of Comparative Psychology* 124:1–13.

Herczeg, G., Gonda, A., & Merilä, J. 2009. The social cost of shoaling covaries with predation risk in nine-spined stickleback, *Pungitius pungitius*, populations. *Animal Behaviour* 77:575–580.

Hering, H., & Sheng, M. 2001. Dendritic spines: Structure, dynamics and regulation. *Nature Reviews Neuroscience* 2:880–888.

Hill, G. E. 1990. Female house finches prefer colourful males: Sexual selection for a condition-dependent trait. *Animal Behaviour* 40:563–572.

Hill, G. E., Inouye, C. Y., & Montgomerie, R. 2002. Dietary carotenoids predict plumage coloration in wild house finches. *Proceedings of the Royal Society B* 269:1119–1124.

Hill, G. E., & Montgomerie, R. 1994. Plumage colour signals nutritional condition in the house finch. *Proceedings of the Royal Society B* 258:47–52.

Hirschenhauser, K., & Oliveira, R. F. 2006. Social modulation of androgens in male vertebrates: Meta-analyses of the challenge hypothesis. *Animal Behaviour* 71:265–277.

Hoare, D. J., Couzin, I. D., Godin, J.-G. J., & Krause, J. 2004. Context-dependent group size choice in fish. *Animal Behaviour* 67:155–164.

Hoefler, C. D., Carlascio, A. L., Persons, M. H., & Rypstra, A. L. 2009. Male courtship repeatability and potential indirect genetic benefits in a wolf spider. *Animal Behaviour* 78:183–188.

Hoekstra, H. 2010. In search of the elusive behavior gene. In *In search of the causes of evolution: From field observations to mechanisms*, ed. P. Grant & R. Grant, 192–210. Princeton, NJ: Princeton University Press.

Höglund, J., & Robertson, J. G. M. 1990. Female preferences, male decision rules and the evolution of leks in the great snipe *Gallinago media*. *Animal Behaviour* 40:15–22.

Hoke, K. L., Ryan, M. J., & Wilczynski, W. 2010. Sexually dimorphic sensory gating drives behavioral differences in túngara frogs. *Journal of Experimental Biology* 213:3463–3472.

Holdo, R. M., Holt, R. D., & Fryxell, J. M. 2009. Opposing rainfall and plant nutritional gradients best explain the wildebeest migration in the Serengeti. *American Naturalist* 173:431–445.

Holland, B., & Rice, W. R. 1998. Chase-away sexual selection: Antagonistic seduction versus resistance. *Evolution* 52:1–7.

Holland, R. A., Wikelski, M., & Wilcove, D. S. 2006. How and why do insects migrate? *Science* 313:794–796.

Hoppitt, W., & Laland, K. N. 2008. Social processes influencing learning in animals: A review of the evidence. *Advances in the Study of Behavior* 38:105–165.

Horn, G. 1998. Visual imprinting and the neural mechanisms of recognition memory. *Trends in Neurosciences* 21:300–305.

Horn, G. 2004. Pathways of the past: The imprint of memory. *Nature Reviews Neuroscience* 5:108–120.

Horowitz, A. 2008. Disambiguating the "guilty look": Salient prompts to a familiar dog behavior. *Behavioral Processes* 81:447–452.

Houck, L. D., & Drickamer, L.C. 1996. *Foundations of animal behavior: Classic papers with commentaries*. Chicago: University of Chicago Press.

Houde, A. E., & Endler, J. A. 1990. Correlated evolution of female mating preference and male color patterns in the guppy *Poecilia reticulata*. *Science* 248:1405–1408.

How, M. J., Hemmi, J. M., Zeil, J., & Peters, R. 2008. Claw waving display changes with receiver distance in fiddler crabs, *Uca perplexa*. *Animal Behaviour* 75:1015–1022.

Howard, E., & Davis, A. K. 2009. The fall migration flyways of monarch butterflies in eastern North America revealed by citizen scientists. *Journal of Insect Conservation* 13:279–286.

Howell, T. F., & Bennett, P. C. 2011. Puppy power! Using social cognition research tasks to improve socialization practices for domestic dogs (*Canis familiaris*). *Journal of Veterinary Behavior* 6:195–204.

Hsu, Y., Earley, R. L., & Wolf, L. L. 2006. Modulation of aggressive behaviour by fighting experience: Mechanisms and contest outcomes. *Biological Reviews* 81:33–74.

Hsu, Y., Lee, S., Chen, M.-H., Yang, S.-Y., & Cheng, K.-C. 2008. Switching assessment strategy during a contest: Fighting in killifish *Kryptolebias marmoratus*. *Animal Behaviour* 75:1641–1649.

Huang, Y.-C., & Hessler, N. A. 2008. Social modulation during songbird courtship potentiates midbrain dopaminergic neurons. *PLoS One* 3:e3281.

Hubbs, A. H., Millar, J. S., & Wiebe, J. P. 2000. Effect of brief exposure to a potential predator on cortisol concentrations in female Columbian ground squirrels (*Spermophilus columbianus*). *Canadian Journal of Zoology* 78:578–587.

Hughes, W. O. H., Oldroyd, B. P., Beekman, M., & Ratnieks, F. L. W. 2008. Ancestral monogamy shows kin selection is key to the evolution of eusociality. *Science* 320:1213–1216.

Hult, C. A. 2002. *Research and writing across the curriculum*. New York: Longman.

Humfeld, S. C. 2008. Intersexual dynamics mediate the expression of satellite mating tactics: Unattractive males and parallel female preferences. *Animal Behaviour* 75:205–215.

Hummel, D. 1983. Aerodynamic aspects of formation flying in birds. *Journal of Theoretical Biology* 104:321–347.

Hunt, J. H., & Amdam, G. V. 2005. Bivoltinism as an antecedent to eusociality in the paper wasp genus *Polistes*. *Science* 308:264–267.

Hunt, S., Low, J., & Burns, K. C. 2008. Adaptive numerical competency in a food-hoarding songbird. *Proceedings of the Royal Society B* 275:2373–2379.

Huntingford, F., & Turner, A. K. 1987. *Animal conflict*. London: Chapman & Hall.

Inouye, D. W., Barr, B., Armitage, K. B., & Inouye, B. 2000. Climate change is affecting altitudinal migrants and hibernating species. *Proceedings of the National Academy of Sciences USA* 97:1630–1633.

Institute for Laboratory Animal Research. 1996. *Guide for the care and use of laboratory animals*. Washington, D.C.: National Academy Press.

James, P. C., & Verbeek, N. A. M. 1984. Temporal and energetic aspects of food storage in Northwestern crows. *Ardea* 72:207–215.

Jaquiéry, J., Broquet, T., Aguilar, C., Evanno, G., & Perrin, N. 2009. Good genes drive female choice for mating partners in the lek-breeding European treefrog. *Evolution* 64:108–115.

Jarvis, J. U. M. 1981. Eusociality in a mammal: Cooperative breeding in naked mole-rat colonies. *Science* 212:571–573.

Jensen, P., & Yngvesson, J. 1998. Aggression between unacquainted pigs—sequential assessment and effects of familiarity and weight. *Applied Animal Behaviour Science* 58:49–61.

Jeon, J. 2008. Evolution of parental favoritism among different-aged offspring. *Behavioral Ecology* 19:344–352.

Johansson, B. G., & Jones, T. M. 2007. The role of chemical communication in mate choice. *Biological Reviews* 82:265–289.

Johansson, B. G., Jones, T. M., & Widemo, F. 2005. Cost of pheromone production in a lekking *Drosophila*. *Animal Behaviour* 69:851–858.

Johnsson, J. I., & Kjallman-Eriksson, K. 2008. Cryptic prey colouration increases search time in brown trout (*Salmo trutta*): Effects of learning and body size. *Behavioral Ecology and Sociobiology* 62: 1613–1620.

Johnstone, R. A. 2008. Sexual selection, honest advertisement and the handicap principle: Reviewing the evidence. *Biological Reviews* 70: 1–65.

Johnstone, R. A., & Grafen, A. 1993 Dishonesty and the handicap principle. *Animal Behaviour* 46: 759–764.

Jones, J. E., & Kamil, A. C. 2001. The use of relative and absolute bearings by Clark's nutcrackers, *Nucifraga columbiana*. *Animal Learning and Behavior* 29: 120–132.

Jones, A. G., & Ratterman, N. L. 2009. Mate choice and sexual selection: What have we learned since Darwin? *Proceedings of the National Academy of Sciences USA* 106:10001–10008.

Jones, A. G., Walker, D., & Avise, J. C. 2001. Genetic evidence for extreme polyandry and extraordinary sex-role reversal in a pipefish. *Proceedings of the Royal Society B* 267:677–680.

Karolinska Institutet. 1973. Nobel Prize in physiology or medicine—1973 [press release]. http://www.nobelprize.org/nobel_prizes/medicine/laureates/1973/press.html (accessed December 4, 2012).

Kasai, H., Fukuda, M., Watanabe, S., Hayashi-Takagi, A., & Noguchi, J. 2010. Structural dynamics of dendritic spines in memory and cognition. *Trends in Neurosciences* 33:121–129.

Kaspari, M., Chang, C., & Weaver, J. 2010. Salted roads and sodium limitation in a northern forest ant community. *Ecological Entomology* 35:543–548.

Keeton, W. T. 1974. The navigational and orientational basis of homing in birds. *Advances in the Study of Behavior* 5:47–132.

Kemp, D. J., & Wiklund, C. 2004. Residency effects in animal contests. *Proceedings of the Royal Society B* 271:1707–1711.

Kennedy, J. S. 1983. Zigzagging and casting as a programmed response to wind-borne odour: A review. *Physiological Entomology* 8:109–120.

Kennedy, P. L., & Ward, J. M. 2003. Effects of experimental food supplementation on movements of juvenile northern goshawks (*Accipiter gentilis atricapillus*). *Oecologia* 134: 284–291.

Kent, C. F., Daskalchuk, T., Cook, L., Sokolowski, M. B., & Greenspan, R. J. 2009. The *Drosophila* foraging gene mediates adult plasticity and gene-environment interactions in behavior, metabolites, and gene expression in response to food deprivation. *PLoS Genetics* 5:e1000609.

Kirschvink, J. L., Walker, M. M., & Diebel, C. E. 2001. Magnetite-based magnetoreception. *Current Opinion in Neurobiology* 11:462–467.

Knapp, R., & Neff, B. D. 2007. Steroid hormones in bluegill, a species with male alternative reproductive tactics including female mimicry. *Biology Letters* 3:628–631.

Kodric-Brown, A. 1986. Satellites and sneakers: Opportunistic male breeding tactics in pupfish (*Cyprinodon pecosensis*). *Behavioral Ecology and Sociobiology* 19:425–432.

Kodric-Brown, A., & Brown, J. H. 1984. Truth in advertising: The kinds of traits favored by sexual selection. *American Naturalist* 124:309–323.

Koh, T. H., Seah, W. K., Yap, L.-M. Y. L., & Daiqin, L. 2009. Pheromone-based female mate choice and its effect on reproductive investment in a spitting spider. *Behavioral Ecology and Sociobiology* 63:923–930.

Köhler, W. 1925. *The mentality of apes.* New York: Harcourt, Brace and Co.

Kokko, H. 1999. Competition for early arrival in migratory birds. *Journal of Animal Ecology* 68: 940–950.

Kokko, H., Brooks, R., Jennions, M. D., & Morley, J. 2002. The evolution of mate choice and mating biases. *Proceedings of the Royal Society B* 270:653–664.

Kokko, H., & Heubel, K. U. 2011. Prudent males, group adaptation, and the tragedy of the commons. *Oikos* 120:641–656.

Kokko, H., & Jennions, M. 2003. It takes two to tango. *Trends in Ecology and Evolution* 18:103–104.

Kokko, H., Jennions, M. D., & Brooks, R. 2006. Unifying and testing models of sexual selection. *Annual Review of Ecology and Systematics* 37: 43–66.

Kokko, H., & Johnstone, R. A. 1999. Social queuing in animal societies: A dynamic model of reproductive skew. *Proceedings of the Royal Society B* 266:571–578.

Kokko, H., López-Sepulcre, A., & Morrell, L.J. 2006. From hawks and doves to self-consistent games of territorial behavior. *American Naturalist* 167: 901–912.

Kokko, H., & Morrell, L. J. 2005. Mate guarding, male attractiveness, and paternity under social monogamy. *Behavioral Ecology* 16:724–731.

Kolm, N., & Ahnesjö, I. 2005. Do egg size and parental care coevolve in fishes? *Journal of Fish Biology* 66:1499–1515.

Kotler, B. P., & Brown, J. S. 1990. Rates of seed harvest by two species of gerbilline rodents. *Journal of Mammalogy* 71:591–596.

Kotler, B. P., Brown, J. S., & Hasson, O. 1991. Factors affecting gerbil foraging behavior and rates of owl predation. *Ecology* 72:2249–2260.

Krakauer, A. H. 2005. Kin selection and cooperative courtship in wild turkeys. *Nature* 434:69–72.

Kramer, G. 1952. Experiments on bird orientation. *Ibis* 94:265–285.

Krause, J., & Ruxton, G. D. 2002. *Living in groups.* Oxford, UK: Oxford University Press.

Krebs, J. R. & Dawkins, R. 1984. Animal signals: Mind-reading and manipulation. In *Behavioural ecology: An evolutionary approach*, 2nd ed, ed. J. R. Krebs & N. B. Davies, 380–402. Oxford, UK: Blackwell.

Krebs, J. R., Sherry, D. F., Healy, S. D., Perry, H., & Vaccarino, A. L. 1989. Hippocampal specialization of foodstoring birds. *Proceedings of the National Academy of Sciences USA* 86:1388–1392.

Kress, S. W., & Nettleship, D. N. 1988. Re-establishment of Atlantic puffins (*Fratercula arctica*) at a former breeding site in the gulf of Maine. *Journal of Field Ornithology* 59:161–170.

Krieger, M. J. B., & Ross, K. G. 2002. Identification of a major gene regulating complex social behavior. *Science* 295:328–332.

Kruuk, L. E. B., Slate, J., Pemberton, J. M., Brotherstone, S., Guinness, F., & Clutton-Brock, T. 2002. Antler size in red deer: Heritability and selection but no evolution. *Evolution* 56:1683–1695.

Kuba, M. J., Byme, R. A., Meisel, D. V., & Mather, J. A. 2006. When do octopuses play? Effects of repeated testing, object type, age, and food deprivation on object play in *Octopus vulgaris*. *Journal of Comparative Psychology* 120:184–190.

Kuchta, S. R., Krakauer, A. H., & Sinervo, B. 2008. Why does the yellow-eyed Ensatina have yellow eyes? Batesian mimicry of pacific newts (genus *Taricha*) by the salamander *Ensatina eschscholtzii xanthoptica*. *Evolution* 62:984–990.

Kulahci, I., Dornhaus, A., & Papaj, D. R. 2008. Multimodal signals enhance decision making in foraging bumblebees. *Proceedings of the Royal Society B* 275:797–802.

Kümmerli, R., & Keller, L. 2007. Reproductive specialization in multiple-queen colonies of the ant *Formica exsecta*. *Behavioral Ecology* 18:375–383.

Kusayama, T., Bischof, H.-J., & Watanabe, S. 2000. Responses to mirror-image stimulation in jungle crows (*Corvus macrorhynchos*). *Animal Cognition* 3:61–64.

Lack, D. 1954. *The natural regulation of animal populations*. London: Oxford University Press.

LaFleur, D.L., Lozano, G.A., & Sclafani, M. 1997. Female mate-choice copying in guppies, *Poecilia reticulata*: A re-evaluation. *Animal Behaviour* 54:579–586.

Lai, C. S., Fisher, S. E., Hurst, J. A., Vargha-Khadem, F., & Monaco A. P. 2001. A forkhead-domain gene is mutated in a severe speech and language disorder. *Nature* 413:519–523.

Laine, C., & Mulrow, C. 2003. Peer review: Integral to science and indispensible to annals. *Annals of Internal Medicine* 139:1038–1040.

Laiolo, P. 2010. The emerging significance of bioacoustics in animal species conservation. *Biological Conservation* 143:1635–1645.

Laland, K. N., & Janik, V. M. 2006. The animal culture debate. *Trends in Ecology and Evolution* 21:542–547.

Laland, K. N., Sterelny, K., Odling-Smee, J., Hoppitt, W., & Uller, T. 2011. Cause and effect in biology revisited: Is Mayr's proximate-ultimate dichotomy still useful? *Science* 334:1512–1516.

Lande, R. 1981. Models of speciation by sexual selection on polygenic traits. *Proceedings of the National Academy of Sciences USA* 78:3721–3725.

Lane, J. E., Gunn, M. R., Slate, J., & Coltman, D. W. 2008. Female multiple mating and paternity in free-ranging North American red squirrels. *Animal Behaviour* 75:1927–1937.

Laufer, G., & Barreneche, J. M. 2008. Re-description of the tadpole of *Pseudopaludicola falcipes* (Anura: Leiuperidae), with comments on larval diversity of the genus. *Zootaxa* 1760:50–58.

Laundré, J. W., Hernández, L., & Altendorf, K. B. 2001. Wolves, elk, and bison: Reestablishing the "landscape of fear" in Yellowstone National Park, U.S.A. *Canadian Journal of Zoology* 79:1401–1409.

Lazarus, J., & Inglis, I. R. 1986. Shared and unshared parental investment, parent-offspring conflict and brood size. *Animal Behaviour* 34:1791–1804.

Lefebvre, L., Reader, S. M., & Sol, D. 2004. Brains, innovations and evolution in birds and primates. *Brain, Behavior and Evolution* 63:233–246.

Lefebvre, L., Whittle, P., Lascaris, E., & Finkelstein, A. 1997. Feeding innovations and forebrain size in birds. *Animal Behaviour* 53:549–560.

Leggett, H. C., Mouden, C. E., Wild, G., & West, S. 2012. Promiscuity and the evolution of cooperative breeding. *Proceedings of the Royal Society B* 279:1405–1411.

Lehner, P. N. 1998. *Handbook of ethological methods*. Cambridge: Cambridge University Press.

Leigh, E. G. Jr. 2010. The group selection controversy. *Journal of Evolutionary Biology* 23:6–19.

Leimar, O., & Hammerstein, P. 2001. Evolution of cooperation through indirect reciprocity. *Proceedings of the Royal Society B* 268:745–753.

Leinfelder, I., de Vries, H., Deleu, R., & Nelissen, M. 2001. Rank and grooming reciprocity among females in a mixed-sex group of captive hamadryas baboons. *American Journal of Primatology* 55:25–42.

Leisler, B., Winkler, H., & Wink, M. 2002. Evolution of breeding systems in Acrocephaline warblers. *Auk* 119:379–390.

Lemish, M. G. 1996. *War dogs: A history of loyalty and heroism*. Dulles, VA: Brassey's Inc.

Levey, D. J., & Stiles, F. G. 1992. Evolutionary precursors of long-distance migration: Resource availability and movement patterns in neotropical landbirds. *American Naturalist* 140:447–476.

Levitis, D. A., Lidicker, W. Z. Jr., & Freund, G. 2009. Behavioural biologists do not agree on what constitutes behavior. *Animal Behaviour* 78:103–110.

Lewis, S. M., Cratsley, C. K., & Rooney, J. 2004. Nuptial gifts and sexual selection in *Photinus* fireflies. *Integrative and Comparative Biology* 44:234–237.

Lifjeld, J. T., & Robertson, R. J. 1992. Female control of extra-pair fertilization in tree swallows. *Behavioral Ecology and Sociobiology* 31:89–96.

Lifjeld, J. T., Slagsvold, T., & Ellegren, H. 1997. Experimental mate switching in pied flycatchers: Male copulatory access and fertilization success. *Animal Behaviour* 53:1225–1232.

Lima, S. L. 1985. Maximizing feeding efficiency and minimizing time exposed to predators: A trade-off in the black-capped chickadee. *Oecologia* 66:60–67.

————. 1998. Stress and decision-making under the risk of predation: Recent developments from behavioral, reproductive and ecological perspectives. *Advances in the Study of Behavior* 27:215–290.

Lima, S. L., Mitchell, W. A., & Roth, T. C. 2003. Predators feeding on behaviorally responsive prey: Some implications for classical models of optimal diet choice. *Evolutionary Ecology Research* 5:1083–1102.

Lima, S. L., Valone, T. J., & Caraco, T. 1985. Foraging-efficiency-predation-risk trade-off in the grey squirrel. *Animal Behaviour* 33:155–165.

Lindburg, D. G., & Fitch-Snyder, H. 1994. Use of behavior to evaluate reproductive problems in captive mammals. *Zoo Biology* 13:433–445.

Linklater, W. L., Cameron, E. Z., Minot, E. O., & Stafford, K. J. 1999. Stallion harassment and the mating system of horses. *Animal Behaviour* 58:295–306.

Linklater, W. L., Cameron, E. Z., Stafford, K. J., & Veltman, C. J. 2000. Social and spatial structure and range use by Kaimanawa wild horses (*Equus caballus*: Equidae). *New Zealand Journal of Ecology* 24:139–152.

Lishman, W. A., Teets, T. L., Duff, J. W., Sladen, W. J. L., Shire, G. G., Goolsby, K. M., Bezner Kerr, W. A., & Urbanke, R. P. 1997. A reintroduction technique for migratory birds: Leading Canada geese and isolation-reared sand-hill cranes with ultralight aircraft. *Proceedings of the North American Crane Workshop* 7:96–104.

Lloyd, J. E. 1975. Aggressive mimicry in *Photuris* fireflies: Signal repertoires by femmes fatales. *Science* 187:452–453.

Lohmann, K. J. 1991. Magnetic orientation by hatchling loggerhead sea turtles (*Caretta caretta*). *Journal of Experimental Biology* 155:37–49.

Lohmann, K. J., Cain, S. D., Dodge, S. A., & Lohmann, C. M. 2001. Regional magnetic fields as navigational markers for sea turtles. *Science* 294:364–366.

Lohmann, K. J., Lohmann, C. M. F., Ehrhart, L. M., Bagley, D. A., & Swing, T. 2004. Animal behaviour: Geomagnetic map used in sea-turtle navigation. *Nature* 428:909–910.

Lohmann, K. J., Luschi, P., & Hays, G. C. 2008. Goal navigation and island-finding in sea turtles. *Journal of Experimental Marine Biology and Ecology* 356:83–95.

Lohmann, K. J., Putnam, N. F., & Lohmann, C. M. F. 2008. Geomagnetic imprinting: A unifying hypothesis of long-distance natal homing in salmon and sea turtles. *Proceedings of the National Academy of Sciences USA* 105:19,096–19,101.

Lorenz, K. 1935. Companions as factors in the bird's environment. *Studies in Animal and Human Behavior* 1:101–258.

———. 1950. The comparative method in studying innate behavior. *Symposium of the Society for Experimental Biology* 4:221–268.

Lorenz, K. Z., & Tinbergen, N. 1957. Taxis and instinct: Taxis and instinctive action in the egg-retrieving behavior of the graylag goose. In *Instinctive behavior: The development of a modern concept*, ed. C. H. Schiller, 176–208. New York: International Universities Press.

Lormée, H., Jouventin, P., Chastel, O., & Mauget, R. 1999. Endocrine correlates of parental care in an Antarctic winter breeding seabird, the Emperor penguin, *Aptenodytes forsteri*. *Hormones and Behavior* 35:9–17.

Loyau, A., Saint Jalme, M., & Sorci, G. 2005. Intra- and intersexual selection for multiple traits in the peacock (*Pavo cristatus*). *Ethology* 111:810–820.

———. 2007. Non-defendable resources affect peafowl lek organization: A male removal experiment. *Behavioural Processes* 74:64–70.

Lozano, G. A. 2009. Multiple cues in mate selection: The sexual interference hypothesis. *Bioscience Hypotheses* 2:37–42.

Lucas, J. R., Brodin, A., de Kort, S. R., & Clayton, N. S. 2004. Does hippocampal size correlate with degree of caching specialization? *Proceedings of the Royal Society B* 271:2423–2429.

Lucas, B. K., Ormandy, C. J., Binart, N., Bridges, R. S., & Kelly, P. A. 1998. Null mutation of the prolactin receptor gene produces a defect in maternal behavior. *Endocrinology* 139:4102–4107.

Luck, N., Dejonghe, B., Fruchard, S., Huguenin, S., & Joly, D. 2007. Male and female effects on sperm precedence in the giant sperm species *Drosophila bifurca*. *Genetica* 130:257–265.

Mace, T. R. 2000. Time budget and pair-bond dynamics in the comb-crested jacana *Irediparra gallinacea*: A test of hypothesis. *Emu* 100:31–41.

Mackey, T. K. C. 2001. The genetic architecture of quantitative traits. *Annual Review of Genetics* 35:303–339.

Magnhagen, C. 1991. Predation risk as a cost of reproduction. *Trends in Ecology & Evolution* 6:183–186.

Mank, J. E., Promislow, D. E. L., & Avise, J. C. 2005. Phylogenetic perspectives in the evolution of parental care in ray-finned fishes. *Evolution* 59:1570–1578.

Mann, J., Stanton, M., Patterson, E. M., Bienenstock, E. J., & Singh, L. O. 2012. Social networks reveal cultural behaviour in tool-using dolphins. *Nature Communications* 3:980.

Manríquez, K. C., Pardo, L. M., Wells, R. J. D., & Palma, A. T. 2008. Crypsis in *Paraxanthus barbiger* (Decapoda: Brachyura): Mechanisms against visual predators. *Journal of Crustacean Biology* 28:473–479.

Manson, J. H., Navarrete, C., Silk, J. B., & Perry, S. 2004. Time-matched grooming in female primates? New analyses from two species. *Animal Behaviour* 67:493–500.

Maple, T., & Hoff, M. P. 1982. *Gorilla behavior*. New York: Van Nostrand Reinhold Company.

Marchetti, K. 1993. Dark habitats and bright birds illustrate the role of the environment in species diversity. *Nature* 362:149–152.

Marks, C., West, T.N., Bagatto, B., & Moore, F. B.-G. 2005. Developmental environment alters conditional aggression in zebrafish. *Copeia* 2005:901–908.

Marler, P., & Peters, S. 1977. Selective vocal learning in a sparrow. *Science* 198:519–521.

Marra, P. P., Francis, C. M., Mulvihill, R. S., & Moore, F. R. 2005. The influence of climate on the timing and rate of spring bird migration. *Oecologia* 142:307–315.

Marriner, L. M., & Drickamer, L. C. 1994. Factors influencing stereotyped behaviors of primates in a zoo. *Zoo Biology* 13:267–275.

Martin, P., & Bateson, P. 1993. *Measuring behaviour: An introductory guide.* 2nd ed. Cambridge: Cambridge University Press.

Martin, R. A., & Pfennig, D. W. 2009. Disruptive selection in natural populations: The roles of ecological specialization and resource competition. *American Naturalist* 174:268–281.

Mas, F., & Kölliker, M. 2008. Maternal care and offspring begging in social insects: Chemical signaling, hormonal regulation and evolution. *Animal Behaviour* 76:1121–1131.

Mason, G. J. 2010. Species differences in response to captivity: Stress, welfare and the comparative method. *Trends in Ecology and Evolution* 25:713–721.

Massaro, M., Setiawan, A. N., & Davis, L. S. 2007. Effects of artificial eggs on prolactin secretion, steroid levels, brood patch development, incubation onset and clutch size in the yellow-eyed penguin (*Megadyptes antipodes*). *General and Comparative Endocrinology* 151:220–229.

Mather, J. 2008. Cephalopod consciousness: Behavioral evidence. *Consciousness and Cognition* 17:37–48.

Mathews, L. M. 2002a. Tests of mate-guarding hypothesis for social monogamy: Does population density, sex ratio, or female synchrony affect behavior of male snapping shrimp (*Alpheus angulatus*). *Behavioral Ecology and Sociobiology* 51:426–432.

————. 2002b. Territorial cooperation and social monogamy: Factors affecting intersexual behaviors in pair-living snapping shrimp. *Animal Behaviour* 63:767–777.

————. 2003. Tests of mate-guarding hypothesis for social monogamy: Male snapping shrimp prefer to associate with high-value females. *Behavioral Ecology* 14:63–67.

Mäthger, L. M., Denton, E. J., Marshall, N. J., & Hanlon, R. T. 2009. Mechanisms and behavioural functions of structural coloration in cephalopods. *Journal of the Royal Society Interface* 6 (suppl 2): S149–S163.

Matos, R. J., & Schlupp, I. 2005. Performing in front of an audience: Signalers and the social environment. In *Animal communication networks*, ed. P. K. McGregor, 63–83. Cambridge: Cambridge University Press.

Maynard Smith, J. M. 1964. Group selection and kin selection. *Nature* 201:1145–1147.

————. 1978. Optimization theory in evolution. *Annual Review of Ecology and Systematics* 9:31–56.

————. 1982. *Evolution and the theory of games.* Cambridge: Cambridge University Press.

Maynard Smith, J., & Harper, D. 2003. *Animal signals.* Oxford, UK: Oxford University Press.

Maynard Smith, J., & Parker, G. A. 1976. The logic of asymmetric contests. *Animal Behaviour* 24:159–175.

Maynard Smith, J., & Price, G. R. 1973. The logic of animal conflict. *Nature* 246:15–18.

Mayr, E. 1961. Cause and effect in biology. *Science* 134: 1501–1506.

————. 1982. *The growth of biological thought.* Cambridge, MA: Harvard University Press.

McDonald, F. 1958. *We the people: The economic origins of the Constitution.* Chicago: University of Chicago Press.

McDonald, D. B. 1993. Delayed plumage maturation and orderly queues for status: A manakin mannequin experiment. *Ethology* 94:31–45.

McEvoy, J. W. 2009. Evolutionary game theory: Lessons and limitations, a cancer perspective. *British Journal of Cancer* 101:2060–2061.

McGlothlin, J. W., Jawor, J. M., & Ketterson, E. D. 2007. Natural variation in a testosterone-mediated trade-off between mating effort and parental care. *American Naturalist* 170:864–875.

McGregor, P., ed. 2005. *Animal communication networks.* Cambridge: Cambridge University Press.

McNamara, J. M., & Houston, A. I. 1987. Starvation and predation as factors limiting population size. *Ecology* 68:1515–1519.

Mello, C. V., Vicario, D. S., & Clayton, D. F. 1992. Song presentation induces gene expression in the songbird forebrain. *Proceedings of the National Academy of Sciences USA* 89:6818–6822.

Meredith, R. M., McCabe, B. J., Kendrick, K. M., & Horn, G. 2004. Amino acid neurotransmitter release and learning: A study of visual imprinting. *Neuroscience* 126:249–256.

Merlin, C., Gegear, R. J., & Reppert, S. M. 2009. Antennal circadian clocks coordinate sun compass orientation in migratory monarch butterflies. *Science* 325:1700–1704.

Mery, F., & Kawecki, T. J. 2002. Experimental evolution of learning ability in fruit flies. *Proceedings of the National Academy of Sciences USA* 99:14,274–14,279.

Mery, F., Varela, S. A. M., Danchin, E., Blanchet, S., Parejo, D., Coolen, I., & Wagner, R. H. 2009. Public versus private information for mate copying in an invertebrate. *Current Biology* 19:730–734.

Meyburg, B. U. 1978. Sibling aggression and cross-fostering of eagles. In *Endangered birds management techniques for threatened species*, ed. S. A. Tempe, 195–200. Madison: University of Wisconsin Press.

Milá, B., Smith, T. B., & Wayne, R. K. 2006. Postglacial population expansion drives the evolution of long-distance migration in a songbird. *Evolution* 60:2403–2409.

Miles, D. B. 2004. The race goes to the swift: Fitness consequences of variation in sprint performance in juvenile lizards. *Evolutionary Ecology Research* 6:63–75.

Milinski, M. 2006. The major histocompatibility complex, sexual selection, and mate choice. *Annual Review of Ecology Evolution and Systematics* 37:159–186.

Milinski, M., Semmann, D., & Krambeck, H. J. 2002a. Donors to charity gain both indirect reciprocity and political reputation. *Proceedings of the Royal Society B* 269:881–883.

————. 2002b. Reputation helps solve the "tragedy of the commons." *Nature* 393:573–577.

Miller, P. 2001. *The power of positive dog training*. New York: Howell.

Miller, L. A., & Surlykke, A. 2001. How some insects detect and avoid being eaten by bats: Tactics and countertactics of prey and predator. *Bioscience* 51:570–581.

Miller, D. A.,Vleck, C. M., & Otis, D. L. 2009. Individual variation in baseline and stress-induced corticosterone and prolactin levels predicts parental effort by nesting mourning doves. *Hormones and Behavior* 56:457–464.

Mitchell, M. D., McCormick, M. I., Ferrari, M. C. O., & Chivers, D. P. 2011. Coral reef fish rapidly learn to identify multiple unknown predators upon recruitment to the reef. *PLoS One* 6:e15764.

Mitchell, W. A., & Valone, T. J. 1990. The optimization research program: Studying adaptations by their functions. *Quarterly Review of Biology* 65:43–52.

Mo, J., Glover, M., Munro, S., & Beattie, G. A. 2006. Evaluation of mating disruption for control of lightbrown apple moth (Lepidoptera: Tortricidae) in citrus. *Journal of Economic Entomology* 99:421–426.

Moller, A. P., & Alatalo, R. V. 1999. Good-genes effects in sexual selection. *Proceedings of the Royal Society B* 266:85–91.

Møller, A. P., & Jennions, M. D. 2001. How important are direct fitness benefits of sexual selection? *Naturwissenschaften* 88:401–415.

Monaghan, P., & Nager, R. G. 1997. Why don't birds lay more eggs? *Trends in Ecology and Evolution* 12:270–274.

Morgan, C. L. 1894. *An introduction to comparative psychology*. London: W. Scott.

————. 1903. *An introduction to comparative psychology*. 2nd ed. London: W. Scott.

Morris, D. W. 1988. Habitat-dependent population regulation and community structure. *Evolutionary Ecology* 2:253–269.

Morton, E. S. 1975. Ecological sources of selection on avian sounds. *American Naturalist* 109:17–34.

Mouritsen, H., & Hore, P. J. 2012. The magnetic retina: Light-dependent and trigeminal magnetoreception in migratory birds. *Current Opinion in Neurobiology* 22:343–352.

Mueller, J. C., Pulido, F., & Kempenaers, B. 2011. Identification of a gene associated with avian migratory behavior. *Proceedings of the Royal Society B* 278:2848–2856.

Muheim, R., Moore, F. R., & Phillips, J. B. 2006. Calibration of magnetic and celestial compass cues in migratory birds: A review of cue-conflict experiments. *Journal of Experimental Biology* 209:2–17.

Muller, K. L. 1995. Habitat settlement in territorial species: The effects of habitat quality and conspecifics. Ph.D. diss., University of California, Davis.

————. 1998. The role of conspecifics in habitat settlement in a territorial grasshopper. *Animal Behaviour* 56:479–485.

Müller, C. A., & Manser, M. B. 2007. "Nasty neighbors" rather than "dear enemies" in a social carnivore. *Proceedings of the Royal Society B* 274:959–965.

Mundinger, P. C. 2010. Behavior genetic analysis of selective song learning in three inbred canary strains. *Behaviour* 147:705–723.

Murray, J. A., Estepp, J., & Cain, S. D. 2006. Advances in the neural bases of orientation and navigation. *Integrative and Comparative Biology* 46:871–879.

Murray, E. A., Kralik, J. D., Wise, S. P. 2005. Learning to inhibit prepotent responses: Successful performance by rhesus macaques, *Macaca mulatta*, on the reversed-contingency task. *Animal Behaviour* 69:991–998.

Murray, C. M., Lonsdorf, E. V., Eberly, L. E., & Pusey, A. E. 2009. Reproductive energetics in free-living female chimpanzees (*Pan troglodytes schweinfurthii*). *Behavioral Ecology* 20:1211–1216.

Murray, C. M., Mane, S. V., & Pusey, A. E. 2007. Dominance rank influences female space use in wild chimpanzees, *Pan troglodytes*: Towards an ideal despotic distribution. *Animal Behaviour* 74:1795–1804.

National Academy of Sciences. 1998. *Teaching about evolution and the nature of science*. Washington, D.C.: National Academies Press.

Neff, B. D. 2003. Decisions about parental care in response to perceived paternity. *Nature* 422: 716–719.

Nicholls, J. A., & Goldizen, A. W. 2006. Habitat type and density influence vocal signal design in satin bowerbirds. *Journal of Animal Ecology* 75:549–558.

Noad, M. J., Cato, D. H., Bryden, M. M., Jenner, M.-N., & Jenner, K. C. S. 2000. Cultural revolution in whale songs. *Nature* 408:537.

Nonacs, P., & Dill, L. M. 1988. Foraging response of the ant *Lasius pallitarsis* to food sources with associated mortality risk. *Insectes Sociaux* 35: 293–303.

––––––. 1990. Mortality risk vs. food quality trade-offs in a common currency: Ant patch preferences. *Ecology* 71:1886–1892.

Nordby, J. C., Campbell, S. E., & Beecher, M. D. 2001. Late song learning in song sparrows. *Animal Behaviour* 61:835–846.

Nordell, S. E. 1994. Observations of the mating behavior and dentition of the round stingray, *Urolophus halleri*. *Environmental Biology of Fishes* 39:219–229.

Nordell, S. E., & Valone, T. J. 1998. Mate choice copying as public information. *Ecology Letters* 1:74–76.

Nordeng, H. 1977. A pheromone hypothesis for homeward migration in anadromous salmonids. *Oikos* 28:155–159.

Nowak, M. A., & Sigmund, K. 1998. Evolution of indirect reciprocity by image scoring. *Nature* 393:573–577.

––––––. 2005. Evolution of indirect reciprocity. *Nature* 437:1291–1298.

Nowak, M. A., Tarnita, C. E., & Wilson, E. O. 2010. The evolution of eusociality. *Nature* 466: 1057–1062.

Ohlsson, T., Smith, H. G., Raberg, L., & Hasselquist, D. 2002. Pheasant sexual ornaments reflect nutritional conditions during early growth. *Proceedings of the Royal Society B* 269:21–27.

Olendorf, R., Getty, T., & Scribner, K. 2004. Cooperative nest defense in red-winged blackbirds: Reciprocal altruism, kinship or by-product mutualism? *Proceedings of the Royal Society B* 271:177–182.

Oliveira, R. F., Lopos, L. A., & Carneiro, A. V. M. 2001. Watching fights raises fish hormone levels. *Nature* 409:784.

Oliveira, R. F., Silva, A., & Canário, A. V. M. 2009. Why do winners keep winning? Androgen mediation of winner but not loser effects in cichlid fish. *Proceedings of the Royal Society B* 276:2249–2256.

Olsson, M., Madsen, T., Nordby, J., Wapstra, E. Ujvari, B., & Wittsell, H. 2003. Major histocompatability complex and mate choice in sand lizards. *Proceedings of the Royal Society B* 270:S254–S256.

O'Riain, M. J., Jarvis, J. U. M., Alexander, R., Buffenstein, R., & Peeters, C. 2000. Morphological casts in a vertebrate. *Proceedings of the National Academy of Sciences USA* 97:13,194–13,197.

Orians, G. H. 1969. On the evolution of mating systems in birds and mammals. *American Naturalist* 103:589–603.

Osorio-Berstain, M., & Drummond, H. 2001. Male boobies expel eggs when paternity is in doubt. *Behavioral Ecology* 12:16–21.

Otte, D. 1974. Effects and function in the evolution of signaling systems. *Annual Review of Ecology and Systematics* 5:385–417.

Outlaw, D. C., Voelker, G., Milá, B., & Girman, D. J. 2003. Evolution of long-distance migration in and historical biogeography of *Catharus* thrushes: A molecular phylogentic approach. *Auk* 120: 299–310.

Owen, P. C., & Perrill, S. A. 1998. Habituation in the green frog, *Rana clamitans*. *Behavioral Ecology and Sociobiology* 44:209–213.

Owen, M. J., Rendell, D., & Ryan, M. J. 2010. Redefining animal signaling: Influence versus information in communication. *Biology & Philosophy* 25:755–780.

Oyegbile, T. O., & Marler, C. A. 2005. Winning fights elevates testosterone levels in California mice and enhances future ability to win fights. *Hormones and Behavior* 48:259–267.

Packer, C., Gilbert, D. A., Pusey, A. E., & O'Brien, S. J. 1991. A molecular genetic analysis of kinship and cooperation in African lions. *Nature* 351:562–565.

Packer, C., Scheel, D., & Pusey, A. E. 1990. Why lions form groups: Food is not enough. *American Naturalist* 136:1–19.

Pak, A. A. 2010. The synergy of laboratory and field studies of dolphin behavior and cognition. *International Journal of Comparative Psychology* 23:538–565.

Panchanathan, K., & Boyd, R. 2004. Indirect reciprocity can stabilize cooperation without the second-order free rider problem. *Nature* 432:499–502.

Parker, G. A. 1970. Sperm competition and its evolutionary consequences in the insects. *Biological Reviews* 45:455–616.

Parker, G. A., Baker, R. R., & Smith, V. G. F. 1972. The origin and evolution of gamete dimorphism and the male-female phenomenon. *Journal of Theoretical Biology* 36:529–553.

Parker, G. A., & Hammerstein, P. 1985. Game theory and animal behavior. In *Evolution: Essays in honour of John Maynard Smith*, ed. J. Maynard Smith, J. Greenwood, P. H. Harvey, & M. Slatkin, 73–94. New York: Cambridge University Press.

Parker, M. W., Kress, S. W., Golightly, R. T., Carter, H. R., Parsons, E. B., Schubel, S. E., Boyce, J. A., McChesney, G. J., & Wisley, S. M. 2007. Assessment of social attraction techniques used to restore a common murre colony in central California. *Waterbirds* 30:17–28.

Parsons, M. H., Lamont, B. B., Kovacs, B. R., & Davies, S. J. J. F. 2007. Effects of novel and historic predator urines on semi-wild Western Grey Kangaroos. *Journal of Wildlife Management* 71:1225–1228.

Patterson, E. M., & Mann, J. 2011. The ecological conditions that favor tool use and innovation in wild bottlenose dolphins (*Tursiops* sp.). *PLoS One* 6:e22243.

Pavey, C. R., & Smyth, A. K. 1998. Effects of avian mobbing on roost use and diet of powerful owls, *Ninox strenua*. *Animal Behaviour* 55:313–318.

Pavey, C. R., Smyth, A. K., & Mathieson, M. T. 1994. The breeding season diet of the powerful owl *Ninox strenua* at Brisbane, Queensland. *Emu* 94:278–284.

Pavlov, I. P. 1927. *Conditioned reflexes: An investigation of the physiological activity of the cerebral cortex*. Trans. and ed. G. V. Anrep. London: Oxford University Press.

Payne, H. F. P., Lawes, M. J., & Henzi, S. P. 2003. Competition and the exchange of grooming among female samango monkeys (*Cercopithecus mitis erythrarchus*). *Behaviour* 140:453–471.

Pennisi, E. 2005. A genomic view of animal behavior. *Science* 307:30–32.

———. 2009. On the origin of cooperation. *Science* 325:1196–1199.

Pepperberg, I. M. 2006. Grey parrot numerical competence: A review. *Animal Cognition* 9:377–391.

Perez, S. M., Taylor, O. R., & Jander, R. 1997. A sun compass in monarch butterflies. *Nature* 387:29.

Persky, H., Smith, K. D., & Basu, G. K. 1971. Relation of psychologic measures of aggression and hostility to testosterone production in man. *Psychosomatic Medicine* 33:265–277.

Peters, R. A., Hemmi, J. M., & Zeil, J. 2007. Signaling against the wind: Modifying motion-signal structure in response to increased noise. *Current Biology* 17:1231–1234.

Petrie, M., Halliday, T., & Carolyne, S. 1991. Peahens prefer peacocks with elaborate trains. *Animal Behaviour* 41:323–331.

Pfennig, D. W. 1992. Polyphenism in spadefoot toads as a locally adjusted evolutionarily stable strategy. *Evolution* 46:1408–1420.

Pfennig, K. S., & Pfennig, D. W. 2005. Character displacement as the "best of a bad situation": Fitness trade-offs resulting from selection to minimize resource and mate competition. *Evolution* 59:2200–2208.

Pfennig, D. W., Rice, A. M., & Martin, R. A. 2007. Field and experimental evidence for competition's role in phenotypic divergence. *Evolution* 61:257–271.

Phelps, S. M., Rand, A. S., & Ryan, M. J. 2007. The mixed-species chorus as public information: Túngara frogs eavesdrop on a heterospecific. *Behavioral Ecology* 18:108–114.

Piep, M., Radespiel, U., Zimmermann, E., Schmidt, S., & Siemers, B. M. 2008. The sensory bias of prey detection in captive born grey mouse lemurs (*Microcebus murinus*). *Animal Behaviour* 75:871–878.

Pietrewicz, A. T., & Kamil, A. C. 1977. Visual detection of cryptic prey by blue jays (*Cyanocitta cristata*). *Science* 195:580–582.

———. 1979. Search image formation in the blue jay (*Cyanocitta cristata*). *Science* 204:1332–1333.

Pitcher, T. J., Magurran, A. E., & Winfield, I. J. 1982. Fish in larger shoals find food faster. *Behavioral Ecology and Sociobiology* 10:149–151.

Platt, J. R. 1964. Strong inference. *Science* 146: 347–353.

Plusquellec, P., & Bouissou, M. 2001. Behavioral characteristics of two dairy breeds of cows selected (Hérens) or not (Brune des Alpes) for fighting and dominance ability. *Applied Animal Behaviour Science* 72:1–21.

Pochron, S. T., & Wright, P. C. 2003. Variability in adult group composition of a prosimian primate. *Behavioral Ecology and Sociobiology* 54:285–293.

Pohlmann, K., Atema, J., & Breithaupt, T. 2004. The importance of the lateral line in nocturnal predation of piscivorous catfish. *Journal of Experimental Biology* 207:2971–2978.

Pohlmann, K., Grasso, F. W., & Breithaupt, T. 2001. Tracking wakes: The nocturnal predatory strategy of piscivorous catfish. *Proceedings of the National Academy of Sciences USA* 98:7371–7374.

Pomfret, J. C., & Knell, R. J. 2006. Sexual selection and horn allometry in dung beetle *Euoniticellus intermedius*. *Animal Behaviour* 71:567–576.

Poole, M. M. 1984. Migration corridors of gray whales along the central California coast, 1980–1982. In *The gray whale, Eschrichotus robustus*, ed. M. L. Jones, S. L. Swartz, & S. Leatherwood, 389–407. Orlando, FL: Academic Press.

Popper, K. R. 1959. *The logic of scientific discovery*. New York: Basic Books.

Pratt, A. E., McLain, D. K., & Lathrop, G. R. 2003. The assessment game in sand fiddler crab contests for breeding burrows. *Animal Behaviour* 65:945–955.

Praw, J. C., & Grant, J. W. A. 1999. Optimal territory size in the convict cichlid. *Behaviour* 136:1347–1363.

Premack, D., & Premack, A. J. 1983. *The mind of an ape.* New York: Norton.

Pribil, S. 2000. Experimental evidence for the cost of polygyny in the red-winged blackbird *Agelaius phoeniceus. Behaviour* 137:1153–1173.

Pribil, S., & Searcy, W. A. 2001. Experimental confirmation of the polygyny threshold model for red-winged blackbirds. *Proceedings of the Royal Society B* 268:1643–1646.

Price, F. E., & Bock, C. E. 1983. Population ecology of the dipper (*Cinclus mexicanus*) in the front range of Colorado. *Studies in Avian Biology* 7:1–84.

Price, E. E., & Stoinski, T. S. 2007. Group size: Determinants in the wild and implications for the captive housing of wild mammals in zoos. *Applied Animal Behaviour Science* 103:255–264.

Pruvost, M., Bellone, R., Benecke, N., Sandoval-Castellanos, E., Cieslak, M., Kuznetsova, T., Morales-Muniz, A., et al. 2011. Genotypes of pre-domestic horses match phenotypes painted in Paleolithic works of cave art. *Proceedings of the National Academy of Science USA* 108:18,626–18,630.

Pulido, F. 2007. The genetics and evolution of avian migration. *Bioscience* 57:165–174.

Pulido, F., & Berthold, P. 2010. Current selection for lower migratory activity will drive the evolution of residency in a migratory bird population. *Proceedings of the National Academy of Sciences USA* 107:7341–7346.

Pulliam, H. R. 1973. On the advantages of flocking. *Journal of Theoretical Biology* 38:419–422.

———. 1974. On the theory of optimal diets. *American Naturalist* 108:59–74.

Purchase, C. F, & Hutchings, J. A. 2008. A temporally stable spatial pattern in the spawner density of a freshwater fish: Evidence for an ideal despotic distribution. *Canadian Journal of Aquatic and Fisheries Sciences* 65:382–388.

Pusey, A. 1987. Sex-biased dispersal and inbreeding avoidance in birds and mammals. *Trends in Ecology and Evolution* 2:295–299.

Pusey, A., & Wolf, M. 1996. Inbreeding avoidance in animals. *Trends in Ecology and Evolution* 11:201–206.

Queller, D. C. 1997. Why do females care more than males? *Proceedings of the Royal Society B* 264:1555–1557.

Queller, D. C., Strassmann, J. E., & Hughes, C. R. 1993. Microsatellites and kinship. *Trends in Ecology and Evolution* 8:285–288.

Radcliffe, J. M., Soutar, A. R., Muma, K. E., Guignion, C., & Fullard, J. H. 2008. Anti-bat flight activity in sound-producing versus silent moths. *Canadian Journal of Zoology* 86:582–587.

Radford, A. N., & Ridley, A. R. 2007. Individuals in foraging groups may use vocal cues when assessing their need for antipredator vigilance. *Biology Letters* 3:249–252.

Radwan, J. 2002. Good genes go Fisherian. *Trends in Ecology and Evolution* 17:539.

Raine, N. E., & Chittka, L. 2007 Adaptive significance of sensory bias in a foraging context: Floral colour preferences in the bumblebee *Bombus terrestris. PLoS ONE* 2:e556.

———. 2008. The correlation of learning speed and natural foraging success in bumble-bees. *Proceedings of the Royal Society B* 275:803–808.

Ramenofsky, M., & Wingfield, J. C. 2007. Regulation of migration. *Bioscience* 57:135–143.

Rankin, S. M., Fox, K. M., & Stotsky, C. E. 1995. Physiological correlates to courtship, mating, ovarian development and maternal behavior in the ring-legged earwig. *Physiological Entomology* 20:257–265.

Rankin, S. M., Palmer, J. O., Yagi, K. J., Scott, G. L., & Tobe, S. S. 1995. Biosynthesis and release of juvenile hormone during the reproductive cycle of the ring-legged earwig. *Comparative Biochemistry and Physiology C: Pharmacology, Toxicology & Endocrinology* 110:241–251.

Reader, S. M., & Laland, K. N. 2002. Social intelligence, innovation and enhanced brain size in primates. *Proceedings of the National Academy of Sciences USA* 99:4436–4441.

Reaney, L. T. 2007. Foraging and mating opportunities influence refuge use in the fiddler crab, *Uca mjoebergi. Animal Behaviour* 73:711–716.

Reaney, L. T., & Backwell, P. R. Y. 2007. Risk-taking behavior predicts aggression and mating success in a fiddler crab. *Behavioral Ecology* 18:521–525.

Rees, P. A. 2009. Activity budgets and the relationship between feeding and stereotypic behaviors in Asian elephants (*Elephas maximus*) in a zoo. *Zoo Biology* 28:79–97.

Reeve, H. K., & Sherman, P. W. 1993. Adaptation and the goals of evolutionary research. *Quarterly Review of Biology* 68:1–32.

Reguera, P., & Gomendio, M. 1999. Predation costs associated with parental care in the golden egg bug *Phyllomorpha laciniata* (Heteroptera: Coreidea). *Behavioral Ecology* 10:541–544.

———. 2002. Flexible oviposition behavior in the golden egg bug (*Phyllomorpha laciniata*) and its implications for offspring survival. *Behavioral Ecology* 13:70–74.

Rendell, D., Owen, M. J., & Ryan, M. J. 2009. What do animal signals mean? *Animal Behaviour* 78:233–240.

Renison, D., Boersma, P. D., & Martella, M. B. 2002. Winning and losing: Causes for variability in outcome of fights in male Magellanic penguins (*Spheniscus magellanicus*). *Behavioral Ecology* 13:462–466.

Renison, D., Boersma, P. D., Van Buren, A. N., & Martella, M. B. 2006. Agonistic behavior in wild male Magellanic penguins: When and how do they interact? *Journal of Ethology* 24:189–193.

Reynolds, S. M., Christman, M. C., Uy, A. C., Patricelli, G. L., Braun, M. J., & Borgia, G. 2009. Lekking satin bowerbird males aggregate with relatives to mitigate aggression. *Behavioral Ecology* 20: 410–415.

Reynolds, J. D., Goodwin, N. B., & Freckleton, R. P. 2002. Evolutionary transition in parental care and live bearing in vertebrates. *Philosophical Transactions of the Royal Society B* 357:269–281.

Reynolds, D. R., & Riley, J. R. 2002. Remote-sensing, telemetric and computer-based technologies for investigating insect movement: A survey of existing and potential techniques. *Computers and Electronics in Agriculture* 35:271–307.

Ribble, D. O. 1991. The monogamous mating system of *Peromyscus californicus* as revealed by DNA fingerprinting. *Behavioral Ecology and Sociobiology* 29:161–166.

Rice, W. R., & Holland, B. 1997. The enemies within: Intergenomic conflict, interlocus evolution (ICE) and the intraspecific red queen. *Behavioral Ecology and Sociobiology* 41:1–10.

Richards, E. L., van Oosterhout, C., & Cable, J. 2010. Sex-specific differences in shoaling affect parasite transmission in guppies. *PLoS ONE* 5:e13285.

Richardson, H., & Verbeek, N. A. M. 1986. Diet selection and optimization by northwestern crows feeding on Japanese littleneck clams. *Ecology* 67:1219–1226.

Riechert, S. E., & Hammerstein, P. 1983. Game theory in the ecological context. *Annual Review of Ecology and Systematics* 14:377–409.

Rieucau, G., & Giraldeau, L.-A. 2011. Exploring the costs and benefits of social information use: An appraisal of current experimental evidence. *Philosophical Transactions of the Royal Society B* 366:949–957.

Riley, J. R., Greggers, U., Smith, A. D., Reynolds, D. R., & Menzel, R. 2005. The flight paths of honeybees recruited by the waggle dance. *Nature* 435:205–207.

Rios-Cardenas, O., & Webster, M. S. 2008. A molecular genetic examination of the mating system of pumpkinseed sunfish reveals high pay-offs for specialized sneakers. *Molecular Ecology* 17:2310–2320.

Ritchie, M. L. 1990. Optimal foraging and fitness for Columbian ground squirrels. *Oecologia* 83:56–67.

Ritz, T., Adem, S., & Schulten, K. 2000. A model for vision-based magnetoreception in birds. *Biophysical Journal* 78:707–718.

Robb, G. N., McDonald, R. A., Chamberlain, D. E., Reynolds, S. J., Harrison, T. J. E., & Bearhop, S. 2008. Winter feeding of birds increases productivity in the subsequent breeding season. *Biology Letters* 4:220–223.

Roberts, J. A., Galbraith, E., Milliser, J., Taylor, P. W., & Uetz, G. W. 2006. Absence of social facilitation of courtship in the wolf spider, *Schizocosa ocreata* (Hentz) (Araneae: Lycosidae). *Acta Ethologica* 9:71–77.

Robinson, G. E., Fernald, R. D., & Clayton, D. F. 2008. Genes and social behavior. *Science* 322:896–900.

Rodd, F. H., Hughes, K. A., Grether, G. F., & Baril, C. T. 2002. A possible non-sexual origin of mate preference: Are male guppies mimicking fruit? *Proceedings of the Royal Society B* 269:475–481.

Rohwer, S., Klein, W. P. Jr., & Heard, S. 1983. Delayed plumage maturation and the presumed pre-alternate molt in American redstarts. *Wilson Bulletin* 95:199–208.

Romanes, G. J. 1882. *Animal intelligence*. London: Kegan Paul, Trench & Co.

———. 1888. *Mental evolution in man: Origin of human faculty*. London: Kegan Paul, Trench & Co.

Ronce, O. 2007. How does it feel to be like a rolling stone? Ten questions about dispersal evolution. *Annual Review of Ecology, Evolution and Systematics* 38:231–253.

Rooney, J., & Lewis, S. M. 1999. Differential allocation of male derived nutrients in two lampyrid beetles with contrasting life history characteristics. *Behavioral Ecology* 10:97–104.

———. 2002. Fitness advantage from nuptial gifts in female fireflies. *Ecological Entomology* 27:373–377.

Rosenzweig, M. L. 2001. Optimality—the biologist's tricorder. *Annales Zoologici Fennici* 38:1–3.

Ross, K. G. 1997. Multilocus evolution in fire ants: Effects of selection, gene flow, and recombination. *Genetics* 145:961–974.

Ross, K. G., & Keller, L. 2002. Experimental conversion of colony social organization by manipulation of worker genotype composition in fire ants (*Solenopsis invicta*). *Behavioral Ecology and Sociobiology* 51:287–295.

Roth, T. C. II, Brodin, A., Smulders, T. V., LaDage, L. D., & Pravosudov, V. V. 2010. Is bigger always better? A critical appraisal of the use of volumetric analysis in the study of the hippocampus. *Philosophical Transactions of the Royal Society B* 365:915–931.

Rowell, J. T., Ellner, S. P., & Reeve, H. K. 2006. Why animals lie: How dishonesty and belief can coexist in a signaling system. *American Naturalist* 168:E180–E204.

Ruefenacht, S., Gebhardt-Henrich, S., Miyake, T., & Gaillard, C. 2002. A behavior test on German shepherd dogs: Heritability of seven different traits. *Applied Animal Behavior Science* 79: 113–132.

Rumbaugh, D. M. 1977. *Language learning by a chimpanzee: The Lana Project.* New York: Academic Press.

Rundus, A. S., Owings, D. H., Joshi, S. S., Chinn, E., & Giannini, N. 2007. Ground squirrels use an infrared signal to deter rattlesnake predation. *Proceedings of the National Academy of Sciences USA* 104:14372–14376.

Russell, Y. I., Call, J., & Dunbar, R. I. M. 2008. Image scoring in great apes. *Behavioral Processes* 78: 108–111.

Russell, A. F., & Hatchwell, B. J. 2001. Experimental evidence for kin-based helping in a cooperatively breeding vertebrate. *Proceedings of the Royal Society B* 268:2169–2174.

Russo, N. F., & Denmark, F. L. 1987. Contributions of women to psychology. *Annual Reviews of Psychology* 38:279–298.

Ruxton, G. D., Sherratt, T. N., & Speed, M. P. 2005. Avoiding attack: The evolutionary ecology of crypsis, warning signals and mimicry. New York: Oxford University Press.

Ryan, M .J. 1998. Principle with a handicap. *Quarterly Review of Biology* 73:477–479.

Ryan, M. J., & Rand, A. S. 1990. The sensory basis of sexual selection for complex calls in the túngara frog, *Physalaemus pustulosus* (sexual selection for sensory exploitation). *Evolution* 44:305–314.

Ryan, M. J., Tuttle, M. D., & Rand, A. S. 1982. Bat predation and sexual advertisement in a neotropical anuran. *American Naturalist* 119: 136–139.

Rypstra, A. L., Weig, C., Walker, S. E., & Persons, M. H. 2003. Mutual mate assessment in wolf spiders: Differences in the cues used by males and females. *Ethology* 109:315–325.

Sachs, J. L., & Bull, J. J. 2005. Experimental evolution of conflict mediation between genomes. *Proceedings of the National Academy of Sciences USA* 102: 390–395.

Salewski, V., & Bruderer, B. 2007. The evolution of bird migration—a synthesis. *Naturwissenschaften* 94:268–279.

Sambandan, D., Carbone, M. A., Anholt, R. R. H., & Mackay, T. F. C. 2008. Phenotypic plasticity and genotype by environment interaction for olfactory behavior in *Drosophila melanogaster. Genetics* 179:1079–1088.

Sánchez, F. 2006. Harvest rates and patch use strategy of Egyptian fruit bats in artificial food patches. *Journal of Mammalogy* 87:1140–1144.

Sandberg, R., Bäckman, J., Moore, F. R., & Lõhmus, M. 2000. Magnetic information calibrates celestial cues during migration. *Animal Behaviour* 60:453–462.

Sapolsky, R. M. 2005. The influence of social hierarchy on primate health. *Science* 308:648–652.

Sapolsky, R. M., Romero, L. M., & Munck, A. U. 2000. How do glucocorticoids influence stress responses? Integrating permissive, suppressive, stimulatory and preparative actions. *Endocrine Reviews* 21:55–89.

Saporito, R. A., Zuercher, R., Roberts, M., Gerow, K. G., & Donnelly, M. A. 2007. Experimental evidence for aposematism in the Dendrobatid poison frog *Oophaga pumlio. Copeia* 2007:1006–1011.

Sargent, B. L., & Mann, J. 2009. Developmental evidence for foraging traditions in wild bottlenose dolphins. *Animal Behaviour* 78: 715–721.

Scheuber, H., Jacot, A., & Brinkof, M. W. G. 2003. The effect of past condition on a multicomponent sexual signal. *Proceedings of the Royal Society B* 270:1779–1784.

Schino, G., & Aureli, F. 2010. The relative roles of kinship and reciprocity in explaining primate altruism. *Ecology Letters* 13:45–50.

Schino, G., Polizzi de Sorrentino, E., & Tiddi, B. 2007. Grooming and coalitions in Japanese macaques (*Macaca fuscata*): Partner choice and the time frame of reciprocation. *Journal of Comparative Psychology* 121:181–188.

Schmidt, K. A., & Ostfeld, R. S. 2008. Eavesdropping squirrels reduce their future value of food under the perceived presence of cache robbers. *American Naturalist* 171:386–393.

Schmidt-Koenig, K. 1979. *Avian orientation and navigation.* London: Academic Press.

———. 1990. The sun compass. *Experimentia* 16:336–342.

Schradin, C., & Anzenberger, G. 1999. Prolactin, the hormone of paternity. *News in Physiological Sciences* 14:223–231.

Scott, J. P. 1943. Effects of single genes on the behavior of *Drosophila. American Naturalist* 77:184–190.

Scott, M. P. 2006. Resource defense and juvenile hormone: The "challenge hypothesis" extended to insects. *Hormones and Behavior* 49:276–281.

Scott-Phillips, T. C. 2008. Defining biological communication. *Journal of Evolutionary Biology* 21:387–395.

Seed, A., & Tomasello, M. 2010. Primate cognition. *Topics in Cognitive Science* 2:407–419.

Seeley, T. D. 1985. *Honey bee ecology: A study of adaptation in social life*. Princeton, NJ: Princeton University Press.

Sefc, K. M., Mattersdorfer, K., Sturmbauer, C., & Koblmüller, S. 2008. High frequency of multiple paternity in broods of a socially monogamous cichlid fish with biparental nest defense. *Molecular Ecology* 17:2531–2543.

Seki, S., Kohda, M., Takamoto, G., Karino, K., Nakashima, Y., & Kuwamura, T. 2009. Female defense polygyny in the territorial triggerfish *Sufflamen chrysopterum*. *Journal of Ethology* 27:215–220.

Semple, S., Gerald, M. S., & Suggs, D. N. 2009. Bystanders affect the outcome of mother-infant interactions in rhesus macaques. *Proceedings of the Royal Society B* 276:2257–2262.

Seyfarth, R. M., Cheney, D. L., Bergman, T., Fischer, J., Züberbühler, K., & Hammerschmidt, K. 2010. The central importance of information in studies of animal communication. *Animal Behaviour* 80:3–8.

Seyfarth, R. M., Cheney, D. L., & Marler, P. 1980. Monkey responses to three different alarm calls: Evidence of predator classification and semantic communication. *Science* 210:801–803.

Sherman, P. W. 1977. Nepotism and the evolution of alarm calls. *Science* 197:1246–1253.

Shier, D. M., & Owings, D. H. 2007. Effects of social learning on predator training and postrelease survival in juvenile black-tailed prairie dogs, *Cynomys ludovicianus*. *Animal Behaviour* 73: 567–577.

Shipman, P. 2010. The animal connection and human evolution. *Current Anthropology* 51:519–538.

Siemers, B. M., Goerlitz, H. R., Robsomanitrandrasana, E., Piep, M., Ramanamamjato, J.-B., Rakotondravony, D., Ramilijaona, O., & Ganzhorn, J. U. 2007. Sensory basis of food detection in wild *Microcebus murinus*. *International Journal of Primatology* 28:291–304.

Sigg, D. P., Goldizen, A. W., & Pople, A. R. 2005. The importance of mating system in translocation programs: Reproductive success of released male bridled nailtail wallabies. *Biological Conservation* 123:289–300.

Sih, A., Bell, A., & Johnson J. C. 2004. Behavioral syndromes: An ecological and evolutionary overview. *Trends in Ecology and Evolution* 19: 372–378.

Sih, A., Bell, A. M., Johnson, J. C., & Ziemba, R. E. 2004. Behavioral syndromes: An integrative overview. *Quarterly Review of Biology* 79:241–277.

Sih, A., & Christensen, B. 2001. Optimal diet theory: When does it work and when and why does it fail? *Animal Behaviour* 61:379–390.

Sih, A., Kats, L. B., & Maurer, E. F. 2003. Behavioural correlations across situations and the evolution of antipredator behavior in a sunfish-salamander system. *Animal Behaviour* 65:29–44.

Sih, A., Krupa, J., & Travers, S. 1990. An experimental study on the effects of predation risk and feeding regime on the mating behavior of the water strider. *American Naturalist* 135:284–290.

Skinner, B. F. 1938. *The behavior of organisms: An experimental analysis*. New York: Appleton-Century.

Slabbekoorn, H., & den Boer-Visser, A. 2006. Cities change the songs of birds. *Current Biology* 16:2326–2331.

Sladen, W. J. L., Lishman, W. A., Ellis, D. H., Shire, G. G., & Rininger, D. L. 2002. Teaching migration routes to Canada geese and trumpeter swans using ultralight aircraft, 1990–2001. *Waterbirds* 25:132–137.

Smith, R. L. 1979. Repeated copulation and sperm precedence: Paternity assurance for a male brooding water bug. *Science* 205:1029–1031.

Smith, B. R., & Blumstein, D. T. 2008. Fitness consequences of personality: A meta-anlaysis. *Behavioral Ecology* 19:448–455.

Smrt, R.D., & Zhao, X. 2010. Epigenetic regulation of neuronal dendrite and dendritic spine development. *Frontiers in Biology* 5:304–323.

Sokal, R. R., & Rohlf, F. J. 2012. *Biometry: The principles and practice of statistics in biological research*. 4th ed. New York: W. H. Freeman and Co.

Sokolowski, M.B. 2001. *Drosophila*: Genetics meets behavior. *Nature Reviews Genetics* 2:879–890.

Sokolowski, M. B., Pereira, H. S., & Hughes, K. 1997. Evolution of foraging behavior in *Drosophila* by density-dependent selection. *Proceedings of the National Academy of Sciences USA* 94:7373–7377.

Sorenson, M. D.1991. The functional-significance of parasitic egg-laying and typical nesting in redhead ducks—an analysis of individual behavior. *Animal Behaviour* 42:771–796.

South, A., & Lewis, S. M. 2011. The influence of male ejaculate quantity on female fitness: A meta-analysis. *Biological Reviews* 86:299–309.

Spady, T. C., & Ostrander, E. A. 2008. Canine behavioral genetics: Pointing out the phenotypes and herding up the genes. *American Journal of Human Genetics* 82:10–18.

Stam, H. J., & Kalmanovitch, T. 1998. E. L. Thorndike and the origins of animal psychology. *American Psychologist* 53:1135–1144.

Stamps. J. A. 1988. Conspecific attraction and aggregation in territorial species. *American Naturalist* 131:329–347.

Stamps, J. A., & Groothuis, T. G. G. 2010. Developmental perspectives on personality: Implications for ecological and evolutionary studies of individual differences. *Philosophical Transactions of the Royal Society B* 365:4029–4041.

Stankowich, T., & Coss, R. G. 2007. Effects of risk assessment, predator behavior and habitat on escape behavior in Columbian black-tailed deer. *Behavioral Ecology* 18:358–367.

Stapput, K., Thalau, P., Wiltschko, R., & Wiltschko, W. 2008. Orientation of birds in total darkness. *Current Biology* 18:602–606.

Stebbins, R. C. 1949. Speciation in salamanders of the plethodontid genus *Ensatina*. University of California Publication in Zoology, vol. 48. Berkeley: University of California Press.

Steinhart, G. A., Marschall, E. A., & Stein, R. A. 2004. Round goby predation on smallmouth bass offspring in nests during simulated catch-and-release angling. *Transactions of the American Fisheries Society* 133:121–131.

Steinhart, G. A., Sandrene, M. E., Weaver, S., Stein, R. A., & Marschall, E. A. 2005. Increased parental care cost for nest-guarding fish in a lake with hyperabundant nest predators. *Behavioral Ecology* 16:427–434.

Stenseth, N. C., & Mysterud, A. 2002. Climate, changing phenology, and other life history traits: Nonlinearity and match-mismatch to the environment. *Proceedings of the National Academy of Sciences USA* 99:13,379–13,381.

Stirling, D. G., Réale, D., & Roff, D. A. 2002. Selection, structure and the heritability of behaviour. *Journal of Evolutionary Biology* 15:277–289.

Stodola, K. W., Linder, E. T., Buehler, D. A., Franzreb, K. E., & Cooper, R. J. 2009. Parental care in the multi-brooded black-throated blue warbler. *Condor* 111:497–502.

Stokke, B. G., Moksnes, A., & Røskaft, E. 2005. The enigma of imperfect adaptations in hosts of avian brood parasites. *Ornithological Science* 4:17–29.

Strathman, R. R., & Strathman, M. F. 1982. The relationship between adult size and brooding in marine invertebrates. *American Naturalist* 119:91–101.

Stutchbury, B. J., & Morton, E. S. 1995. The effect of breeding synchrony on extra-pair mating system in songbirds. *Behaviour* 132:675–690.

Stutchbury, B. J. M., Tarof, S. A., Done, T., Gow, E., Kramer, P. M., Tautin, J., Fox, J. W., & Afanasyev, V. 2009. Tracking long-distance songbird migration by using geolocators. *Science* 323:896.

Sullivan, K. A. 1988. Age specific profitability and prey choice. *Animal Behaviour* 36:613–615.

Summers, K., Sea, C. S., & Heying, H. 2006. The evolution of parental care and egg size: A comparative analysis in frogs. *Proceedings of the Royal Society B* 273:687–692.

Suttie, J. M. 1979. The effect of antler removal on dominance and fighting behaviour in farmed red deer stags. *Journal of Zoology* 190:217–224.

Suzuki, S., Nagano, M., & Kobayashi, N. 2006. Mating competition and parentage assessment in *Ptomascopus morio* (Coleoptera: Silphidae): A case for resource defense polygyny. *European Journal of Entomology* 103:751–755.

Suzuki, S., Nagano, M., & Trumbo, S. T. 2005. Intrasexual competition and mating behavior in *Ptomascopus morio* (Coleoptera: Silphidae Nicrophorinae). *Journal of Insect Behavior* 18:233–242.

Swingland, I. R., & Greenwood, P. J. 1983. *The ecology of animal movement*. Oxford, UK: Clarendon Press.

Switzer, P. V. 1993. Site fidelity in predictable and unpredictable habitats. *Evolutionary Ecology* 7:533–555.

————. 1997. Past reproductive success affects future habitat selection. *Behavioral Ecology and Sociobiology* 40:307–312.

Székely, T., & Cuthill, I. 2000. Trade-off between mating opportunities and parental care: Brood desertion by female Kentish plovers. *Proceedings of the Royal Society B* 267:2087–2092.

Taborsky, M., Oliveira, R. F., & Brockmann, H. J. 2008. The evolution of alternative reproductive tactics: Concepts and questions. In *Alternative reproductive tactics*, ed. R. F. Oliveira, M. Taborsky, & H. J. Brockmann, 1–21. Cambridge: Cambridge University Press.

Tack, E. J., Robson, T. E., Putland, D. A., & Goldizen, A. W. 2005. Geographic variation in vocalisations of satin bowerbirds, *Ptilonorhynchus violaceus*, in south-eastern Queensland. *Emu* 105:27–31.

Tallamy, D. W. 2005. Egg dumping in insects. *Annual Review of Entomology* 50:347–370.

Tanaka, Y. 2000. Realized heritability of behavioral responsiveness to an oviposition pheromone in the azukibean weevil *Callosobruchus chinensis*. *Entomologia Experimentalis et Applicata* 96:239–243.

Tang-Martínez, Z., & Ryder, T. B. 2005. The problem with paradigms: Bateman's worldview as a case study. *Integrative and Comparative Biology* 45:821–830.

Tay, W. T., Miettinen, M., & Kaitala, A. 2003. Do male golden egg bugs carry eggs they have fertilized? A microsatellite analysis. *Behavioral Ecology* 14:481–485.

Taylor, M. L., Wigmore, C., Dodgson, D. J., Wedell, N., & Hosken, D. J. 2008. Multiple mating increases female fitness in *Drosophila simulans*. *Animal Behaviour* 76:963–970.

Temeles, E. J. 1994. The role of neighbours in territorial systems: When are they "dear enemies"? *Animal Behaviour* 47:339–350.

Templeton, J. J., & Giraldeau, L.-A. 1996. Vicarious sampling: The use of personal and public information by starlings foraging in a simple patchy environment. *Behavioral Ecology and Sociobiology* 38:105–114.

Thacker, M., Lima, S. L., & Hews, D. L. 2009. Alternative antipredator tactics in tree lizard morphs: Hormonal and behavioural responses to a predator encounter. *Animal Behaviour* 77:395–401.

Thorndike, E. L. 1911. *Animal intelligence: Experimental studies.* New York: MacMillan.

Thornhill, R. 1983. Cryptic female choice and its implications in the scorpionfly *Harpobittacus nigricepts. American Naturalist* 122:765–788.

Thorup, K., & Holland, R.A. 2009. The bird GPS— long-range navigation in migrants. *Journal of Experimental Biology* 212:3597–3604.

Tibbetts, E. A., & Dale, J. 2007. Individual recognition: It is good to be different. *Trends in Ecology and Evolutionary Biology* 22:529–537.

Tinbergen, N. 1951. *The study of instinct.* New York: Oxford University Press.

_____. 1960. The natural control of insects in pinewoods. I. Factors influencing the intensity of predation by songbirds. *Archives Néerlandaises de Zoologie* 13:265–343.

_____. 1963. On aims and methods in ethology. *Zeitschrift für Tierpsychologie* 20:410–433.

Tome, M. W. 1988. Optimal foraging: Food patch depletion by ruddy ducks. *Oecologia* 76:27–36.

Tracy, B. F., & McNaughton, S. J. 1995. Elemental analysis of mineral lick soils from the Serengeti national park, the Konza prairie, and Yellowstone national park. *Ecography* 18:91–94.

Tregenza, T., & Wedell, N. 2000. Genetic compatibility, mate choice and patterns of parentage: Invited review. *Molecular Ecology* 9:1013–1027.

Trillmich, F., & Wolf, J. B. W. 2008. Parent–offspring and sibling conflict in Galápagos fur seals and sea lions. *Behavioral Ecology and Sociobiology* 62:363–375.

Trivers, R. L. 1971. The evolution of reciprocal altruism. *Quarterly Review of Biology* 46:35–57.

Trivers, R. L. 1972. Parental investment and sexual selection. In: *Sexual selection and the descent of man, 1871–1971,* ed. B. Campbell, 136–179. Chicago: Aldine.

_____. 1974. Parent-offspring conflict. *American Zoologist* 14:249–264.

Trumbo, S. T. 2002. Hormonal regulation of parental behavior in insects. In *Hormones, brain and behavior,* vol. 3, ed. D. W. Pfaff, A. P. Arnold, A. M. Etgen, S. E. Fahrbach, R. L. Moss, & R. R. Rubin, 115–139. New York: Academic Press.

Vadas, R. L. Jr. 1994. The anatomy of an ecological controversy: Honey-bee searching behavior. *Oikos* 69:158–166.

Vahed, K. 2007. All that glitters is not gold: Sensory bias, sexual conflict and nuptial feeding in insects and spiders. *Ethology* 113:105–127.

Valone, T. J. 1989. Group foraging, patch estimation, and public information. *Oikos* 56:357–363.

_____. 2006. Are animals capable of Bayesian updating? An empirical review. *Oikos* 112: 252–259.

_____. 2007. From eavesdropping on performance to copying the behavior of others: A review of public information use. *Behavioral Ecology and Sociobiology* 62:1–14.

Valone, T. J., & Lima, S. L. 1987. Carrying food items to cover for consumption: The behavior of ten bird species feeding under the risk of predation. *Oecologia* 71:286–294

van den Berghe, E. P., & Gross, M. R. 1989. Natural selection resulting from female breeding competition in a pacific salmon (Coho: *Oncorhynchus kisutch*). *Evolution* 43:125–140.

van der Staay, F. J. 2006. Animal models of behavioral dysfunctions: Basic concepts and classification, and an evaluation strategy. *Brain Research Reviews* 52:131–159.

Varricchio, D. J., Moore, J. R., Erickson, G. M., Norell, M. A., Jackson, F. D., & Borkowski, J. J. 2009. Avian paternal care had dinosaur origin. *Science* 322:1826–1828.

Vehrencamp, S. C. 1983. Optimal degree of skew in cooperative societies. *American Zoologist* 23:327–335.

Verbeek, M. E. M., Drent, P. J., & Wiepkema, P. R. 1994. Consistent individual differences in early exploratory behaviour of male great tits. *Animal Behaviour* 48:1113–1121.

Verner, J. 1964. Evolution of polygamy in the long-billed marsh wren. *Evolution* 18:252–261.

Verner, J., & Wilson, M. F. 1966. The influence of habitats on mating systems of North American passerine birds. *Ecology* 47:143–147.

Verrell, P. A. 1986. Wrestling in the red-spotted newt (*Notophthalmus viridescens*): Resource value and contestant asymmetry determine contest duration and outcome. *Animal Behaviour* 34:398–402.

Vila-Gispert, A., Moreno-Amich, R., & García-Berthou, E. 2002. Gradients of life-history variation: An intercontinental comparison of fishes. *Reviews in Fish Biology and Fisheries* 12:417–427.

Vincent, A. C. J. 1995a. A role for daily greetings in maintaining seahorse pair bonds. *Animal Behaviour* 49:258–260.

_____. 1995b. Trade in seahorses for traditional Chinese medicines, aquarium fishes, and curios. *TRAFFIC Bulletin* 15:125–128.

_____. 2008. Conservation on coral reefs: The world as an onion. *American Fisheries Society Symposium* 49:1435–1467.

Vincent, A., Evans, K., & Marsden, A. D. 2005. Home range behaviour of the monogamous Australian seahorse, *Hippocampus whitei*. *Environmental Biology of Fishes* 72:1–12.

Vitt, L. J., & Price, H. J. 1982. Ecological and evolutionary determinants of relative clutch mass in lizards. *Herpetologica* 38:237–255.

Vleck, C. M. 2001. Hormonal control of incubation behaviour. In *Avian incubation: Behaviour, environment and evolution*, ed. D. C. Deeming, 54–62. Oxford, UK: Oxford University Press.

von Frisch, K. 1967. *The dance language and orientation of bees*. Cambridge, MA: Harvard University Press.

_____. 1971. *Bees: Their vision, chemical senses, and language*. 2nd ed. Ithaca, NY: Cornell University Press.

von Frisch, K., Wenner, A. M., & Johnson, D. L. 1967. Honeybees: Do they use direction and distance information provided by their dances? *Science* 158:1072–1077.

Vos, P., Hogers, R., Bleeker, M., Reijans, M., van de Lee, T., Hornes, M., Frijters, A., et al. 1995. AFLP: A new technique for DNA fingerprinting. *Nucleic Acids Research* 23:4407–4414.

Walcott, C. 1996. Pigeon homing: Observations, experiments and confusions. *Journal of Experimental Biology* 199:21–27.

_____. 2005. Multimodal orientation cues in homing pigeons. *Integrative and Comparative Biology* 45:574–581.

Wallace, A. R. 1878. *Tropical nature, and other essays*. London: Macmillan.

Ward, M. P., & Schlossberg, S. 2004. Conspecific attraction and the conservation of territorial songbirds. *Conservation Biology* 18:519–525.

Warner, R. R. 1988. Traditionality of mating-site preferences in a coral reef fish. *Nature* 335:719–721.

_____. 1990. Male versus female influences on mating-site determination in a coral reef fish. *Animal Behaviour* 39:540–548.

Warren, W. C., Clayton, D. F., Ellegren, H., Arnold, A. P., Hillier, L. W., Künstner, A., Searle, S., et al. 2010. The genome of a songbird. *Nature* 464:757–762.

Washburn, M. F. 1908. *The animal mind*. New York: MacMillan.

Watson, J. B. 1913. Psychology as the behaviorist views it. *Psychological Review* 20:158–177.

_____. 1924. *Behaviorism*. New York: People's Institute.

Watts, C. R., & Stokes, A. W. 1971. The social order of turkeys. *Scientific American* 224:112–118.

Weber, J. N., & Hoekstra, H. E. 2009. The evolution of burrowing behavior in deer mice (genus *Peromyscus*). *Animal Behaviour* 77:603–609.

Wehner, R. 1998. Navigation in context: Grand theories and basic mechanisms. *Journal of Avian Biology* 29:370–386.

_____. 2001. Polarization vision—a uniform sensory capacity? *Journal of Experimental Biology* 204:2589–2596.

Weimerskirch, H., Martin, J., Clerquin, Y., Alexandre, P., & Jiraskova, S. 2001. Energy saving in flight formation. *Nature* 413:697–698.

Welke, K., & Schneider, J. M. 2009. Inbreeding avoidance through cryptic female choice in the cannibalistic orb-web spider *Argiope lobata*. *Behavioral Ecology* 20:1056–1062.

Wenner, A. M. 2002. The elusive honey bee dance "language" hypothesis. *Journal of Insect Behavior* 15:859–878.

Wenner, A. M., Well, P. H., & Johnson, D. L. 1969. Honey bee recruitment to food sources: Olfaction or language? *Science* 164:84–86.

West-Eberhard, M. J. 1984. Sexual selection, competitive communication and species-specific signals in insects. In *Insect communication*, ed. T. Lewis, 283–324. New York: Academic Press.

Westneat, D. F., & Stewart, I. R. K. 2003. Extra-pair paternity in birds: Causes, correlates, and conflict. *Annual Review of Ecology and Systematics* 34:365–396.

Wheeler, B. C. 2009. Monkeys crying wolf? Tufted capuchin monkeys use anti-predator calls to usurp resources from conspecifics. *Proceedings of the Royal Society B* 276:3013–3018.

Whiten, A., Goodall, J., McGrew, W. C., Nishida, T., Reynolds, V., Sugiyama, Y., Tutin, C. E. G., Wrangham, R. W., & Boesch, C. 1999. Cultures in chimpanzees. *Nature* 399:682–685.

Whiten, A., & van Schaik, C. P. 2007. The evolution of animal "cultures" and social intelligence. *Philosophical Transactions of the Royal Society B* 362:603–620.

Wilkinson, G. S. 1984. Reciprocal food sharing in the vampire bat. *Nature* 308:181–184.

_____. 1990. Food sharing in vampire bats. *Scientific American* 262:76–82.

Williams, G. C. 1966. *Adaptation and natural selection: A critique of some current evolutionary thought*. Princeton, NJ: Princeton University Press.

Williams, T. M., Selegue, J. E., Werner, T., Gompel, N., Kopp, A., & Carroll, S. B. 2008. The regulation and evolution of a genetic switch controlling sexually dimorphic traits in *Drosophila*. *Cell* 134:610–623.

Wilson, A. B., Ahnesjö, I., Vincent, A. C. J., & Meyer, A. 2003. The dynamics of male brooding, mating patterns, and sex roles in pipefishes and seahorses (family Syngnathidae). *Evolution* 57:1374–1386.

Wilson, D. S., Van Vugt, M., & O'Gorman, R. 2008. Multilevel selection theory and major evolutionary transitions. *Current Directions in Psychological Science* 17:6–9.

Wilson, D. S., & Wilson, E. O. 2007. Rethinking the theoretical foundation of sociobiology. *Quarterly Review of Biology* 82:327–348.

Wilson, E. O. 2008. One giant leap: How insects achieved altruism and colonial life. *BioScience* 58:17–25.

Wilsson, E., & Sundgren, P.-E. 1997. The use of a behavior test for the selection of dogs for service and breeding. I. Methods of testing and evaluating test results in the adult dog, demands on different kinds of service dogs, sex and breed differences. *Applied Animal Behavior Science* 53:279–295.

Wiltschko, R., & Wiltschko, W. 1981. The development of sun compass orientation in young homing pigeons. *Behavioral Ecology and Sociobiology* 9:135–141.

Wiltschko, R., & Wiltschko, W. 1989. Pigeon homing: Olfactory orientation—a paradox. *Behavioral Ecology and Sociobiology* 24:163–173.

Wiltschko R., & Wiltschko, W. 2009. Avian navigation. *Auk* 126:717–743.

Wiltschko, W., Munro, U., Ford, H., & Wiltschko, R. 1998. Effect of a magnetic pulse on the orientation of Silvereyes, *Zosterops l. lateralis*, during spring migration. *Journal of Experimental Biology* 201:3257–3261.

Wiltschko, W., Traudt, J., Güntürkün, O., Prior, H., & Wiltschko, R. 2002. Lateralization of magnetic compass orientation in a migratory bird. *Nature* 419:467–470.

Wiltschko, W., & Wiltschko, R. 2002. Magnetic compass orientation in birds and its physiological basis. *Naturwissenschaften* 89:445–452.

Wiltschko, W., & Wiltschko, R. 2005. Magnetic orientation and magnetoreception in birds and other animals. *Journal of Comparative Physiology A* 191:675–693.

Wingfield, J. C., Hegner, R. E., Dufty, A. M. Jr., & Ball, G. F. 1990. The "challenge hypothesis": Theoretical implications for patterns of testosterone secretion, mating systems, and breeding strategies. *American Naturalist* 136:829–846.

Wingfield, J. C., Jacobs, J. D., Tramontin, A. D., Perfito, N., Meddle, S., Maney, D. L., & Soma, K. 2000. Toward an ecological basis of hormone-behavior interactions in reproduction of birds. In *Reproduction in context*, ed. K. Wallen & J. Schneider, 85–128. Cambridge, MA: M.I.T. Press.

Wingfield, J. C., Lynn, S. E., & Soma, K. K. 2001. Avoiding the "costs" of testosterone: Ecological bases of hormone-behavior interactions. *Brain, Behavior and Evolution* 57:239–251.

Wisby, W. J., & Haesler, A. D. 1954. Effect of olfactory occlusion on migrating silver salmon (*O. kisutch*). *Journal of the Fisheries Research Board of Canada* 11:472–478.

Wisenden, B. D. 1999. Alloparental care in fishes. *Reviews in Fish Biology and Fisheries* 9:45–70.

Witzgall, P., Kirsch, P., & Cork, A. 2010. Sex pheromones and their impact on pest management. *Journal of Chemical Ecology* 36:80–100.

Wolf, M., van Doorn, G. S., Leimar, O., & Weissing, F. J. 2007. Life-history trade-offs favour the evolution of animal personalities. *Nature* 447:581–584.

Wolff, J. O., & Macdonald, D. W. 2004. Promiscuous females protect their offspring. *Trends in Ecology and Evolution* 19:127–134.

Wong, B. M., & McCarthy, M. 2009. Prudent male mate choice under perceived sperm competition risk in the eastern mosquito fish. *Behavioral Ecology* 20:278–282.

Wright, T. F., Dahlin, C. R., & Salinas-Melgoza, A. 2008. Stability and change in vocal dialects of the yellow-naped amazon. *Animal Behaviour* 76:1017-1027.

Wu, L-Q., & Dickman, J. D. 2012. Neural correlates of magnetic sense. *Science* 336:1054–1057.

Wynne, C. D. L. 2004a. *Do animals think?* Princeton, NJ: Princeton University Press.

Wynne, C. D. L. 2004b. The perils of anthropomorphism. *Nature* 428:606.

Wynne, C. D. L. 2007. What are animals? Why anthropomorphism is still not a scientific approach to behavior. *Comparative Cognition & Behavior Reviews* 2:125–135.

Wynne-Edwards, V. C. 1962. *Animal dispersion in relation to social behavior*. Edinburgh: Oliver and Boyd.

Wynne-Edwards, K. E., & Timonin, M. E. 2007. Paternal care in rodents: Weakening support for hormonal regulation of the transition to behavioral fatherhood in rodent animal models of biparental care. *Hormones and Behavior* 52:114–121.

Yamamoto, Y., Hino, H., & Ueda, H. 2010. Olfactory imprinting of amino acids in lacustrine sockeye salmon. *PLoS ONE* 5:e8633.

Yamaguchi, N., Dugdale, H. L., & MacDonald, D. W. 2006. Female receptivity, embryonic diapause, and superfetation in the European badger (*Meles meles*): Implications for the reproductive tactics of males and females. *Quarterly Review of Biology* 81:33–48.

Yang, G., Pan, F., & Gan, W-B. 2009. Stably maintained dendritic spines are associated with lifelong memories. *Nature* 462:920–925.

Yanoviak, S. P. 2006. Steve Yanoviak research—gliding ants—FAQ. http://www.canopyants.com/glide_faq.html (accessed November 28, 2012).

Yanoviak, S. P., Dudley, R., & Kaspari, M. 2005. Directed aerial descent in canopy ants. *Nature* 433:624–626.

Yasué, M., Quinn, J. L., & Cresswell, W. 2003. Multiple effects of weather on the starvation and predation risk trade-off in choice of feeding location in Redshanks. *Functional Ecology* 17:727–736.

Yerkes, R. M. 1913. The heredity of savageness and wildness in rats. *Journal of Animal Behavior* 3:286–296.

Yin, S., Fernandez, E. J., Pagan, S., Richardson, S. L., & Snyder, G. 2008. Efficacy of a remote-controlled, positive-reinforcement, dog-training system for modifying problem behaviors exhibited when people arrive at the door. *Applied Animal Behaviour Science* 113:123–138.

Yom-Tov Y. 2001. An updated list and some comments on the occurrence of intraspecific nest parasitism in birds. *Ibis* 143:133–143.

Zahavi, A. 1975. Mate selection: A selection for a handicap. *Journal of Theoretical Biology* 53: 205–214.

Zarrow, M. X., Gandelman, R., & Denenberg, V. H. 1971. Prolactin: Is it an essential hormone for maternal behavior in the mammal? *Hormones and Behavior* 2:343–354.

Zink, A. 2003. Quantifying the costs and benefits of parental care in female treehoppers. *Behavioral Ecology* 14:687–693.

Zuberbühler, K. 2008. Audience effects. *Current Biology* 18:R189–R190.

Zuk, M., Thornhill, R., Ligon, D. J., & Johnson, K. 1990. Parasites and mate choice in red jungle fowl. *American Zoologist* 30:235–244.

Answers to Selected Questions

Chapter 1: The Science of Animal Behavior

2. (a) Behavior. (b) Behavior. (c) Behavior. (d) Not a behavior, because the mane moved as a result of the breeze and was not internally coordinated by the animal. (e) Behavior. (f) Behavior.

4. A hypothesis is an explanation that makes a prediction. An educated guess does not make a prediction.

6. This is a hypothesis based on anthropomorphism and therefore is very difficult to test. However, to begin to address this question, you could examine the levels of stress hormones in birds at bird feeders and birds not at bird feeders and compare the two. This test would indicate stress levels, but not happiness.

Chapter 2: Evolution and the Study of Animal Behavior

2. The conclusion is premature because we do not know whether aggressive behavior is heritable in this population. If aggression in this species is heritable, then the conclusion is valid. However, selection cannot affect the frequency of a trait in a population if the trait is not heritable.

4. Male success in territory acquisition is based on individuals' ability to compete for territories and mates. There are only a finite number of territories available, which limits the number of successful male breeders.

Chapter 3: Methods for Studying Animal Behavior

4. European wildcat and domestic cat, cheetah and cougar, jaguar and lion.

Chapter 4: Behavioral Genetics

2. Conduct a parent-offspring regression analysis of boldness behavior. For example, you could measure boldness in adult females and their offspring to see if there was a positive correlation between the two. If so, you would conclude that boldness is heritable.

4. In the study of social organization in fire ants, you could reasonably conclude that variation in the genotype at the *Gp-9* locus affects behavioral organization more strongly than do environmental factors.

6. Lower heritability for a behavioral trait means that more of the variation in the behavior is due to environmental than genetic variation. This suggests that behaviors are more affected by environmental variation than morphology, perhaps because behaviors develop in more variable environments.

Chapter 5: Learning and Cognition

2. This mutant should have no ability to learn: it can't benefit from experiences because it can't remember them.

4. Local enhancement occurs when individuals focus their attention on others and use their behavior to learn about the location of a resource such as a food patch. Public information is the use of information derived from the behavior of others to learn about the quality of a resource like a food patch.

6. There were likely two changes: (1) an increased number of neurotransmitters were released when Max learned the new behavior; and (2) new dendritic spines formed when the behavior was learned, some of which still persist. Only the second change is associated with lifelong memory.

Chapter 6: Communication

2. If only the largest, most dominant males can produce low-frequency sounds, then females can use male vocalizations to assess male quality (i.e., body size). If females prefer to mate with dominant males, then we can predict that females will exhibit a mating preference for males that produce the lowest frequency vocalizations.

4. Allow half the test females to observe a fight between two males and the other half to observe nothing. Immediately after the male contest, allow the female to exhibit a mating preference for one of the two males. If females prefer the dominant male without seeing him win a fight, it is likely he produces chemical pheromones that advertise his status. If females select the dominant male only after watching him win, they are gaining information about relative fighting ability via eavesdropping. For an example, see Alquiloni, L., Buřič, M., & Gherardi, F., Crayfish females eavesdrop on fighting males before choosing the dominant male, *Current Biology* 18 (2008): R462–R463.

6. The squirrels eavesdrop on blue jay vocal signals to learn about the presence of potential cache robbers, which changes their behavior. A rival male or female fighting fish was a bystander observing fighting behavior of a focal male. The focal male changed its behavior because of the presence of a female but did not change its behavior in response to the presence of a male bystander. Thus, in the first example, the eavesdropper's behavior changed, while in the latter example, the focal male changed its behavior in the presence of a female bystander.

Chapter 7: Foraging Behavior

2. Experiment A: The patch with heavy sand has a higher energetic cost of foraging and is predicted to have a higher GUD. Experiment B: The jays will have higher GUDs in the cold, shady patch than the warm, sunny patch, because the energetic cost of foraging is higher in the cold, shady patch. For a comparable study, see A. M. Kilpatrick, The impact of thermal regulatory costs on foraging behavior: A test with American crows (*Corvus brachyrhynchos*) and eastern grey squirrels (*Sciurus carolinensis*), *Evolutionary Ecology Research* 5 (2003):781–786.

4. The patch-use model predicts that increases in travel time will increase patch-use time. You can predict that shrews will spend more than one minute and harvest more than nine items from patches when travel distance (time) is increased.

6. Both models assume that individuals maximize their fitness by maximizing energy intake rate.

Chapter 8: Antipredator Behavior

2. If a predator is not present and predation risk is low, then territory size will be large. If a predator is present and predation risk is high, then territory size will be small, because individuals will want to minimize the amount of time spent defending the territory and susceptible to predators. You could set up aquaria with predator cues (visual or chemical) or predators both present and absent and measure the size of territories defended by males.

4. The dilution effect states that the probability of being killed declines as group size increases. The group size effect is not applicable to this situation, because the antipredator benefit is that larger groups spot predators sooner than smaller groups (even though each individual scans less and feeds

more), and here the predator has already been spotted. The selfish herd is also not applicable, because this antipredator behavior requires simply moving toward the center of a group.

6. Scientific Process 8.1: Null hypothesis (1): There is no difference in food in the mudflat and saltmarsh habitats. Null hypothesis (2): Ambient temperature does not affect the feeding decisions of redshanks. Scientific Process 8.2: Null hypothesis: The presence of a predator will not affect the mating behavior of water striders. Scientific Process 8.3: Null hypothesis: Ground squirrels do not increase blood flow to their tail when attacked by rattlesnakes.

Chapter 9: Dispersal and Migration

2. Natal dispersal and proximate mechanisms: The density of individuals, hunger levels, and the number of aggressive interactions could all be proximate mechanisms for determining the level of competition. Pheromones and visual recognition could be used to determine the relatedness of others as a proximate mechanism to initiate natal dispersal (e.g., inbreeding avoidance). Natal dispersal and ultimate mechanisms: Competition reduces fitness, and so individuals that disperse to areas of lower competition should have higher fitness. Inbreeding depression reduces fitness. Individuals that avoid breeding with close kin will have higher fitness than those that breed with relatives.

4. There must be a positive year-to-year correlation in breeding site quality. A good (or poor) site this year must have a high probability of also being a good (or poor) site next year.

6. As climate warms, overwinter survivorship of residents should increase. This change may result in higher fitness for residents than migrants and may lead to a reduction in the proportion of migrants in the population.

Chapter 10: Habitat Selection, Territoriality, and Aggression

2. The predation risk pattern described suggests that parrots exhibit an Allee effect: fitness increases as group size increases. Therefore, you might predict that parrots will exhibit conspecific attraction.

4. Patch A = 20 ducks, Patch B = 10 ducks, and Patch C = 6 ducks. To solve the problem, first convert each patch to the number of items per minute. This yields 10/min, 5/min, and 3/min, respectively, for a total

of 18 food items offered each minute in the patches. There are 36 ducks, and all must obtain food at the same rate, which can be calculated to be 2/min.

6. For humans, a short line is a rich patch, while a long line is a poor patch. Humans will sort themselves between these lines according to the IFD model. Individuals must be free to move between lines and must be able to assess the quality (or speed) of each line. Students could view lines at local stores or their university bookstore and see what parameters are important for moving through lines quickly. For example, is it the number of items in a cart or the number of individuals in each line that is most important? Recent work indicates that it is best to choose the line with the fewest individuals, no matter how much they have in their carts, because the paying process requires so much more time than scanning objects.

Chapter 11: Mating Behavior

2. Because females can take up to four days to become sexually mature and males mature in one day, some females likely had less opportunities for mating than others, including males. This would bias the results toward males having higher reproductive success, since they were all sexually mature at the time of the experiments.

4. Observations demonstrated that males varied naturally in terms of exhibiting mate guarding. Those that had the highest rate of paternity in their nest also exhibited high levels of mate guarding. The experiment manipulated the amount of time a male was away from his mate. Males that were experimentally removed suffered higher levels of cuckoldry, again supporting the hypothesis that mate guarding minimizes cuckoldry.

6. The fitness of the territorial strategy appears to be higher than the fitness of the roving tactic because it results in more offspring produced. This holds true even in a population with similar numbers of territorial and roving males. It is therefore likely that the roving tactic is a conditional strategy.

Chapter 12: Mating Systems

2. In an extra-pair mating system, individuals typically form pair-bonded social associations, and social and genetic mating systems differ. In a promiscuous mating system, pair bonds are not formed.

4. Polygynous species have higher variance in male mating success than do monogamous species. Common crossbills should exhibit lower variance in male mating success than Gunnison's sagegrouse.

6. If individual females independently seek out tents, this behavior could be an example of resource defense polygyny, in which the resource needed by females is the tent structure. If females associate in harems that a male defends from rivals, this system would be described as female defense polygyny.

Chapter 13: Parental Care

2. The individual with the increased flight costs should reduce its level of care. This experiment has been conducted several times—see, for example, B. I. Tieleman, T. H. Dijkstrab, K. C. Klasing, G. H. Visser, & J. B. Williams, Effects of experimentally increased costs of activity during reproduction on parental investment and self-maintenance in tropical house wrens, *Behavioral Ecology* 19 (2008): 949–959.

4. When environmental variation is large, resources may be very low in some years. When resources are low, an entire synchronous clutch may fail because of insufficient food if all chicks are equal competitors. Asynchronous clutches establish a competitive hierarchy so that the first chick to hatch is a superior competitor for food. In years of low resources, there may be enough food to successfully raise one offspring, which results in higher parent fitness compared to that which would result from a synchronous clutch.

6. Males with small clutches face higher predation risk when defending a clutch. As a result, the benefits of defense may be smaller than the costs, and so the males will not provide parental care. This prediction could be tested by examining the effect of predation risk on frogs defending egg clutches and by manipulating predation risk for males. Those exposed to higher predation risk would be predicted to exhibit lower levels of care.

Chapter 14: Social Behavior

2. Groups find patches faster, so you would predict larger groups for the treatment with hidden patches.

4. Subordinate individuals that stay in a group obtain other fitness benefits (e.g., reduced predation risk). Subordinate individuals may stay in a group because they may gain future benefits via social queuing.

6. Two individuals must interact repeatedly.

8. Females are more susceptible to pathogens than males.

Credits

Brandon Wheeler; **6.36 inset** © Pontier, John / Animals Animals; **6.37** Michael & Patricia Fogden / Getty Images

Chapter 7   7.1 © Chappell, Mark / Animals Animals; **p143 thumbnails, left to right,** © Dick, Michael / Animals Animals; Photo by Burt Adrinenssens; Benoit Guénard photo; © Chappell, Mark / Animals Animals; **7.2a** © Vasapolli, Salvatore / Animals Animals; **7.2b** Ipek G. Kuhlaci; **7.5b** Ipek G. Kuhlaci; **p149 Scientific Process 7.1** © Dick, Michael / Animals Animals; **p151 Scientific Process 7.2** Photo by Burt Adrinenssens; **7.9** Copyright by David W. Jamison; **7.11** Benoit Guénard photo; **7.15** © Chappell, Mark / Animals Animals; **p162 Scientific Process 7.3** © Ardea / Bahr, Chris Martin / Animals Animals; **7.18** © Nussbaumer, Rolf / Animals Animals; **p165 Applying the Concepts 7.1** Lorna Patrick, U.S. Fish and Wildlife Service; **7.20a** Photo by Jay M. Biernaske; **7.20b** Photo by Jay M. Biernaske

Chapter 8   8.1 Tony Campbell / Shutterstock.com; **p171 thumbnails, left to right,** Aaron Corcoran; Dr Ajay Kumar Singh / Shutterstock.com; © Kent, Breck P. / Animals Animals; Ron Rowan Photography / Shutterstock.com; **8.5a** Maria Eifler; **8.5b** Maria Eifler; **8.7** Aaron Corcoran; **p176 Applying the Concepts 8.1** © Miller, Steven David / Animals Animals; **8.8** Nordell; **8.9a** © Stone, Lynn / Animals Animals; **8.10c** © Michael Francis Photography / Animals Animals; **8.11a** © Lightwave Photography, Inc. / Animals Animals; **p181 Scientific Process 8.1** Martin Fowler / Shutterstock.com; **8.12** Alex Wild/Visuals Unlimited, Inc.; **p185 Scientific Process 8.2** © Degginger, Phil / Animals Animals; **8.13** Dr Ajay Kumar Singh / Shutterstock.com; **8.15** © Kent, Breck P. / Animals Animals; **8.16** Shawn E. Nordell; **8.18** Pete Oxford / DRKPhoto; **p191 Scientific Process 8.3** © John A.L. Cooke / Animals Animals; **p191 Scientific Process 8.3.1a** © Dennis, David M. / Animals Animals; **p191 Scientific Process 8.3.1b** © Gerlach Nature Photography / Animals Animals; **8.21 inset** Jason Edwards / Getty Images; **8.22a** © Ardea / Labat, Ferrero / Animals Animals; **8.22b** Tom Willard / Shutterstock.com; **8.23** Ron Rowan Photography / Shutterstock.com

Chapter 9   9.1 © Nussbaumer, Rolf / Animals Animals; **p197 thumbnails, left to right,** © Lacz, Gerard / Animals Animals; Paul V. Switzer; Henry M. Wilbur; © Chappell, Mark / Animals Animals; © Azure Computer & Photo Services / Animals Animals; **9.3** © Lacz, Gerard / Animals Animals; **9.5a** Photo by E. K. Bollinger; **9.5b** Photo by E. K. Bollinger; **p203 Scientific Process 9.1** Paul V. Switzer; **9.6a** Colours-ringed kittiwakes on their nests (Picture Thierry Boulinier); **9.6b** Black-legged kittiwakes on their nests (Picture Thierry Boulinier); **9.8** Gail Johnson / Shutterstock.com; **9.10a** Alice Boyle; **9.10b** Alice Boyle; **9.12a** Kristine L. Grayson; **9.12b** Henry M. Wilbur; **9.13 inset** Centre for Wildlife Ecology; **9.14a** © Cranston, Bob / Animals Animals; **9.15** Jan S. / Shutterstock.com; **9.17** © Garber, Howie / Animals Animals; **p217 Scientific Process 9.2** © Chappell, Mark /

Animals Animals; **9.21a** © Azure Computer & Photo Services / Animals Animals; **9.21b** © Whitehead, Fred / Animals Animals; **9.24** James L Amos / Getty Images; **9.30a** Photo by N. Chernetsov; **9.30b** Florian Andronache / Shutterstock.com; © Azure Computer & Photo Services / Animals Animals; **9.32a** © Azure Computer & Photo Services / Animals Animals; **p224 Toolbox 9.1b** © Lightwave Photography, Inc. / Animals Animals

Chapter 10   10.1 Peggy Sherman; **p227 thumbnails, left to right,** Photograph by J. Lill; mitalpatel / Shutterstock.com; © Degginger, Phil / Animals Animals; © Welling, David / Animals Animals; **p230 Scientific Process 10.1** Shawn E. Nordell; **10.4** © Maier, Robert / Animals Animals; **10.7a** Josh Engel; **10.7b** Photograph by J. Lill; **10.8** Dhalusa / Wikimedia Commons; **p235 Scientific Process 10.2** Brad Fiero, Pima Community College; **10.10** Steve Byland / Shutterstock.com; **p238 Applying the Concepts 10.1.1a** Michael Ward; **p238 Applying the Concepts 10.1.1b** Michael Ward; **p238 Applying the Concepts 10.1.1c** Michael Ward; **10.14a** mitalpatel / Shutterstock.com; **10.14b** © Adrea / Roberts, Sid / Animals Animals; **p244 Scientific Process 10.3** (NOTE: The photo referenced in the Excel sheet is different from the one in the text. The credit for the photo referenced in the Excel sheet is given here.) Tim Laman / National Geographic / Aurora Photos; **10.17** © W.A.N.T. Photography / Animals Animals; **10.20** © Degginger, Phil / Animals Animals; **10.22a** © Mc Donald Wildlife Photog. / Animals Animals; **10.22b** © Welling, David / Animals Animals; **10.24 inset** © Mark A. Chappell; **10.26 inset** Mark Bowler/Getty Images

Chapter 11   11.1 © Sheldon, Allen Blake / Animals Animals; **p255 thumbnails, left to right,** © Richardson, Ray / Animals Animals; © Leszczynski, Zigmund / Animals Animals; Agnès Horr; Raleigh J. Robertson; © Lacz, Gerard / Animals Animals; © Specker, Donald / Animals Animals; **11.2** © Lemker, John / Animals Animals; **11.3a** Francois Loubser / Shutterstock.com; **11.3b** MartinMaritz / Shutterstock.com; **11.3c** photowind / Shutterstock.com; **11.4** © Watkins, Bruce / Animals Animals; **11.5** Dr. Gerald Schatten / Science Source; **11.6** kurt_G / Shutterstock.com; **11.7a** © Richardson, Ray / Animals Animals; **11.9** topten22photo / Shutterstock.com; **11.11a** Adeline Loyau; **11.12** Shawn E. Nordell; **p264 Scientific Process 11.1** photo Anders Berglund; **11.15a** Trevor Jinks. Queensland. Australia; **11.15b** Trevor Jinks. Queensland. Australia; **11.15c** Trevor Jinks. Queensland. Australia; **11.16** © Robinson, James / Animals Animals; **11.19** © Leszczynski, Zigmund / Animals Animals; **11.21** Emmanuelle Pouivré and Glenn Yannic; **11.22** Agnès Horr; **11.24a** mooinblack / Shutterstock.com; **11.24b** © Murray, Patti / Animals Animals; **11.26** © Willis, Gladden Williams / Animals Animals; **11.29** Tania Thomson / Shutterstock.com; **11.32** Raleigh J. Robertson; **11.34** Buquet Christophe / Shutterstock.com; **11.37** © Leszczynski, Zigmund / Animals Animals; **11.38** © Lacz, Gerard / Animals Animals; **p282 Scientific**

Index